AN INTRODUCTION TO THE
THEORY OF
PROBABILITY

AN INTRODUCTION TO THE
THEORY OF PROBABILITY

Parimal Mukhopadhyay

Indian Statistical Institute, India

 **World Scientific**

NEW JERSEY · LONDON · SINGAPORE · BEIJING · SHANGHAI · HONG KONG · TAIPEI · CHENNAI

Published by

World Scientific Publishing Co. Pte. Ltd.

5 Toh Tuck Link, Singapore 596224

USA office: 27 Warren Street, Suite 401-402, Hackensack, NJ 07601

UK office: 57 Shelton Street, Covent Garden, London WC2H 9HE

British Library Cataloguing-in-Publication Data
A catalogue record for this book is available from the British Library.

AN INTRODUCTION TO THE THEORY OF PROBABILITY

ISBN-13 978-981-4313-42-1
ISBN-10 981-4313-42-4

Printed in Singapore by World Scientific Printers.

To

Promit (Piku)

and

Pritish (Tipu)

Preface

The theory of probability is the branch of pure mathematics, concerned with chance phenomena. The theory has its origin in the gambling houses of the 16th/17th century Europe and since then has traversed many paths, grown in depth and scope and today it finds applications in practically all aspects of natural and social sciences, engineering, business, even the humanities.

A myriad number of books on probability theory are currently available. Some of these books are written at elementary level, others at a little higher level, and some at a very high level using sophisticated tools of analysis and/or measure theory. The last mentioned group of books can be comprehended only by very advanced-level students of probability. The books which give a balanced mixture of classical approach and modern day approach to the theory appear to be limited in number. And so, the present book is an attempt to find a place in this area.

Modern probability theory cannot be approached without using some concepts and results from measure theory. In this treatise we have visited the needed measure-theoretical arena as only a part of our journey to the subject of probability. The book is, therefore, intended mainly for the students, who want to be aware of the classical approach to the theory as well as applications of measure theory to probability theory, without shedding too much sweat to understand the intricacies of measure. The basic requirement in reading the book is the knowledge in mathematics up to pre-graduation level and an elementary course in real analysis and calculus.

Since the theory of probability originated from the games of chance, we have developed the classical theory, elucidated with a number of examples and exercises, in Chapter 2. The short-comings of the classical theory have been noted and the statistical or empirical definition of probability has been traced out.

Chapter 3 veers the discussion to the axiomatic approach to the theory, as developed by Kolmogorov, Bernstein, Khintchein and others. This chapter introduces the set-algebra, concepts of fields, σ-fields, point-functions, set-functions, measures, measurable sets and measurable functions. Against this backdrop, probability measure, conditional probability measures, concepts of independent trials and product-measures are introduced and examined in depth.

Chapter 4 ushers in the concepts of random variables, distribution functions and their properties. This chapter also elucidates the Lebesgue-integration of a Borel-measurable function with special reference to the same of a random variable and the Lebesgue-Stieltjes integral and the Riemann-Stieltjes integral.

Discrete random variables being sometimes easier to deal with, the next chapter considers the expectation of discrete random variables and its associated properties. Chapter 6 addresses the properties of the distribution of a random variable in a general set-up. Different probability inequalities are also considered.

The generating functions, - probability generating function, moment generating function, factorial moment generating function, cumulant generating function, characteristic functions - are the subjects of interest of Chapter 7. Chapters 8 and 9, respectively, bring together various standard discrete and continuous random variables. The generalized power series distribution has found its place.

The next chapter extends the discussion to the properties of distributions in $\mathcal{R}^n$, the n-dimensional Euclidean space. Two important probability distributions, - multinomial distribution and bivariate normal distribution - have been explored. Chapter 11 focusses on the computation of probability distributions of functions of random variable(s).

The next chapter addresses the different modes of convergence of sequences of random variables and the celebrated central limit theorems. The concluding chapter discusses different discrete and continuous stochastic processes with special reference to Markov chains.

The readers interested in solving problems with finite number of possible cases, would find Chapters 2 and 5 useful.

The author of a textbook is indebted to almost every writer who has touched the subject. In writing this treatise, I have unhesitatingly taken the help of many publications, which have been cited at the proper places and have been listed in the bibliography.

Theorems, Lemmas, Equations, Tables, Notes, Remarks, Figures are numbered sequentially within a section. Corollaries are numbered sequentially within the theorems or lemmas they correspond to. In preparing the subject index, only the pages, where the subjects have been defined, have been cited. The list is, however, not exhaustive.

Thanks are due to Mr. Arunava Sen of Books and Allied, Kolkata and Dr. K. Vijayan of the University of Western Australia for their valuable support and encouragement in the preparation of the manuscript.

I am indebted to my readers of other books in Statistics (ten in number) who have given me encouragement to attempt this project. I will be thankful if this book is also well-accepted by them.

Two gentlemen, both aged less than four years, have given me invaluable impetus to complete this endeavor; my thanks are due to them.

Parimal Mukhopadhyay

Kolkata

8th December, 2010

Contents

Glossary of Some Frequently Used Symbols and Abbreviations

$\Rightarrow$	:	implies
$\Leftrightarrow$	:	implies and is implied by
$\rightarrow$	:	converges to
$\uparrow, \downarrow$	:	increasing, decreasing
$\Gamma(x)$	:	gamma function
$\overline{\lim}, \underline{\lim}, \lim$	:	lim superior, lim inferior, limit
$\mathcal{R}, \mathcal{R}^n$	:	real line, n-dimensional Euclidean space
$\mathcal{B}, \mathcal{B}^n$	:	Borel σ-field on $\mathcal{R}$, Borel σ-field on $\mathcal{R}^n$
I_A	:	indicator function of set A
μ_k	:	$E(X - E(X))^k, k \geq 0$ integral
f', f'', f'''	:	first, second and third derivative of f
$\cap$	:	distributed as
$\approx$	:	asymptotically (or approximately) equal to
$a.s$	:	almost surely
$a.e.$	:	almost everywhere
$\rightarrow^L$	:	convergence in law
$\rightarrow^P$	:	convergence in probability
$\rightarrow^{a.s.}$	:	convergence almost surely
$\rightarrow^{L_r}$	:	convergence in rth mean
pdf	:	probability density function
pmf	:	probability mass function
pgf	:	probability generating function
mgf	:	moment generating function
cgf	:	cumulant generating function
cf	:	characteristic function
$i.o.$	:	infinitely often
iid	:	independently and identically distributed
WLLN	:	weak law of large numbers
SLLN	:	strong law of large numbers
CLT	:	central limit theorem
$N(\mu, \sigma^2)$	:	normal with mean μ, variance σ^2

1

Chapter 1

Preliminaries

1.1 Introduction

If a coin is tossed, nobody knows for certain whether a head or a tail will turn up, though it is sure that either a head or a tail will occur. We are, however, ignoring the possibility that the coin stands on the edge, or rolls away and is lost. Similarly, if a dice is rolled, either of the faces $1, 2, \ldots, 6$ will turn up, though it is not known beforehand, which one will exactly occur. Experiments of these types where the outcomes cannot be predicted exactly are called 'random' experiments as opposed to the 'deterministic' type of experiments where the results of the experiments can be exactly foretold. Examples of deterministic experiments are throwing a stone upwards where it is known that the stone will definitely fall to the ground (due to force of gravitation), throwing a missile with a specified velocity and at a given angle with the ground, where the position of the missile at any given time can be exactly calculated. The problems of deterministic experiments are dealt with in subjects like physics, statics, dynamics. Random experiments are subjects of study of the theory of probability. The theory of probability analyzes the results of such experiments, states which one is more 'probable' than the other, which one is impossible under the conditions of the given experiment, relations between different kinds of outcomes, so on and so forth.

'Probability' or 'chance' is a word we often encounter in our day-to-day life. We often say that it is very probable that it will rain tonight meaning thereby we very much expect to have downpour this night. This expectation, of course, comes from our general knowledge about the conditions of weather in this month of the season. One may say that the chances of England and West Indies winning the World Cup in cricket, before the start of

3

the game, are in the ratio 50:50, meaning thereby that one expects to see either of the teams coming out winner with equal expectation. Again, this expectation is based on one's present knowledge and belief about the relative performance of the teams. In short, the word, 'probability' or 'chance' here measures one's degree of belief about the occurrence of certain events or truth of certain statements. This degree of belief cannot be verified by making repeated experiments under identical conditions as detailed below. The 'probability' of this type may be called 'subjective probability'.

If a symmetric coin (that is, one, which is devoid of any asymmetry in physical properties like distribution of weight, materials between two sides, and along the edges, etc.) is tossed, we say that a head and a tail are equally likely to occur. In this case, our belief that both the faces have the same chance to turn up can be objectively verified. If we repeat the experiment a large number of times (say, 100, 200, 500 times) under identical conditions, we will notice that on an average 50% of the results are 'heads' and 50% 'tails'. Similarly, in rolling a six-faced 'balanced' dice, we say that each face is equally likely to turn up. Repetition of this experiment a large number of times under identical conditions will show on an average $\frac{1}{6}$ proportion of experiments resulting in each of the faces $1, \ldots, 6$. This type of probability may be called 'objective probability'. The 'objective probability' is very much a property of an experiment as 'volume', 'mass', 'temperature' are properties of a substance. Throughout this treatise we shall be concerned with this objective measurement of probability.

The above process of verifying the probability of an outcome is based on the 'statistical' or 'relative frequency' interpretation of probability, a concept explained in Section 2.8.

However, the mathematical theory often starts from the assumption that we can assign probabilities to various possible outcomes of a random experiment. What precisely these probabilities mean are not part of the mathematical aspect of the subject. The mathematical theory aims at developing the properties of these probabilities, whatever their interpretation may be.

1.2 Historical Perspective

The principal impetus for the development of the theory of probability originated from an interest in games of chance, which involve an element of randomness. The rudiments of a mathematical theory to describe these elements of randomness, probably, took shape in the sixteenth century as is evidenced from a short note written in the early seventeenth century by the famous astronomer and mathematician Galileo Galilei. A French nobleman Chevalier de Me're' asked Blaise Pascal (1623 - 62), a great mathematician to explain some apparent contradictions between his reasonings and experience in gambling. Pascal solved these problems and some other problems brought to him by de Me're'. Pierre de Fermat (1601 - 65), another mathematician, subsequently became interested in the subject and in the private correspondence between Pascal and Fermat in 1654 was laid down the foundation of the theory of probability. Subsequently, Christianus Huygenes (1629 - 95), a great Dutch scientist published in 1657 the first treatise on probability, entitled *De Ratiociniis in Aleae Ludo*. Another pioneering contribution is the work, *Ars Conjectandi*, by James Bernoulli (1654 - 1705) [published posthumously in 1713, but written sometime around 1690]. Mathematicians like A. de Moivre (1667 - 1754), Laplace (1749 - 1827) enriched the science of probability through their important pioneering contributions.

The development of the subject took an important turn with the works of Chebychev (1821 - 1894), A. Markov (1856 - 1922) and A. Liapounoff (1858 - 1918). The fundamental definition of probability which was accepted in the earlier period remained the classical definition of probability until in the recent times mathematicians like S. Bernstein, A. Khintchein, A. Kolmogorov, E. Borel, H. Steinhaus, P. Le'vy developed axiomatic formulations providing important extensions and generalizations to earlier results. The first systematic presentation of probability theory was made in 1935 by Kolmogorov in a monograph available in English translation entitled *Foundations of Theory of Probability*, Chesla, New York.

1.3 Plan of the Book

Since the theory of probability originated from the games of chance, we have developed the classical theory, elucidated with a number of examples and exercises, in the pioneering chapter. The problems of geometric probability have also been considered. The short-comings of the classical theory have been noted and the statistical or empirical definition of probability has been traced out.

Chapter 3 veers the discussion to the axiomatic approach to the theory, as developed by Kolmogorov, Bernstein, Khintchein and others. The modern probability theory cannot be approached without using some concepts and results from measure theory. In this treatise we have visited the needed measure-theoretical arena as only a part of the journey to the developments of the subject of probability. Chapter 3 introduces the set-algebra, concepts of fields, σ-fields, point-functions, set-functions, measures, measurable sets and measurable functions. Against this backdrop, probability measure and its extension, conditional probability measures, concepts of independent trials and product-measures are introduced and examined in depth.

Chapter 4 ushers in the concepts of random variables, distribution functions and their properties. This chapter also elucidates the Lebesgue-integration of a Borel-measurable function with special reference to the same of a random variable and the Lebesgue-Stieltjes integral and the Riemann-Stieltjes integral.

Discrete random variables being sometimes easier to deal with, the next chapter considers the expectation of discrete random variables and its associated properties. Chapter 6 addresses the properties of the distribution of a random variable in a general set-up. Different probability inequalities are also considered.

The generating functions, - probability generating function, moment generating function, factorial moment generating function, cumulant generating function, characteristic functions - are the subjects of interest of Chapter 7. Chapters 8 and 9, respectively, bring together various standard discrete and continuous random variables. The generalized power series distribution has also been accommodated

The next chapter extends the discussion to the properties of distributions in $\mathcal{R}^n$, the n-dimensional Euclidean space. Two important probability distributions, - multinomial distribution and bivariate normal distribution - have been explored. Chapter 11 focusses on the computation of probability distributions of functions of random variable(s).

The next chapter addresses the different modes of convergence of sequences of random variables and the celebrated central limit theorems.

The concluding chapter addresses different stochastic processes. Markov chains, martingales, discrete branching process, simple random walk, continuous time processes including Poisson process, birth process, birth and death process and normal process are considered.

The concept of a random variable as the limit of a non-decreasing sequence of simple functions, Lebesgue integral of a Borel Measurable function including that of a random variable and the associated convergence theorems, Lebesgue-Stieltjes integral and Riemann-Stieltjes integral have been discussed explicitly in the Appendix.

The ideas have been explained lucidly and supplemented by many illuminating examples and innovative exercises.

Chapter 2

The Classical Approach

2.1 Introduction

This chapter elucidates the classical approach to the theory of probability. Section 2.2 introduces some elementary definitions, like those of cases or events, mutually exclusive, exhaustive and equally likely cases, etc. This section also represents the algebra of events in terms of set theory. The next section ushers in the definition of probability of an event. Section 2.4 gives examples of some simple problems which can be solved by using results of combinatorial algebra only. Models of statistical mechanics are the subject matter of the next section. Different theorems and concepts, - probability of union of events, concept of conditional probability, theorem of compound probability, concept of independence of events, theorem of total probability, Bayes theorem, binomial probability are subsequently introduced. Section 2.7 examines the limitations of the classical approach to the theory. The next section addresses the statistical or empirical definition of probability. Computation of geometric probability is the subject-matter of the next section. Finally varied problems of finding the probabilities of events have been reviewed from the classical perspective.

2.2 Some Definitions

We first consider a few definitions:

(a) *Random Experiment*: Experiments of the type whose outcomes cannot be predicted exactly will be called random experiments. Thus, if a coin is tossed, nobody knows for sure whether a 'head' (H) or a 'tail' (T) will turn up, though it is certain that either a head or a tail will turn up. Similarly,

9

if a dice is rolled, either of the faces, $1, 2, \ldots, 6$ will turn up, though it is not known beforehand which one will actually occur. If a card is randomly drawn out of a well-shuffled pack of cards, the result cannot be foretold exactly, though it is sure that one of the 52 cards will turn up.

In this chapter (except for Section 2.9) we shall consider experiments of this type only (vide note 2.2.1).

(b) *Event*: An event is the result of an experiment. Thus, in throwing a dice, the results - face 3 turns up, an odd number turns up, - are both events. In drawing a card from a well-shuffled pack, ace of spade appears or a red card appears, are both events. Events may be of two types.

(b.1) *Elementary Event* or *Case*: It is an event which cannot be decomposed into smaller events. In tossing a coin, both H and T are elementary events. In tossing two coins, each of $(HH), (HT), (TH), (TT)$, where in $(XY), X$ and Y represent the result of first and second toss respectively, is an elementary event. In throwing a dice, each of the events, $1, \ldots, 6$ is an elementary event. Symbolically, an elementary event will be represented by small case letters, a, b, etc. or by $e_1, e_2, \ldots$.

(b.2) *Compound (Contingent) Event*: A compound (contingent) event is obtained through a combination of several elementary events. In throwing two dice, the event that the sum of the face values is 8 is composed of five elementary events: $(2, 6), (3, 5), (4, 4), (5, 3), (6, 2)$ where (x, y) represents the pair of the face values of the first dice and the second dice respectively. In tossing two coins, the event that at least one head occurs is composed of three elementary events, $(HH), (HT), (TH)$.

Symbolically, a compound event will often be represented by large case letters A, B, etc. or by $E_1, E_2, \ldots$.

We shall often simply use the word 'event', whether it is an elementary event or a compound event will be clear from the context.

(c) *Favorable Cases*: A case 'a' is said to be favorable to an event 'A' if whenever 'a' occurs, 'A' occurs. In throwing two dice, each of the cases, $(2, 6), (3, 5), (4, 4), (5, 3)$ and $(6, 2)$ is favorable to the event, sum of the face values is 8. In tossing two coins, each of the cases, $(HH), (HT), (TH)$ is favorable to the event, at least one head turns up.

(d) *Mutually Exclusive Events*: Events are said to be mutually exclusive with respect to an experiment, if when one of them occurs, the others can not occur; i.e., any two events cannot occur simultaneously. Thus, in tossing a dice, the events (cases), '2 occurs' and '5 occurs' are mutually exclusive. In tossing two coins, the two events, 'both are heads (HH)' and 'both are tails (TT)' are mutually exclusive.

Again, in tossing a dice, the events - 'an odd face turns up' and '3 occurs' are not mutually exclusive, because, if 3 occurs, an odd face also turns up.

(e) *Exhaustive Events*: A set of events is said to be exhaustive in relation to a random experiment, if one of them necessarily occurs, every time the experiment is performed. In throwing two dice, the events (cases), $(1,1), \ldots (1,6), \ldots, (6,6)$ form an exhaustive set of events. There are 36 elementary events in this case. In drawing a card from a well-shuffled pack of cards, the events, 'a black card turns up', 'a red card turns up' form an exhaustive set of events.

(f) *Equally Likely Cases*: Cases are said to be equally likely when we have no reason to believe that one is more likely to occur than the other. In tossing a fair coin, H and T are equally likely cases. In drawing a card at random from a well-shuffled pack of cards, one may believe that each of the cards is equally likely to appear and thus each of the 52 cases is equally likely. Here, we are making the simplest assumption regarding the possibility of occurrences of different cases, namely, each case is equally likely to occur.

EXAMPLE 2.2.1: In tossing a fair coin thrice, elementary events are: $(HHH), (HHT), (HTH), (HTT), (THH), (THT), (TTH), (TTT)$. These cases are mutually exclusive, equally likely and they form an exhaustive set.

EXAMPLE 2.2.2: The experiment consists of forming a three digited number out of digits $1, 2, \ldots, 9$, each digit being allowed to be repeated any number of times. A typical outcome is (x, y, z) where each of x, y, z can take any value between 1 and 9. The number of (elementary) events is, therefore, $9^3 = 729$. All these cases are mutually exclusive, exhaustive and equally likely.

Note 2.2.1: We note that in all the above experiments, the number of elementary events N is finite.

2.2.1   *Representation in set theory*

Since the classical theory of probability depends heavily on set theory we recall some rudiments of theory of sets. We shall also use different features of sets to represent events and relations among events.

(i) *Set*: A set is a collection of some elements which are its members and is often pictured as an oval-shaped space with various points within the space denoting its elements. The diagram representing a set is called a *Venn Diagram*.

Depending on how many elements it has, a set may be finite or infinite. The set of all natural numbers $N_1 = \{1, 2, \ldots, n, \ldots\}$ is infinite. However, its elements are countable. Hence, N_1 is a countably infinite set. Similarly, the set of all even numbers $N_2 = \{2, 4, \ldots, 2n, \ldots\}$ is countably infinite. The members of the set $A = \{\omega : a < \omega < b\}$ where $a < b$ are some real numbers are infinite in number and not countable. Hence, the set A is uncountably infinite.

(ii) *Universal Set*: It is the largest possible set of interest in any given situation. All other sets are parts of this set and hence are subsets of this universal set.

If we are interested in real numbers, the real line $\mathcal{R}^1 = \mathcal{R} = \{\omega : -\infty < \omega < \infty\}$ is the universal set. Any interval on the real line is a subset of this universal set.

A universal set is often represented by Ω.

(iii) *Null Set*: It is a set containing no element. It is also called an *empty set* and is often represented as ϕ. It is a member of all other sets.

(iv)*Disjoint Sets*: Two sets A_1, A_2 are disjoint, if they have no common element. Similarly, sets $A_1, A_2, \ldots, A_m$ are mutually disjoint, if no two of them have any common element.

In the theory of probability, an elementary event or case resulting from an experiment is represented as a point in a set. Thus, in throwing a dice, each of the cases, $1, 2, \ldots, 6$ is denoted as a point in a set.

The set containing all the possible points representing all the elementary events which can result from the given random experiment is called the *Sample Space* for this experiment and is often represented as $\mathcal{S}$. Therefore, the universal set is the sample space in probability theory. In tossing a dice, $\mathcal{S} = \{1, 2, \ldots, 6\}$. In tossing two dice, $\mathcal{S} = \{(1, 1), (1, 2), \ldots, (6, 6)\}$.

The total number of possible cases is $6^2 = 36$, since each of y, and z in the point (y, z) can take any of the values $1, 2, \ldots, 6$. In tossing two coins $\mathcal{S} = \{(HH), (HT), (TH), (TT)\}$. Clearly, a sample space consists of all the exhaustive set of points.

However, since in the above examples, we have considered only experiments with finite number of cases, sample spaces for these experiments are finite sets.

A compound event A is represented as a set containing all the points which represent the elementary events of which A is made up (i.e., which are favorable to A). In throwing a dice, the compound event, an odd face turns up, is represented by the set containing points 1, 2, 3 only.

The events $A, B, C, \ldots$ are mutually exclusive, if the sets representing them are disjoint.

EXAMPLE 2.2.3: A coin is tossed until a head appears. Here the sample space $\mathcal{S} = \{H, TH, TTH, TTTH, \ldots\}$. The number of points in the sample space is countable and infinite. The sample space, here, is a countably infinite set.

EXAMPLE 2.2.4: The experiment consists of dropping the pointed end of a needle on a circular surface of radius 5". The outcome is the coordinate of the point of contact. The sample space, here, is the set $\mathcal{S} = \{(x, y) : x^2 + y^2 \leq 5\}$. The number of points in $\mathcal{S}$, here, is uncountably infinite.

We shall, however, in this chapter (except in Section 2.9) consider only those experiments whose sample space is (countably) finite.

(v) *Union*: Let A and B be two sets. The union of A and B, denoted as $A \cup B$ is the set of all elements contained in either A or B or both. In the terminology of events, $A \cup B$ represents the event that at least one of the events A and B occurs.

Similarly, the union of m sets $A_1, \ldots, A_m$, namely $\cup_{i=1}^m A_i$ is the set of al elements contained in at least one of $A_1, \ldots, A_m$. When the sets $A_1, \ldots, A_m$ are mutually disjoint, we shall sometimes write $\cup_{i=1}^m A_i$ as $A_1 + \ldots + A_m$. The set $\cup_{i=1}^m A_i$ represents the event that at least one of the events $A_1, \ldots, A_m$ occurs. $\cup_{i=1}^m A_i$ is also called a *total event*.

(vi) *Intersection*: The intersection of two sets A and B, denoted as $A \cap B$ or simply AB is the set of all elements contained in both A and B. The set

$A \cap B$ denotes the event that both A and B occur.

Similarly, the intersection of m sets $A_1, \ldots, A_m$, denoted as $\cap_{i=1}^m A_i$ or simply, $A_1 A_2 \ldots A_m$ is the set of all elements contained in all of $A_1, \ldots, A_m$. The set $A_1 \ldots A_m$ represents the event that each of the events $A_1, \ldots, A_m$ occur simultaneously. The event $\cap_{i=1}^m A_i$ is also called a *compound event*. If the sets $A_1, \ldots A_m$ are mutually disjoint, $A_i \cap A_j = \phi$ for all $i \neq j = 1, \ldots, m$.

In the next chapter, we shall define the union $\cup_{i=1}^\infty A_i$ and intersection $\cap_{i=1}^\infty A_i$ of a countable number of sets $A_1, A_2, \ldots$.

(vii) *Complementation*: The complement of set A, denoted as A^c or A' or $\bar{A}$ is the set of all elements not contained in A. Hence, $A^c = \mathcal{U} - A$. The set A^c represents the event that A does not occur.

Clearly, $\Omega^c = \phi, \phi^c = \Omega, (A^c)^c = A$.

(viii) *Subset*: A set C is a subset of a set D if every member of C is necessarily a member of D. This is denoted as $C \subseteq D$. When D contains at least one element not contained in C, C is a proper subset of D and we write $C \subset D$.

For two sets A, B, clearly, $(A \cap B) \subset (A \cup B)$. For three sets $A, B, C, (ABC) \subset (AB), ABC \subset A \cup (BC)$, etc.

If $C \subset D$, event D is implied by C but not vice-versa. In this case, as noted in Section 2.3, $P(C) \leq P(D)$.

(ix) *Difference*: The difference $A - B$ is the set of all elements contained in A but not in B. Similarly $B - A$ is the set of all elements contained in B but not in A. Clearly $A - B \neq B - A$. Also $A \supset (A - B), (B - A) \subset B$. Clearly, $A - B$ and $B - A$ are disjoint sets.

It follows that $A - B = A \cap B^c, B - A = B \cap A^c$.

(x) The operations of union and intersection satisfy the following properties:

Commutativity: $A \cup B = B \cup A$; $\ A \cap B = B \cap A$.

Associativity: $A \cup (B \cup C) = (A \cup B) \cup C$; $\ A \cap (B \cap C) = (A \cap B) \cap C$.

Distributivity: $A \cap (B \cup C) = (A \cap B) \cup (A \cap C)$;

$$A \cup (B \cap C) = (A \cup B) \cap (A \cup C).$$

Idempotency: $A \cap A = A$; $\ A \cup A = A$.

Lemma 2.2.1: The sets obey de Morgan's laws of complementation:

For m sets $A_i, i = 1, \ldots, m$,

(a)
$$(\cup_{i=1}^m A_i)^c = \cap_{i=1}^m A_i^c; \qquad (2.2.1)$$

(b)
$$(\cap_{i=1}^m A_i)^c = \cup_{i=1}^m A_i^c. \qquad (2.2.2)$$

Proof. We have

$$\begin{aligned}
(\cup_{i=1}^m A_i)^c &= \{\omega : \omega \text{ does not belong to even one } A_i, i = 1, \ldots, m\} \\
&= \{\omega : \omega \notin A_i \text{ for each one of } i = 1, \ldots, m\} \\
&= \cap_{i=1}^m A_i^c.
\end{aligned}$$

Also,

$$\begin{aligned}
(\cap_{i=1}^m A_i)^c &= \{\omega : \omega \text{ does not belong to each one of } A_i, i = 1, \ldots, m\} \\
&= \{\omega : \omega \text{ belongs to at least one } A_i^c\} \\
&= \cup_{i=1}^m A_i^c.
\end{aligned}$$

$\square$

Field of Events: We now construct a field of events F which has the following properties:

(a) F contains the set S as one of its elements.

(b) If A, B which are subsets of S are elements of F, so are $A \cup B, A \cap B, \bar{A}, \bar{B}$. Thus the null set is contained in F, because $S \in F$. The union, intersection and complements of a finite number of events belong to F (for further details on this concept see Chapter 3).

EXAMPLE 2.2.5: A dice is thrown. The sample space $S = \{E_1, \ldots, E_6\}$, E_i denoting the elementary event, roll of point i. The field F consists of 2^6 (=64) sets, $F = \{(\phi); (E_1) \ldots (E_6); (E_1, E_2), \ldots, (E_5, E_6); (E_1, E_2, E_3), \ldots, (E_4, E_5, E_6); (E_1, \ldots, E_4), \ldots, (E_3, \ldots, E_6); (E_1, \ldots, E_5), \ldots, (E_2, \ldots, E_6); (E_1, \ldots, E_6)\}$.

2.3 The Classical Definition of Probability

We assume that the total number of elementary events N in the sample space of the experiment is finite. We also assume that all the elementary events are equally likely to occur i.e. the circumstances of the experiment are such that no elementary event is more likely to occur than the other. Under these assumptions we have the following classical or *a priori* definition of probability.

If an experiment can result in N mutually exclusive, exhaustive and equally likely cases, of which $n(A)$ cases are favorable to an event A, then the probability of occurrence of A is

$$P(A) = \frac{n(A)}{N}. \tag{2.3.1}$$

It follows, therefore, that probability of A is the sum of the probabilities of all elementary events favorable to A.

EXAMPLE 2.3.1: Find the probability that among three-digited numbers formed by $1, 2, \ldots, 6$, there is no repetition.

Let (x, y, z) take any of the values $1, 2, \ldots, 6$. All these cases are mutually exclusive and equally likely. We note that the required event A occurs if any of the following cases occur: $(x, y, z), x \neq y \neq z = 1, 2, \ldots, 6$. Suppose x is chosen first, then y and z in that order. x may be anything between 1 to 6 and hence x may be chosen in 6 ways. After x is chosen, y may be chosen in any of the remaining 5 ways. Since, with each choice of x, there are 5 ways of choosing y, the total number of ways in which x and y can be chosen is $6.5 = 30$. Thus the total number of ways in which (x, y, z) can be chosen satisfying the above-mentioned condition is $6.5.4 = 120$. Therefore, $N = 216, n(A) = 120$ and $P(A) = \frac{120}{216}$.

The following properties follow immediately from the above-mentioned definition.

(i) Since $0 \leq n(A) \leq N$, probability $P(A)$ of an event satisfies

$$0 \leq P(A) \leq 1. \tag{2.3.2}$$

If $P(A) = 0$, then $n(A) = 0$, i.e., there is no favorable case for A and hence A is impossible under the given circumstances. If $P(A) = 1, n(A) = N$, i.e., all the cases are favorable to A and hence A is sure to occur. A is a certain event.

(ii) The number of cases which are not favorable for A is $N - n(A)$. Hence, probability that A does not occur is $1 - P(A)$.

(iii) If occurrence of an event A implies occurrence of an event B, then the set of the points representing cases favorable for A is a subset of the points representing cases favorable for B. Hence $n(A) \leq n(B)$. Therefore, $P(A) \leq P(B)$.

Clearly, the classical definition fails, if at least one of the assumptions, (i) N is finite (ii) all the cases are equally likely, is not satisfied. This has been discussed in Section 2.7.

However, most of the problems on finding probabilities of simple events can be calculated using a priori definition of probability. As such we shall consider the development of classical theory in further detail.

2.4 Some Examples

We shall consider in this section some illustrative examples.

EXAMPLE 2.4.1: If n men, among whom are A and B stand in a row, what is the probability there will be exactly r men between A and B? (b) If they stand in a ring instead of in a row, show that the probability is independent of r and is $1/(n-1)$. In the circular position consider only the cases where there will be r persons between A and B in the arc leading from A to B in the clockwise direction.

(i)The total number of mutually exclusive, exhaustive and equally likely cases is $n!$.

Suppose A and B are kept fixed; r persons can be chosen out of $(n-2)$ persons and arranged in $(n-2)_r = (n-2)(n-3)\ldots(n-r-1)$ ways[1]. The remaining $(n-r-2)$ persons can be arranged in $(n-r-2)!$ ways. Again, if A occupies a position on the left of B, then A may choose any of the following positions: first, second, ..., $(n-r-1)$th from the left. Similarly, B may also occupy a position on the left of A. Hence, probability is

$$\frac{2.(n-r-1).(n-2)_r.(n-r-2)!}{n!} = \frac{2(n-r-1)}{n(n-1)}.$$

(ii) If they form a ring, the total number of possible arrangements is obtained by keeping the position of any person fixed and arranging the remaining positions in all possible ways. This is $(n-1)!$.

Now, A can occupy any position, allowing only r persons between A and

---

[1]We shall use $(a)_b$ to denote $a(a-1)\ldots(a-b+1), (b \leq a)$.

B. Hence, probability is

$$\frac{(n-2)_r(n-r-2)!}{(n-1)!} = \frac{1}{n-1}.$$

EXAMPLE 2.4.2: If n distinct balls are tossed into m boxes so that each ball is equally likely to fall in any box, show that the probability that a specified box will contain r balls is

$$p_r = \binom{n}{r}\frac{(m-1)^{n-r}}{m^n}. \tag{2.4.1}$$

Show that the most probable number of balls in a cell is r which satisfies

$$\frac{(m-1)^2}{(n-r)(n-r+1)} < \frac{(m-1)}{r(n-r)} > \frac{1}{r(r+1)}. \tag{2.4.2}$$

Also, show that

$$\frac{1}{r!}(\frac{n}{m})^r(1-\frac{1}{m})^{n-r}(1-\frac{r}{n})^{(r-1)/2} \le p_r \le \frac{1}{r!}(\frac{n}{m})^r(1-\frac{1}{m})^{n-r}(1-\frac{r}{2n})^{r-1}. \tag{2.4.3}$$

Show also that for $m = n$ with $m \to \infty$,

$$p_r \to \frac{e^{-1}}{r!}.$$

Total number of mutually exclusive, exhaustive and equally likely cases is m^n, because, each ball can go to any of the m cells, and we are assuming that a cell can contain all the n balls.

r balls can be chosen out of n balls in $\binom{n}{r}$ ways. The remaining $(n-r)$ balls can go into $(m-1)$ boxes in $(m-1)^{n-r}$ ways. Hence, the expression p_r.

The most probable number of balls is r which satisfies $p_{r-1} < p_r < p_{r+1}$. (2.4.2) is obtained after simplification.

(iii)Now,

$$p_r = (1-\frac{1}{m})^{n-r}\frac{Q}{r!}(\frac{n}{m})^r$$

where

$$Q = (1-\frac{1}{n})(1-\frac{2}{n})\dots(1-\frac{r-1}{n}).$$

For

$$1 \le k \le r-1, (1-\frac{r}{n}) < (1-\frac{k}{n})(1-\frac{r-k}{n}) < (1-\frac{r}{2n})^2.$$

Therefore,

$$(1 - \frac{r}{n})^{(r-1)/2} < Q < (1 - \frac{r}{2n})^{r-1}.$$

Hence (2.4.3) holds. For $m = n$, with $n \to \infty, Q \to 1, (1 - \frac{1}{m})^{m-r} \to e^{-1}$, when $p_r \to \frac{e^{-1}}{r!}$.

EXAMPLE 2.4.3: What is the probability that two throws with three dice each will show the same configuration if (i) the dice are distinguishable, (ii) they are not.

(i) The first throw of 3 dice may occur in 6^3 ways; the second throw must show the same configuration. Hence, probability $= \frac{6^3}{6^6} = \frac{1}{216}$.

(ii) The first three throws may show results of the following three types: (a) all face values are different, (b) two face values are same, (iii) all face values are same.

(a) Total number of such cases $= (6)_3.3! = 720$, because, the first three throws may occur in $(6)_3$ ways and therefore the same face values may occur in the second three throws in 3! ways.

(b) Total number of possible cases is $6.5.\binom{3}{2}^2 = 270$, because two different face values can be chosen and permuted, - one to account for the two dice and the other the remaining, - in 6.5 ways; two different throws to show the same face value out of the first three throws in $\binom{3}{2}$ ways and similarly, for the second set of three throws.

(c) The total number of possible cases is 6.

Hence probability $= \{720 + 270 + 6\}/6^6 = 996/6^6$.

EXAMPLE 2.4.4: What is the probability that (a) the birthdays of twelve randomly selected persons will fall in twelve calendar months; (ii) the birthdays of six given persons will fall in exactly two calender months.

Total number of mutually exclusive, exhaustive and equally likely cases is 12^{12}, because, each person can choose his birthday in any of 12 months.

(i)The birthday of each person has to fall in different month. Hence, probability is $12!/(12)^{12}$.

(ii) The two months can be chosen in $\binom{12}{2}$ ways, the chosen six persons can have their birthdays in at least one of these months in $(2^6 - 2)$ ways and the remaining six persons can have their birthday in 12^6 ways. Hence, the probability is

$$\frac{\binom{12}{2}(2^6 - 2)}{12^6}.$$

EXAMPLE 2.4.5: From n balls numbered $1, 2, \ldots, n$, m balls are drawn. Let X be the largest number drawn. Find $P(X = k)$ if the sampling is done: (i) with replacement, (ii) without replacement.

(i) Here $P(X \le k) = (\frac{k}{n})^m$. Hence,

$$P(X = k) = P(X \le k) - P(X \le k - 1) = \frac{k^m - (k-1)^m}{n^m}.$$

(ii) Here $P(X \le k) = \binom{k}{m}/\binom{n}{m}$. Hence,

$$P(X = k) = \frac{\binom{k-1}{m-1}}{\binom{n}{m}}.$$

EXAMPLE 2.4.6: A number a is chosen at random from the set $1, \ldots, N$. Find the probability p_N that when divided by the integer $r(> 1), a$ will have a remainder q.

a must be one of the following integers: $q, q + r, q + 2r, \ldots$. Hence $p_N = \frac{1}{N}\{[\frac{N-q}{r}+1]\}$, where $[z]$ is the highest integer contained in z; $\lim_{N\to\infty} p_N = \frac{1}{r}$.

EXAMPLE 2.4.7: What is the probability of obtaining a given sum s of points with n dice?

The total number of possible cases with 6 dice is 6^n. The total number of favorable cases is the number of solutions of the equation $a_1 + a_2 + \ldots a_n = s$, where a_i represents the number turned up by the ith draw and hence $1 \le a_i \le 6, i = 1, \ldots, n$. This number is given by the coefficient of x^s in the expansion of

$$T = (x + x^2 + \ldots + x^6)^n,$$

because coefficient of x^s is $\frac{n!}{b_1! b_2! \ldots b_n!}$ where $b_1 + 2b_2 + \ldots + 6b_6 = s, b_i = $ number of throws with result $i, i = 1, \ldots, 6$. Now, $T = x^n(1-x^6)^n(1-x)^{-n}$;

$$x^n(1 - x^6)^n = \sum_{r=0}^{n}(-1)^r \binom{n}{r} x^{n+6r}, \qquad (i)$$

$$(1 - x)^{-n} = \sum_{k=0}^{\infty} \binom{n+k-1}{n-1} x^k. \qquad (ii)$$

Multiplying (i) and (ii), the coefficient of x^s in T is

$$\sum_{r=0}^{[(s-n)/6]} (-1)^r \binom{n}{r} \binom{s-6r-1}{n-1} = N_n \text{ (say)}.$$

For $n = 3, s = 15, N_n = 10$. The required probability $= \frac{N_n}{6^n}$.

EXAMPLE 2.4.8: An urn contains a white and b black balls. Balls are drawn one by one until only those of the same color are left. What is the probability that they are all white.

All the black balls should have been exhausted. The sequence of balls drawn contain x white balls and b black balls, $x = 0, 1, \ldots, a - 1$. The probability of drawing x white balls to precede the last black ball is $\binom{x+b-1}{x} / \binom{a+b}{a}$, because the total number of possible cases is the number of permutations of a white and b black balls and is $\binom{a+b}{a}$. Hence probability $= \sum_{x=0}^{a-1} \binom{x+b-1}{x} / \binom{a+b}{a}$.

Now,

$$\sum_{x=0}^{a-1} \binom{x+b-1}{x} = \binom{b-1}{0} + \binom{b}{1} + \binom{b+1}{2} + \ldots + \binom{a+b-2}{a-1},$$

is the term independent of x in

$$[(1+x)^{b-1} + \frac{1}{x}(1+x)^b + \frac{1}{x^2}(1+x)^{b+1} + \ldots + \frac{1}{x^{a-1}}(1+x)^{b+a-2}]$$

$$= (1+x)^{a+b-1} x^{1-a},$$

which is $\binom{a+b-1}{a-1}$. Hence, probability $= \frac{a}{a+b}$.

2.5 Models in Statistical Mechanics

The model of distributing balls over cells is an appropriate one for various problems in statistical mechanics. The n balls are physical particles and the m cells represent m equal parts into which a spacial region is divided.

So far we have assumed that when the n distinct balls are distributed in m cells, each of the m^n sample points (elementary events) in the sample space has the probability $1/m^n$.

When the balls are indistinguishable, an outcome of an experiment is described by a sequence of m numbers, $(k_1, \ldots, k_m), k_1 + \ldots + k_m = n$, giving the number of balls in each cell. The probability of such an outcome, using the assumption of equal likelihood of each of the m^n sample points (as in the case of distinguishable balls as above) is

$$P(k_1, \ldots, k_m) = \frac{n!}{k_1! \ldots k_m!} \frac{1}{m^n}.$$

This is called the *Maxwell-Boltzman statistic*. (If the balls were distinct, there would be $\frac{n!}{k_1! \ldots k_m!}$ distinct arrangements for the event $(k_1, k_2, \ldots, k_m)$, all of which have now become indistinguishable.)

Under *Bose-Einstein statistic* only distinguishable arrangements of n indistinguishable balls in m cells are considered and each is given the equal probability. There are $\binom{m+n-1}{n}$ such distinguishable arrangements and probability of each distinct arrangement is $[\binom{m+n-1}{n}]^{-1}$.

Proof: Suppose m cells are represented by a sequence of $(m+1)$ bars ($|$), each bar representing an wall of a cell. Within these cells, the balls (crosses) are placed. For example, $|\times|\times\times|||$ shows a configuration of 3 balls in 4 cells with $k_1 = 1, k_2 = 2, k_3 = k_4 = 0$. Any arrangement of n balls in m cells is thus represented by a sequence of $(m+1)$ bars and n crosses, provided the sequence must start with a bar and end with a bar. Hence the number of distinguishable arrangements equal the number of choices of n positions for crosses out of $(m+n-1)$ available positions. This number is $\binom{m+n-1}{n}$ $\square$.

The *Fermi-Dirac statistic* is based on the following hypotheses: (a) it is impossible for two or more balls to be in the same cell; (b) and all distinguishable arrangements satisfying (a) have equal probability. There are $\binom{m}{n}$ such arrangements and each is given equal probability $1/\binom{m}{n}, n \le m$.

Bose-Einstein statistic hold good for photons, nuclei and atoms containing an even number of elementary particle, Fermi-Dirac statistic apply to electrons, neutrons and protons. However, different situations fit different models and it is impossible to select or justify probability models by a priori arrangements.

EXAMPLE 2.5.1: n indistinguishable balls are distributed without replacement among $M(> 1)$ urns numbered 1 to M. Determine the probability that each urn numbered 1 to n contains exactly one ball.

The total number of possible arrangements is $\binom{n+M-1}{n}$ by Bose-Einstein statistic, all of which are assumed equally likely. Only one arrangement satisfies the desired event. Hence, probability is $[\binom{n+M-1}{n}]^{-1}$.

2.6 Some Theorems on Probability of Events

We now consider some basic theorems on probability of events.

2.6.1 Theorems on probability of union of events

Theorem 2.6.1: If $A_1, \ldots, A_n$ are mutually exclusive events, then the probability that at least one of the events $A_1, \ldots, A_n$ occur is

$$P(\cup_{i=1}^{n} A_i) = P(A_1) + P(A_2) + \ldots + P(A_n). \tag{2.6.1}$$

Proof. Let M be the total number of mutually exclusive and equally likely cases in the sample space of which m_i are favorable to the event $A_i, i = 1, \ldots, n$. Since $A_1, \ldots, A_n$ are mutually exclusive events, the sets representing them are mutually disjoint. Hence, the total number of cases favorable to the event $\cup_{i=1}^{n} A_i (= \sum_{i=1}^{n} A_i)$ is $\sum_{i=1}^{n} m_i$. Therefore,

$$P(\cup_{i=1}^{m} A_i) = \frac{m_1 + \ldots + m_n}{M} = \frac{m_1}{M} + \ldots + \frac{m_n}{M} = P(A_1) + \ldots P(A_n).$$

Corollary 2.6.1.1: If an event A can occur only in m mutually exclusive ways, so that A occurs if and only if any of $A_1, \ldots, A_m$ occurs, then

$$P(A) = P(A_1) + \ldots + P(A_m).$$

Corollary 2.6.1.2: Let the event B imply A, but not vice-versa. Then the set $(A - B)$ represents the event $A \cap B^c$, that is, the event A occurs, but not B. Then

$$P(A - B) = P(A) - P(B). \tag{2.6.2}$$

Proof. The set B is a proper subset of $A, B \subset A$. Now, the set A can be decomposed into two disjoint subsets, B and $A - B$. Hence,

$$P(A) = P(B) + P(A - B).$$

Therefore,

$$P(A - B) = P(A) - P(B), B \subset A.$$

EXAMPLE 2.6.1: 3 cards are drawn randomly from a well-shuffled pack. What is the probability that all are diamonds or spades.

Let A_1 be the event that cards drawn are all diamonds and A_2 be the event that they are all spades. The events A_1, A_2 are mutually exclusive. Then $A = A_1 + A_2$ is the event that the cards drawn are all diamonds or spades. Now, $P(A_1) = \binom{13}{3}/\binom{52}{3} = a$ (say). Also $P(A_2) = a$. Hence, $P(A) = 2a$.

We now consider the case, where $A_1, \ldots, A_n$ are not necessarily mutually disjoint. First we introduce the following notations.

$$\sum_{i=1}^{n} P(A_i) = S_1, \quad \sum_{i<j=1}^{n} \sum P(A_i A_j) = S_2,$$

$$\sum \sum \sum_{i<j<k=1}^{n} P(A_i A_j A_k) = S_3,$$

$$\cdots \cdots \cdots$$

$$P(A_1 A_2 \ldots A_n) = S_n.$$

Thus, S_1 is the sum of n terms, each of probability of an event A_i; S_2 is the sum of $\binom{n}{2}$ terms, each of probability of joint occurrence of a pair A_i, A_j; S_3 is the sum of $\binom{n}{3}$ terms, each of probability of joint occurrence of a triplet of events A_i, A_j, A_k, etc.

Theorem 2.6.2: Let $A_1, \ldots, A_n$ be the events not necessarily mutually exclusive. The probability that at least one of $A_1, \ldots A_n$ occurs is

$$P(\cup_{i=1}^{n} A_i) = S_1 - S_2 + S_3 + \ldots + (-1)^{n-1} S_n. \qquad (2.6.3)$$

Proof. (By method of induction) First consider the events $A_1, A_2 (n = 2)$. Now

$$A_1 \cup A_2 = A_1 + (A_2 - A_1 \cap A_2).$$

Hence, by Theorem 2.6.1,

$$P(A_1 \cup A_2) = P(A_1) + P(A_2 - A_1 \cap A_2).$$

Since $A_1 \cap A_2 \subset A_2$, by Corollary 2.6.1.2,

$$P(A_1 \cup A_2) = P(A_1) + P(A_2) - P(A_1 \cap A_2). \qquad (2.6.4)$$

Thus, the theorem is true for $n = 2$. Assume that the result (2.6.3) holds for $n = m(> 2)$. Therefore,

$$P(\cup_{i=1}^{m} A_i) = \sum_{i=1}^{m} P(A_i) - \sum_{i<j=1}^{m} \sum P(A_i A_j) + \sum \sum \sum_{i<j<k=1}^{m} P(A_i A_j A_k)$$

$$- \ldots + (-1)^{m-1} P(A_1 \ldots A_m). \qquad (2.6.5)$$

Now,

$$\cup_{i=1}^{m+1} A_i = \cup_{i=1}^{m} A_i + (A_{m+1} - (\cup_{i=1}^{m} A_i) \cap A_{m+1})$$
$$= \cup_{i=1}^{m} A_i + (A_{m+1} - \cup_{i=1}^{m} (A_i \cap A_{m+1})).$$

Hence,

$$P(\cup_{i=1}^{m+1}) = P(\cup_{i=1}^m A_i) + P\left(A_{m+1} - \cup_{i=1}^m (A_i \cap A_{m+1})\right)$$
$$= P\left(\cup_{i=1}^m A_i\right) + P(A_{m+1}) - P\left(\cup_{i=1}^m (A_i \cap A_{m+1})\right). \tag{2.6.6}$$

Since, by assumption, the theorem holds for m sets,

$$P\left(\cup_{i=1}^m (A_i \cap A_{m+1})\right) = \sum_{i=1}^m P(A_i \cap A_{m+1})$$

$$- \sum\sum_{i<j=1}^m P\{(A_i \cap A_{m+1})(A_j \cap A_{m+1})\}$$

$$+ \sum\sum\sum_{i<j<k=1}^m P\{(A_i \cap A_{m+1})(A_j A_{m+1})(A_k A_{m+1})\}$$

$$- \dots + (-1)^{m-1} P\{(A_1 \cap A_{m+1}) \dots (A_m \cap A_{m+1})\}. \tag{2.6.7}$$

Now,

$$(A_i \cap A_{m+1})(A_j \cap A_{m+1}) = A_i A_j A_{m+1},$$

$$(A_i \cap A_{m+1})(A_j \cap A_{m+1})(A_k \cap A_{m+1}) = A_i A_j A_k A_{m+1}, \text{ etc.}$$

Using in (2.6.6) the expressions in (2.6.5) and (2.6.7), and after rearrangements of terms, we have

$$P\left(\cup_{i=1}^{m+1} A_i\right) = \sum_{i=1}^{m+1} P(A_i) - \sum\sum_{i<j=1}^{m+1} P(A_i A_j)$$
$$+ \sum_{i<j<k=1}^{m+1} P(A_i A_j A_k) - \dots + (-1)^m P(A_1 \dots A_{m+1}).$$

Hence, if the theorem holds for $n = m$, it holds for $n = m+1$. The theorem has been shown to hold good for $n = 2$. Hence, it holds for all integer values of n.

An Alternative Proof.

To calculate $P(\cup_{i=1}^n A_i)$ one should add the probabilities of all elementary events (points) contained in at least one $A_i (i = 1, \dots, m)$, each point being taken only once. Consider an arbitrary point E contained in exactly t of the sets A_i, say, $A_1, \dots, A_t$ and find its contribution to the right side of (2.6.3). $P(E)$ occurs t times in S_1, since S_1 is the sum of the probabilities $P(A_1), \dots, P(A_t)$. Similarly it occurs $\binom{t}{2}$ times among the $\binom{n}{2}$ terms in S_2, $\binom{t}{3}$ times in $S_3, \dots, \binom{t}{j}$ times in $S_j, \dots, \binom{t}{t}$ times in S_t. Its contribution to $S_{t+1}, \dots, S_n$ is zero, since E occurs only in t of the sets $A_1, \dots, A_n$. Therefore, the coefficient of $P(E)$ in the right side of (2.6.3) is

$$\binom{t}{1} - \binom{t}{2} + \dots + - \binom{t}{t}$$

which equals 1. Thus $P(E)$ occurs only once in the right side of (2.6.3). Hence the proof.

Corollary 2.6.2.1: For n events $A_1, \ldots, A_n$,

$$P\left(\cap_{i=1}^n A_i\right) = \sum_{i=1}^n P(A_i) - \sum \sum_{i<j=1}^n P(A_i \cup A_j) + \sum \sum \sum_{i<j<k=1}^n P(A_i \cup A_j \cup A_k)$$

$$- \ldots + (-)^{n-1} P(A_1 \cup A_2 \cup \ldots \cup A_n). \tag{2.6.8}$$

Proof. We have,

$$P\left(\cap_{i=1}^n A_i\right) = 1 - P\left(\cup_{i=1}^n A_i^c\right).$$

Now, by Theorem 2.6.2,

$$P\left(\cup_{i=1}^n A_i^c\right) = \sum_{i=1}^n P(A_i^c) - \sum \sum_{i<j=1}^n P(A_i^c A_j^c) + \sum \sum \sum_{i<j<k=1}^n P(A_i^c A_j^c A_k^c)$$

$$- \ldots + (-1)^{n-1} P(A_1^c A_2^c \ldots A_n^c). \tag{2.6.9}$$

By de Moivre's laws,

$$\cap A_i^c = (\cup A_i)^c.$$

Hence,

$$P(A_i^c A_j^c) = 1 - P(A_i \cup A_j),$$

$$P(A_i^c A_j^c A_k^c) = 1 - P(A_i \cup A_j \cup A_k), \text{ etc.}$$

Therefore, from (2.6.9),

$$
\begin{aligned}
1 - P\left(\cap_{i=1}^n A_i\right) &= \sum_{i=1}^n \{1 - P(A_i)\} - \sum \sum_{i<j=1}^n \{1 - P(A_i \cup A_j)\} \\
&+ \sum \sum \sum_{i<j<k=1}^n \{1 - P(A_i \cup A_j \cup A_k)\} - \ldots \\
&+ (-1)^{n-1} \{1 - P(\cup_{i=1}^n A_i)\} \\
&= \binom{n}{1} - \binom{n}{2} + \binom{n}{3} + \ldots + (-1)^{n-1}\binom{n}{n} - \sum_{i=1}^n P(A_i) \\
&+ \sum \sum_{i<j=1}^n P(A_i \cup A_j) - \sum \sum \sum_{i<j<k=1}^n P(A_i \cup A_j \cup A_k) \\
&+ \ldots + (-1)^n P(\cup_{i=1}^n A_i)
\end{aligned}
$$

when the result follows on simplification.

EXAMPLE 2.6.2: A sample of r individuals are taken from a population of n people with replacement. Find the probability p_r that m specified individuals are selected. Show that as $n \to \infty$ and $r \to \infty$, so that $\frac{r}{n} \to p$, a constant, $p_r \to (1 - e^{-r})^m$.

Let A_i be the event that the ith individual in the set of m specified individuals is not selected. Then

$$p_r = 1 - P(\cup_1^m A_i) = 1 - \binom{m}{1}\left(\frac{n-1}{n}\right)^r + \binom{m}{2}\left(\frac{n-2}{n}\right)^r - \cdots + (-1)^m \left(\frac{n-m}{n}\right)^r$$

$$= \sum_{k=0}^m (-1)^k \binom{m}{k}\left(1 - \frac{k}{n}\right)^r.$$

Now, $\left(1 - \frac{kp}{r}\right)^r \to e^{-kp}$ as $r \to \infty$. Hence, $p_r \to \sum_{k=0}^m (-1)^k \binom{m}{k} e^{-kp} = (1 - e^{-p})^m$ as $n \to \infty, r \to \infty$.

EXAMPLE 2.6.3 (*Matching Problem*): Suppose n cards marked $1, \ldots, n$ are shuffled and laid out against n positions, also marked $1, 2, \ldots, n$. Consider the number of matches, that is, the number of times a card marked i is in position $i (i = 1, \ldots, n)$. Find the probability that (a) there is at least one match (b) there are exactly r matches.

(a) Let A_i be the event that there is a match at the ith position. Then

$$P(A_i) = \frac{(n-1)!}{n!} = \frac{1}{n},$$

because, if the entry at the ith position is kept fixed at i, the remaining positions can be filled up in $(n-1)!$ ways out of the total number of $n!$ equally likely ways of distributing the cards.

Similarly, let $A_i A_j$ be the event that there is a match at the ith position and a match at the jth position. Then

$$P(A_i A_j) = \frac{(n-2)!}{n!} = \frac{1}{n(n-1)}.$$

Thus,

$$P(A_i A_j A_k) = \frac{(n-3)!}{n!} = \frac{1}{n(n-1)(n-2)},$$

$$\cdots\cdots\cdots$$

$$P(A_{i_1} A_{i_2} \ldots A_{i_r}) = \frac{1}{n(n-1)\ldots(n-r+1)}.$$

Hence,

$$P(\cup_{i=1}^n A_i) = S_1 - S_2 + S_3 - \ldots + (-1)^{n-1} S_n$$
$$= 1 - \frac{1}{2!} + \frac{1}{3!} - \ldots + (-1)^{n-1}\frac{1}{n!}$$
$$= P_1 \text{ (say)}.$$

$1 - P_1 = p_n$ (say) is the first $(n+1)$ terms in the expansion of e^{-1}. Thus, the probability of no matches in n positions, $p_n \to e^{-1} = 0.36788$ as $n \to \infty$. Hence, $P_1 \to 0.63212$ as $n \to \infty$. We have $|p_n - e^{-1}| \leq \frac{1}{(n+1)!}$. The probability p_n is almost independent of n.

(b) Let $N_{r,n}$ be the number of sample points with exactly r matches out of n positions. Required probability is $N_{r,n}/n!$.

To evaluate $N_{r,n}$ let us assume first that the r positions are fixed. Number of sample points with no matches in the remaining $(n-r)$ positions is $N_{0,(n-r)}$. Again these r positions can be chosen in $\binom{n}{r}$ ways. Thus, $N_{r,n} = \binom{n}{r} N_{0,(n-r)}$. Hence the probability is

$$\frac{\binom{n}{r} N_{0,(n-r)}}{n!} = \frac{N_{0,(n-r)}}{r!(n-r)!} = \frac{p_{n-r}}{r!}$$
$$= \frac{1}{r!}[1 - \frac{1}{1!} + \frac{1}{2!} - \ldots + (-1)^{n-r}\frac{1}{(n-r)!}].$$

Theorem 2.6.3 (*Boole's Inequality*): Given $n(> 1)$ events $A_1, \ldots, A_n$,

$$P(\cup_{i=1}^n A_i) \leq \sum_{i=1}^n P(A_i). \qquad (2.6.10)$$

Proof. Consider first A_1 and A_2. Now

$$A_1 \cup A_2 = A_1 + (A_2 - A_1),$$

the events $A_1, A_2 - A_1$ being disjoint. Hence,

$$P(A_1 \cup A_2) = P(A_1) + P(A_2 - A_1) \leq P(A_1) + P(A_2)$$

since $(A_2 - A_1) \subset A_2$. Thus the inequality is proved for $n = 2$. Now,

$$\cup_{i=1}^n P(A_i) = \left(\cup_{i=1}^{n-1} A_i\right) \cup A_n$$

$$\Rightarrow P\left(\cup_{i=1}^n A_i\right) \leq P\left(\cup_{i=1}^{n-1} A_i\right) + P(A_n) \leq P\left(\cup_{i=1}^{n-2} A_i\right) + P(A_{n-1}) + P(A_n)$$

$$\cdots\cdots\cdots$$

$$\leq \sum_{i=1}^n P(A_i).$$

Corollary 2.6.3.1: Given n events $A_1, \ldots, A_n$,

$$P\left(\cap_{i=1}^n A_i\right) \geq 1 - \sum_{i=1}^n P(A_i^c) = \sum_{i=1}^n P(A_i) - (n-1). \qquad (2.6.11)$$

Proof. Since $\cap_{i=1}^{n} A_i = (\cup_{i=1}^{n} A_i^c)^c$,

$$P\left(\cap_{i=1}^{n} A_i\right) = 1 - P\left(\cup_{i=1}^{n} A_i^c\right) \geq 1 - \sum_{i=1}^{n} P(A_i^c)$$

by (2.6.10). The second expression follows by putting $P(A_i^c) = 1 - P(A_i)$.

Theorem 2.6.4: *(Bonferroni's Inequality)* Given $n(> 1)$ events $A_1, \ldots, A_n$,

$$\sum_{i=1}^{n} P(A_i) - \sum \sum_{i<j=1}^{n} P(A_i A_j) \leq P\left(\cup_{i=1}^{n} A_i\right) \leq \sum_{i=1}^{n} P(A_i). \qquad (2.6.12)$$

We only prove the left side. The right side follows by (2.6.11). For $n = 2$

$$P(A_1) + P(A_2) - P(A_1 A_2) = P(A_1 \cup A_2)$$

and thus the inequality is true for $n = 2$ (by equality). For $n = 3$,

$$P\left(\cup_{i=1}^{n} A_i\right) = \sum_{i=1}^{3} P(A_i) - \sum \sum_{i<j=1}^{3} P(A_i A_j) + P(A_1 A_2 A_3)$$

and the result holds. Assuming that it holds for m events $(3 < m \leq n-1)$, we show that it holds for $m + 1$.

$$
\begin{aligned}
P\left(\cup_{i=1}^{m+1} A_i\right) &= P\left(\cup_{i=1}^{m} A_i \cup A_{m+1}\right) \\
&= P\left(\cup_{i=1}^{m} A_i\right) + P(A_{m+1}) - P\left\{A_{m+1} \cap \left(\cup_{i=1}^{m} A_i\right)\right\} \\
&\geq \sum_{i=1}^{m} P(A_i) - \sum \sum_{i<j=1}^{m} P(A_i A_j) + P(A_{m+1}) - \\
& \quad P\left(\cup_{i=1}^{m} (A_i \cap A_{m+1})\right) \\
&\geq \sum_{i=1}^{m+1} P(A_i) - \sum \sum_{i<j=1}^{m} P(A_i A_j) - \sum_{i=1}^{m} P(A_i A_{m+1}) \\
& \quad \text{by Boole's inequality} \\
&= \sum_{i=1}^{m+1} P(A_i) - \sum \sum_{i<j=1}^{m+1} P(A_i A_j).
\end{aligned}
$$

Theorem 2.6.5: The probability that exactly r of the events $A_1, \ldots, A_n$ occur, is $(1 \leq r \leq n)$

$$P_{[r]} = S_r - \binom{r+1}{r} S_{r+1} + \binom{r+2}{r} S_{r+2} - \ldots +_- \binom{n}{r} S_n. \qquad (2.6.13)$$

Proof. To calculate $P_{[r]}$ we should add the probabilities of all points which occur in exactly r of the sets $A_1, \ldots, A_n$, each point taken only once. Let E be an arbitrary point contained in exactly m of the events $A_1, \ldots, A_n$. We calculate its contribution to the expression in the right side of (2.6.13).

If $m < r, P(E)$ does not contribute. If $m = r, P(E)$ occurs only in S_m and its coefficient in the right side of (2.6.13) is 1. For $m > r$, coefficient of $P(E)$ in the right side of (2.6.13) is

$$\binom{m}{r} - \binom{r+1}{r}\binom{m}{r+1} + \binom{r+2}{r}\binom{m}{r+2} - \ldots +_- \binom{m}{r}\binom{m}{m}, \qquad (2.6.14)$$

since $P(E)$ occurs $\binom{m}{r}$ times in S_r, $\binom{m}{r+1}$ times in S_{r+1}, etc.

Now, (2.6.14)

$$= \binom{m}{r} \left\{ 1 - \binom{m-r}{1} + \binom{m-r}{2} - \ldots + \binom{m-r}{m-r} \right\} = 0.$$

Hence the proof.

Corollary 2.6.5.1: The probability that at least r of the n events $A_1, \ldots, A_n$ occur is

$$P_{[r]} + P_{[r+1]} + \ldots + P_{[n]}. \tag{2.6.15}$$

EXAMPLE 2.6.4 : Considering the matching problem of Example 2.6.3, probability of exactly r matches is

$$\frac{1}{r!} - \binom{r+1}{r} \frac{1}{(r+1)!} + \ldots + \binom{n}{r} \frac{1}{n!}$$

$$= \frac{1}{r!} [1 - \frac{1}{1!} + \frac{1}{2!} - \ldots + \frac{1}{(n-r)!}]$$

as obtained earlier. Similarly, probability of at least r matches can be obtained.

2.6.2 *Conditional probability: Theorem of compound probability*

Suppose in the sample space of a random experiment there are N mutually exclusive, exhaustive and equally likely cases. Of these, N_A cases are favorable to an event A. Out of these N_A cases, N_{AB} cases are favorable to B. Therefore, N_{AB} cases are favorable to both A and B. The proportion N_{AB}/N_A is the proportion of cases among the cases favorable to A which are favorable to B and is defined as the conditional probability of occurrence of B given that the event A has occurred. This is denoted as $P(B|A)$. The proportion N_{AB}/N is the probability of joint occurrence of the events A and B and is $P(AB)$.

Now,

$$\frac{N_{AB}}{N} = \left(\frac{N_A}{N} \right) \left(\frac{N_{AB}}{N_A} \right),$$

i.e.,

$$P(AB) = P(A)P(B|A). \tag{2.6.16}$$

Hence we have the following theorem.

Theorem 2.6.6: The probability of simultaneous occurrence of two events A and B is equal to the probability of A multiplied by the conditional probability of B given that A has occurred (It is also equal to the probability of B multiplied by the conditional probability of A given that B has occurred).

Clearly, (2.6.16) can also be written as

$$P(AB) = P(B)P(B|A). \tag{2.6.17}$$

Whether one should use (2.6.16) or (2.6.17) depends on the physical condition of the experiment, if it is possible for A to occur before B or otherwise.

From (2.6.16), we have the conditional probability

$$P(B|A) = \frac{P(AB)}{P(A)} \tag{2.6.18}$$

provided $P(A) > 0$. The conditional probability $P(B|A)$ is not defined if $P(A) = 0$.

Corollary 2.6.6.1: For three events, A_1, A_2, A_3,

$$\begin{aligned} P(A_1 A_2 A_3) &= P(A_1 A_2)P(A_3|A_1 A_2) \\ &= P(A_1)P(A_2|A_1)P(A_3|A_1 A_2), \quad \text{provided } P(A_1 A_2) > 0. \end{aligned} \tag{2.6.19}$$

Similarly, for four events A_1, A_2, A_3, A_4,

$$\begin{aligned} P(A_1 A_2 A_3 A_4) &= P(A_1 A_2 A_3)P(A_4|A_1 A_2 A_3) \\ &= P(A_1)P(A_2|A_1)P(A_3|A_1 A_2)P(A_4|A_1 A_2 A_3), \end{aligned} \tag{2.6.20}$$

provided $P(A_1 A_2 A_3 A_4) > 0$.

In general, for m events $A_1, \ldots, A_m$

$$P(A_1 \ldots A_m) = P(A_1)P(A_2|A_1) \ldots P(A_m|A_1 \ldots A_{m-1}), \tag{2.6.21}$$

provided $P(A_1 \ldots A_m) > 0$.

EXAMPLE 2.6.5: An urn contains 6 red balls and 4 black balls. Two balls are drawn without replacement. What is the probability that the second ball is red, if it is known that the first is red.

Let A and B be the events that the first and the second ball are red respectively. Hence, $P(AB) = \binom{6}{2}/\binom{10}{2} = \frac{1}{3}, P(A) = \frac{6}{10}$. Thus, $P(B|A) = \frac{5}{9}$.

EXAMPLE 2.6.6: Two unbiased dice are thrown. Find the conditional probability that the two fives occur if it is known that the total is divisible by 5?

Let A be the event that the total is divisible by 5, B the event that the two fives occur; then AB is the event that two fives occur. $P(A) = 7/36, P(AB) = 1/36$. Hence, $P(B|A) = 1/7$.

EXAMPLE 2.6.7: A die is loaded in such a way that the probability of a given number turning up is proportional to that number (e.g., a 6 is twice as probable as 3). What is the probability of a 3, given that an odd number is rolled?

Here, probability of a given number i is $\frac{i}{21}$, since $\sum_{i=1}^{6} i = 21$. Let A be the event, an odd number turns up, B the event 3 is obtained. Thus $P(B|A) = P(AB)/P(A) = \frac{1}{3}$.

Theorem 2.6.7: Let $P(AC) > 0$. Then

$$P(AB|C) = P(A|C)P(B|AC). \tag{2.6.22}$$

Proof. We note that the conditional probabilities involved in (2.6.22) are all defined, since $AC \subset C, P(AC) \leq P(C)$ and thus $P(C) > 0$, because $P(AC) > 0$. Now,

$$P(AB|C) = \frac{P(ABC)}{P(C)} = \frac{P(B|AC)P(AC)}{P(C)} = P(B|AC)P(A|C).$$

We now consider the theorem of total probability.

Theorem 2.6.8: Suppose the event A can occur only along with the event B. Suppose also that B can occur only in m mutually exclusive ways, $B_1, \ldots, B_m$. Then

$$P(A) = \sum_{i=1}^{m} P(B_i)P(A|B_i) \tag{2.6.23}$$

provided $P(B_i) > 0, i = 1, \ldots, m$.

Proof. The events $AB_1, \ldots, AB_m$ are mutually exclusive. Thus $P(A) = P(AB) = \sum_{i=1}^{m} P(AB_i) = \sum_{i=1}^{m} P(B_i)P(A|B_i)$. (Actually, (2.6.23) remains valid even if $P(B_i) = 0$ for some i, when its contribution to the right side of (2.6.23) is zero.)

Corollary 2.6.8.1: For any event B which can occur along with (not necessarily always) $A, P(A) = P(AB) + P(AB^c)$.

EXAMPLE 2.6.8: An urn contains 'a' white balls and 'b' black balls; another contains 'c' white balls and 'd' black balls. One ball is transferred from the first into the second and then one ball is drawn from the later. What is the probability that it is a white ball?

The ball transferred from the first urn may be a black (B) one or a white (W) one. Thus the possible events are BW, WW. Now,

$$P(BW) = \frac{bc}{(a+b)(c+d+1)}, \quad P(WW) = \frac{a(c+1)}{(a+b)(c+d+1)}.$$

Hence, the required probability $= P(BW) + P(WW) = \frac{ac+bc+a}{(a+b)(c+d+1)}$.

EXAMPLE 2.6.9: Two players play a game as follows. Taking turns, they draw the balls out of an urn containing 'a' white and 'b' black balls, one ball at a time. He who extracts the first white one wins the game. What is the probability that the player who starts the game will win the game.

Let A, B be two players. Probability of A to win is

$$= P(W) + P(BBW) + P(BBBBW) + \ldots$$

$$= \frac{a}{a+b} \left[1 + \frac{b(b-1)}{(a+b-1)(a+b-2)} + \right.$$

$$\left. \frac{b(b-1)(b-2)(b-3)}{(a+b-1)(a+b-2)(a+b-3)(a+b-4)} + \ldots \right].$$

2.6.3 Independence of events

We have seen in the previous subsection that $P(B|A)$ does not generally equal $P(B)$. Thus, the information whether A has happened changes the probability of occurrence of B. If, however, $P(B|A) = P(B)$, probability of B does not depend on whether A has or has not happened. Similarly, if $P(A|B) = P(A)$, probability of A does not depend on whether B has happened or not. In this case, events A, B are said to be stochastically independent. Hence we have the following definition.

DEFINITION 2.6.1.*Independence of a pair of events*: Two events A and B are said to be independent if the conditional probability of occurrence of B given the occurrence of the A is the same as its unconditional probability, i.e. if

$$P(B|A) = P(B) \tag{2.6.24}$$

provided the $P(B|A)$ is well-defined.

It follows that

$$P(B|A) = P(B) \Rightarrow P(A|B) = P(A), \qquad (2.6.25)$$

because,

$$P(A)P(B|A) = P(AB) = P(B)P(A|B).$$

Therefore, for independent events A, B,

$$P(AB) = P(A)P(B). \qquad (2.6.26)$$

Again, if (2.6.26) holds, (2.6.24) will also hold.

Therefore, events A and B are stochastically independent if and only if (2.6.26) holds.

The definition of independence can easily be extended to any finite number of events.

Given n events $A_1, \ldots, A_n, P(A_i|A_1, \ldots A_{i-1}A_{i+1} \ldots A_n)$ denotes the conditional probability of A_i given that the events $A_1, \ldots, A_{i-1}, A_{i+1}, \ldots, A_n$ have occurred. Now

$$P(A_i|A_1, \ldots, A_{i-1}, A_{i+1}, \ldots, A_n) = \frac{P(A_1 \ldots A_n)}{P(A_1 \ldots A_{i-1}A_{i+1} \ldots A_n)}, \quad (2.6.27)$$

provided $P(A_1 \ldots A_{i-1}A_{i+1} \ldots A_n) > 0$. We have the following definition of independence of n events.

DEFINITION 2.6.2.*Independence of a number of events*: n events $A_1, \ldots, A_n$ are said to be mutually independent if the conditional probabilities of any of these events, say A_i, given one or more of the remaining events, is equal to the unconditional probability of A_i.

Theorem 2.6.9: (*Compound probability for independent events*) If $A_1, \ldots, A_n$ are independent events,

$$P(A_1 \ldots A_n) = P(A_1)P(A_2) \ldots P(A_n). \qquad (2.6.28)$$

Proof. We have, for any set of events $A_1, \ldots, A_n$,

$$P(A_1 \ldots A_n) = P(A_1)P(A_2|A_1) \ldots P(A_n|A_1, \ldots, A_{n-1}).$$

Since $A_1, \ldots, A_n$ are independent, we have $P(A_2|A_1) = P(A_2), \ldots,$ $P(A_n|A_1, \ldots, A_{n-1}) = P(A_n)$. Thus, the theorem follows. $\square$

By definition, three events A, B, C are independent, if

$$P(A|BC) = P(A|B) = P(A|C) = P(A),$$

$$P(B|AC) = P(B|A) = P(B|C) = P(B),$$

$$P(C|AB) = P(C|A) = P(C|B) = P(C).$$

It is easy to verify that three events A, B, C are stochastically independent if and only if the following conditions hold:

(i) they are pairwise independent, i.e.,

$$P(AB) = P(A)P(B), \ P(AC) = P(A)P(C), \ P(BC) = P(B)P(C)$$

(ii)

$$P(ABC) = P(A)P(B)P(C). \tag{2.6.29}$$

Similarly, m events $A_1, \ldots, A_m$ are stochastically independent if the following conditions hold.

$$
\begin{aligned}
P(A_i A_j) &= P(A_i)P(A_j), \ \ i < j = 1, \ldots, m \\
P(A_i A_j A_k) &= P(A_i)P(A_j)P(A_k), \ \ i < j < k = 1, \ldots, m \\
&\cdots \cdots \\
P(A_1 A_2 \ldots A_m) &= P(A_1)P(A_2) \ldots P(A_m).
\end{aligned}
\tag{2.6.30}
$$

The conditions in (2.6.30) are $\binom{m}{1} + \binom{m}{2} + \ldots \binom{m}{m} = 2^m - m - 1$ in number.

EXAMPLE 2.6.10: Consider families with 3 children. Sample space, here, consists of 8 points $\{(bbb), (bbg), (bgb), (bgg), (gbb), (gbg), (ggb), (ggg)\}$, where 'b' stands for a boy and 'g' for a girl. Assume that the cases are all equally likely. Let A be the event that the family has children of both sexes and B be the event there is at least one girl. Examine if A and B are stochastically independent.

Here $P(A) = 3/4, P(B) = 1/2$. The event AB means the family has exactly one girl. $P(AB) = 3/8 = P(A)P(B)$. Therefore, A, B are independent events.

EXAMPLE 2.6.11: The events A, B are independent. Determine whether the events (a) A and $\bar{B}$ (b) $\bar{A}$ and $\bar{B}$ are independent.

(a) We have $(AB) \cup (A\bar{B}) = A$; also, $A\bar{B}$ and AB are disjoint sets. Thus, $P(A) = P(A\bar{B}) + P(AB)$. Now

$$
\begin{aligned}
P(A)P(\bar{B}) = P(A)\,(1 - P(B)) &= P(A) - P(AB) \\
&= P(A\bar{B}).
\end{aligned}
$$

Thus, A and $\bar{B}$ are independent.

(b) We have $(A\bar{B}) \cup (\bar{A}\bar{B}) = \bar{B}$; also, $P(\bar{B}) = P(A\bar{B}) + P(\bar{A}\bar{B})$. Now

$$P(\bar{A})P(\bar{B}) = P(\bar{B})\{1 - P(A)\} = P(\bar{B}) - P(\bar{B}A) \text{ by } (a)$$
$$= P(\bar{A}\bar{B}).$$

Hence, $\bar{A}, \bar{B}$ are independent.

Similarly, $\bar{A}, B$ are independent.

EXAMPLE 2.6.12: Show that if any of the pairs $\{A, B\}, \{A, B^c\}$, $\{A^c, B\}, \{A^c, B^c\}$ is an independent pair, then all pairs are independent.

We have

$$P(AB^c) = P(A) - P(AB) = P(A)[1 - P(B)] \quad \textit{iff} \text{ the pair}$$
$$\{A, B\} \text{ is independent}$$
$$= P(A)P(B^c).$$

By interchanging the role of A and B in the above result, we obtain $P(A^c B) = P(A^c)P(B)$ iff $\{A, B\}$ is an independent pair. On replacing B by B^c in the last result, we obtain $P(A^c B^c) = P(A^c)P(B^c)$ *iff* $\{A, B^c\}$ is an independent pair. However, this pair is independent *iff* $\{A, B\}$ is independent.

EXAMPLE 2.6.13: Show that any event A is independent of the impossible event ϕ and the sure event S.

We have $P(A\phi) = P(\phi) = 0 = P(A)P(\phi)$ and $P(AS) = P(A) = P(A)P(S)$.

EXAMPLE 2.6.14: Two dice are tossed. Let A denote the event of an odd total, B that of an ace on the first die, C that of having a total of seven. Are A, B, C mutually independent?

Here $P(A) = 1/2, P(B) = 1/6, P(C) = 1/6, P(AB) = 1/12, P(AC) = 1/6, P(BC) = 1/36, P(ABC) = 1/36$. Thus, A, B, C are not mutually independent.

EXAMPLE 2.6.15: A total of n shells is fired at at a target. The probability of the i th cell hitting the target is $p_i, i = 1, \ldots, n$. Find the probability that at least two cells out of n find the target.

The required probability

$$P = 1 - \text{Prob. [at most one shell hits the target]}$$
$$= 1 - \text{Prob. [no shell hits the target]} - \text{Prob. [exactly one shell hits the target]}$$
$$= 1 - P_1 - P_2 \text{ (say)}.$$

Now,

$$P_1 = q_1 \ldots q_n, \quad q_i = 1 - p_i, i = 1, \ldots, n$$

$$P_2 = q_1 \ldots q_n \left[\frac{p_1}{q_1} + \frac{p_2}{q_2} + \ldots + \frac{p_n}{q_n} \right].$$

Hence,

$$P = 1 - \Pi_{i=1}^n q_i \left\{ 1 + \sum_{k=1}^n \frac{p_k}{q_k} \right\}.$$

EXAMPLE 2.6.16: In each of a set of games it is 2 to 1 in favor of the winner of the previous game. What is the chance that the player who wins the first game shall win at least 3 of the next 4 games?

Required probability = prob. of winning 3 games + prob. of winning 4 games.

$$= \binom{4}{3} (\frac{2}{3})^3 (\frac{1}{3}) + (\frac{2}{3})^4 = \frac{16}{27},$$

since 3 games to be won can be chosen in $\binom{4}{3}$ ways, probability of a win is $\frac{2}{3}$, a failure is $\frac{1}{3}$ and the trials are independent.

2.6.4 Bayes theorem

Suppose an event A can occur if and only if one of the hypotheses $B_1, \ldots, B_k$ is true. The probability $P(B_i)$ of occurrence of B_i is known for each $i, i = 1, \ldots, k$. Also, known is the conditional probability $P(A|B_i)$ of occurrence of A given B_i has already occurred, $i = 1, \ldots, k$. We want to find the probability $P(B_i|A)$ of occurrence of B_i given that A has already occurred. This is given in Bayes Theorem as follows.

To fix the idea suppose that a scientist observes a certain event A. He considers that A can happen only if one of the hypotheses $B_1, \ldots, B_k$ holds good. Before observing A, he assigns certain probabilities $P(B_1), \ldots, P(B_k)$ of these hypotheses to be true. He also knows the conditional probability $P(A|B_i)$ of occurrence of A when B_i is true. Now that he has observed the event A, how is he going to change his probabilities to different hypotheses $B_1, \ldots, B_k$? This is obtained by the conditional probability $P(B_i|A)$.

Generally, $P(B_i)$ and $P(B_i|A)$ will not be the same. Thus the occurrence of A generally changes one's assessment of probabilities to different hypotheses. The probabilities $P(B_i)$ which are assigned to B_i without any reference to A are called 'prior' probabilities, $i = 1, \ldots, k$. The probabilities $P(B_i|A)$ which are calculated after A has been observed are called 'posterior' probabilities, $i = 1, \ldots, k$. Our main interest lies here in the hypotheses $B_1, \ldots, B_k$.

Theorem 2.6.10 (*Bayes Theorem*): Let an event A occur only if one of the hypotheses $B_1, \ldots, B_k$ is true. Known are the probabilities $P(B_1), \ldots, P(B_k)$ of the occurrence of $B_1, \ldots, B_k$ respectively. The conditional probabilities $P(A|B_i), i = 1, \ldots, k$ are also known. Then the posterior probability $P(B_i|A)$ is given by

$$P(B_i|A) = \frac{P(B_i)P(A|B_i)}{\sum_{j=1}^{k} P(B_j)P(A|B_j)}, \qquad (2.6.31)$$

provided at least one $P(B_i) > 0$.

Proof. We have

$$P(AB_i) = P(B_i)P(A|B_i) = P(A)P(B_i|A).$$

Hence,

$$P(B_i|A) = \frac{P(B_i)P(A|B_i)}{P(A)}.$$

Now,

$$P(A) = \sum_{j=1}^{k} P(AB_j) = \sum_{j=1}^{k} P(B_j)P(A|B_j)$$

by (2.6.23). Hence the theorem.

EXAMPLE 2.6.17: The first of three urns contains 7 white and 10 black balls, the second contains 5 white and 12 black balls and the third contains 17 white balls (and no black ball). A person chooses an urn at random and draws a ball from it. The ball is white. Find the probabilities that the ball came from (i) the first (ii) the second (iii) the third urn.

Let H_i be the hypothesis that the ith urn was chosen and E be the event a white ball is drawn.

$$P(H_i) = \frac{1}{3}, i = 1, \ldots, 3; P(E|H_1) = \frac{7}{17}, P(E|H_2) = \frac{5}{17}, P(E|H_3) = 1,$$

$$P(H_1|E) = \frac{7/17}{7/51 + 5/51 + 1/3} = \frac{7}{29}, P(H_2|E) = \frac{5}{29}, P(H_3|E) = \frac{17}{29}.$$

EXERCISE 2.6.18: Two production lines manufacture the same types of items. In a given time line 1 turns out n_1 items of which $n_1 p_1$ are defective; in the same time, line 2 turns out n_2 items of which $n_2 p_2$ are defective. Suppose a unit is selected at random from the combined lot produced by the two lines. Let D be the event of a defective item, A the event the unit was produced by the line 1 and B the event it was produced by line 2. Determine $P(A|D), P(B|D)$. Here

$$P(A) = \frac{n_1}{N}, \ P(B) = \frac{n_2}{N}, \ \text{where } N = n_1 + n_2;$$

$$P(D|A) = p_1, P(D|B) = p_2.$$

Hence,

$$P(A|D) = \frac{P(A)P(D|A)}{P(A)P(D|A) + P(B)P(D|B)} = \frac{n_1 p_1}{n_1 p_1 + n_2 p_2}.$$

Similarly,

$$P(B|D) = \frac{n_2 p_2}{n_1 p_1 + n_2 p_2}.$$

If $n_1 p_1 > n_2 p_2, P(A|D) > P(B|D)$.

2.6.5 *Binomial probability*

Consider an experiment having only two outcomes, success (S) and failures (F) occurring with probabilities p and $q(= 1 - p)$ respectively. Such a trial is called a Bernoulli trial. Consider n such experiments conducted independently. We have thus n independent experiments each having two outcomes S and F with probabilities p and q respectively. An example is that of throwing a fair dice n times, turning up a 6 being reckoned aa S and any other point a F when $p = \frac{1}{6}, q = \frac{5}{6}$. We are interested in finding the probability of having r successes out of n such trials. The probability is

$$p_r = \binom{n}{r} p^r q^{n-r} \qquad (2.6.32)$$

because the r trials to result in successes may be chosen in $\binom{n}{r}$ ways. Again, any sequence of S's and F's with rS's and $(n-r)F$'s has the probability $p^r q^{n-r}$, because of independence of trials. The probability (2.6.32) is called the *binomial probability*.

EXAMPLE 2.6.19: A die is tossed 8 times. What is the probability of having 2 sixes? At least 2 sixes?

In this case, the probability of a success is $\frac{1}{6}$ and that of a failure is $\frac{5}{6}$. Hence, probability of having 2 sixes is $\binom{8}{2}(\frac{1}{6})^2(\frac{5}{6})^6$ and at least two sixes is $\sum_{x=2}^{8}\binom{8}{x}(\frac{1}{6})^x(\frac{5}{6})^{8-x}$.

2.7   Limitations of Classical Definition

Though the classical definition of probability is easy to comprehend, it suffers from some serious limitations as follows.

It is limited to situations in which there is only a finite number of possible outcomes. Consider an experiment in which trials are performed until a particular event occurs, say, tossing a coin until a head occurs (Example 2.2.3). Here, as has been seen, the sample space contains a countably infinite number of points rendering N infinite. Thus, the probability of this event cannot be defined by the classical definition. In practice, such an experiment will terminate in a finite number of trials, but there is no 'a priori' assurance that this will happen. It is desirable, both for theoretical

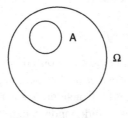

Fig. 2.7.1

and practical reasons, to extend the theory to situations where there is a continuum of possibilities. Suppose a physical variable like the height of an individual, the value of electric current in a wire, etc. are observed. Here, each of the continuum of possible values of these variables is to be regarded as a possible outcome. The number of such outcomes being infinite in number, probability of an event, say, the height of a man lies between 5'2"and 5'3", can not be defined in classical theory.

Consider again the following problem. A point object is thrown at random on the space Ω. We want to find the probability that the object lies within a subspace A in Ω (Fig. 2.7.1). Here, the sample space Ω is a domain on a plane and the elementary events ω are points within the domain. The conditions of the experiment are such that all the points ω are equally likely. Hence, both the total number of points in Ω and A are uncountably infinite and as such probability can not be defined in classical theory. (The above problem is an example of geometric probability dealt with in Section 2.9.)

The classical definition is based on the assumption that the cases are equally likely. This assumption may not be fulfilled in many cases. A simple example is the loaded dice. For a dice, which is asymmetrical in mass or shape, it is not intuitively expected that each side has the same probability of turning up. The classical definition fails to answer the question like the probability of obtaining a six.

In the classical definition, the value of probability comes necessarily as a rational number.

2.8 Statistical or Empirical Definition

We have noted in Section 1.2 that the equal likelihood of each of six faces turning up in rolling a balanced dice can be verified by rolling the dice a large number of times under identical conditions and observing that each of the faces occurs approximately $\frac{1}{6}$ proportion of times. This verifies that the probability of any particular face showing up is $\frac{1}{6}$. Such findings form the basis for empirical or statistical definition of probability.

In summary we observe that the experiments conducted under identical conditions a large number of times show a statistical regularity, namely, the relative frequency of an outcome in several sets of sequences of trials is more or less constant, provided each set consists of a large number of trials. The rate of convergence of relative frequencies to this particular value increases rapidly as the number of trials increases. This constant value may be taken as the probability of the outcome. The basic requirements (assumptions) for this definition is that the experiments must be conducted under identical conditions and the number of trials must be large.

DEFINITION 2.8.1: Let $f_n(A)$ be the number of times in which an event A, the outcome of an experiment, occurs in a series of n repetitions of the trial conducted under identical conditions. The relative frequency of A is $r_n(A) = f_n(A)/n$. The probability of the event A is defined as

$$P(A) = \lim_{n \to \infty} \frac{f_n(A)}{n}$$

provided the limit exists and is unique.

Note that even if the conditions of the experiment are such that the elementary events are not equally likely, the probability can be defined in the statistical sense, though, however, it remains undefined in the classical sense.

2.9 Geometric Probability

Geometric approach to the calculation of probabilities is employed when the sample space Ω includes an uncountable set of elementary events ω, none of which is more likely to occur than the other. Suppose, as in figure 2.7.1, the sample space Ω is a domain in a plane and the elementary events ω are points within Ω. If an event A is represented by a region A within Ω, so that all ω belonging to the region A are favorable to the event A, then the probability of A is

$$P(A) = \frac{\text{area of subdomain } A}{\text{area of domain } \Omega}. \tag{2.9.1}$$

In general,

$$\text{Probability of an event } A = \frac{S_A}{S_\Omega} \tag{2.9.2}$$

where S_A and S_Ω are measures of the specific parts of the region representing the event A and the sample space Ω respectively. The measures may be length, area, volume, etc. according as the region is in one, two, three, .., dimensions.

Formula (2.9.2) is a generalization of the classical formula (2.3.1) to an uncountable set of elementary events. The symmetry of the experimental conditions with respect to the elementary events ω is usually formulated by the assumption of randomness.

The theory of geometric probabilities is often criticized for arbitrariness in determining the probability of events. Many authors point out that for an infinite number of outcomes, the probability cannot be determined objectively, i.e., independently of the mode of computation.

EXAMPLE 2.9.1: A straight line of unit length, say, (interval) $(0, 1)$ is divided into three intervals by choosing two points at random. What is the probability that the three line segments form a triangle?

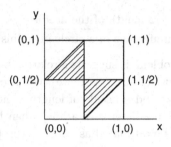

Fig. 2.9.1

Let x, y be the abscissas of any two points chosen at random on $(0, 1)$. A set of necessary and sufficient conditions for the three segments to form a triangle is that the length of any one of the segments be less than the sum of the other two. Hence, we should have either

$$0 < x < \frac{1}{2} < y < 1 \text{ and } y - x < \frac{1}{2},$$

or

$$0 < y < \frac{1}{2} < x < 1 \text{ and } x - y < \frac{1}{2}.$$

This is represented by the shaded area in the figure 2.9.1. Hence, probability is $\frac{1}{4}$.

EXAMPLE 2.9.2: *Buffon's needle problem*: A plane is ruled with parallel straight lines at distance L from each other. A needle of length $l < L$ is thrown at random on the plane (Fig. 2.9.2(a)). Find the probability that it will hit one of the lines.

We characterize the outcomes of the experiment by the numbers: the abscissa x of the center of the needle with respect to the nearest line on the left and by the angle θ the needle makes with the direction of the lines (Fig. 2.9.2(b)). Since the needle is thrown at random, all values of x and θ are equiprobable. Without any loss of generality we consider only the possibilities of the needle hitting the nearest line on the left when $0 \le x \le \frac{L}{2}$ and $0 \le \theta \le \frac{\pi}{2}$ (the probability is same for hitting the nearest line on the right). The sample space Ω is thus a rectangle of area $S_\Omega = \frac{L\pi}{4}$ (Fig. 2.9.2(c)). The needle will hit the line if $x < \frac{l}{2} \sin \theta$. We are thus interested in the event $A = \{x < \frac{l}{2} \sin \theta\}$. Area S_A of $A = \int_0^{\pi/2} \frac{l}{2} \sin \theta d\theta = \frac{l}{2}$. Hence, $P(A) = \frac{S_A}{S_\Omega} = \frac{2l}{\pi L}$

$$= \frac{2 \text{ length of the needle}}{\text{circumference of a circle of radius } \frac{L}{2}}.$$

[This is an interesting problem; it suggests a relation between a pure chance experiment and a famous number π. If we take a graph paper ruled by parallel lines 1 inch apart and a needle of length 1 inch and keep track of the fraction of times the needle crosses a line, when thrown randomly on the graph paper, π may be estimated as about 2/(proportion of crosses)].

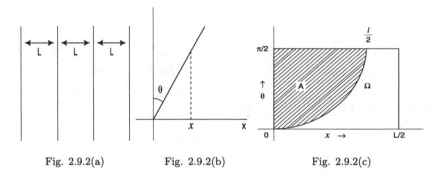

Fig. 2.9.2(a)   Fig. 2.9.2(b)   Fig. 2.9.2(c)

EXAMPLE 2.9.3: A plane is ruled in rectangles with sides L and M. A needle of length $l (l < L, l < M)$ is thrown at random on the plane (Fig. 2.9.3). Find the probability that the needle cuts at least one line.

Let A, B be the events that the needle cuts a vertical line and a horizontal line respectively. The events A, B are independent. Hence,

$$P(A \cup B) = P(A) + P(B) - P(A)P(B).$$

On the basis of Buffon's problem, $P(A) = \frac{2l}{\pi L}, P(B) = \frac{2l}{\pi M}$. Hence, $P(A \cup B) = \frac{2l}{\pi L} + \frac{2l}{\pi M} - \frac{4l^2}{\pi^2 LM}$.

EXAMPLE 2.9.4: In a game a player tosses a coin on the surface of a table ruled one inch square. If the coin (of $\frac{3}{4}''$ in diameter) falls entirely inside a square, the player receives a prize; otherwise, he looses the coin. If the coin lands on the table, what is his chance to win?

We assume that the center of the coin may fall on any region of the table with equal probability. Since the coin is $\frac{3}{8}''$ in radius, its center must lie within $\frac{3}{8}''$ from any edge, for the coin to fall entirely inside a square (Fig. 2.9.4). The restriction generates a square A of side $\frac{1}{4}''$ within which the center of the coin must lie for the player to have a prize. Hence

$$\text{probability} = \frac{\text{area of square } A}{\text{area of the table}} = \frac{1}{16}.$$

We have ignored the thickness of the lines making the square.

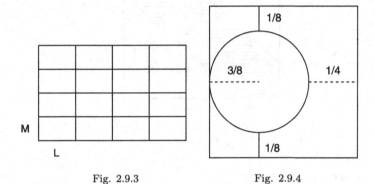

Fig. 2.9.3   Fig. 2.9.4

EXAMPLE 2.9.5: Two persons A and B come to the club between 6 P.M. and 7 P.M. at random points of time and each stays for 10 minutes. What is the chance they will meet?

Let A and B arrive respectively after x and y minutes from 6 P.M. They will meet if $(x - 10) \le y \le (x + 10)$, i.e., if (x, y) falls in the shaded region (Fig. 2.9.5). Area of the unshaded portion is 2500. Hence, the probability they meet is $\frac{11}{36}$.

EXAMPLE 2.9.6: If a stick is broken at random in two parts, (a) what is the average length of the smaller part? (b) what is the average ratio of the smaller length to the larger?

(a) The breaking point may be anywhere on the stick. If the breaking point is on the left half, the smaller half is on the left; its average size is half of that half, i.e. $\frac{1}{4}$ of the total length. Similar argument applies when the breaking point is on the right half.

(b) Suppose that the point falls on the right-hand half. Then $(1-x)/x$ is the ratio of the smaller length to the larger, assuming that the stick is of unit length and x is the length of the breaking point from the left-end. Since x is a random point in the interval $(\frac{1}{2}, 1)$ (and since a similar argument applies if $x \in (0, \frac{1}{2})$), the average value is

$$2 \int_{1/2}^{1} \frac{1-x}{x} dx \approx 2 \log_e 2 - 1 \approx 0.386.$$

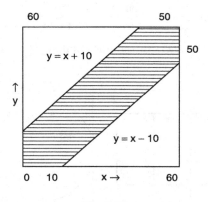

Fig. 2.9.5

2.10   Further Examples

EXAMPLE 2.10.1: A company manufacturing biscuits put a card numbered $1, 2, \ldots, r$ at random in each packet, all numbers being equally likely to occur. If $n(> r)$ packets of biscuits are purchased, find the probability that there is at least one complete set of cards.

Let A_i be the event that the number i is represented, $i = 1, 2, \ldots, r$. Now,

$$\cap_{i=1}^{r} A_i = (\cup_{i=1}^{r} A_i^c)^c.$$

$$P(\cap_{i=1}^{r} A_i) = 1 - P(\cup_{i=1}^{r} A_i^c)$$

$$= 1 - \sum_i P(A_i^c) + \sum\sum_{i<j} P(A_i^c A_j^c)$$
$$\quad - \sum\sum\sum_{i<j<k} P(A_i^c A_j^c A_k^c) + \ldots + (-1)^r P(A_1^c \ldots A_r^c)$$

$$= 1 - \binom{r}{1}\left(\frac{r-1}{r}\right)^n + \binom{r}{2}\left(\frac{r-2}{r}\right)^n - \ldots + (-1)^{r-1}\left(\frac{1}{r}\right)^n$$

$$= \sum_{k=0}^{r}(-1)^k \binom{r}{k}\left(\frac{r-k}{r}\right)^n = \frac{\Delta^r 0^n}{r^n}$$

where $\sum_{k=0}^{r}(-1)^k \binom{r}{k}(r-k)^n = \Delta^r 0^n = \Delta^r x^n]_{x=0}$, Δ being the operator defined by $\Delta f(x) = f(x+1) - f(x)$.

For $n = 10, r = 6$, probability ≈ 0.27181.

EXAMPLE 2.10.2: n balls are distributed among m cells ($n > m$). Find the probability that (i) no cell remains empty, (ii) k cells remain empty.

(i) Let A_i be the event that the ith cell remains empty. Then $P(A_i) = (\frac{m-1}{m})^n$, $P(A_iA_j) = (\frac{m-2}{m})^n$, etc.

The required probability $= 1 - P(\cup_{i=1}^{m} A_i)$

$$= 1 - m(\tfrac{m-1}{m})^n + \binom{m}{2}(\tfrac{m-2}{m})^n - \ldots + (-1)^{m-1}(\tfrac{1}{m})^n$$

$$= \sum_{k=0}^{m}(-1)^k \binom{m}{k}(\tfrac{m-k}{m})^n = \frac{\Delta^m 0^n}{m^n} = p_{n,m} \text{ (say)}.$$

(ii) If n balls are distributed among n cells, let $N_{n,m}$ be the number of sample points leaving no cell empty. Thus, $p_{n,m} = \frac{N_{n,m}}{m^n}$. Suppose first that the k cells to remain empty are specified. The remaining $(m-k)$ cells must remain occupied. The number of sample points in the distribution of n balls in $(m-k)$ cells, in which all these cells remain occupied is $N_{n,m-k}$. Again, the k cells may be chosen in $\binom{m}{k}$ ways. Hence, probability $=$

$$\frac{\binom{m}{k}N_{n,m-k}}{m^n} = \binom{m}{k}(\frac{m-k}{m})^n p_{n,m-k} = \binom{m}{k}\frac{\Delta^{m-k}0^n}{m^n}.$$

EXAMPLE 2.10.3: A number $'a'$ is chosen at random from the numbers $\{1, 2, \ldots, N\}$. Find the probability that (a) the number a is not divisible either by a_1 or by a_2, where a_1, a_2 are fixed natural coprime numbers; (b) the number a is not divisible by either of the numbers $a_1, \ldots, a_k$, where a_i are natural pairwise coprime numbers. Find $\lim_{N\to\infty} p_N$ in case (a) and (b).

We have $p_N = 1 - P['a'$ is divisible by one of a_1 or $a_2] = 1 - q_N$ (say). Then $q_N = P_1 + P_2 - P_3$ where $P_i = $ probability $'a'$ is divisible by $a_i (i = 1, 2)$, $P_3 = $ probability $'a'$ is divisible by both. Thus

$$p_N = 1 - \frac{1}{N}\{[\frac{N}{a_1}] + [\frac{N}{a_2}] - [\frac{N}{a_1 a_2}]\}, \quad \lim_{N\to\infty} p_N (1 - \frac{1}{a_1})(1 - \frac{1}{a_2}).$$

(b)

$$p_N = 1 - \frac{1}{N}\{\sum_{1}^{k}[\frac{N}{a_i}] - \sum\sum_{i<j=1}^{k}[\frac{N}{a_i a_j}] + \sum\sum\sum_{i<j<l=1}^{k}[\frac{N}{a_i a_j a_l}] +$$

$$\ldots + (-1)^{k-1}[\frac{N}{a_1 \ldots a_k}]\};$$

$$\lim_{N \to \infty} p_N = \Pi_{i=1}^k (1 - \frac{1}{a_i}).$$

EXAMPLE 2.10.4: From an urn containing M white and $(N - M)$ black balls, all the balls are drawn one by one and without replacement. Find the probabilities: (i) $A_k = \{$ the kth ball is white $\}$, (ii) $A_{k,l} = \{$ the kth ball and the lth ball are white $\}$, (iii) $B_{k,l} = \{$ the kth ball is black and the lth ball is white$\}$.

Consider any two sequences, say, $BBWWB\ldots, WBBWB\ldots$;

$$P(BBWWB\ldots) = \frac{(N - M)(N - M - 1)M(M - 1)(N - M - 2)\ldots}{N(N - 1)(N - 2)(N - 3)(N - 4)\ldots},$$

$$P(WBBWB\ldots) = \frac{M(N - M)(N - M - 1)(M - 1)(N - M - 2)\ldots}{N(N - 1)(N - 2)(N - 3)(N - 4)\ldots}.$$

Thus, all sequences are equally likely. Hence, $P(A_k) = P(A_1) = \frac{M}{N}, P(A_{kl}) = P(A_{12}) = \frac{M(M-1)}{N(N-1)}, P(B_{kl}) = P(B_{12}) = \frac{(N-M)M}{N(N-1)}$.

EXAMPLE 2.10.5: The probability that a family will have r children is βp^r, where $r = 1, \ldots$. Assuming that the sex ratio at birth is 1:1, (a) show that the probability that the family will have m sons is $2\beta p^m/(2 - p)^{m+1}$; (b) given that a family has at least one boy, what is the probability that there are two or more?

(a) Let B_n be the event that the family has n children and A_k, the event that the family has k boys. Now,

$$P(A_m) = \sum_{n=m}^{\infty} P(B_n)P(A_m|B_n) = \beta \sum_{n=m}^{\infty} p^n \binom{n}{m}(\frac{1}{2})^n$$

$$= \beta \sum_{n=m}^{\infty} \binom{n}{m}(\frac{p}{2})^n = \beta(\frac{p}{2})^m \sum_{n=m}^{\infty} \binom{n}{m}(\frac{p}{2})^{n-m}$$

$$= \beta(\frac{p}{2})^m (1 - \frac{p}{2})^{m+1} = \frac{2\beta p^m}{(2-p)^{m+1}},$$

since,

$$\sum_{n=m}^{\infty} \binom{n}{m}(\frac{p}{2})^{n-m} = 1 + (m + 1)\frac{p}{2} + \binom{m + 2}{2}(\frac{p}{2})^2 + \ldots$$

$$= (1 - \frac{p}{2})^{m+1}.$$

(b) We have

$$P(\cup_{k=2}^{\infty} A_k | \cup_{k=1}^{\infty} A_k) = \frac{P\{(\cup_2^{\infty} A_k) \cap (\cup_{k=1}^{\infty} A_k)\}}{P\{\cup_1^{\infty} A_k\}} = \frac{P(\cup_2^{\infty} A_k)}{P(\cup_1^{\infty} A_k)}$$

$$= \frac{\sum_2^{\infty} P(A_k)}{\sum_1^{\infty} P(A_k)}.$$

Now,

$$\sum_2^{\infty} P(A_k) = \frac{2\beta}{(2-p)} \sum_2 (\frac{p}{2-p})^k = \frac{\beta p^2}{(2-p)^2(1-p)},$$

$$\sum_1^{\infty} P(A_k) = \frac{\beta p}{(2-p)(1-p)}.$$

Hence, probability $= \frac{p}{2-p}$.

EXAMPLE 2.10.6: A player tosses a coin and is to score one point for every head turned up and two for every tail. He is to play on until his score reaches or passes n. Find the probability of his attaining exactly n.

The player attains n, if (i) he tosses a head after attaining $(n-1)$ points or (ii) he tosses a tail after attaining $(n-2)$ points. Hence the difference equation is $p_n = \frac{1}{2} p_{n-1} + \frac{1}{2} p_{n-2}$. Thus,

$$p_n - p_{n-1} = -\frac{1}{2}(p_{n-1} - p_{n-2}) = (-\frac{1}{2})^2 (p_{n-2} - p_{n-3}) = \ldots = (-\frac{1}{2})^{n-2}(p_2 - p_1).$$

Now, $p_1 = \frac{1}{2}, p_2 = \frac{3}{4}$. Hence, $p_n = p_{n-1} + (-1)^n \frac{1}{2^n} = p_{n-1} + (-1)^n \frac{1}{2^n}(\frac{2}{3} + \frac{1}{3})$, i.e., $p_n - \frac{1}{3}(-1)^n \frac{1}{2^n} = p_{n-1} - \frac{1}{3}(-1)^{n-1} \frac{1}{2^{n-1}}$.

Thus, $p_n - \frac{1}{3}(-1)^n \frac{1}{2^n} = p_1 - \frac{1}{3}(-1)\frac{1}{2}$ when $p_n = \frac{1}{3}(2 + (-1)^n \frac{1}{2^n})$.

EXAMPLE 2.10.7: A and B play a series of games with \$a and \$b respectively. The stake is \$1 on each game and no game can be drawn. If the probability of A winning a game is p (a constant), find the initial probability of A's ultimate win (winning B's capital). Show that if b is infinitely large, (i) A is sure to be ruined for $p = \frac{1}{2}$; (ii) A's initial probability of ruin is $\frac{1}{2}$ if $p = 2^{\frac{1}{a}}/(1 + 2^{\frac{1}{a}})$.

Let p_n be the probability of A's ultimate win when his capital is n. Then

$$p_n = p p_{n+1} + (1-p) p_{n-1}, \tag{i}$$

because, when his capital is n, if he wins a game, his capital will be $n + 1$ and if he loses, his capital will be $n - 1$. Initial conditions are: $p_0 = 0$, $p_{a+b} = 1$. From (i),

$$p_{n+1} - p_n = \frac{1-p}{p}(p_n - p_{n-1}) = \ldots = (\frac{1-p}{p})^n p_1 \text{ (by repeated application)}.$$

$$\tag{ii}$$

Hence,

$$p_{n+1} = p_1[1 - (\frac{1-p}{p})^{n+1}]\frac{p}{2p-1}.$$

Using $p_{a+b} = 1$, we have,

$$p_n = \frac{1 - (\frac{1-p}{p})^n}{1 - (\frac{1-p}{p})^{a+b}}.$$

Initial probability of A's win is

$$p_a = \frac{p^a - (1-p)^a}{p^{a+b} - (1-p)^{a+b}} \cdot p^b. \qquad (iii)$$

Initial probability of A's ruin $= 1 - p_a$.

For $p = \frac{1}{2}, p_a = \frac{a}{a+b}$ (from (i) by similar analysis) $\to 0$ as $b \to \infty$.

For $p \neq \frac{1}{2}$ and $b \to \infty$, (from (iii)), $p_a = \frac{1}{2}$ when $p = \frac{2^{1/a}}{1+2^{1/a}}$.

EXAMPLE 2.10.8: Each of the n urns contains 'a' white balls and 'b' black balls. A ball is selected from the first urn and then transferred into the second, another one from the second to the third and so on. Finally, a ball is drawn at random from the nth urn. What is the probability that it is white? How does the probability change when the first ball transferred is known to be white?

Let the urns be numbered $1, 2, \ldots, n$ and p_r the probability of drawing a white ball from the rth urn. Then

$$p_r = p_{r-1}\frac{a+1}{a+b+1} + (1 - p_{r-1})\frac{a}{a+b+1}, \quad r = 2, 3, \ldots, n.$$

We have $p_1 = \frac{a}{a+b}$.

$$p_2 = \frac{a(a+1)}{(a+b)(a+b+1)} + \frac{ba}{(a+b)(a+b+1)} = \frac{a}{a+b}.$$

Thus $p_n = \frac{a}{a+b}$.

In the second case, $p_1 = 1, p_2 = \frac{a+1}{a+b+1} = \frac{a}{a+b} + \frac{b}{(a+b)(a+b+1)}, p_3 = \frac{(a+1)^2}{(a+b+1)^2} + \frac{ab}{(a+b+1)^2} = \frac{a}{a+b} + \frac{b}{(a+b)(a+b+1)^2}$. Proceeding in this way,

$$p_n = \frac{a}{a+b} + \frac{b}{(a+b)(a+b+1)^{n-1}} \to \frac{a}{a+b} \text{ as } n \to \infty.$$

EXAMPLE 2.10.9: Two urns contain respectively 'a' white and 'b' black balls and 'b' white and 'a' black balls. A series of random drawings is made as follows: (i) each time only one ball is drawn and it is returned to the same urn from which it was obtained; (ii) if the ball drawn is white, the next draw is made from the first urn; otherwise, the next draw is made from the second urn.

Assuming that the first ball drawn is drawn from the first urn, find the probability p_n that the nth ball drawn is white and obtain its limiting value as $n \to \infty$.

We have

$$p_n = p_{n-1}\frac{a}{a+b} + (1 - p_{n-1})\frac{b}{a+b} = \frac{a-b}{a+b}p_{n-1} + \frac{b}{a+b}$$
$$= \alpha + \beta p_{n-1} \text{ (say), i.e. } \frac{p_n}{\beta^n} = \frac{\alpha}{\beta^n} + \frac{p_{n-1}}{\beta^{n-1}}.$$

This is a difference equation in p_n. Replacing n by $n-1, n-2, \ldots, 2$ and adding

$$\frac{p_n}{\beta^n} = \alpha\left(\frac{1}{\beta^n} + \ldots + \frac{1}{\beta^2}\right) + \frac{p_1}{\beta}$$
$$\text{or } p_n = \frac{\alpha(1-\beta^{n-1})}{1-\beta} + p_1\beta^{n-1} = \frac{1}{2} + \frac{1}{2}\left(\frac{a-b}{a+b}\right)^n \text{ (since } p_1 = 1\text{)}$$
$$\to \frac{1}{2} \text{ as } n \to \infty.$$

EXAMPLE 2.10.10: A unbiased dice is tossed $(r + s)$ times. Find the probability of getting $r(> s)$ consecutive sixes.

As $r > s$, only one sequence of r consecutive sixes can occur. Let p_a be the probability of getting r consecutive sixes in $(r + a)$ tosses. Hence, $p_s = p_{s-1} +$ probability of getting a 6 at the $(r + s)$th throw immediately after obtaining $(r-1)6$'s preceded by a number other than $6 = p_{s-1}+(\frac{1}{6})^r.\frac{5}{6}$.

Similarly,

$$p_{s-1} = p_{s-2} + \left(\frac{1}{6}\right)^r.\frac{5}{6}, \ldots\ldots\ldots, p_1 = p_0 + \left(\frac{1}{6}\right)^r.\frac{5}{6}.$$

Adding

$$p_s = p_0 + \frac{5s}{6^{r+1}} = \frac{1}{6^r} + \frac{5s}{6^{r+1}} = \frac{6 + 5s}{6^{r+1}}.$$

EXAMPLE 2.10.11: A safety device is designed to have a high conditional probability of operating when there is a failure (dangerous condition) and a high conditional probability of not operating when a failure does not occur. For a particular band of safety device both these probabilities are $(1 - \alpha)$. Given that a dangerous condition occurs when probability is β, find the conditional probability that there was a failure when the safety device worked.

Let W = the event the safety device works; F = there is a failure (dangerous condition). Then $P(F) = \beta, P(W|F) = 1 - \alpha, P(W^c|F^c) = 1 - \alpha$. Hence,

$$
\begin{aligned}
P(F|W) &= \frac{P(F)P(W|F)}{P(F)P(W|F)+P(F^c)P(W^c|F^c)} \\
&= \frac{\beta(1-\alpha)}{\beta(1-\alpha)+(1-\beta)(1-\alpha)} \\
&= 0.04676 \text{ for } 1 - \alpha = 0.98, \beta = .001.
\end{aligned}
$$

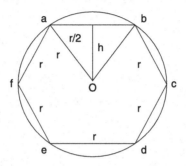

Fig. 2.10.1

EXAMPLE 2.10.12: If a chord is selected at random on a fixed circle, find the probability that its length exceeds the radius of the circle.

Depending on the manner, how the term 'random' is explained, we have three different solutions. Let the radius of the circle be r.

Solution 1 Assume that the distance of the chord from the center of the circle is a random value in $[0, r]$. A regular hexagon $abcdef$ of side r can be inscribed in the circle (Fig. 2.10.1). Any chord lying within this polygon will have length greater than r. Hence, the required probability is $\frac{h}{r}, h$ being the distance of the chord ab from the center 0. Now, $h = r\frac{\sqrt{3}}{2}$. Hence, probability = $\frac{\sqrt{3}}{2} \approx 0.866$.

Solution 2 Assume that the mid-point of the chord is evenly distributed over the interior of the circle. The chord is larger than the radius when the mid-point of the chord is within distance h from the center of the circle. Thus all points within the circle of radius h with O as center can serve as the midpoint of a chord of length greater than r. Hence, probability = $\frac{\pi h^2}{\pi r^2} = \frac{3}{4}$.

Solution 3 Assume that the chord is determined by two points chosen randomly on the circumference of the circle. Assume without loss of generality that one point is fixed at a. The other point must fall anywhere on the arc fab to have the length of the chord shorter than r. The length of the arc fab is $\frac{1}{3}$(length of the circumference of the circle). Hence, probability is $\frac{2}{3}$.

2.11 Exercises and Complements

2.1 Considering each of the following sets of events, state whether they are sets of : (i) elementary events, (ii) mutually exclusive events (iii) exhaustive events, (iv) equally likely events.

(a) Two coins are tossed. The events are: A_1: two heads; A_2: two tails; A_3: one head and one tail.

[(i) No, (ii)Yes, (iii) Yes, (iv) No]

(b) A dice is rolled. The events are: $B_1 : \{1 \text{ or } 2\}$, $B_2 : \{2 \text{ or } 3\}$, $B_3 : \{3 \text{ or } 4\}$, $B_4 : \{4 \text{ or } 5\}$, $B_5 : \{5 \text{ or } 6\}$.

[(i) No, (ii) No, (iii) Yes, (iv) Yes]

(c) A gun is fired at a target. The events are: C_1: a hit, C_2: a miss.

[(i) No, (ii) Yes, (iii) Yes, (iv) No]

2.2 The last three digits of a telephone number beginning 584 are missing. Assuming that all combinations and arrangements of the last three digits are equally likely, find the probability of the following events. A: distinct digits different from 5,8,4 are missing; B: identical digits are missing; C: two of the missing digits coincide.

$[P(A) = (7)_3/10^3, \; P(B) = 0.001, \; P(C) = 0.27]$

2.3 What is the probability that the birthdays of r randomly selected persons will be all different?

$[(365)_r/365^r; \text{ for } r = 30, \text{ probability} = 0.294]$

2.4 If n balls are placed at a random order into n cells, find the probability that exactly one cell remains empty. $[\binom{n}{2}.n!/n^n]$

2.5 From a population of n elements, a sample of size r is taken. Find the probability that none of the N prescribed elements will be included in the sample assuming the sample to be (a) without replacement (b) with replacement.

$[(a)\ \binom{n-N}{r}/\binom{N}{r},\ (b)\ (n-N)^r/n^r]$

2.6 Two boxes each have r balls, labeled $1, 2, \ldots, r$. A random sample of size $n(\leq r)$ is drawn without replacement from each box. Find the probability that the sample contains exactly k balls having the same number in common.

$[\ \binom{r}{n}\binom{n}{k}\binom{r-n}{n-k}/\binom{r}{n}^2\ ]$

2.7 From the population of 5 symbols a, b, c, d, e, a sample of size 25 is taken. Find the probability that the sample will contain 5 symbols of each kind.

$[\ \frac{25!}{(5!)^5 5^{25}}\ ]$

2.8 Calculate the probability $p_0(r), p_1(r), \ldots, p_4(r)$ that among r bridge cards drawn at random from a well-shuffled full pack of cards, there are $0, 1, \ldots, 4$ aces respectively. Verify that $p_0(r) = p_4(52 - r)$.

[Hints: $p_1(r) = 4\binom{48}{r-1}/\binom{52}{r} = 4r(52 - r)_3/(52)_4.$]

2.9 Find the probability p_N that a natural number chosen at random from the set $1, \ldots, N$ is divisible by a fixed natural number k. Find $\lim_{N\to\infty} p_N$.

$[p_N = \frac{1}{N}\{[\frac{N}{k}]\}$ where $[z]$ is the integer part of z; $\lim_{N\to\infty} p_N = \frac{1}{k}.]$

2.10 One sequence is chosen at random from the set of all sequences of length n consisting of numbers 0, 1, 2. Find the probabilities of the following events: $A = \{$ the sequence begins with 0$\}$, $B = \{$ the sequence contains exactly $(m + 2)$ zeroes, two of them being at the end points of the sequence$\}$, $C = \{$ the sequence contains exactly m units$\}$, $D = \{$ the sequence contains exactly m units$\}$, $D = \{$ the sequence contains exactly m_0 zeroes, m_1 units and m_2 twos.$\}$

$$[P(A) = \tfrac{1}{3}, P(B) = \binom{n-2}{m}2^{n-m-2}/3^n\ (0 \leq m \leq n - 2),$$
$$P(B) = \binom{n}{m}2^{n-m}/3^n,$$
$$P(D) = \binom{n}{m_0}\binom{n-m_0}{m_1}\tfrac{1}{3^n} = \frac{n!}{m_0!m_1!m_2!3^n},\ m_0 + m_1 + m_2 = n.]$$

2.11 A box has r balls labeled $1, 2, \ldots r$. $N(\leq r)$ balls are selected at random from the box, their numbers noted and are then returned to the box. If the procedure is repeated r times, what is the probability that none of the original N balls are duplicated?

$[\{\binom{r-N}{N}/\binom{r}{N}\}^{r-1}]$

2.12 (*Coupon Collector's problem*) In every box of cornflakes is a picture of a baseball player. A full set is represented by m players. Find the probability that n boxes of cereals must be purchased to obtain a full set of pictures.

[Let Z denote the number of boxes that must be purchased to obtain a full set of m pictures. If the pictures are taken as cells, boxes as balls, the problem reduces to finding the number of balls that must be distributed so that all the cells remain occupied. By the notation of Example 2.10.2, $P(Z \le n) = p_{n,m}$. Hence,

$$P(Z = n) = P(Z \le n) - P(Z \le n - 1) = p_{n,m} - p_{n-1,m}$$
$$= \sum_{k=1}^{m-1}(-1)^{k-1}\binom{m-1}{k-1}(\frac{m-k}{m})^{n-1}.]$$

2.13 In a lottery m tickets are drawn at a time out of the total number of n tickets and returned before the next drawing is made. What is the probability that in k drawings, each of the numbers $1, 2, \ldots, n$ will appear at least once?

[Let A_i : the event, number i does not occur. The required probability $= 1 - P(\cup_{i=1}^n A_i)$. Now,

$$P(A_i) = [\binom{n-1}{m}/\binom{n}{m}]^k = (1 - \frac{m}{n})^k$$
$$P(A_i A_j) = [(1 - \frac{m}{n})(1 - \frac{m}{n-1})]^k, \quad \text{etc.}]$$

2.14 (*Polay's urn model*) An urn contains b black balls and r red balls. A ball is drawn at random and then replaced with c additional balls of the same color. The process is repeated. Let B_k denote the event that the kth ball drawn is black and R_k, the event that the kth ball drawn is red. Determine (a) $P(R_2|R_1)$, (b) $P(B_1 B_2 B_3)$. Show by induction that the probability of selecting a red ball at any trial is $\frac{r}{b+r}$.

[(a) $\frac{r+c}{b+r+c}$ (b) $\frac{b(b+c)(b+2c)}{(b+r)(b+r+c)(b+r+2c)}$]

2.15 An urn contains a white balls and b black balls. After a ball is drawn, it is to be returned to the urn if it is white; if it is black, it is replaced by a white ball from another urn. Show that the probability of drawing a white ball, after the operation has been repeated n times is

$$p_n = 1 - \frac{b}{a+b}[1 - \frac{1}{a+b}]^n.$$

[Let A_r denote the expected number of white balls, after r operations (for definition of expectation, see Chapter 5). Then

$$A_{r+1} = A_r + 1(1 - \frac{A_r}{a+b}) = A_r(1 - \frac{1}{a+b}) + 1, \ A_0 = a.$$

Thus,

$$A_n = (a+b) - b(1 - \frac{1}{a+b})^n \quad \text{(by repeated application).}$$

Now, $p_n = \frac{A_n}{a+b}$.]

2.16 Two urns contain respectively ('a' white, 'b' black), (c white, d black) balls. One ball is taken from the first urn and transferred into the second and simultaneously, one ball is taken from the second urn and transferred into the first. Find the probability p_n of drawing a white ball from the first urn after the operation has been repeated n times.

[Let A_r, C_r denote the expected number of white balls in the first and the second urn respectively, after the operation has been repeated r times, $A_r + C_r = a + c, A_0 = a, C_0 = c$.

$A_{r+1} = A_r + ($ prob. a white ball is transferred from the second urn$)$
$\quad\quad - $ (prob. a white ball is transferred from the first urn)
$\quad\quad = A_r + \frac{C_r}{c+d} - \frac{A_r}{a+b}$,

when

$$p_n = \frac{A_n}{a+b} = \frac{a+c}{a+b+c+d} + \frac{ad-bc}{(a+b)(a+b+c+d)}[1 - \frac{1}{a+b} - \frac{1}{c+d}]^n$$

$$\rightarrow \frac{a+c}{a+b+c+d} \quad \text{as } n \rightarrow \infty.]$$

2.17 Using a probabilistic argument, prove the identity $(A > a)$,

$$1 + \frac{A-a}{A-1} + \frac{(A-a)_{(2)}}{(A-1)_{(2)}} + \ldots + \frac{(A-a)!}{(A-1)_{(A-a-1)}.a} = \frac{A}{a}.$$

[Suppose an urn contains A balls of which a are drawn. The balls are drawn at random without replacement. The probability that a white ball will be drawn ultimately is one. Thus,

$$\frac{a}{A} + \frac{(A-a)a}{A(A-1)} + \frac{(A-a)(A-a-1)a}{A(A-1)(A-2)} + \ldots + \frac{(A-a)!}{A(A-1)\ldots(a+1)} = 1]$$

2.18 Let the events $A_1, \ldots, A_n$ be mutually independent. If $P(A_i) = p_i$, find the probability that (i) none of the events will occur; (ii) at least one of the events will occur; (iii) at most one of the events will occur.

[(i) $\Pi_{i=1}^n (1 - p_i) = Q$ (say), (ii) $1 - Q$, (iii) $\Pi_{i=1}^n (1 - p_i)[1 + \sum_{k=1}^n \frac{p_k}{1-p_k}]$]

2.19 Show that if A and B are mutually exclusive events and each has positive probability, they cannot be independent. If they have positive probabilities and are independent, they cannot be mutually exclusive.

2.20 In an objective type question there are n alternatives. Let A be the event of a correct choice from among these alternatives. Let B be the event of knowing which of the alternatives is correct. Suppose $P(B) = p, 0 < p < 1$. (a) Determine $P(B|A)$; (b) Show that $P(B|A) \geq P(B)$ and $P(B|A)$ increases with increasing n for fixed p.

$$\begin{aligned}
[P(B) &= p, P(B^c) = 1 - p, P(A|B) = 1, P(A|B^c) = \tfrac{1}{n}. \\
P(B|A) &= \frac{P(B)P(A|B)}{P(B)P(A|B)+P(B^c)P(A|B^c)} = \frac{p}{p+(1-p)\frac{1}{n}} \\
&= \frac{p}{\frac{n-1}{n}p+\frac{1}{n}}, \quad \text{because } \tfrac{n-1}{n}p + \tfrac{1}{n} < 1.
\end{aligned}$$

Also,

$$\begin{aligned}
P_{n+1}(B|A) - P_n(B|A) &= \frac{(n+1)p}{np+1} - \frac{np}{(n-1)p+1} \\
&= \frac{p(1-p)}{(np+1)\{(n-1)p+1\}} > 0.]
\end{aligned}$$

2.21 An urn contains n balls, each of different colors, of which one is white. Two independent observers, each with probability 0.1 of telling the truth, assert that a ball drawn at random from the urn is white. Prove that the probability that the ball is, in fact, white is $\frac{n-1}{n+80}$. Also, show that if $n < 20$, this probability is less than the probability that at least one of the observers is telling the truth.

2.22 A noisy communication channel is transmitting a message which is a sequence of 0's and 1's, a 1 being sent with probability p. A transmitted symbol may or may not be perturbed into the opposite symbol in the process of transmission. It is known that a 1 is converted into 0 with probability p_1 and a 0 is converted into 1 with probability p_2 during the process of transmission. Calculate

(i) the conditional probability that a 1 was sent given that a 0 is received at the reception;

    (ii) the conditional probability that a 0 was sent given that a 1 is
received at the reception.

$$[(i)\ P(A|B) = \frac{pp_1}{pp_1 + (1-p)(1-p_2)}; \ (ii)\ P(B|A) = \frac{(1-p)p_2}{(1-p)p_2 + p(1-p_1)}.]$$

2.23 A and B have, respectively, $n+1$ and n coins. If they toss their coins
simultaneously, what is the probability that (a) A will have more heads
than B? (b) A and B will have equal number of heads? (c) B will have
more heads than A?

[Let P_n be the probability of A having more heads than B.

$$P_n = \sum_{x=1}^{n+1} \sum_{a=0}^{n} \binom{n+1}{x+a}\binom{n}{a}\frac{1}{2^{n+1}}.$$

Now, considering the coefficient of t^x in

$$(1+t)^{n+1}(1+\frac{1}{t})^n = \frac{(1+t)^{2n+1}}{t^n}, \quad \sum_{a=0}^{n}\binom{n+1}{x+a}\binom{n}{a} = \binom{2n+1}{n+x}.$$

Hence,

$$P_n = \frac{\sum_{x=1}^{n+1}\binom{2n+1}{n+x}}{2^{n+1}} = \frac{1}{2}\cdot\frac{2^{n+1}}{2^{n+1}} = \frac{1}{2}.$$

(b)

    The probability $= \frac{1}{2^{2n+1}}\sum_{x=1}^{n}\binom{n+1}{x}\binom{n}{n-x} = \frac{\binom{2n+1}{n}}{2^{2n+1}} = Q_n$ (say) .

(c) The probability $= 1 - P_n - Q_n = \frac{1}{2} - Q_n$]

2.24: Let the probability that the weather on any day is of the same type
(rain or no rain) as the previous day be p. Let P_1 be the probability of no
rain on the first day of the year. What is the probability P_n of rain on the
nth day?

$$[P_n = P_{n-1}p + (1 - P_{n-1})(1-p) = (2p-1)P_{n-1} + 1 - p$$
$$= (2p-1)\{(2p-1)P_{n-2} + (1-p)\} + (1-p)$$
$$= \ldots\ldots\ldots\ldots$$
$$= (2p-1)^{n-1}P_1 + (1-p)\{(2p-1)^{n-2} + (2p-1)^{n-3} + \ldots + 1\}$$
$$= (2p-1)^{n-1}P_1 + \frac{1}{2}\{1 - (2p-1)^{n-1}\}$$
$$= \frac{1}{2}\{1 + (2P_1 - 1)(2p-1)^{n-1}\}.]$$

2.25 The train R-passenger and C-express arrive at B junction on their way to Kolkata between 8-50 AM and 9-15 AM and each of the trains wait for the other for at most 10 minutes. Supposing both the trains arrive at the junction at random moments between 8-50 AM and 9-15 AM, what is the probability that one person traveling by the passenger on an urgent business will be able to get the express train at the junction to reach Kolkata in time?

$$\left[\frac{9}{25} \right]$$

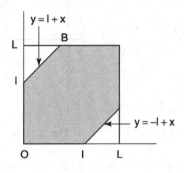

Fig. 2.11.1

2.26 Two points x, y are tossed at random and independently on a line-segment of length L. What is the probability that the distance between the two points does not exceed l?

[Suppose x falls in the interval $(0, L)$ of the x-axis, y in the interval $(0, L)$ of the y-axis. The required probability is the probability that a point (x, y) tossed at random into the square will fall in the region bounded by the lines $y = l + x$ and $y = x - l$ (Fig. 2.11.1). Hence, probability $=$ $\frac{\text{area of the shaded region}}{L^2} = \frac{2L.l - l^2}{L^2} = \frac{l(2L - l)}{L^2}.$]

2.27 A rod of length l is broken into three parts. What is the probability that a triangle can be formed from these parts?

[Let the length of these three parts be $a, b - a, l - b (a < b < l)$. A set of necessary and sufficient conditions for the three segments to form a triangle is that the length of any of the segments be less than the sum of the other two. Hence, we must have

$$a + (b - a) > l - b \quad \text{or} \quad b > (l/2)$$
$$(b - a) + (l - b) > a \quad \text{or} \quad a < (l/2)$$
$$(l - b) + a > b - a \quad \text{or} \quad b - a < (l/2).$$

Hence, $0 \leq a \leq \frac{l}{2}$ and $\frac{l}{2} < b \leq \frac{l}{2} + a$. The required probability is

$$\left(\int_0^{l/2} \int_{l/2}^{(l/2)+a} dadb \right) / \left(\int_0^l \int_0^a dadb \right) = \frac{1}{4}.$$

2.28 (*Bertrand's paradox*) A chord is chosen at random in a circle of radius r. What is the probability that its length will exceed the length of the side of the equilateral triangle inscribed in the circle?

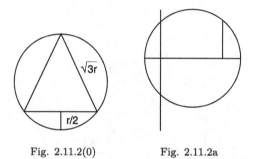

Fig. 2.11.2(0)   Fig. 2.11.2a

[**Solution 1** Consider any chord. Draw a diameter perpendicular to it. It is seen that only chords that intersect the diameter between $\frac{1}{4}$ and $\frac{3}{4}$ of its length exceeds the side of the regular triangle inscribed in the circle. Hence, probability is $\frac{1}{4}$ (Fig. 2.11.2(a)).

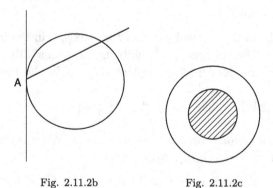

Fig. 2.11.2b   Fig. 2.11.2c

Solution 2 Fix one end of the chord on the circle, say A, in advance. The tangent to the circle at A and the two sides of the regular triangle with vertex at A form three 60° angles. Clearly, chords falling in the middle angle satisfy the conditions of the event. Hence, probability is $\frac{1}{3}$ (Fig. 2.11.2(b)).

Solution 3 To define the position of the chord, it is sufficient to specify its mid-point. The chord will satisfy the conditions of the problem if its mid-point lies within distance $\frac{r}{2}$ from the center, i.e., within a circle concentric with the given one, but with radius $\frac{r}{2}$ (Fig. 2.11.2(c)). Hence probability $= \frac{1}{4}$.]

Chapter 3

Axiomatic Approach

3.1 Introduction

In the previous chapter we have noted the short-comings of the classical definition. The vagueness of the classical definition sometimes led to paradoxical results (example: Bertrand's paradox). We shall consider here axiomatic approach to the theory of probability.

The chapter is organized as follows. Section 3.2 considers set-algebra and introduces the concepts of fields and σ-fields. The next section discusses point functions, set functions, including inverse functions. The concepts of measure and measurable sets are also addressed. Section 3.4 incorporates the notion of measurable functions.

Against this background axiomatic definition of probability is introduced in Section 3.5. The next section deals with conditional probability measure. The concepts of independent trials and the product-space are examined in the concluding section.

3.2 Set Algebra, Fields, σ-Fields

3.2.1 *Algebra of sets*

In this chapter we shall denote by Ω the universal set or the whole space.

DEFINITION 3.2.1: The union of an arbitrary, possibly uncountable number of sets is given by

$$\cup_{i \in I} A_i = \{\omega \in \Omega : \omega \in A_i \text{ for some } i \in I\} \qquad (3.2.1)$$

where I is an arbitrary index set assumed non-empty. If I is finite we

63

have a finite union. If I is countable and is given by $\{1, 2, \ldots, \}$, we have a countable union denoted by $\cup_{i=1}^{\infty} A_i$.

Similarly, the intersection of $A_i (i \in I)$ is

$$\cap_{i \in I} A_i = \{\omega \in \Omega : \omega \in A_i \ \forall \ i \in I\}. \tag{3.2.2}$$

As before, the operations of $\cup$ and $\cap$ are reflexive, commutative, associative and distributive for sets $A_i, i \in I$. Also, de Morgan's rule of complements hold, i.e.

$$\begin{aligned} (\cup_{i \in I} A_i)^c &= \cap_{i \in I} A_i^c \\ (\cap_{i \in I} A_i)^c &= \cup_{i \in I} A_i^c. \end{aligned} \tag{3.2.3}$$

This can be proved as in Lemma 2.7.1.

Lemma 3.2.1: Given n sets, $A_1, A_2, \ldots, A_n$, there exists a class of disjoint sets $\{B_i, i = 1, \ldots, n\}$ such that

$$\cup_{i=1}^{n} A_i = B_1 + B_2 + \ldots + B_n \tag{3.2.4}$$

where $B_i \cap B_j = \phi, i \neq j = 1, 2, \ldots, n$.

Proof: The lemma will be proved by induction. Evidently,

$$A_1 \cup A_2 = A_1 + A_1^c A_2 = B_1 + B_2 \text{ (say)}, \tag{3.2.5}$$

where B_1, B_2 are disjoint sets. Thus, the lemma is true for $n = 2$. Suppose, it is true for $n = m (\geq 2)$. Then

$$\cup_{i=1}^{m+1} A_i = (\cup_{i=1}^{m} A_i) \cup A_{m+1}$$

$$= (\textstyle\sum_{i=1}^{m} B_i) \cup A_{m+1}$$

$$= (\textstyle\sum_{i=1}^{m} B_i) + (\textstyle\sum_{i=1}^{m} B_i)^c \cap A_{m+1} \quad \text{(by (3.2.5))} \tag{3.2.6}$$

$$= \textstyle\sum_{i=1}^{m} B_i + B_{m+1} \quad \text{(say)}$$

where B_{m+1} and $\sum_{i=1}^{m} B_i$ are disjoint and hence $B_1, \ldots, B_{m+1}$ are mutually disjoint. Hence if the lemma holds for $n = m$, it holds for $n = m + 1$. Therefore, by induction, the lemma holds for any arbitrary n. $\qquad\square$

Corollary 3.2.1.i: Given n sets, $A_1, \ldots, A_n$ we can write

$$\cup_{i=1}^{n} A_i = A_1 + A_1^c A_2 + A_1^c A_2^c A_3 + \ldots + A_1^c A_2^c \ldots A_{n-1}^c A_n \tag{3.2.7}$$

as the sum of n disjoint sets $A_1, A_1^c A_2, A_1^c A_2^c A_3, \ldots$, etc.

Corollary 3.2.1.ii: For a countably infinite number of sets $A_1, A_2, \ldots$,

$$\cup_{i=1}^{\infty} A_i = A_1 + A_1^c A_2 + A_1^c A_2^c A_3 + \ldots \tag{3.2.8}$$

Proof. Suppose $\omega \in \cup_{i=1}^{\infty} A_i$. Then $\omega \in A_i$ for some i. Then $\omega \in A_1$ or $\omega \in A_1^c$. In the later case, $\omega \in A_1^c A_2$ or $\omega \in A_1^c A_2^c$. In the later case, $\omega \in A_1^c A_2^c A_3$ or $\omega \in A_1^c A_2^c A_3^c$. Continuing in this manner, ω has to belong to either of $A_1, A_1^c A_2, A_1^c A_2^c A_3, \ldots, A_1^c A_2^c \ldots A_{n-1}^c A_n$, etc. Hence ω belongs to the right side of (3.2.8).

Conversely, suppose ω belongs to the right side of (3.2.8). Then $\omega \in A_1^c A_2^c \ldots A_{k-1}^c A_k$ for some k. But $A_1^c A_2^c \ldots A_{k-1}^c A_k \subset A_k$. Hence $\omega \in A_k$ for some k. Thus, $\omega \in \cup_{i=1}^{\infty} A_i$. Therefore, the sets on the two sides of (3.2.8) in which ω belongs are equivalent. Hence the proof. □

DEFINITION 3.2.2: *Sequence and limits of sets*: To every integer $n = 1, 2, \ldots$, we assign a set. The ordered class of sets $A_1, A_2, \ldots$ is a sequence $\{A_n\}$ of sets.

The sequence $\{A_n\}$ is monotonically increasing if $A_n \subseteq A_{n+1}$. In this case $\cup_{k=1}^{n} A_k = A_n; \cup_{k=1}^{\infty} A_k = A$ is called the limit of $\{A_n\}$. Symbolically $A_n \uparrow A$.

The sequence $\{A_n\}$ is monotonically decreasing if $A_n \supseteq A_{n+1}$. Here $\cap_{k=1}^{n} A_k = A_n; \cap_{k=1}^{\infty} A_k = A$ is called the limit of $\{A_n\}$. Symbolically, $A_n \downarrow A$.

EXAMPLE 3.2.1: Let $A_n = \{\omega : 0 < \omega < 1 - \frac{1}{n}\} \subset (-\infty, \infty) = \Omega$. Then $A_n \uparrow A = \{\omega : 0 < \omega < 1\}$.

Let $B_n = \{\omega : 0 < \omega < 1 + \frac{1}{n}\} \subset (-\infty, \infty) = \Omega$. Then $B_n \downarrow B = \{\omega : 0 < \omega \leq 1\}$. □

Now we define the limit of an arbitrary sequence of sets.

For any sequence $\{A_n\}$, define

$$B_n = \inf_{k \geq n} A_k = \cap_{k=n}^{\infty} A_k$$

$$= \{\omega : \omega \text{ belong to all } A_k \text{ except perhaps for } A_1, \ldots, A_{n-1}\}.$$
$$C_n = \sup_{k \geq n} A_k = \cup_{k=n}^{\infty} A_k$$

$$= \{\omega : \omega \text{ belongs to at least one of } A_n, A_{n+1}, \ldots\}.$$

Thus, B_n is a monotonically increasing sequence with limit

$$B = \cup_{n=1}^{\infty} \cap_{k=n}^{\infty} A_k = \liminf_{n} A_n = \underline{\lim} A_n. \tag{3.2.9}$$

Thus $\omega \in \liminf_n A_n$ *iff* for some $n, \omega \in A_k$ for all $k \geq n$; in other words, $\omega \in \liminf_n A_n$ *iff* $\omega \in A_n$ eventually, that is, for all but finitely many n.

Hence B is the set of all points which belong to almost all A_n (all but any finite number of sets). We shall also call $\liminf_n A_n$ as the lower limit of the sequence of sets A_n.

C_n is a monotonically decreasing sequence with limit

$$C = \cap_{n=1}^{\infty} \cup_{k=n}^{\infty} A_k = \limsup_n A_n = \overline{\lim} A_n. \qquad (3.2.10)$$

Thus, $\omega \in \limsup_n A_n$, *iff* for any $n, \omega \in A_k$ for some $k \geq n$; in other words, $\omega \in \limsup_n A_n$ *iff* $\omega \in A_n$ for infinitely many n.

Again,

$$
\begin{aligned}
C^c &= (\cap_{n=1}^{\infty} \cup_{k=n}^{\infty} A_k)^c \\
&= \cup_{n=1}^{\infty} (\cup_{k=n}^{\infty} A_k)^c = \cup_{n=1}^{\infty} \cap_{k=n}^{\infty} A_k^c \text{ (by de Morgan's rule)} \\
&= \liminf A_k^c \\
&= \{\omega : \omega \text{ belong to almost all } A_n^c\text{'s}\} \\
&= \{\omega : \omega \text{ belong to only finite number of } A_n\text{'s}\}.
\end{aligned}
$$

Thus, C is the set of all those points which belong to infinitely many A_n's. We shall sometimes call $\limsup_n A_n$ as the upper limit of the sequence of sets A_n.

Since, every point which belongs to almost all A_n belongs to infinitely many A_n's,

$$\underline{\lim} A_n \subseteq \overline{\lim} A_n. \qquad (3.2.11)$$

If $\underline{\lim} A_n = \overline{\lim} A_n = A$ (say), the limit of $\{A_n\}$ is said to exist and A is called the limit of $\{A_n\}$.

EXAMPLE 3.2.2: Let A_n be the set of points (x, y) of the Cartesian plane $(\mathcal{R}^2)$ bounded by the two axes and the lines $x = n$ and $y = 1/n$. Thus,

$$A_n = \{(x, y) : 0 \leq x < n; 0 \leq y < \frac{1}{n}\}, (x, y) \in \mathcal{R}^2.$$

Hence,

$$B_n = \cap_{k=n}^{\infty} A_k \subset \{(x, y) : 0 < x < n, y = 0\}$$

and

$$B = \underline{\lim} A_n = \{(x, y) : 0 \leq x < \infty, y = 0\}.$$

Also,

$$C_n = \cup_{k=n}^{\infty} A_k \subset \{(x.y) : 0 \leq x < \infty; 0 \leq y < 1/n\}$$

and

$$C = \overline{\lim} A_n = \{(x,y) : 0 \le x < \infty, y = 0\}.$$

Therefore, $B = C$ is the limit of $\{A_n\}$. □

DEFINITION 3.2.3. *Class of sets*: A class of sets is a set of sets. The elements of a class are sets. An example is $C = \{\{a\}, \{b\}, \{a, b\}\}$.

DEFINITION 3.2.4. *Power set.* The class of all subsets of Ω is called the power set of Ω.

3.2.2 Fields

First we consider a basic criterion for defining a field.

DEFINITION 3.2.5. *Closure*: Suppose C is a class of sets. If by performing a certain operation on one or more members of C, we obtain an element which is also a member of C, then C is said to be closed under that operation.

Closure under complementation: If

$$C \in C \Rightarrow C^c \in C, \tag{3.2.12}$$

then C is closed under complementation.

Closure under union: If

$$A, B \subset C \Rightarrow A \cup B \subset C \tag{3.2.13}$$

then C is said to be closed under union. We note that, if (3.2.13) holds, then, by induction it follows that

$$A_1, A_2, \ldots, A_n \subset C \Rightarrow \cup_{i=1}^{n} A_i \subset C, \ \forall \, n < \infty \tag{3.2.14}$$

i.e. C is closed under finite union.

Closure under intersection: If

$$A, B \subset C \Rightarrow A \cap B \subset C, \tag{3.2.15}$$

then C is closed under intersection. We note that, if (3.2.15) holds, then by induction, it follows that

$$A_1, A_2, \ldots, A_n \subset C \Rightarrow \cap_{i=1}^{n} A_i \subset C, \ \forall \, n < \infty \tag{3.2.16}$$

i.e. C is closed under finite intersection.

EXAMPLE 3.2.3: Let C be the class of all intervals of the form $(x, \infty), x \in \mathcal{R}$. Then, for two sets (x, ∞) and (y, ∞),

$$(x, \infty) \cup (y, \infty) = (u, \infty),$$
$$(x, \infty) \cap (y, \infty) = (v, \infty)$$

where $u = \min. (x, y)$ and $v = \max. (x, y)$. Hence C is closed under union and intersection.

However, $(x, \infty)^c = (-\infty, x] \notin C$. Hence, C is not closed under complementation.

EXAMPLE 3.2.4: Let C be the class of all open intervals of the form $(a, b), a < b; a, b \in \mathcal{R}$. Assume that C also contains the empty set ϕ. Then for two intervals (a, b) and $(c, d), c < d; c, d \in \mathcal{R}$,

$$\begin{aligned}
(a, b) \cap (c, d) &= \phi, \text{ if } a < b < c < d \text{ or } c < d < a < b, \\
&= (c, b) \quad \text{if } a < c < b < d, \\
&= (a, d) \quad \text{if } c < a < d < b, \\
&= (c, d) \quad \text{if } a < c < d < b, \\
&= (a, b), \quad \text{if } c < a < b < d.
\end{aligned}$$

Thus, C is closed under finite intersection. It is, however, not closed under complementation, because $(a, b)^c = (-\infty, a] \cup [b, \infty) \notin C$. Also, it is not closed under union, since $(a, b) \cup (c, d)$ is not an interval, if $a < b < c < d$, or $c < d < a < b$.

DEFINITION 3.2.6. *Field*: A non-empty class of sets, $\mathcal{A}$, closed under complementation and finite intersections, is called a field.

A field is also called a *finitely additive class of sets* or *Boolean field of sets* or *Boolean algebra*.

Lemma 3.2.2: A field is closed under finite unions. Conversely, a class closed under complementation and finite unions is a field.

Proof. Suppose $\mathcal{A}$ is a field. Then,

(i) $A \in \mathcal{A} \Rightarrow A^c \in \mathcal{A}$, $\qquad\qquad\qquad\qquad\qquad$ (3.2.17)

(ii) $A_1, A_2, \ldots, A_n \in \mathcal{A} \Rightarrow \cap_{i=1}^n A_i \in \mathcal{A}$. $\qquad\qquad$ (3.2.18)

But

$$A_1, A_2, \ldots, A_n \in \mathcal{A} \Rightarrow A_1^c, A_2^c, \ldots, A_n^c \in \mathcal{A} \quad \text{(by (3.2.17))}$$

$$\Rightarrow \cap_{i=1}^n A_i^c \in \mathcal{A} \quad \text{(by (3.2.18))}$$

$$\Rightarrow (\cap_{i=1}^n A_i^c)^c \in \mathcal{A} \quad \text{(by (3.2.18))}$$

$$\Rightarrow \cup_{i=1}^n A_i \in \mathcal{A} \quad \text{(by de Morgan's rule)}.$$

Hence, $\mathcal{A}$ is closed under finite unions also.

Conversely, suppose $\mathcal{A}$ is a class such that (3.2.17) holds and

$$A_1, A_2, \ldots, A_n \in \mathcal{A} \Rightarrow \cup_{i=1}^n A_i \in \mathcal{A}.$$

Then, $\cup_{i=1}^n A_i^c \in \mathcal{A} \Rightarrow (\cup_{i=1}^n A_i^c)^c \in \mathcal{A} \Rightarrow \cap_{i=1}^n A_i \in \mathcal{A}$, by de Morgan's rule. Hence the proof. $\qquad \square$

Note 3.2.1: A field is, therefore, a class, which is closed under complementation and finite intersections or finite unions.

Lemma 3.2.3: Every field contains the empty set ϕ and the whole space Ω.

Proof. Let A be a set in $\mathcal{A}$. Then $A^c \in \mathcal{A}$. Hence,

$$A \cap A^c = \phi \in \mathcal{A} \qquad (3.2.19)$$

and

$$A \cup A^c = \Omega \in \mathcal{A}. \qquad (3.2.20)$$

Hence the proof. $\qquad \square$

Note 3.2.2: The power set (definition 3.2.3) is also a field.

EXAMPLE 3.2.5: The class containing only ϕ and Ω is a field. It is the smallest field and is contained in every other field. It is called the *degenerate field* or *trivial field*.

EXAMPLE 3.2.6: The largest field of all subsets of Ω is the collection of all subsets of Ω. The power set consisting of every subset of Ω is the largest field.

EXAMPLE 3.2.7: Suppose a field contains a set A. Then it has to contain A^c also. Hence, the field

$$\mathcal{A} = \{A, A^c, \phi, \Omega\}$$

is the smallest field containing A. It is contained in all the fields containing A.

EXAMPLE 3.2.8: Let A, B, C be three mutually exclusive and exhaustive sets. It can be verified that the smallest field containing A, B, C is

$$\mathcal{A} = \{A, B, C, A+B, A+C, B+C, \phi, \Omega\}.$$

DEFINITION 3.2.7. *Minimal Field*: Consider an arbitrary class $\mathcal{C}$ of sets. The smallest field containing $\mathcal{C}$ is called the minimal field containing $\mathcal{C}$ or the *field generated by* $\mathcal{C}$. It is contained in every field containing $\mathcal{C}$. Thus, if each of the fields, $\mathcal{C}_i, i \in I$, where I is a non-empty index set, contains $\mathcal{C}$, then $\mathcal{C} \subset \cap_{i \in I}\mathcal{C}_i = \mathcal{C}_0$ (say). The following theorem, stated without proof, shows that $\mathcal{C}_0$ is a field. Therefore, $\mathcal{C}_0$ is the minimal field containing $\mathcal{C}$ and is often denoted as $\mathcal{F}(\mathcal{C})$.

Theorem 3.2.1: The intersection of an arbitrary number of fields is a field.

Note 3.2.3: Though the intersection of two or more fields is a field, the union of two fields may not be a field. For example, $\{A, A^c, \phi, \Omega\}$ and $\{B, B^c, \phi, \Omega\}$ are fields. But their union, $\{A, B, A^c, B^c, \phi, \Omega\}$ is not a field.

EXAMPLE 3.2.9: Let $\{A_i\}, i = 1, \ldots, n$ be a class of mutually exclusive and exhaustive sets, so that $A_i \cap A_j = \phi, i \neq j = 1, \ldots, n; \cup_{i=1}^n A_i = \Omega$. Then the minimal field containing $\{A_1, \ldots, A_n\}$ is

$$\mathcal{A} = \{\phi;\ A_1, \ldots, A_n;\ A_1 + A_2, \ldots, A_1 + A_n, \ldots, A_{n-1} + A_n;$$

$$A_1 + A_2 + A_3, \ldots, A_{n-2} + A_{n-1} + A_n;\ \ldots; \Omega\}. \tag{3.2.21}$$

In (3.2.21), unions of A_i's are taken one set at a time, two sets at a time, $\ldots, n$ sets at a time. Total number of sets in (3.2.21) is, therefore, $\binom{n}{0} + \binom{n}{1} + \ldots \binom{n}{n} = 2^n$.

EXAMPLE 3.2.10: If $A_1, \ldots A_n$ are arbitrary subsets of Ω, the smallest field containing $A_1, \ldots, A_n$ may be described as follows. The minimal field $\mathcal{C}$ consists of the collection of all finite unions of sets of the form $B_1 \cap B_2 \cap \ldots \cap B_n$, where B_i is either A_i or A_i^c.

DEFINITION 3.2.8. *Partition*: Let $\{A_i\}$ be a class of mutually exclusive and exhaustive class of sets, so that $A_i A_j = \phi, i \neq j = 1, \ldots, n$ and $\cup_{i=1}^n A_i = \Omega$. Then the class $\{A_i\}$ is said to be a partition of Ω.

Note 3.2.4. A method of constructing a minimal field: We have noted in Example 3.2.9, a rule for constructing a minimal field $\mathcal{F}(\mathcal{P})$ if $\mathcal{P}$ is a partition of Ω.

Let $\mathcal{C}$ be a class of sets in Ω, not necessarily forming a partition of Ω. To form a minimal field $\mathcal{F}(\mathcal{C})$, we proceed as follows.

(i) First form a class $\mathcal{A}_1$ containing all the sets in $\mathcal{C}$ and their complements and ϕ and Ω.

(ii) Form a class $\mathcal{A}_2$ containing all the sets in $\mathcal{A}_1$ and the intersections of all the sets in $\mathcal{A}_1$, the sets taken two at a time, three at a time, ..., all the sets taken at a time.

(iii) Form a class $\mathcal{A}_3$ containing all the sets in $\mathcal{A}_2$ and the unions of all the disjoint sets in $\mathcal{A}_2$, the sets taken two at a time, three at a time, all sets taken together.

(iv) Then $\mathcal{A}_3$ is a field and is the minimal field containing $\mathcal{C}$.

EXAMPLE 3.2.11: Let $\mathcal{C} = \{A, B\}$. Then

$$\mathcal{A}_1 = \{\phi, \Omega, A, A^c, B, B^c\};$$

$$\mathcal{A}_2 = \{\mathcal{A}_1; AB, AB^c, A^cB, A^cB^c\},$$

$$\mathcal{A}_3 = \{\mathcal{A}_2; AB + A^cB^c, AB^c + A^cB^c;$$

$$AB+AB^c+A^cB, AB+A^cB+A^cB^c, AB+A^cB+A^cB^c, AB^c+A^cB+A^cB^c\}.$$

$\mathcal{A}_3$ is the minimal field containing $\{A, B\}$. It can be verified that it coincides with the minimal field containing the partition $\mathcal{P} = \{AB, A^cB, AB^c, A^cB^c\}$.

3.2.3   σ-Field

DEFINITION 3.2.9. *σ-Field:* A non-empty class of sets which is closed under complementation, countable intersections (or countable unions) is called a σ-field.

A σ-field is also called a *completely additive class of sets* or *Borel field* or *σ-algebra* .

A σ-field is, of course, a field, as it is closed under complementation and finite intersections (unions).

If $\mathcal{C}$ contains only a finite number of sets and is a field, it is also a σ-field; however, a field containing an infinite number of sets may not be a σ-field.

EXAMPLE 3.2.12: In Example 3.2.3, it was shown that $\mathcal{C}$, the class of intervals of the form (x, ∞) is closed under finite intersections. However, it is not closed under countable intersections. Consider

$$\cap_{n=1}^{\infty}(x - \frac{1}{n}, \infty) = [x, \infty) \notin \mathcal{C}.$$

Therefore, though $\mathcal{C}$ contains an infinite number of sets, it is not a σ-field.

EXAMPLE 3.2.13: Let $\Omega = \{1, 2, 3, 4, \ldots\}$ and $\mathcal{C}$ be the class of all sets A of Ω such that either A contains a finite number of points or A^c contains a finite number of points. Then, obviously, $\mathcal{C}$ is closed under complementation.

If each one of A, B contains only a finite number of points, $A \cup B$ will also contain a finite number of points, and $(A \cup B)^c$ will contain an infinite number of points. If either A or B is infinite, then $(A \cup B)^c = A^c \cap B^c$ will contain a finite number of points. If both A and B contain an infinite number of points, then of course, $(A \cup B)^c$ will contain only a finite number of points. Hence, $\mathcal{C}$ is a field.

Consider now the sets, $A_1 = \{3\}, A_2 = \{6\}, \ldots, A_k = \{3k\}, \ldots$. The class $\mathcal{C}$ is till closed under complementation. But $\cup_{k=1}^{\infty} A_k$ is neither finite, nor its complement is finite. Hence, $\mathcal{C}$ is not closed under countable unions. Thus, $\mathcal{C}$ is not a σ-field of such sets.

EXAMPLE 3.2.14: Let Ω be $\mathcal{R}$, the set of all real numbers and let $\mathcal{F}$ consists of all finite disjoint unions of right-semi-closed intervals $(a, b]$. It may be verified that $\mathcal{F}$ is a field; but it is not a σ-field. For example, $A_n = (0, 1 - \frac{1}{n}] \in \mathcal{F}, n = 1, 2, \ldots$ and $\cup_{n=1}^{\infty} A_n = (0, 1) \notin \mathcal{F}$.

If Ω is the extended real line $\bar{\mathcal{R}} = [-\infty, \infty]$, then as above, the collection of finite disjoint unions of right- semi-closed intervals form a field, but not a σ-field. Here, the intervals are sets of the form $(a, b] = \{x : a < x \leq b\}, -\infty \leq a < b \leq \infty$.

Theorem 3.2.2: The intersection of an arbitrary number of σ-fields is a σ-field.

Proof. Omitted.

DEFINITION 3.2.10. *Minimal σ-field:* Given a class $\mathcal{C}$ of sets, the minimal σ-field containing $\mathcal{C}$ is the intersection of all σ-fields containing $\mathcal{C}$. It is also called the σ-field generated by $\mathcal{C}$ and is denoted by $\sigma(\mathcal{C})$.

If C is finite, then $\sigma(C)$ coincides with the minimal field $\mathcal{F}(C)$ containing C.

To obtain $\sigma(C)$ from a given C, the same procedure as in the Note 3.2.4 can be followed. Here, however, n is infinite.

EXAMPLE 3.2.15: Consider $\Omega = \mathcal{R}$, the real line. There are eight intervals of the form,

$$(-\infty, a), \ (-\infty, a], \ (b, \infty), \ [b, \infty),$$

$$(a, b), \ [a, b), \ [a, b], \ (a, b], \ a < b, a, b \in \mathcal{R}.$$

Consider the class C of all intervals of the form $(-\infty, a)$. Let $\mathcal{B} = \sigma(C)$ be the minimal σ-field containing C. Then

$$[a, \infty) = (-\infty, a)^c \subset \mathcal{B} \ \text{(by complementation)};$$

$$(-\infty, a] = \cap_{n=1}^{\infty} (-\infty, a + \frac{1}{n}) \subset \mathcal{B} \ \text{(by countable intersection)},$$

$$(a, \infty) = (-\infty, a]^c \subset \mathcal{B},$$

$$(a, b) = (-\infty, b) \cap (-\infty, a) \subset \mathcal{B}, \ a < b,$$

$$[a, b) = \cap_{n=1}^{\infty} (a - \frac{1}{n}, b) \subset \mathcal{B},$$

$$[a, b] = \cap_{n=1}^{\infty} (a - \frac{1}{n}, b + \frac{1}{n}) \subset \mathcal{B},$$

$$(a, b] = \cap_{n=1}^{\infty} (a, b + \frac{1}{n}) \subset \mathcal{B}.$$

Thus $\mathcal{B}$ is generated by the class C of all intervals of the form $(-\infty, a)$. The class $\mathcal{B}$ is called the *Borel field* of subsets of the real line $\mathcal{R}$ and its sets, the *Borel sets*. □

In the same manner, it can be proved that $\mathcal{B}$ is the minimum σ-field, containing any one of the other remaining seven classes of intervals listed in Example 3.2.15.

DEFINITION 3.2.11. *Borel Field*: If $\Omega = \mathcal{R}$, the real line, C is the class of all real open intervals of the form (a, b), minimal σ-field containing C is called the Borel field $\mathcal{B}$ of subsets of the real line and its sets are called Borel sets of $\mathcal{R}$. The same Borel field is obtained if C is taken as the class of all intervals of the type $[a, b), [a, b], (a, b], (-\infty, a), (-\infty, a], (b, \infty), [b, \infty)$.

Note 3.2.5: If $x \in \mathcal{R}$, the set $\{x\}$ is a Borel set, because

$$\{x\} = (-\infty, x] \cap [x, \infty)$$

and hence lies in $\mathcal{B}$. Hence, any countable subset of $\mathcal{R}$, containing a countable union of singletons, is a Borel set. Thus, $\{0, 1, 2, \ldots\}$ is a Borel set.

However, an uncountable union of such singletons, namely, a subset containing an uncountable number of points of $\mathcal{R}$ may not be a Borel set.

Note 3.2.6: However, there are other subsets of $\mathcal{R}$ which are not Borel sets and hence are not contained in $\mathcal{B}$ (Halmos, 1958). However, non-Borel sets of $\mathcal{R}$ do not naturally arise in the application of probability theory.

DEFINITION 3.2.12: The *Borel field $\mathcal{B}_2$ of subsets of the real plane $\mathcal{R}^2$* is defined as the minimal σ-field containing the class of all open intervals of the form

$$C_2 = \{(x, y) : a < x < b,\ c < y < d\},\ a < b, c < d, a, b, c, d \in \mathcal{R},$$

a, b, c, d arbitrary. Now, the set

$$C_n = \{(x, y) : a < x < b + \frac{1}{n},\ c < y < d\} \subset C_2.$$

Hence,

$$\cap_{n=1}^{\infty} C_n = \{(x, y) : a < x \le b,\ a < y < d\} \subset \mathcal{B}_2.$$

Thus, all half-open rectangles with one side open and one side closed lie in $\mathcal{B}_2$ and are *Borel sets*. Similarly, all closed rectangles are also Borel sets.

The infinite rectangles of the form

$$\{x < b, c < y < d\} = \cup_{n=1}^{\infty} \{-n < x < b, c < y < d\},$$

$$\{a < x, c < y < d\} = \cup_{n=1}^{\infty} \{a < x < n, c < y < d\},$$

$$\{a < x < b, c < y\} = \cup_{n=1}^{\infty} \{a < x < b, c < y < n\},$$

$$\{a < x < b, y < d\} = \cup_{n=1}^{\infty} \{a < x < b, -n < y < d\}$$

and they also generate $\mathcal{B}_2$. It can be shown that all these rectangles belong to $\mathcal{B}_2$. It is apparent that $\mathcal{B}_2$ contains all rectangle sets whose coordinate sets are Borel sets on the real line.

Similarly, *Borel fields $\mathcal{B}_n$ of subsets of $\mathcal{R}^n$*, the n-dimensional Euclidean space (the space of n-tuples of real numbers $(x_1, x_2, \ldots, x_n)$), is the minimal σ-field containing the class of all n-dimensional open rectangles

$$C_n = \{(x_1, x_2, \ldots, x_n) : a_i < x_i < b_i, i = 1, 2, \ldots, n\},$$

$a_i \in \mathcal{R}, b_i \in \mathcal{R}(i = 1, \ldots, n)$, arbitrary. The same Borel field is obtained by starting with rectangles, closed or open on any of the $2n$ sides. Clearly, there are 2^n types of such rectangles.

Remark 3.2.1: The two most important completely additive class of sets considered in this book are the *class of events* and the class of Borel sets on the real line $\mathcal{R}$ or on Euclidean spaces of higher dimensions.

EXAMPLE 3.2.16: An open rectangle is a subset $A \subseteq \mathcal{R}^n$ of the form $A = \{\omega : a_i < \omega_i < b_i, i = 1, \ldots, n\}$. Show that an open rectangle is a Borel set.

Let $A_n = \{\omega : a_i < \omega_i \leq b_i - \frac{1}{n}, i = 1, \ldots, n\}$. The union $\cup_n A_n$ of the rectangles A_n is the open interval $A = \{\omega : a_i < \omega_i < b_i, i = 1, \ldots, n\}$. Since each A_n is a Borel set and the Borel sets are closed under countable unions, it follows that A is a Borel set.

DEFINITION 3.2.13. *Monotone Field*: A field $\mathcal{F}$ is said to be a monotone field, if it is closed under monotone operations, i.e., if $\{A_n\}$ is a monotone sequence of sets in $\mathcal{F}$, then $\lim A_n \subset \mathcal{F}$. Thus,

$$A_n \subset \mathcal{F}, \ A_n \uparrow A \Rightarrow A \subset \mathcal{F},$$

$$A_n \subset \mathcal{F}, \ A_n \downarrow A \Rightarrow A \subset \mathcal{F}.$$

Theorem 3.2.3: A σ-field is a monotone field and conversely.

Proof. Omitted.

Note 3.2.7: Field, σ-field of events: We have seen in Chapter 2 that an event is a collection of results of a random experiment and is represented by a set A. The collection of all possible elementary events (singleton sets, $\{e\}$) of an experiment is the sample space, $\mathcal{S}$.

In the language of algebra of sets, the sample space is, therefore, the whole space Ω, the events are different sets of Ω. Since A is an event A^c is also an event, viz., the event that A does not occur. The union $\cup_{i=1}^{n} A_i$ is an event that at least one of $A_1, \ldots, A_n$ occurs. The intersection $\cap_{i=1}^{n} A_i$ is an event that each of $A_1, \ldots, A_n$ occurs. Hence, if the number of possible outcomes n is finite, the class of all events is a *field of events*. If n is countably infinite, the sample space is countable and the class of events is a *σ-field of events*.

The space $\mathcal{S}$ of all outcomes of an experiment together with the specification of a σ-field of events is sometimes called a *measurable space* and is denoted as $(\mathcal{S}, \mathcal{A})$.

3.3   Point Function, Set Function

3.3.1   *Point function*

DEFINITION 3.3.1: *Real-valued point function*: A real-valued point function f defined on Ω is a rule that associates with each point ω of Ω, a real number $f(\omega)$, called the value of f for the point ω. The set Ω is called the *domain of f* and the set of values $\{f(\omega) : \omega \in \Omega\}$ is called the range of f. We denote the set containing the range of f as Ω'.

If for each ω, $f(\omega)$ is different, the function is one-to-one $(1 : 1)$. If two or more ω's have the same value, the function is many to one. An example of a $2 : 1$ function is $f(\omega) = \omega^2$.

Thus, a function f which maps Ω to Ω' (we denote it as $\Omega \to^f \Omega'$) assigns to each $\omega \in \Omega$, a value $f(\omega) \in \Omega'$. $f(\omega)$ is called the image of ω under f and ω is called the pre-image of $f(\omega)$.

If $\Omega' = \{f(\omega) : \omega \in \Omega\}$, f is said to be mapping from Ω *onto* Ω'. In this case, Ω' is said to be *strict range* of the function f. If $\{f(\omega) : \omega \in \Omega\} \subset \Omega'$, f is said to be mapping from Ω *into* Ω'.

The function f is a numerical function, if $\Omega' = \mathcal{R}$, the real line $(-\infty < y < \infty)$. Real-valued point functions f_1 and f_2 are equal, if

$$f_1(\omega) = f_2(\omega) \; \forall \; \omega \in \Omega$$

and we write $f_1 = f_2$. Similarly, we write $f_1 < (>)f_2$ *iff* $f_1(\omega) < (>)f_2(\omega) \; \forall \; \omega$.

If $f(\omega) = c$, a constant for all ω, f is a degenerate function.

EXAMPLE 3.3.1: Consider the experiment of tossing two balanced dice. Here, the sample space S consists of 36 points (e_1, e_2) where $e_i = 1, 2, \ldots, 6; i = 1, 2$. Let $X(e_i, e_2) = e_1 + e_2$. Then X takes values $2, 3, \ldots, 12$. Thus $X(4, 6) = 10$. Here $\Omega = \{\omega_1, \omega_2), \omega_1, \omega_2 = 1, \ldots, 6; \Omega' = \{2, 3, \ldots, 12\}, \Omega$ is mapped onto Ω' through the many-to-one function X.

3.3.2   *Set function*

DEFINITION 3.3.2 *Set Function*: If a function is defined on the sets of a certain class, then the function is called a set function. If $\mathcal{C}$ is a class of sets and with each $A \subset \mathcal{C}$, we associate a value $\mu(A)$, say, then μ is a set function. For example, with an interval (a, b), we may associate a value

$\mu(a,b) = b - a$, the length of the interval. For the union of two disjoint sets (a,b) and (c,d) we may associate a value $\mu((a,b)\cup(c,d)) = (b-a)+(d-c)$, etc. We shall denote

$$f(B) = [f(\omega) : \omega \in B], B \subset \Omega. \qquad (3.3.1)$$

The function f has, therefore, its domain $\mathcal{B}$ (say), the class of all subsets of Ω and range $\mathcal{B}'$, the class of all subsets of Ω'. Evidently,

$$f(\Omega) = [\omega' : f(\omega) = \omega'] = \Omega'.$$

DEFINITION 3.3.3. *Finitely Additive Real-Valued Set Function*: Let $\mathcal{C}$ be a field. A real-valued set function defined on $\mathcal{C}$ is finitely additive if (i) $\mu(\phi) = 0$ (ii) if $C_i \in \mathcal{C}, i = 1, \ldots, n$ are disjoint sets for any n, then $\mu(\cup_{i=1}^{n} C_i) = \sum_{i=1}^{n} \mu(C_i)$.

DEFINITION 3.3.4. *Completely Additive Real-Valued Set Function*: Let $\mathcal{C}$ be a σ-field of sets. A real-valued set function μ defined on $\mathcal{C}$ is called completely additive if (i) $\mu(\phi) = 0$ (ii) if $C_i \in \mathcal{C}, i = 1, 2, \ldots$ are disjoint, then $\mu(\cup_{i=1}^{\infty} C_i) = \sum_{i=1}^{\infty} \mu(C_i)$.

3.3.3 *Measure and measurable sets*

DEFINITION 3.3.5. *Measure and Measurable Sets*: A non-negative, additive, real-valued set function defined on a field or σ-field is called a *measure*. Thus if $\mathcal{C}$ is a field or σ-field of subsets of Ω, a real-valued set function defined on $\mathcal{C}$ is a measure if

  (i) $\mu(C) \geq 0 \ \forall \ C \in \mathcal{C}$,
  (ii) $\mu(\phi) = 0$,
  (iii) if $C_i \in \mathcal{C}(i = 1, 2, \ldots)$ are disjoint such that $\cup_{i=1}^{\infty} C_i \in \mathcal{C}$, then $\mu(\cup_{i=1}^{\infty} C_i) = \sum_{i=1}^{\infty} \mu(C_i)$.

The measure μ is finitely additive if (iii) is valid for a finite number of disjoint sets; it is completely additive if (iii) also extends to a countable number of disjoint sets. The sets C are called *measurable sets* with respect to measure μ. The measure μ is called bounded if $\mu(\Omega) < \infty$. The measure μ is finite *iff* $\mu(A) < \infty \ \forall \ A \in \mathcal{C}$.

A non-negative finitely additive set function μ defined on the field $\mathcal{C}$ is said to be σ-*finite iff* Ω can be written as $\cup_{n=1}^{\infty} A_n$, where the A_n's belong to $\mathcal{C}$ and $\mu(A_n) < \infty \ \forall \ n$.

DEFINITION 3.3.6. *Counting Measure*: Suppose Ω is the real line $\mathcal{R}$ and $\mathcal{B}$ is the class of all Borel sets in $\mathcal{R}$. Let μ be a set function such that

$$\mu(B) = \text{ number of integers (positive, negative or zero) in } B.$$

Then,

$$\mu(\phi) = 0; \; \mu(\sum_i B_i) = \sum_i \mu(B_i), \; B_i \in \mathcal{B}.$$

Therefore, μ is a measure on $\mathcal{B}$. It is called the counting measure. The measure associates unit mass with each of the points $0, +_-1, +_-2, \dots$. Note that μ takes only values $0, 1, 2, \dots$ only. However, $\mu(B) = 0$ does not imply $B = \phi$.

Counting measure may also be defined in $(\mathcal{R}^k, \mathcal{B}_k)$ as follows. For $B_k \in \mathcal{B}_k$, define $\mu(B_k) = $ number of points $(x_1, \dots, x_k)$ with integral (positive, negative or zero) coordinates lying in B_k.

DEFINITION 3.3.7: *Lebesgue Measure and Lebesgue Measurable Sets*: Let $\mathcal{R}$ be the real line and $\mathcal{I}$ the class of all bounded semi-closed intervals $(a, b], (a < b)$. On $\mathcal{I}$ we define a set function μ such that

$$
\begin{aligned}
\mu(a, b] &= b - a, \\
\mu(\phi) &= 0, \\
\mu(\sum I_i) &= \sum \mu(I_i),
\end{aligned}
\tag{3.3.2}
$$

where $I_1, I_2, \dots$ are disjoint set of $\mathcal{I}$. Thus, μ is a measure on $\mathcal{I}$.

The measure μ can be extended to all Borel sets. This extended measure on $(R, \mathcal{B})$ is called the Lebesgue measure.

Sometimes, one may take the σ-field obtained by adjoining to $\mathcal{B}$ all subsets of measure zero and here μ is defined over this larger σ-field, $\mathcal{B}'$, say. The members of $\mathcal{B}'$ are called Lebesgue measurable sets. μ defined over $\mathcal{B}$ or $\mathcal{B}'$ is called the Lebesgue measure.

DEFINITION 3.3.8 *Lebesgue-Stieltjes Measure*: Let F be any monotone non-decreasing, real-valued function on the whole real line $\mathcal{R}$, which is finite or have bounded variation.

Without loss of generality, it may be taken to be continuous on the right. Thus the value of the function at the jump point is the limit of its values on the right side of the jump point. Thus,

$$\lim_{\epsilon \downarrow 0} F(x + \epsilon) = F(x), \text{ i.e. } F(x + 0) = F(x) \forall \; x \in \mathcal{R}. \tag{3.3.3}$$

(Similarly, we may also take F to be continuous on the left.) On all sets $(a, b], (a < b \in \mathcal{R})$ of the class $\mathcal{I}$, we define a measure μ_F as follows:

(i) $\mu_F(a, b] = F(b) - F(a) =$ increment of F in the interval $(a, b]$;

(ii) $\mu_F(\{a\}) = F(a) - F(a-) =$ magnitude of jump of F at a;

(iii) $\mu_F(\sum I_i) = \sum_i \mu_F(I_i)$,

where $\{I_i\}$ is a class of disjoint sets in $\mathcal{I}$.

μ_F can be extended to all Borel sets and hence it is a measure on $(\mathcal{R}, \mathcal{B})$. This measure is called the Lebesgue-Stieltjes (LS) measure induced by the function F. The function F is called a *distribution function* for $\mu_F(.)$. The relation $\mu_F(a, b] = F(b) - F(a)$ sets up a one-to-one correspondence between the Lebesgue-Stieltjes measure and distribution function, where however, the two distribution functions which differ by a constant, are considered identical.

Clearly, the measure μ_F depends on the increment of F and not on its absolute value. A class of functions F having the same increment will give rise to the same value of μ_F.

It follows that the measure of any interval, right-semiclosed or not, may be expressed in terms of F. For, if $F(x - 0) = \lim_{\epsilon \downarrow 0} F(x - \epsilon)$, then

(i) $\mu_F(a, b] = F(b) - F(a)$;

(ii) $\mu_F(a, b) = F(b - 0) - F(a)$;

(iii) $\mu_F[a, b] = F(b) - F(a - 0)$;

(iv) $\mu_F[a, b) = F(b - 0) - F(a - 0)$.

The following results are obtained from (i) - (iv) when a approaches $-\infty$ and b approaches ∞.

(v) $\mu_F(-\infty, x] = F(x) - F(-\infty)$;

(vi) $\mu_F(-\infty, x) = F(x - 0) - F(-\infty)$;

(vii) $\mu_F(x, \infty) = F(\infty) - F(x)$;

(viii) $\mu_F[x, \infty) = F(\infty) - F(x - 0)$;

(ix) $\mu_F(R) = F(\infty) - F(-\infty)$.

DEFINITION 3.3.9: *Almost Everywhere (a.e.)*: If a certain property can hold for all ω except for a set of measure zero, the property is said to hold almost everywhere (abbreviated to 'a.e.').

3.3.4   *Inverse function*

In this section we shall consider another type of set functions, inverse functions.

DEFINITION 3.3.10. *Inverse Function*: The set of all points $\omega \in \Omega$ whose image under f is ω' is called the inverse image of $\{\omega'\}$, denoted as $f^{-1}(\{\omega'\})$. Thus,

$$f^{-1}(\{\omega'\}) = \{\omega : f(\omega) = \omega'\}. \tag{3.3.4}$$

In general, let $B' \subseteq \Omega'$. Then $\{\omega \in \Omega : f(\omega) \in B'\}$ is the inverse image of B' under f and is denoted as $f^{-1}(B')$,

$$f^{-1}(B') = [\omega : f(\omega) \in B']. \tag{3.3.5}$$

Clearly,

$$f^{-1}(\Omega') = [\omega : f(\omega) \in \Omega] = \Omega. \tag{3.3.6}$$

Thus, with each point function f (whose domain is Ω and range Ω') there is a set function f^{-1}, whose domain is a class $\mathcal{B}'$ of subsets of Ω' and range is a class $\mathcal{B}$ of subsets of Ω. f^{-1} is called the inverse function (or mapping) of f. We shall denote

$$f^{-1}(\mathcal{B}') = [f^{-1}(B') : B' \subset \mathcal{B}']. \tag{3.3.7}$$

EXAMPLE 3.3.2: Consider the function $f(\omega) = \omega^2$ from $\mathcal{R}$ to $\mathcal{R}$. Let $B' = (2,3)$. Then $f^{-1}(B') = (-\sqrt{3}, -\sqrt{2}) \cup (\sqrt{2}, \sqrt{3})$.

EXAMPLE 3.3.3: Consider the real-valued point function

$$I_A(\omega) = \begin{cases} 1 & \text{if } \omega \in A \\ 0 & \text{if } \omega \notin A. \end{cases} \tag{3.3.8}$$

I_A is called the *indicator function* or *characteristic function* of A. The strict range of I_A is $\{I_A(\omega) : \omega \in \Omega\} = \{0, 1\}$. Clearly, $I_\Omega = 1 \ \forall \ \omega \in \Omega$.

Let B' be a interval on the real line $\mathcal{R}$. Then

$$I_A^{-1}(B') = \begin{cases} \phi & \text{if } B' \text{ does not contain '0' or '1'}, \\ A & \text{if } B' \text{ contains '1', but not '0'}, \\ A^c & \text{if } B' \text{ contains '0', but not '1'}, \\ \Omega & \text{if } B' \text{ contains both '0' and '1'}. \end{cases} \tag{3.3.9}$$

Hence,

$$I_A^{-1}(\mathcal{B}') = \{\phi, A, A^c, \Omega\} = \sigma(A),$$

the σ-field generated by A. Clearly, I_A^{-1} is a set function.

Let $J_A = c.I_A$ where c is a constant. Then $J_A = c(0)$, if $\omega \in A$ (otherwise); the strict range of J_A is $\{0, c\}$. Also $J_A^{-1}(\mathcal{B}') = I_A^{-1}(\mathcal{B}') = \sigma(A)$. □

Lemma 3.3.1 *Properties of Indicator Functions*: The following relations are easy to prove.

(a) (i) $I_A = I_A^2 = \ldots I_A^n$; (ii)$A = B \Rightarrow I_A = I_B$; (iii) $A \subseteq B \Rightarrow I_A \leq I_B$

(b) (i) $I_{A^c} = 1 - I_A$; (ii) $I_{B-A} = I_B - I_A$.

(c) (i) $I_{AB} = I_A.I_B = \min.(I_A, I_B)$; (ii) $I_{\cap_{i=1}^n A_i} = \Pi_{i=1}^n I_{A_i} = \min.\{I_{A_i}, i = 1, \ldots, n\}$.

(d) (i)$I_{A+B} = I_A + I_B$; (ii) $I_{A \cup B} = I_A + I_B - I_A.I_B = \max.(I_A, I_B)$; (iii)

$$I_{\cup_{i=1}^n} = \sum_i I_{A_i} - \sum\sum_{i<j} I_{A_i A_j} + \sum\sum\sum_{i<j<k} I_{A_i A_j A_k} - \cdots$$

$$+(-1)^{n-1} I_{\Pi_{i=1}^n A_i}. \tag{3.3.10}$$

To establish the relation in (diii), we note that if ω lies in k of the A_i's, then its contribution to the left side of (3.3.10) is one. The first term on the right contributes k, the second term $\binom{k}{2}$, the third term $\binom{k}{3}, \ldots$, the kth term 1 and the remaining terms each zero. Hence contribution of ω on the right side of (3.3.8) is

$$\binom{k}{1} - \binom{k}{2} + \binom{k}{3} - \cdots + (-1)^{k-1}\binom{k}{k} = 1.$$

Hence the proof. $\qquad\qquad\square$

The following lemma gives the properties of the inverse function.

Lemma 3.3.2: (i) Let $B \subset C \subset \Omega'$. Then

$$f^{-1}(B) = [\omega : f(\omega) \in B] \subset [\omega : f(\omega) \in C] = f^{-1}(C). \tag{3.3.11}$$

(ii)

$$f^{-1}(\cap_k B_k) = \cap_k f^{-1}(B_k). \tag{3.3.12}$$

(iii)

$$f^{-1}(\cup_k B_k) = \cup_k f^{-1}(B_k). \tag{3.3.13}$$

(iv)

$$f^{-1}(B^c) = (f^{-1}(B))^c. \tag{3.3.14}$$

Proof. The relation (i) follows readily.

(ii) Let $B_k \subset \Omega', k = 1, 2, \ldots$. Then

$$\omega \in f^{-1}(\cap B_k) \Leftrightarrow f(\omega) \in \cap B_k$$
$$\Leftrightarrow f(\omega) \in B_k \; \forall \; k$$
$$\Leftrightarrow \omega \in f^{-1}(B_k) \; \forall \; k$$
$$\Leftrightarrow \omega \in \cap_k f^{-1}(B_k).$$

(iii) Let $B_k \subset \Omega', k = 1, 2, \ldots$. Then

$$
\begin{aligned}
\omega \in f^{-1}(\cup_k B_k) &\Leftrightarrow f(\omega) \in \cup B_k \\
&\Leftrightarrow f(\omega) \in B_k \text{ for at least one } k \\
&\Leftrightarrow \omega \in f^{-1}(B_k) \text{ for at least one } k \\
&\Leftrightarrow \omega \in \cup_k f^{-1}(B_k).
\end{aligned}
$$

(iv)

$$
\omega \in f^{-1}(B^c) \Leftrightarrow f(\omega) \in B^c \Leftrightarrow f(\omega) \notin B
$$

$$
\Leftrightarrow \omega \notin f^{-1}(B) \Leftrightarrow \omega \in [f^{-1}(B)]^c.
$$

Note that the unions and intersections considered above need not be countable. □

Corollary 3.3.2.i:

$$
\begin{aligned}
f^{-1}(\Omega') &= [\omega : f(\omega) \in \Omega'] = \Omega, \\
f^{-1}(\phi) &= f^{-1}[(\Omega')^c] \qquad = [f^{-1}(\Omega')]^c = \Omega^c = \phi.
\end{aligned}
$$

Corollary 3.3.2.ii: If

$$
\cap_k B_k = \phi,
$$

then

$$
f^{-1}(\cap_k B_k) = \cap_k f^{-1}(B_k) = \phi.
$$

Thus, if B_k's are disjoint, then the sets $f^{-1}(B_k)$'s are also disjoint.

Inverse function and σ-field

Let $\mathcal{B}$ be a certain class of subsets of Ω' and be a σ-field. The question arises, if $f^{-1}(\mathcal{B}) = [f^{-1}(B) : B \subset \mathcal{B}]$ is a σ-field.

The answer is affirmative as given in Lemma 3.3.4(i). In this direction we state two results without proof.

Lemma 3.3.3: Let $\mathcal{C}$ be a certain class of subsets of Ω and be a σ-field. Then the class $\mathcal{B}$ of subsets of Ω' whose members have inverse images in $\mathcal{C}$ is a σ-field.

Lemma 3.3.4: (i) Let $\mathcal{B}$ be a class of subsets of Ω' and be a field (σ-field). Then $f^{-1}(\mathcal{B})$ is a field (σ-field) over subsets of $f^{-1}(\mathcal{B})$ in Ω.

(ii) Let $\sigma(\mathcal{B})$ be the minimal σ-field generated by $\mathcal{B}$ over Ω'. Then $f^{-1}(\sigma(\mathcal{B}))$ is the minimal σ-field generated by $f^{-1}(\mathcal{B})$ in Ω.

3.4 Measurable Functions

DEFINITION 3.4.1. *Measurable Function*: If f maps $\Omega_1 \to \Omega_2$, f is a measurable function relative to the σ-fields $\mathcal{F}_j$ of subsets of $\Omega_j (j = 1, 2)$ *iff* $f^{-1}(A) \in \mathcal{F}_1$ for each $A \in \mathcal{F}_2$. The notation $f : (\Omega_1, \mathcal{F}_1) \to (\Omega_2, \mathcal{F}_2)$ will mean $f : \Omega_1 \to \Omega_2$, measurable relative to $\mathcal{F}_1$ and $\mathcal{F}_2$.

It is sufficient that $f^{-1}(A) \in \mathcal{F}_1$ for each $A \in \mathcal{C}$, where $\mathcal{C}$ is a class of subsets of Ω_2 such that the minimal σ-field over $\mathcal{C}$ is $\mathcal{F}_2$.

Note that measurability of f does not imply that $f(A) \in \mathcal{F}_2$ for each $A \in \mathcal{F}_1$.

If $\mathcal{F}$ is a σ-field of subsets of Ω, $(\Omega, \mathcal{F})$ is sometimes called a *measurable space* and the sets of $\mathcal{F}$ are called *measurable sets*.

If $(\Omega, \mathcal{F})$ is a measurable space and $f : \Omega \to \mathcal{R}^n$ (or $\bar{\mathcal{R}}^n$), f is said to be *Borel measurable* [on $(\Omega, \mathcal{F})$] *iff* f is measurable relative to the σ-fields $\mathcal{F}$ and $\mathcal{B}_n$, the class of all Borel sets. Here, $\bar{\mathcal{R}} = \mathcal{R} \cup \{\infty\} \cup \{-\infty\}$.

If $(\Omega, \mathcal{F}, \mu)$ is a measure space, the terminology f is Borel measurable on $(\Omega, \mathcal{F}, \mu)$ would mean f is Borel measurable on $(\Omega, \mathcal{F})$ and μ is a measure on $\mathcal{F}$.

Let $\Omega_2 = \mathcal{R}$, the real line and $\mathcal{B}$, the Borel field of subsets of $\mathcal{R}$. If $f^{-1}(B) \subset \mathcal{C}$ for all Borel sets B of $\mathcal{B}$, then f is said to be a *$\mathcal{C}$-measurable function* or a *function measurable with respect to $\mathcal{C}$*.

If Ω is also $\mathcal{R}$ and if f is measurable with respect to $\mathcal{B}$, the Borel field of all subsets of $\mathcal{R}$, then f is called a *Borel function* (or some tine *Baire function*).

Thus, f is a Borel function, if $f^{-1}(B) \subset \mathcal{B}$ for all Borel sets B in the range space of f and the Borel field $\mathcal{B}$ in the domain of f. Thus the Borel function is measurable with respect to the class of al Borel sets.

Note 3.4.1: The class of Borel functions is quite broad and includes most of the functions ordinarily encountered in practice. This class includes all the continuous functions. It also includes sectionally continuous functions if reasonable care is taken to define the functions at discontinuities. For most practical purposes the class is so broad that there is little need to verify that a given function under consideration is in fact a Borel function.

We note below some properties of Borel measurable functions (measurable functions).

(a) If f_1 and f_2 are Borel measurable functions from Ω to $\bar{R}$, then $f_1 + f_2, f_1 - f_2, f_1 f_2$ and f_1/f_2 are also Borel measurable, assuming that these are well-defined (i.e., $f_1 + f_2$ is not of the type $\infty - \infty$ and f_1/f_2 is not of the type a/∞ or $a/0$.)

(b) A non-negative Borel measurable function f is the limit of an increasing sequence of non-negative finite-valued simple functions f_n.

(c) An arbitrary Borel Measurable function f is the limit of a sequence of finite-valued simple functions f_n, with $|f_n| \leq |f| \ \forall \ n$. (Simple functions have been defined in definition 3.4.3).

(d) If $f_1, f_2, \ldots$ are Borel measurable functions from Ω to $\bar{R}$, and $f_n(\omega) \to f(\omega) \ \forall \ \omega \in \Omega$, then f is Borel measurable.

(f) A composition of measurable functions is measurable. Thus if $f : (\Omega_1, \mathcal{F}_1) \to (\Omega_2, \mathcal{F}_2)$ and $g : (\Omega_2, \mathcal{F}_2) \to (\Omega_3, \mathcal{F}_3)$ and $f \circ g : (\Omega_1, \mathcal{F}_1) \to (\Omega_3, \mathcal{F}_3)$, then $f \circ g$ is measurable.

EXAMPLE 3.4.1: Consider the indicator function I_A. We have seen

$$I_A^{-1}(\mathcal{B}) = \{\phi, \Omega, A, A^c\} = \sigma(A).$$

Therefore, if $A \subset \Omega$ and $\mathcal{C} = \sigma(A)$, the I_A is $\mathcal{C}$-measurable.

EXAMPLE 3.4.2: Suppose $f(\omega) = c$, a constant $\forall \ \omega \in \Omega$. Then $f^{-1}(B) = \Omega(\phi)$ if $c \in B(\notin B)$, where B is a Borel set in $\mathcal{B}$. Now, the degenerate σ-field $\{\phi, \Omega\}$ belongs to any σ-field. Hence, $f(\omega)$ is measurable with respect to any σ-field.

EXAMPLE 3.4.3: Suppose $f(\omega)$ is such that it can take only three distinct values,

$$f(\omega) = \begin{cases} c_0 \text{ if } \omega \in A_0 \\ c_1 \text{ if } \omega \in A_1 \\ c_2 \text{ if } \omega \in A_2, \end{cases}$$

where $\{A_0, A_1, A_2\}$ is a partition of Ω. Then

$$f^{-1}(B) = \begin{cases} \phi \text{ if } B & \text{does not contain any of } c_0, c_1, c_2 \\ A_0 \text{ if } B & \text{contains } c_0 \text{ only} \\ A_1 \text{ if } B & \text{contains } c_1 \text{ only} \\ A_2 \text{ if } B & \text{contains } c_2 \text{ only} \\ A_0 + A_1 \text{ if } B & \text{contains } c_0 \text{ and } c_1 \\ A_0 + A_2 \text{ if } B & \text{contains } c_0 \text{ and } c_2 \\ A_1 + A_2 \text{ if } B & \text{contains } c_1 \text{ and } c_2 \\ \Omega \text{ if } B & \text{contains } c_0, c_1, c_2. \end{cases} \quad (3.4.1)$$

Hence,

$$f^{-1}(\mathcal{B}) = \{\phi, A_0, A_1, A_2, A_0 + A_1, A_0 + A_2, A_1 + A_2, \Omega\} = \sigma(A_0, A_1, A_2),$$

the minimal σ-field containing the partition $\{A_0, A_1, A_2\}$. Hence, f is measurable with respect to $\mathcal{C} \supset \sigma\{A_1, A_2, A_3\}$.

EXAMPLE 3.4.4: Let $\Omega = \mathcal{R}$. Then the indicator function of any Borel set B is $\mathcal{B}$-measurable and is a Borel function.

EXAMPLE 3.4.5: Let $\Omega = \mathcal{R}$ and $f(\omega) = \omega + 4$. Then for $(a, b) \in \mathcal{B}, f^{-1}(a, b) = (a - 4, b - 4) \in \mathcal{B}$. In general, $f^{-1}(B) \in \mathcal{B}$ for any Borel set B in the range space and $\mathcal{B}$ in the domain of f and hence f is a Borel function.

If $f(\omega) = \omega^2, \omega \in \mathcal{R}, f^{-1}(a, b)(b > a > 0)$ equals $[\omega : \omega^2 \in (a, b)] = (-\sqrt{b}, -\sqrt{a}) \cup (\sqrt{a}, \sqrt{b}) \in \mathcal{B}$. Thus f is a Borel function.

It can be shown that if $\Omega = \mathcal{R}$, any 1:1 monotone function $f(\omega)$, which has an inverse point function is a Borel function.

DEFINITION 3.4.2. Let f be a real-valued function on Ω and let $\Omega' = \mathcal{R}$ and $\mathcal{B}$ be the Borel field. By Lemma 3.3.4, it is known that $f^{-1}(\mathcal{B})$ is a σ-field on Ω (i.e., a σ-field containing some or all of subsets of Ω). It is called the σ-field *induced by* f and is written as $\mathcal{B}(f)$.

The function f will be $\mathcal{C}$-measurable *iff* it induces a σ-field which is a subset of $\mathcal{C}$. Clearly, if $\mathcal{C} \subset \mathcal{C}'$ and if f is $\mathcal{C}$-measurable, it is also $\mathcal{C}'$-measurable. However, if f is $\mathcal{C}'$-measurable, it may not be $\mathcal{C}$-measurable. It will be $\mathcal{C}$-measurable, if $f^{-1}(\mathcal{B}) \subset \mathcal{C} \ \forall \ B \subset \mathcal{B}$.

EXAMPLE 3.4.6: If $f(\omega) = c \ \forall \ \omega \in \Omega, \mathcal{B}(f) = \{\phi, \Omega\}$.

EXAMPLE 3.4.7: Let $f(\omega) = c_i$, if $\omega \in A_i, i = 1, 2, \ldots, n$, where $A_1 + \ldots A_n = \Omega$. Then, the σ-field induced by f is the smallest σ-field containing the partition $\{A_1, A_2, \ldots, A_n\}$. $\square$

It follows that any single-valued function of the form (3.4.1) taking only a finite number of distinct values, $c_1, \ldots, c_n$, may be written as

$$f = \sum_{k=1}^{n} c_k I_{A_k} \tag{3.4.2}$$

where the sets A_k's are mutually disjoint and exhaustive,

$$A_k = \{\omega : f(\omega) = c_k\} = f^{-1}(c_k). \tag{3.4.3}$$

We therefore, consider the following definition.

DEFINITION 3.4.3. *Simple Function*: Let $(\Omega, \mathcal{F})$ be a fixed measure space. If $f : \Omega \rightarrow \mathcal{R}$ (or $\bar{\mathcal{R}}$), f is said to be a simple function *iff* f is Borel measurable and takes on finitely many distinct values. Equivalently, f is simple if it can be written as

$$f = \sum_{k=1}^{m} c_k I_{B_k} \qquad (3.4.4)$$

where the sets B_k are not necessarily disjoint, $\cup_{k=1}^{m} B_k = \Omega$.

Lemma 3.4.1: Any simple function (3.4.4) can be written as

$$f = \sum_{k=1}^{n} x_k I_{A_k} \qquad (3.4.5)$$

where the class $\{A_1, \ldots, A_n\}$ form a partition of Ω and x_k's are the distinct values assumed by the function f on each point in the set A_k.

Proof. Omitted.

Note 3.4.2: We note that a simple function takes only a finite number n of distinct values. Conversely, any function, taking only a finite number of values, is a simple function.

The partition $\{A_1, \ldots, A_n\}$ is called the *partition induced by* f. We note that the partition induced by f and $cf(c \neq 0)$ are the same.

Henceforward, we shall assume that a simple function is of the form (3.4.5) with A_k's forming a partition of Ω and x_k's distinct.

Lemma 3.4.2: The σ-field induced by the simple function f is the minimal σ-field containing the partition $\{A_1, \ldots, A_n\}$.

Proof. Omitted.

DEFINITION 3.4.4. *Elementary Function*: A linear combination of a countable number of indicator functions is called an elementary function. Without loss of generality we shall write an elementary function as

$$f = \sum_{k=1}^{\infty} x_k I_{A_k}, \qquad (3.4.6)$$

where x_k's are distinct values and $\{A_1, A_2, \ldots\}$ is a partition of Ω.

We have seen that if $\Omega = \mathcal{R}$, the indicator function of a Borel set B is a Borel function. A linear combination of indicator functions of Borel sets B_i is also $\mathcal{B}$-measurable and hence is a Borel function. This follows from the fact that if $B_i, i = 1, 2, \ldots \subset \mathcal{B}$, then $\sigma(B_1, B_2, \ldots) \subset \mathcal{B}$.

Therefore, if $\Omega = \mathcal{R}$, all the results which are true for $\mathcal{C}$ measurable functions, also hold for Borel functions.

3.5 Axiomatic Definition of Probability

With this background we now introduce the axiomatic definition of probability.

Axioms are propositions that are regarded as true and are not proved within the framework of the subject. All other propositions of the theory are derived from the accepted axioms in a logical fashion. The axioms are formulated on the basis of prolonged accumulation of facts and their logical analysis. The problem of an axiomatic definition of probability was first posed and solved by S. N. Bernstein in the first quarter of the last century. A.N. Kolmogorov (1933) considered a different approach which closely relates the probability theory with the set theory and the metric theory of functions. We shall follow here Kolmogorov's approach.

Let (i) $\mathcal{S}$ be the sample space underlying a fixed random experiment, (ii)$\mathcal{A}$, the σ-field of subsets of $\mathcal{S}$ and (iii)P, a real-valued set function defined on $\mathcal{A}$ which satisfies the following axioms: For any sets $A, A_1, A_2, \ldots \in \mathcal{A}$,

  (a) $P(A) \geq 0$ (non-negativity);
  (b) $P(\mathcal{S}) = 1$ (normed);
  (c) if $A = \sum_{i=1}^{\infty} A_i, P(A) = \sum_{i=1}^{\infty} P(A_i)$($\sigma$-additivity).

$$(3.5.1)$$

Then the function P is called a *probability measure* defined on $\mathcal{A}$ and the triplet $(\mathcal{S}, \mathcal{A}, P)$ the *probability space*. Sets belonging to $\mathcal{A}$ are called events.

It is obvious that if axiom (c) is satisfied, the result in (c) also holds for a finite number of disjoint events, $A_1, \ldots, A_n$; i.e.,

(d) $P(\cup_{i=1}^{n} A_i) = \sum_{i=1}^{n} P(A_i)$ (finite additivity). (A proof of this result is given below).

In the classical definition of probability, there was no need to postulate the axioms (b) and (c)', since these properties were proved by us. The assertion in axiom (a) was contained in the classical definition.

Clearly, the probability function P is a finite measure (vide definition 3.3.5). Since $P(\Omega) = 1$, the probability is a normed or scaled measure. The pair

$(S, \mathcal{A})$ is a measurable space and the sets in $\mathcal{A}$ are P-measurable sets.

If the set S contains only a finite number of points (elementary events), we have a finite sample space. If S contains at most a countable number of points, we have a discrete sample space. If S contains uncountably many points, we have a uncountable sample space. In particular, if $S = \mathcal{R}^k$ or some rectangles in $\mathcal{R}^k$, we have a continuous sample space.

In case the sample space S is finite, the σ-field $\mathcal{A}$ in the axiomatic definition coincides with all subsets of S. In case S is infinite, our primary interest would usually be a well-defined family $\mathcal{C}$ of events (subsets of S), which may not form an additive system. It is usual to take the σ-field $\mathcal{A}$ in the axiomatic approach as the (unique) minimal σ-field containing $\mathcal{C}$.

It may be noted that all the events may not be of our interest. We take $\mathcal{A}$ as the σ-field of all events of interest.

EXAMPLE 3.5.1: Suppose S is finite containing N elementary points ω_i. We assign to each ω_i an arbitrary non-negative number p_i such that $\sum_{i=1}^{N} p_i = 1$. As the σ-field, we take the class of sets of all elements in S, the total number of such sets (events) being 2^N. If A is an event, $P(A) = \sum_{i:\omega_i \in A} p_i$.

In particular, if $p_i = 1/N \ \forall \ i$, we have the situation of equally likely cases.

EXAMPLE 3.5.2: Suppose S consists of a countable number of elementary events $\omega_i, i = 1, 2, \ldots$. To each ω_i, attach a non-negative quantity $p_i, \sum_{i=1}^{\infty} p_i = 1$. As the σ-field we take the class of all sets (events) of S. For an event $A, P(A) = \sum_{i:\omega_i \in A} p_i$.

Note that it is not possible in this case to have $p_i = c$ (a constant) $\forall \ i$.

EXAMPLE 3.5.3: Let S be infinite and uncountable, say, $\mathcal{R}^1 = \mathcal{R}$. Such cases occur when the experiment consists in taking measurements of characters, such as the length, breadth, height, weight of an article. Here, $\mathcal{A}$ is the class of all Borel sets of $\mathcal{R}$. As a probability function, one may take arbitrary non-negative, countably additive set function on the Borel sets such that $P(S) = 1$. Such a function is uniquely defined by its value for the specific interval (∞, x) where x is any finite real number.

The similar method applies when S is a real plane $\mathcal{R}^2$ or the real hyperplane $\mathcal{R}^n$.

3.5.1 *Some simple properties*

The following simple properties follow from axiom (a), (b), (c) ((d)).

(i) Suppose $\{A_i\}(i = 1, \ldots, n)$ form a partition of $\mathcal{S}$ and $P(A_i)$ are given. Then, $\sum_{i=1}^{k} A_i(k \leq n) = A$ (say) $\in \mathcal{A}$. By (d) $P(A) = \sum_{i=1}^{k} P(A_i)$.

Hence, probability of all sets belonging to $\sigma(A_1, \ldots, A_n)$ is defined.

If $\mathcal{S}$ is finite and is $\{\omega_1, \omega_2, \ldots, \omega_N\}$, then the singletons $\{\omega_i\}(i = 1, \ldots, N)$ form a partition of $\mathcal{S}$. If $P(\{\omega_i\}) = p_i$ is known $\forall\ i$ and if A contains $\omega_{j_1}, \ldots, \omega_{j_k}$, then $P(A) = \sum_{i=1}^{k} p_{j_i}$. In particular, if $p_i = 1/N\ \forall\ i$, (the so-called equally likely case), $P(A) = k/N$.

(ii) For any event $A, P(A^c) = 1 - P(A)$, because, $\mathcal{S} = A + A^c$ and $P(\mathcal{S}) = P(A) + P(A^c)$ by (d).

(iii) Probability of an impossible event $\phi, P(\phi) = 0$, because $\Omega = \Omega + \phi$. (Alternatively, $\phi = \cup_{i=1}^{n}\phi$, union of disjoint sets ϕ; hence, $P(\phi) = P(\phi) + \ldots + P(\phi)$, which implies $P(\phi) = 0$.)

(iv) If $A \subset B, B = A + BA^c$ and hence, $P(B) = P(A) + P(BA^c)$. By (a), $P(BA^c) \geq 0$. Therefore,

$$A \subset B \Rightarrow P(B) \geq P(A). \tag{3.5.2}$$

This exhibits *monotonicity* of the probability function P. Also, we have

$$P(BA^c) = P(B - A) = P(B) - P(A), \tag{3.5.3}$$

(v) From (a) and (3.5.2), for any event $A \in \mathcal{A}$,

$$0 \leq P(A) \leq 1. \tag{3.5.4}$$

(vi) Suppose $A_i(i = 1, 2, \ldots) \in \mathcal{A}$ and are disjoint. Then $B = \sum_{i=1}^{\infty} A_i \in \mathcal{A}$. Obviously, $\sum_{i=1}^{n} A_i = B_n$ (say) $\in B \in \mathcal{A}$. Hence,

$$P(B_n) = \sum_{i=1}^{n} P(A_i) \leq P(B) \leq 1(\text{ by (d), (3.5.2) and (3.5.4)}).$$

Now, $P(B_n) \uparrow P(B)$. This gives an unique way of defining probability of the event of the form $\sum_{i=1}^{\infty} A_i$.

Also, it implies

$$P(A_n) \to 0 \text{ as } n \to \infty. \tag{3.5.5}$$

(vii) Let A and B be two arbitrary events. Now, $A \cup B = A + (B - AB)$ and $B = AB + (B - AB)$. By axiom (d),

$$P(A \cup B) = P(A) + P(B - AB)$$

and

$$P(B) = P(AB) + P(B - AB).$$

Hence,

$$P(A \cup B) = P(A) + P(B) - P(AB)$$
$$\leq P(A) + P(B) \quad [\text{ as } P(AB) \geq 0].$$

By induction,

$$P(\cup_{i=1}^n A_i) \leq \sum_{i=1}^{n} P(A_i). \tag{3.5.6}$$

All the properties $(i) - (v)$ and (vii) were proved in the classical approach.

(viii) We note that (d) follows from (c) by taking $A_i = \phi$ for $i = n+1, n+2, \ldots$ and by applying the result (i).

(ix) All the axioms (a)-(c) satisfy the properties of the relative frequency which provide the basis for the statistical definition of probability. If $f_n(A)$ is the number of times an event A occurs in n repetitions of an experiment, then $\lim_{n \to \infty} r_n(A)$ has been defined as the probability of occurrence of A, where $r_n(A) = f_n(A)/n$, the relative frequency of A. Obviously, (i) $r_n(A) \geq 0$, (ii) if A is a sure event, $r_n(A) = 1$. Axioms (a) and (b) are based on the idea that the probability which is the long term relative frequency should have the above-mentioned properties. Also, if the events $A_i(i = 1, \ldots, k)$ are mutually disjoint and occur with relative frequency $r_n(A_i)$, then the relative frequency with which at least one of the events occurs is

$$r_n(A_1) + r_n(A_2) + \ldots + r_n(A_k).$$

Axiom (c) represents a generalization of this property to the probability of occurrence of at least one of these countable number of events.

Note 3.5.1: An event A such that $P(A) = 1$ is called an *almost sure* event and A is said to happen *almost surely (a.s.)*. Its complement A^c has probability zero and is called a *null event*. For example, in tossing a coin, if p is the probability of having a 'head' and $1 - p$ is the probability of having a 'tail', the event that the coin falls on its edge is a null event.

If a certain property can hold for all ω, except those belonging to a null set, it is said to hold *almost surely (a.s.)* or for *almost all (a.a.)* ω or *almost everywhere (a.e.)*.

We consider in more detail a property of the general probability space $(S, \mathcal{A}, P)$ where the number of outcomes in S may be uncountable and P is a probability function satisfying the axioms (a), (b), (c) of (3.5.1) and (d).

Lemma 3.5.1: If the condition (d) is satisfied, condition (c) is equivalent to the following condition:

(c)':

$$B_n \in \mathcal{A}, B_n \uparrow B \text{ as } n \to \infty \Rightarrow P(B_n) \to P(B) \text{ as } n \to \infty. \tag{3.5.7}$$

Proof. We take $A_n = B_n - B_{n-1}, B_0 = \phi$. Then, the sets A_n are disjoint and $B_n = \sum_{i=1}^{n} A_i$. Now,

$$B_n \uparrow \sum_{i=1}^{\infty} A_i; \text{ hence } B = \sum_{i=1}^{\infty} A_i.$$

Suppose (c)' holds. Then

$$\sum_{i=1}^{n} A_i \uparrow \sum_{i=1}^{\infty} A_i \Rightarrow P(\sum_{i=1}^{n} A_i) \uparrow P(\sum_{i=1}^{\infty} A_i). \tag{3.5.8}$$

But by (d), $P(\sum_{i=1}^{n} A_i) = \sum_{i=1}^{n} P(A_i)$. Again,

$$\sum_{i=1}^{n} P(A_i) \uparrow \sum_{i=1}^{\infty} P(A_i). \tag{3.5.9}$$

Therefore, from the right side of (3.5.8) and (3.5.9), $P(\sum_{i=1}^{\infty} A_i) = \sum_{i=1}^{\infty} P(A_i)$, i.e. (c) holds.

Alternatively, suppose (c) holds. Then $P(\sum_{i=1}^{\infty} A_i) = \sum_{i=1}^{\infty} P(A_i)$. This means

$$P(B) = \sum_{i=1}^{\infty} P(A_i) = \lim \sum_{i=1}^{n} P(A_i) = \lim P(\sum_{i=1}^{n} A_i) \text{ (by (d))} = \lim P(B_n).$$

Condition (c) therefore implies (and is implied by) the condition that if B_n is a sequence of monotonically increasing sets, which converges to B, then $P(B_n)$ is also a monotonically increasing sequence of values which converges to $P(B)$. □

Corollary 3.5.1.i: If $B_n \in \mathcal{A}$ and $B_n \downarrow B$, then $P(B_n) \downarrow P(B)$.

Proof.

$$B_n \downarrow B \Rightarrow B_n^c \uparrow B^c \Rightarrow P(B_n^c) \uparrow P(B^c)$$
$$\Rightarrow (1 - P(B_n)) \uparrow (1 - P(B))$$
$$\Rightarrow P(B_n) \downarrow P(B).$$

The Lemma 3.5.1 and its corollary imply that probability is continuous from below and from above. The following theorem shows that probability is continuous.

Theorem 3.5.1 (*Continuity Theorem*): If $A_n \in \mathcal{A}$ and $A_n \to A$, then $P(A_n) \to P(A)$.

Proof. We have

$$\cap_{k \geq n} A_k \subset A_n \subset \cup_{k \geq n} A_k$$

$$\Rightarrow P(\cap_{k \geq n} A_k) \leq P(A_n) \leq P(\cup_{k \geq n} A_k). \tag{3.5.10}$$

But

$$\cap_{k \geq n} A_k \uparrow \underline{\lim} A_n, \quad \cup_{k \geq n} A_k \downarrow \overline{\lim} A_n. \tag{3.5.11}$$

Now,

$$A_n \to A \Rightarrow \underline{\lim} A_n = \overline{\lim} A_n = A. \tag{3.5.12}$$

From Lemma 3.5.1 and from (3.5.11),

$$P(\cap_{k \geq n} A_k) \uparrow P(A). \tag{3.5.13}$$

Again, from Corollary 3.5.1.i and from (3.5.11),

$$P(\cup_{k \geq n} A_k) \downarrow P(A). \tag{3.5.14}$$

Hence, from (3.5.10),

$$P(A_n) \to P(A).$$

$$\square$$

Thus, σ-additivity condition is sometimes referred to as the continuity condition of the probability function.

3.6   Conditional Probability Measure

Consider the probability space $(\mathcal{S}, \mathcal{A}, P)$. Let A_1, A_2 be two events in $\mathcal{A}$ such that $P(A_i) > 0, i = 1, 2$. The ratio

$$P(A_2|A_1) = \frac{P(A_1 \cap A_2)}{P(A_1)} \tag{3.6.1}$$

is called the conditional probability of the event A_2 given that A_1 has occurred. When $P(A_1) = 0, P(A_1 \cap A_2) = 0$ and $P(A_2|A_1)$ remains undefined.

For fixed A_1 in $\mathcal{A}$ such that $P(A_1) > 0$, $(\mathcal{S}, \mathcal{A}, P(.|A_1))$ is a probability space where $P(.|A_1)$ is a probability measure which takes the value $P(A_2|A_1)$ on the event $A_2 \in \mathcal{A}$. To prove this it is enough to prove

Theorem 3.6.1: $P(A_2|A_1)$ given in (3.6.1) satisfies axioms (a), (b), (c) provided $P(A_1) > 0$.

Proof. (a) $P(A_2|A_1) \geq 0$ since $P(A_1 \cap A_2) \geq 0$.

(b) $P(\mathcal{S}|A_1) = \frac{P(A_1)}{P(A_1)} = 1$.

(c) Let $\{B_i\}$ be a sequence of disjoint events in $\mathcal{A}$. We have

$$\{\cup_{i=1}^{\infty} B_i\} \cap A_1 = \cup_{i=1}^{\infty}(B_i \cap A_1).$$

Since $\{B_i\}$ is a sequence of disjoint events in $\mathcal{A}$, $\{B_i \cap A_1\}$ is a sequence of disjoint events in $\mathcal{A}$. Hence

$$P[\{\cup_{i=1}^{\infty} B_i\} \cap A_1] = \sum_{i=1}^{\infty} P(B_i \cap A_1).$$

Dividing both sides by $P(A_1)$,

$$P\left((\cup_{i=1}^{\infty} B_i | A_1)\right) = \sum_{i=1}^{\infty} P(B_i|A_1).$$

Hence the proof.□

The equation (3.6.1) can also be written as

$$P(A_1 \cap A_2) = P(A_1)P(A_2|A_1). \tag{3.6.2}$$

If $P(A_2|A_1) = P(A_2)$, then (3.6.2) reduces to the case of independent events A_1, A_2 and we have

$$P(A_1 \cap A_2) = P(A_1)P(A_2).$$

More generally, it can be shown that if $A_1, A_2, \ldots$ is a finite or countably infinite sequence of events in $\mathcal{A}$ having probability measure P such that $P(A_1) > 0, P(A_1 \cap A_2) > 0, P(A_1 \cap A_2 \cap A_3) > 0, \ldots$ then

$$P(A_1 \cap A_2 \cap \ldots) = P(A_1)P(A_2|A_1)P(A_3|A_1 \cap A_2) \ldots$$
$$P(A_n|A_1 \cap A_2 \cap \ldots \cap A_{n-1}) \ldots \tag{3.6.3}$$

In the case of mutual independence of $A_1, A_2, \ldots$ (3.6.3) can be written as

$$P(A_1 \cap A_2 \cap \ldots) = P(A_1)P(A_2)P(A_3) \ldots \tag{3.6.4}$$

DEFINITION 3.6.1: A class of events $\mathcal{C} = \{A_i : i \in J\}$, where J is a finite or an infinite index set, is said to be an independent class iff the product rule (3.6.4) holds for any finite subset of $\mathcal{C}$.

Thus, if J_0 is any finite subset of the index set J, then, we must have

$$P(\cap_{i \in J_0} A_i) = \Pi_{i \in J_0} P(A_i).$$

We note that this product rule must hold for any and hence every finite subclass.

EXAMPLE 3.6.1: Suppose $\{A, B, C, D, E\}$ is an independent class of events. Then the following classes are also independent classes of events: $\{ABC, E\}, \{AC, BC, D\}, \{B, C, DE\}$.

EXAMPLE 3.6.2: Show that if $C = \{A, B, C, D\}$ is an independent class, so also all the classes $\{A, B^c, C^c, D^c\}, \{A, B^c, C, \phi\}, \{S, B, \phi, D^c\}$, etc.

Theorem 3.6.2: If $A_1, A_2, \ldots$ is a finite or countably infinite sequence of disjoint events in $\mathcal{A}$ having non-zero probabilities, where $\cup_{i=1}^{\infty} A_i = S$ (sample space), and if A is any event in $\mathcal{A}$, then

$$P(A) = P(A_1)P(A|A_1) + P(A_2)P(A|A_2) + \ldots$$

Proof. Since $A_1, A_2, \ldots$ is a sequence of disjoint events in $\mathcal{A}$ and $\cup A_i = S, A \cap A_1, A \cap A_2, \ldots$ is a sequence of disjoint events in $\mathcal{A}$ and $\cup_{i=1}^{\infty}(A \cap A_i) = A$. Hence,

$$P(A) = \sum_{i=1}^{\infty} P(A \cap A_i) = \sum_{i=1}^{\infty} P(A_i)P(A|A_i).$$

EXAMPLE 3.6.3: Suppose $\{B_i : i \in J\}$ is a partition of S and $P(A|B_i) = p \; \forall \; i \in J$. Then show that A is independent of each B_i and $P(A) = p$.

We have

$$P(A) = \sum_{i \in J} P(AB_i) = \sum_{i \in J} P(A|B_i)P(B_i) = p \sum_{i \in J} P(B_i) = p.$$

Remark: Theorems of the previous chapter were proved for a finite number of events in $\mathcal{A}$. The axiomatic approach extends the validity of these theorems to countably infinite sequence of events in $\mathcal{A}$. The Theorem 2.6.2 can now be modified as

Theorem 3.6.3: If $\{A_i\}$ is a sequence of events in $\mathcal{A}$,

$$P(\cup_{i=1}^{\infty} A_i) = \sum_{i=1}^{\infty} P(A_i) - \sum \sum_{i<j=1}^{\infty} P(A_i \cap A_j) + \sum \sum \sum_{i<j<k=1}^{\infty} P(A_i \cap A_{\cap} A_k) - \ldots.$$

Theorem 2.6.8 (theorem of total probability) is a particular case of Theorem 3.6.2 for a finite number of events.

3.7 Independent Trials and Product Space

Consider two random experiments e_1, e_2 which are independent - results of e_1 do not depend on those of e_2 and vice-versa - with sample spaces S_1, S_2 respectively. The cartesian product S of the sample spaces S_1 and S_2 is the set of all ordered pairs of elementary events $(\omega^{(1)}, \omega^{(2)})$ where $\omega^{(i)} \in S_i, i = 1, 2$. We write

$$S = S_1 \times S_2.$$

Similarly, if $A_1^{(1)}, A_2^{(2)}$ are events in (subsets of) S_1, S_2, respectively, the cartesian product of $A^{(1)}$ and $A^{(2)}$ is the event $A = \{(\omega^{(1)}, \omega^{(2)}); (\omega^{(i)} \in A_i, i = 1, 2\}$ and we write

$$A = A_1 \times A_2.$$

The event A is called the joint occurrence of $A^{(1)}$ and $A^{(2)}$. A is empty if and only if $A^{(1)}$ or $A^{(2)}$ is an empty set.

S_1, S_2 are called the components or marginal sample spaces of S. If we take all points $(\omega^{(1)}, \omega^{(2)})$ in S such that $\omega^{(1)} \in A^{(1)}$ we obtain a cylindrical set in S which is the cartesian product $A^{(1)} \times S_2$. Thus

$$A^{(1)} \times S_2 = \{(\omega^{(1)}, \omega^{(2)}) : \omega^{(1)} \in A^{(1)}\}.$$

Similarly, $S_1 \times A^{(2)}$ is the cylindrical set $\{(\omega^{(1)}, \omega^{(2)}) : \omega^{(2)} \in A^{(2)}\}$. Thus $A^{(1)} \times A^{(2)}, A^{(1)} \times S_2, S_1 \times A^{(2)}$ are all events in the sample space $S = S_1 \times S_2$ such that the cartesian product $A^{(1)} \times A^{(2)}$ is the intersection of the cylindrical sets $A^{(1)} \times S_2$ and $S_1 \times A^{(2)}$,

$$A^{(1)} \times A^{(2)} = (A^{(1)} \times S_2) \cap (S_1 \times A^{(2)}).$$

The notation extends readily to any finite or countably infinite number of events $A^{(1)}, A^{(2)}, \ldots$ in sample spaces $S_1, S_2, \ldots$.

If $\mathcal{A}^{(1)}, \mathcal{A}^{(2)}$ are σ-field of events of S_1, S_2 respectively, then the minimal σ-field of subsets of $S, \mathcal{A}$ can be generated out of the sets of the type $A^{(1)} \times A^{(2)}, A^{(1)} \in F^{(1)}, A^{(2)} \in F^{(2)}$, where $F^{(1)}, F^{(2)}$ are the fields generating $\mathcal{A}^{(1)}, \mathcal{A}^{(2)}$ respectively.

Let $P^{(1)}, P^{(2)}$ be probability measures on S_1, S_2 respectively. For any set $A = A^{(1)} \times A^{(2)}$ where $A^{(1)} \in \mathcal{A}^{(1)}, A^{(2)} \in \mathcal{A}^{(2)}$, let us assign a probability measure by the formula

$$P(A) = P^{(1)}(A^{(1)}) \times P^{(2)}(A^{(2)}). \tag{3.7.1}$$

In this case, we say that the component probability spaces $(S_1, A^{(1)}, P^{(1)})$ and $(S_2, A^{(2)}, P^{(2)})$ are statistically independent. The probability space (S, A, P) where P is defined by (3.7.1) is the cartesian product of the two component probability spaces, is

$$(S_1, A^{(1)}, P^{(1)}) \times (S_2, A^{(2)}, P^{(2)}).$$

The notion extends readily to more than two probability spaces.

EXAMPLE 3.7.1: Consider n Bernoulli trials, $e_1, e_2, \ldots, e_n$. The probability space for e_i is $(S_i, A^{(i)}, P^{(i)})$ where $S_i = \{S', F\}$ [$S' = $ success, $F = $ failure, $A^{(i)} = \{\phi, \{S'\}, \{F\}, S_i\}, P^{(i)}(S') = p, P^{(i)}(F) = 1 - p$. The product sample space $S = S_1 \times \ldots \times S_n = \{(\omega^{(1)}, \ldots, \omega^{(n)}), \omega^{(i)} = S'$ or $F, i = 1, \ldots, n\}$ consists of 2^n points, A is the minimal σ-field over S, the probability measure P over A is such that

$$P\{(\omega^{(1)}, \ldots, \omega^{(n)})\} = p^r (1-p)^{n-r}$$

where $\omega^{(i_1)} = \ldots \omega^{(i_r)} = S'$ and the remaining ω's are each F.

EXAMPLE 3.7.2: A and B each draws two cards from a full pack without replacement. The probability space for A's experiments is $(S_1, A^{(1)}, P^{(1)})$ where S_1 consists of $\binom{52}{2}$ elementary events $\omega, A^{(1)}$ is the minimal σ-field of subsets of S_1 and $P^{(1)}$ is such that $P^{(1)}(\omega_i) = 1/\binom{52}{2} \; \forall \; i = 1, \ldots, 52$. The sample space for B is identical. The joint occurrence of the event $A^{(1)}$ - A has two aces, and $A^{(2)}$, - B has two aces is represented by the event $A = A^{(1)} \times A^{(2)}$. The probability measure P on the product probability space $(S, A, P) = (S_1, A^{(1)}, P^{(1)}) \times (S_2, A^{(2)}, P^{(2)})$ is such that $P(E) = P^{(1)}(E_1) \times P^{(2)}(E_2)$ for each $E_1 \in A^{(1)}, E_2 \in A^{(2)}$ and $E \in A$ where $E = E_1 \times E_2$. Now, $P^{(1)}(A^{(1)}) = \binom{4}{2}/\binom{52}{2} \left(= \sum_{\omega_i \in A^{(1)}} P^{(1)}(\omega_i)\right) = P^{(2)}(A^{(2)})$ and hence $P(A) = \left[\binom{4}{2}/\binom{52}{2}\right]^2$.

3.8   Exercises and Complements

3.1 If $A_n \downarrow A$, then, by using de Morgan's rule, show that $A_n^c \uparrow A^c$.

3.2 Verify the following results:

(i) $(\limsup_n A_n)^c = \liminf_n A_n^c$.

(ii) $(\liminf_n A_n)^c = \limsup_n A_n^c$.

(iii) $\liminf_n A_n \subset \limsup_n A_n$.

(iv) If $A_n \uparrow A$ or $A_n \downarrow A$, then $\liminf_n A_n = \limsup_n A_n = A$.

3.3 Let

$$A_n = \begin{array}{l} (-\frac{1}{n}, 1] \text{ if n is odd;} \\ (-1, \frac{1}{n}] \text{ if n is even.} \end{array}$$

Find $\limsup_n A_n$ and $\liminf_n A_n$.

3.4 Let $A = (a, b)$ and $B = (c, d)$ be disjoint intervals of $\mathcal{R}$ and let $C_n = A$ if n is odd and $C_n = B$ if n is even. Find $\limsup_n C_n$ and $\liminf_n C_n$.

3.5 Prove that $\cup_{n=1}^{\infty} A_n = \cup_{n=1}^{\infty}(A_1^c \cap \ldots \cap A_{n-1}^c \cap A_n)$.

3.6 Let $\mathcal{F}$ be a collection of subsets of Ω. Suppose that $\Omega \in \mathcal{F}$ and $A, B \in \mathcal{F} \Rightarrow A \cap B^c \in \mathcal{F}$. Show that $\mathcal{F}$ is a field.

3.7 Suppose $\mathcal{F}$ is a field and $\mathcal{F}$ is closed under disjoint countable unions. Show that $\mathcal{F}$ is a σ-field.

3.8 Show that the minimal field containing $\{A, B, C\}$ is the minimal field containing the partition $\{ABC, A^cBC, AB^cC, ABC^c, A^cB^cC, A^cBC^c, AB^cC^c, A^cB^cC^c\}$.

3.9 Let $\mathcal{C}$ be a class of subsets of Ω and $A \subset \Omega$. We denote by $\mathcal{C} \cap A$ the class $\{B \cap A : B \subset \mathcal{C}\}$. If the minimal σ-field over $\mathcal{C}$ is $\sigma(\mathcal{C}) = \mathcal{F}$, then show that

$$\sigma_A(\mathcal{C} \cap A) = \mathcal{F} \cap A$$

where $\sigma_A(\mathcal{C} \cap A)$ is the minimal σ-field of subsets of A over $\mathcal{C} \cap A$.

3.10 Let $\mathcal{F}_1$ and $\mathcal{F}_2$ be two σ-fields of subsets of Ω. Show that $\mathcal{F}_1 \cup \mathcal{F}_2$ need not be a σ-field.

3.11 Show that (i) an open set is a Borel set (ii) a closed set is a Borel set (iii) an open rectangle is a Borel set.

3.12 Let μ be a finitely additive set function on the field $\mathcal{F}$. Then show that

(a) $\mu(\phi) = 0$;
(b) $\mu(A \cup B) + \mu(A \cap B) = \mu(A) + \mu(B)$ for $A, B \in \mathcal{F}$;
(c) If $A, B \in \mathcal{F}$, and $B \subset A$, then $\mu(A) = \mu(B) + \mu(A - B)$.

3.13 Let μ be a finitely additive set function on the field $\mathcal{F}$.

(a) Assume that μ is continuous from below at each $A \in \mathcal{F}$; that is, if $A_1, A_2, \ldots \in \mathcal{F}, A = \cup_{n=1}^{\infty} A_n \in \mathcal{F}$ and $A_n \uparrow A$, then $\mu(A_n) \to \mu(A)$. Then prove that μ is countably additive on $\mathcal{F}$.

(b) Assume that μ is continuous from above at the empty set; that is, if $A_1, A_2, \ldots \in \mathcal{F}$, and $A_n \downarrow \phi$, then $\mu(A_n) \to \mu(\phi) = 0$. Then show that μ is countably additive on $\mathcal{F}$.

Hence, observe that for a set function, finite additivity plus continuity implies countable additivity. This result generalizes the results in Lemma 3.5.1 and its corollary.

3.14 Let $\mathcal{F}$ be the field of finite disjoint unions of right-semiclosed intervals of $\mathcal{R}$. Consider the set function μ on $\mathcal{F}$ as follows:

  (i) $\mu(-\infty, a] = a, \ a \in \mathcal{R}$,
  (ii) $\mu(a, b] = b - a, \ a, b \in \mathcal{R}, \ a < b$,
  (iii) $\mu(b, \infty) = -b, \ b \in \mathcal{R}$,
  (iv) $\mu(\mathcal{R}) = 0$,
  (v) $\mu(\cup_{i=1}^{n} I_i) = \sum_{i=1}^{n} \mu(I_i)$, if $I_1, \ldots, I_n$ are disjoint right-semiclosed intervals.

(a) Show that μ is finitely additive, but not countably additive on $\mathcal{F}$. (b) show that μ is finite but unbounded on $\mathcal{F}$.

3.15 Let Ω be a countably infinite set and let $\mathcal{F}$ be the field containing all finite subsets of Ω and their complements. Let μ be a set function such that $\mu(A) = 0$, if A is finite and $\mu(A) = 1$ if A^c is finite. (a) Show that μ is finitely additive but not countably additive on $\mathcal{F}$. (b) Show that Ω is the limit of an increasing sequence of sets A_n in $\mathcal{F}$ with $\mu(A_n) = 0$ for all n, but $\mu(\Omega) = 1$.

3.16 Let f and g be extended real-valued Borel measurable functions on $(\Omega, \mathcal{F})$ and define

$$h(\omega) = \begin{cases} f(\omega) & \text{if } \omega \in A \\ g(\omega) & \text{if } \omega \in A^c, \end{cases}$$

where A is a set in $\mathcal{F}$. Prove that h is Borel measurable.

3.17 If $f_1, f_2, \ldots$ are extended real-valued Borel measurable functions on $(\Omega, \mathcal{F}), n = 1, 2, \ldots$, show that $\sup_n f_n$ and $\inf_n f_n$ are Borel measurable. Hence show that $\limsup_{n \to \infty} f_n$ and $\liminf_{n \to \infty} f_n$ are also Borel measurable.

3.18 Let $A_0 = \phi$ and $A_1, \ldots, A_n$ be different sets. Show that $B_j = A_j - \cup_{i=0}^{j-1} A_i, j = 1, \ldots, n$ are disjoint and $B_1 + \ldots + B_n = \cup_{i=1}^{n} A_i$.

3.19 Suppose that $\sum_n P(A_n) < \infty$. Show that $P(\overline{\lim}_n A_n) = 0$.

3.20 Prove that if $\mathcal{C} = \{A_i : i \in J\}$ is an independent class of events, then the class $\mathcal{C}'$ obtained by arbitrarily replacing the A_i in any subclass of $\mathcal{C}$ by either of $\phi, \mathcal{S}$ or A_i^c is also an independent class of events. (This generalizes the result in Example 3.6.2.)

3.21 Suppose $\mathcal{C}_1 = \{A_i : i \in I\}$ and $\mathcal{C}_2 = \{B_j : j \in J\}$ are finite or countably infinite disjoint classes whose members have the property that any A_i is independent of any B_j, that is, $P(A_i B_j) = P(A_i)P(B_j) \, \forall \, i \in I$ and $j \in J$. Then, show that the events $A = \cup_{i \in I} A_i$ and $B = \cup_{j \in J} B_j$ are independent.

3.22 Let F be a distribution function on $\mathcal{R}$ given by $F(x) = 0$, $x < -1, F(x) = x + 2$, $-1 \le x < 0; F(x) = 3 + x^2$, $0 \le x < 2; F(x) = 10, x \ge 2$. If μ is the Lebesgue-Stieltjes measure corresponding to F, compute the measures of each of the following sets. (i) $\{3\}$, (ii) $[-\frac{1}{2}, 4)$, (iii) $(-1, 0] \cup (1, 2)$ (iv) $[0, \frac{3}{4}) \cup (1, 3]$, (v) $\{x : |x| + 2x^2 > 1.5\}$.

Chapter 4

Random Variables and Probability Distributions

4.1 Introduction

This chapter examines the concept of probability distribution from a measure-theoretic point of view. Section 4.2 introduces the concept of a random variable and its properties. The next section traces out the probability space induced by a random variable. The concept of a probability distribution function and its properties are then visited. The notions of discrete, continuous and mixed random variables are examined subsequently. The notion of independence among random variables is the subject-matter of the next section. The concluding section introduces the Lebesgue-integration of a Borel measurable function with special reference to the same over a random variable and the Lebesgue-Stieltjes integral and the Riemann-Stieltjes integral as special cases. Detail discussion on the concept of random variable as the limit of a non-decreasing sequence of simple functions, different types of integration of Borel measurable functions and associated convergence theorems have been relegated to the Appendix to the treatise.

4.2 Random Variables

In Section 3.4 we have introduced the concept of a C-measurable function. The concept of a random variable is very similar.

DEFINITION 4.2.1 *Random Variable*: Consider the measurable space $(\mathcal{S}, \mathcal{A})$, $\mathcal{S}$ being the sample space of a random experiment and $\mathcal{A}$ being the σ-field of all events. Note that all the events may not be of our interest and $\mathcal{A}$ is the σ-field of events of our interest. We shall consider only the events in $\mathcal{A}$.

A finite single-valued point function which maps $\mathcal{S}$ into the real-line $\mathcal{R}$ is called a random variable if the inverse image under X of all Borel sets in $\mathcal{B}$ are events, i.e. if

$$X^{-1}(B) = \{\omega : X(\omega) \in B\} \in \mathcal{A}, \ \forall \ B \in \mathcal{B}, \qquad (4.2.1)$$

i.e., if X is a $\mathcal{A}$-measurable function. Any $\mathcal{A}$-measurable function is a random variable on $(\mathcal{S}, \mathcal{A})$.

A random variable X on a probability space $(\mathcal{S}, \mathcal{A}, P)$ is therefore a $\mathcal{A}$-measurable function (also a Borel measurable function) from $\mathcal{S}$ to $\mathcal{R}$. Thus, $X : (\mathcal{S}, \mathcal{A}) \to (\mathcal{R}, \mathcal{B})$. In many situations, it is convenient to allow X to take on the values $+_-\infty$; X is said to be an *extended random variable iff* X is a $\mathcal{A}$-measurable function from $\mathcal{S}$ to $\bar{\mathcal{R}}$, that is, $X : (\mathcal{S}, \mathcal{A}) \to (\bar{\mathcal{R}}, \mathcal{B}(\bar{\mathcal{R}}))$ where $\bar{\mathcal{R}} = \mathcal{R} \cup \{\infty\} \cup \{-\infty\}$ and $\mathcal{B}(\bar{\mathcal{R}})$ is the minimal σ field over $\bar{\mathcal{R}}$.

To verify whether a certain function is a random variable it is not necessary to verify if $X^{-1}(B) \in \mathcal{A}$ for every $B \in \mathcal{B}$. It is sufficient if we verify if $X^{-1}(\mathcal{C}) \in \mathcal{A}$, where $\mathcal{C}$ is any class of subsets of $\mathcal{R}$ given in Example 3.2.15, which generate $\mathcal{B}$. As stated in Example 3.2.15, there are eight such classes of subsets of $\mathcal{R}$.

Lemma 4.2.1: X is a random variable *iff* $X^{-1}(\mathcal{C}) \subset \mathcal{A}$, where $\mathcal{C}$ is any class of subsets of $\mathcal{R}$, which generate $\mathcal{B}$.

Proof. Suppose X is a random variable so that $X^{-1}(\mathcal{B}) \subset \mathcal{A}$. Now, since $\mathcal{C} \subset \mathcal{B}$,

$$X^{-1}(\mathcal{C}) \subset X^{-1}(\mathcal{B}) \subset \mathcal{A}.$$

To prove the converse, suppose $X^{-1}(\mathcal{C}) \subset \mathcal{A}$. Now, since $\mathcal{A}$ is a σ-field,

$$\begin{aligned}
X^{-1}(\mathcal{C}) \subset \mathcal{A} &\Rightarrow \sigma(X^{-1}(\mathcal{C})) \subset \mathcal{A} \\
&\Rightarrow X^{-1}(\sigma(\mathcal{C})) \subset \mathcal{A} \ \text{(by Lemma 3.3.3 (ii))} \\
&\Rightarrow X^{1}(\mathcal{B}) \subset \mathcal{A}.
\end{aligned}$$

Hence the lemma.   □

We have, therefore, the following theorem.

Theorem 4.2.1: X is a random variable *iff* for each $x \in \mathcal{R}$, the set $X^{-1}(-\infty, x] \in \mathcal{A}$, i.e., if the set $\{\omega : X(\omega) \le x\} \in \mathcal{A}$.

Note 4.2.1: Sometimes it is more convenient to verify if the set $X^{-1}(x, \infty) \in \mathcal{A}$, i.e., if $\{\omega : X(\omega) > x\} \in \mathcal{A}$.

EXAMPLE 4.2.1: A coin is tossed three times. Here $\mathcal{S} = \{\omega_1 = (HHH), \omega_2 = (HHT), \omega_3 = (HTH), \omega_4 = (HTT), \omega_5 = (THT), \omega_6 =$

$(THH), \omega_7 = (TTH), \omega_8 = (TTT)\}$. Let X denote the number of heads. Then $X(\omega_1) = 3, X(\omega_2) = X(\omega_3) = X(\omega_5) = 2, X(\omega_4) = X(\omega_6) = X(\omega_7) = 1, X(\omega_8) = 0$. Therefore,

$$X^{-1}(-\infty, x] = \begin{cases} \phi & x < 0, \\ \{(TTT)\}, & 0 \le x < 1, \\ \{(HTT), (THT), (TTH)\}, & 1 \le x < 2, \\ \{(HHT), (HTH), (THH)\}, & 2 \le x < 3, \\ \{(HHH)\}, & 3 \le x < 4, \\ S, & 4 \le x. \end{cases}$$

Hence, X is a random variable.

EXAMPLE 4.2.2: A coin is tossed until a head appears. Let X be the number of tosses, including the toss in which the head occurs. Here $S = \{\omega_1 = (H), \omega_2 = (TH), \omega_3 = (TTH), \ldots\}$. Here X assumes the countably infinite number of values, $X(\omega_i) = i, i = 1, 2, \ldots$. Hence,

$$X^{-1}(-\infty, x] = \begin{cases} \phi, & x < 1, \\ \{(H)\}, & 1 \le x < 2, \\ \{(TH)\}, & 2 \le x < 3, \\ \{(TTH)\}, & 3 \le x < 4, \\ \ldots \end{cases}$$

Thus X is a random variable.

Theorem 4.2.2: A constant function $X(\omega)$ defined on $(S, \mathcal{A})$ is a random variable.

Proof. Let $X(\omega) = c$ (a real quantity), $\forall \, \omega \in S$. Let $B \in \mathcal{B}$ be any Borel set. Then, either $c \in B$ or $c \notin B$. If $c \in B, X^{-1}(B) = S$. If $c \notin B, X^{-1}(B) = \phi$. Since the class $\{\phi, S\} \in \mathcal{A}, X$ is a $\mathcal{A}$-measurable function and hence is a random variable.

Theorem 4.2.3: If $X(\omega)$ is a random variable on $(S, \mathcal{A}), cX(\omega)$ is also a random variable on $(S, \mathcal{A})$.

Let x be any arbitrary but fixed real number. Then $(-\infty, x] \in \mathcal{B}$. For $c = 0$, the theorem is trivial (by Theorem 4.2.2). Assume $c > 0$. Then

$$(cX)^{-1}(-\infty, x] = \{\omega : cX(\omega) \le x\} = \{\omega : X(\omega) \le \frac{x}{c}\} = X^{-1}(-\infty, \frac{x}{c}] \in \mathcal{A}$$

(since X is a random variable).

Hence cX is a $\mathcal{A}$-measurable function and hence is a random variable on $(S, \mathcal{A})$. The case $c < 0$ can be similarly proved.

The following theorems hold.

Theorem 4.2.4: If $\{c_i\}, i = 1, 2, \ldots, n$ are real numbers and $X_1, X_2, \ldots, X_n$ are random variables on $(\mathcal{S}, \mathcal{A})$, then $\sum_{i=1}^{n} c_i X_i$ is a random variable on $(\mathcal{S}, \mathcal{A})$.

Theorem 4.2.5: Any polynomial function of a random variable is a random variable.

Theorem 4.2.6: If X and Y are random variables on $(\mathcal{S}, \mathcal{A})$ and if $\{\omega : Y(\omega) = 0\} = \phi$, then $\frac{X}{Y}$ is a random variable on $(\mathcal{S}, \mathcal{A})$.

Theorem 4.2.7: If X, Y are random variables on $(\mathcal{S}, \mathcal{A})$, then $\max(X, Y)$ and $\min(X, Y)$ are also random variables.

Theorem 4.2.8: The indicator function $I_A(\omega)$ (vide Example 3.3.3) is a random variable on $(\mathcal{S}, \mathcal{A})$ if and only if $A \in \mathcal{A}$. (If $A \notin \mathcal{A}, I_A$ need not be a random variable.

We have seen in Example 3.3.3 that $I_A^{-1}(\mathcal{B}) = \{\phi, A, A^c, \mathcal{S}\} = \sigma(A)$. Hence, if $A \in \mathcal{A}$ (i.e. if A is an event), then $\sigma(A) \subset \mathcal{A}$ and I_A is a random variable.

Theorem 4.2.9: Let $X(\omega)$ be a random variable on $(\mathcal{S}, \mathcal{A})$ and $g(X(\omega))$ be a Borel measurable function of $g(X(\omega))$. Then $g(X(\omega))$ is a random variable.

Proof. The function g maps $\mathcal{R} \to \mathcal{R}$. Let B be any Borel set. Then

$$g(X)^{-1}(B) = \{\omega : g(X(\omega)) \in B\} = \{\omega : X(\omega) \in g^{-1}(B)\} = X^{-1}[g^{-1}(B)].$$

Since g is Borel measurable, $g^{-1}(B) \in \mathcal{B}$. Thus $g^{-1}(B)$ is a Borel set. Therefore, $X^{-1}[g^{-1}(B)] \in \mathcal{A}$. Hence, $g(X)^{-1}(B) \in \mathcal{A}$.

Corollary 4.2.9.1: For a given random variable $X(\omega)$ define two non-negative functions,

$$X^+(\omega) = \begin{cases} X(\omega), & \text{if } X(\omega) \geq 0, \\ = 0 \text{ if } & X(\omega) < 0, \end{cases}$$

and

$$X^-(\omega) = \begin{cases} -X(\omega) \text{ if } X(\omega) < 0, \\ = 0 \text{ if } \quad X(\omega) \geq 0. \end{cases}$$

X^+ and X^-, respectively, are called the *positive part* and *negative part* of X.

X^+ and X^- are Borel measurable functions of X.

If X is a random variable on $(\mathcal{S}, \mathcal{A})$, then X^+ and X^- are also random variables on $(\mathcal{S}, \mathcal{A})$.

We have

$$|X(\omega)| = X^+(\omega) + X^-(\omega).$$

If $X(\omega)$ is a random variable, then $|X(\omega)|$ is also a random variable.

Corollary 4.2.9.2: If $X(\omega)$ is a random variable, then $g(X(\omega))$ is also a random variable, where g is a continuous function from $\mathcal{R}$ to $\mathcal{R}$.

Proof. Continuity of g implies g is Borel measurable.

Corollary 4.2.9.3: If $X(\omega)$ is a random variable, then $g(X(\omega))$ is also a random variable, where g is a monotonic function from $\mathcal{R}$ to $\mathcal{R}$.

Proof. Monotonicity of g implies g is a Borel measurable function.

We note that the concept of probability is not required in the definition of a random variable.

EXAMPLE 4.2.3: Let $A_1, \ldots, A_n$ be disjoint sets with $\cup_1^n A_i = \mathcal{S}; X$ is a function defined on $(\mathcal{S}, \mathcal{A})$ such that $X(\omega) = a_i$ for each $\omega \in A_i (i = 1, \ldots, n)$. Thus one can write

$$X(\omega) = \sum_{i=1}^{n} a_i I_{A_i}(\omega) \qquad (4.2.2)$$

where A_i is the indicator function of the set A_i. As noted in Definition 3.4.3, a function of the type (4.2.2) is called a simple function.

Consider any Borel set $B \in \mathcal{B}$. B includes either some of a_i's, say, $a_1, \ldots, a_r (r \le n)$ or none at all. In the former case $X^{-1}(B)$ is $\cup_1^r A_i$ and in the later case ϕ. If each of $A_i (i = 1, \ldots, n) \in \mathcal{A}, X$ is a random variable.

EXAMPLE 4.2.4: In Definition 3.4.4, we have defined an elementary function

$$X(\omega) = \sum_{1}^{\infty} a_n I_{A_n}(\omega). \qquad (4.2.3)$$

Here $\{A_n, n = 1, 2, \ldots\}$ is a sequence of disjoint subsets of $\mathcal{S}, \cup_1^{\infty} A_n = \mathcal{S}, A_n \in \mathcal{A}(n = 1, 2, \ldots)$. X is a function on $(\mathcal{S}, \mathcal{A})$ such that $X(\omega) = a_n$ whenever $\omega \in A_n (n = 1, 2, \ldots)$. As in Example 4.2.3, it can be proved that X is $\mathcal{A}$-measurable and hence is a random variable.

EXAMPLE 4.2.5: Suppose X is a random variable on $(\mathcal{S}, \mathcal{A})$. Let $A_t = \{\omega : X(\omega) \le t\} \in \mathcal{A}, t \in R$. Consider several sets defined by $X(.)$.

(i) $\{\omega : X(\omega) \in (a, b]\} = A_b A_a^c$, which is an event, since

$$\{\omega : X(\omega) \in (a, b]\} = \{\omega : X(\omega) \le b\} \cap \{\omega : X(\omega) > a\}.$$

(ii) Noting that $t > a$ *iff* $t > a - \frac{1}{n}$ for every positive integer n, we have

$$\{\omega : X(\omega) \in [a, b]\} = \{\omega : a \le X(\omega) \le b\}$$

$$= \cap_{n=1}^{\infty} \{\omega : a - \tfrac{1}{n} < X(\omega) \le b\}$$

$$= \cap_{n=1}^{\infty} A_b A_{a-\frac{1}{n}}^c$$

which is an event.

(iii) $\{\omega : X(\omega) = a\} = \{\omega : X(\omega) \in [a, a]\} = \cap_{n=1}^{\infty} A_a A_{a-\frac{1}{n}}^c$ is also an event.

4.3   Induced Probability Space

We now consider the concept of probability distribution of a random variable X.

Theorem 4.3.1: The random variable X defined on $(\mathcal{S}, \mathcal{A}, P)$ induces a probability space $(\mathcal{R}, \mathcal{B}, P_X)$. The probability measure P_X is given by

$$P_X(B) = P\{X^{-1}(B)\} = P\{\omega : X(\omega) \in B\} \qquad (4.3.1)$$

for all $B \in \mathcal{B}$. The function P_X is called the probability distribution of X.

Proof. Clearly $P_X(B) \ge 0 \; \forall \; B \in \mathcal{B}$. $P_X(R) = P(X \in R) = 1$.

Let $B_i \in \mathcal{B}, i = 1, 2, \ldots$ with $B_i \cap B_j = \phi (i \ne j)$. Now, $X^{-1}(\cup_{i=1}^{\infty} B_i) = \cup_{i=1}^{\infty} X^{-1}(B_i)$ and $X^{-1}(B_i)$ and $X^{-1}(B_j)$ are disjoint sets in $\mathcal{A}$.

$$P_X \left(\cup_1^{\infty} B_i \right) = P\left\{ X^{-1}\left(\cup_1^{\infty} B_i \right) \right\} = P\left\{ \left(\cup_{i=1}^{\infty} X^{-1}(B_i) \right) \right\}$$
$$= \sum_{i=1}^{\infty} P\{X^{-1}(B_i)\} = \sum_{i=1}^{\infty} P_X(B_i).$$

Thus P_X satisfies all the three axioms and hence is a probability measure. $(R, \mathcal{B}, P_X)$ is, therefore, a probability space.

It can be shown that the continuity property of the probability measure (vide Theorem 3.5.1) also holds for P_X.

EXAMPLE 4.3.1: A unbiased dice is tossed. Let X be a random variable which takes the value 1 if the result is an odd number, 2 if the result is an even number.

Here $\mathcal{S} = \{1, 2, \ldots, 6\}$, where i denotes the event i comes up. X is a function on $\mathcal{S}$ such that its value on i is $X(i) = 1$, for $i = 1, 3, 5$ and 2 for $i = 2, 4, 6$. Let $\mathcal{A}$ denote the σ-field of events in $\mathcal{S}$. Therefore

$$X^{-1}(B) = \begin{cases} \{1, 3, 5\} & \text{if } B = B_1 \text{ (say), contains 1 but not 2,} \\ \{2, 4, 6\} & \text{if } B = B_2 \text{ (say), contains 2 but not 1,} \\ \mathcal{S} & \text{if } B = B_3 \text{ (say), contains both 1 and 2,} \\ \phi & \text{if } B = B_4 \text{ (say), contains neither 1 nor 2.} \end{cases}$$

Hence $X^{-1}(B) \in \mathcal{A}$ for all $B \in \mathcal{B}$. Therefore, X is a random variable. Again,

$$P_X(B_1) = P\{X^{-1}(B_1)\} = P\{i : X_i \in B_1\} = P\{i : X_i = 1\}$$

$$= P(\{1\}) + P(\{3\}) + P(\{5\}) = \frac{1}{2}.$$

Similarly, $P_X(B_2) = \frac{1}{2}, P_X(B_3) = 1, P_X(B_4) = P(\phi) = 0$. If the sets B_1 and B_2 are single-member sets, containing only 1 and 2 respectively, then they are disjoint and

$$P_X(B_1 \cup B_2) = P_X(B_1) + P_X(B_2) = 1.$$

We have noted in Theorem 4.2.9 that a Borel measurable function $Z = g(X)$ of a random variable X is a random variable. We will now trace out the probability space induced by the random variable Z.

Let $A \in \mathcal{A}$ be an event of the original sample space $\mathcal{S}$. Let $A = X^{-1}(N) = Z^{-1}(M) = X^{-1}[g^{-1}(M)]$ where $N \in \mathcal{R}^{(1)}$ and $M \in \mathcal{R}^{(2)}$, say, and $(\mathcal{S}, \mathcal{A}) \to^X (\mathcal{R}^{(1)}, \mathcal{B}) \to^g (\mathcal{R}^{(2)}, \mathcal{B})$. Then, probability measure $P_Z(.)$ on the Borel sets in $\mathcal{B}$ is

$$P_Z(M) = P_X(N) = P(A) = P\{\omega : Z(\omega) \in M\} = P\{\omega : X(\omega) \in N\}.$$
$$(4.3.2)$$

4.4 Probability Distribution Function of a Random Variable

We have seen in Theorem 4.2.1, X is a random variable *iff* for each $x \in \mathcal{R}, \{\omega : X(\omega) \leq x\} \in \mathcal{A}$. Also, we have seen that a random variable induces a set function P_X on $(\mathcal{R}, \mathcal{B})$. Therefore, $P_X(-\infty, x]$ becomes a very important function of study.

We shall therefore consider a point function

$$P(X \leq x) = P_X((-\infty, x]) = F_X(x) \text{ (say)}$$

on $(R, \mathcal{B})$. $F_X(x)$ is called *the distribution function* or *cumulative distribution function* of the random variable X.

Clearly

$$P_X(a, b] = P[a < X \le b]$$
$$= F_X(b) - F_X(a) \; (b > a).$$

EXAMPLE 4.4.1: Let $X = I_A$ with $P(A) = p$. Here $X(\omega) = 1(0)$ for $\omega \in A(A^c)$. $P(X = 1) = p$, $P(X = 0) = 1 - p$. Therefore,

$$F_X(x) = P(-\infty < X \le x) = \begin{cases} 0, & x < 0, \\ 1 - p, & 0 \le x < 1, \\ 1 & x \ge 1. \end{cases}$$

The function $F_X(x)$ has jumps of magnitudes $(1 - p)$ and p at 0 and 1 respectively.   □

Theorem 4.4.1: The following are the properties of a distribution function:

   (i) F is monotonically non-decreasing;
   (ii) $F(-\infty) = 0$;
   (iii) $F(\infty) = 1$;
   (iv) F is continuous to the right, i.e., $F(x + 0) = F(x)$.

Proof: (i) Let $x < y$ be two real numbers. Denote the events $[X \le x], [X \le y]$ by B_x, B_y respectively. Clearly, $B_x, B_y \in \mathcal{B}$. Now $B_x \subseteq B_y$. Hence, by the continuity property of probability function, $P(B_x) \le P(B_y)$, i.e. $F_X(x) \le F_X(y)$.

(b) Let for integral $n(> 0), B_n$ be the event $[X \le -n]$. We have $B_1 \supset B_2 \supset B_3 \ldots$ and $\lim_{n\infty} B_n = \phi$. Also, each of $B_1, B_2, \ldots \in \mathcal{B}$, i.e. they are measurable events under X. Hence, by continuity theorem (Theorem 3.5.1), $\lim_{n\to\infty} P(B_n) = P(\lim_{n\to\infty} B_n) = P(\phi) = 0$, i.e. $F(\infty) = 0$.

(iii) Let for integral $n(> 0), B_n$ be the event $[X \le n]$. Then $B_1 \subset B_2 \subset B_3 \subset \ldots$ and $\lim_{n\to\infty} B_n = \mathcal{S}$. Also, each of $B_1, B_2, \ldots \in \mathcal{B}$, i.e., they are measurable events under X. By the continuity theorem, therefore, $\lim_{n\to\infty} P(B_n) = P(\mathcal{S}) = 1$, i.e., $F(\infty) = 1$.

(iv) Let for integral $n(> 0), C_n = \left[X \le x + \frac{1}{n}\right]$. Then $C_1 \supset C_2 \supset C_3 \ldots$ and $\lim_{n\to\infty} C_n = [X \le x]$. Hence, by the continuity theorem,

$$\lim_{n\to\infty} P[C_n] = P(X \le x) \text{ i.e. } \lim_{n\to\infty} P[X \le x + \frac{1}{n}] = F_X(x)$$

i.e.

$$\lim_{\epsilon \downarrow o} F(x + \epsilon) = F_x(x) \quad \text{i.e.} \quad F_X(x+0) = F_X(x).$$

We now prove that $F(x - 0)$ is not necessarily equal to $F_X(x)$. Consider $D_n = [X \leq x - \frac{1}{n}] \rightarrow]$. Then $D_1 \subset D_2 \subset \ldots$ and $\lim_{n \to \infty} D_n = [X < x]$. Thus

$$\lim_{n \to \infty} P\left[X \leq x - \frac{1}{n}\right] = P(X \leq x)$$
$$\text{i.e.} \quad \lim_{n \downarrow 0} F_X(x - 0) = P(X < x)$$
$$\text{i.e.} \quad F_X(x - 0) = P(X < x).$$

Thus $F_X(x - 0)$ is not necessarily equal to $F_X(x)$ and hence $F_X(x)$ is not continuous to the left.

The quantity $F_X(x) - F_X(x - 0) = P[X \leq x] - P[X < x] = P[X = x]$ is called the *saltus* (or *jump*) of the distribution function $F_X(x)$. If $P[X = x] > 0, F_X(x)$ has a discontinuity at x with the value $P(X = x)$. If $P(X = x) = 0, F_X(x)$ is continuous at x (Fig. 4.4.1).

Note 4.4.1: In fact, any real-valued function F defined on R that satisfies the properties (i) - (iv) of Theorem 3.4.1 is called a distribution function.

Theorem 4.4.2 (*Jordan Decomposition Theorem*): The points of discontinuity of $F(x)$ form an at most countable set. Moreover, $F = F_c + F_d$, where F_c is a continuous function and F_d is a step function. Again, this decomposition is unique.

Proof. Let $(a, b]$ be a finite interval of $\mathcal{R}$ with at least n discontinuity points $x_1, \ldots, x_n; a_1 < x_1 < x_2 < \ldots < x_n \leq b$. Then $F(a) < F(x_1-) < F(x_1) \leq \ldots \leq F(x_n-) < F(x_n) \leq F(b)$.

Let $p_k = F(x_k) - F(x_{k-1})(= \text{saltus of } F_X(x) \text{ at } x_k), k = 1, \ldots, n$. Then

$$\sum_{k=1}^{n} p_k \leq F(b) - F(a). \tag{4.4.1}$$

Let ν_ϵ be the number of discontinuity points with jump $> \frac{1}{m} = \epsilon$. Then

$$\sum_{k=1}^{n} p_k \geq \epsilon \nu_\epsilon + \sum' p_k, \tag{4.4.2}$$

$\sum'$ being summation over other points. Hence, from (4.4.1) and (4.4.2),

$$\epsilon \nu_\epsilon \leq F(b) - F(a) \quad \text{i.e.} \quad \nu_\epsilon \leq \frac{F(b) - F(a)}{\epsilon} \quad \text{(a finite quantity)}.$$

Thus for every integer n, number of discontinuity points with jump $> \frac{1}{m}$ is finite. Therefore, there are not more than a countable number of discontinuity points in any finite interval $(a, b]$. Again, $\mathcal{R}$ is union of countable number of such intervals. Hence, the number of discontinuity points is at most countable. This proves the first part.

Define

$$F^d(x) = \sum_{x_k \leq x} p(x_k), \ x \in \mathcal{R}, \qquad (4.4.3)$$

$$F^c(x) = F(x) - F^d(x). \qquad (4.4.4)$$

As in Theorem 4.4.1, $F^d(x)$ is non-negative, non-decreasing and continuous on the right. Also, by (4.4.4), the function F^c is also non-negative, and continuous on the right.

It is also continuous on the left and non-decreasing, since for $x < x'$,

$$\begin{aligned}
F^c(x') - F^c(x) &= F(x') \textstyle\sum_{x_k \leq x'} p(x_k) - F(x) + \sum_{x_k \leq x} p(x_k) \\
&= F(x') - F(x) - \textstyle\sum_{x < x_k \leq x'} p(x_k) \\
&= F(x' - 0) - F(x) - \textstyle\sum_{x < x_k < x'} p(x_k),
\end{aligned}$$

is non-negative and tends to zero as $x \uparrow x'$. This decomposition is unique, because, otherwise, if

$$F = F^c + F^d = F^{c'} + F^{d'}$$

then

$$F^c - F^{c'} = F^{d'} - F^d, \qquad (4.4.5)$$

the left side of which is continuous, while the right hand side is a step function. This is a contradiction, unless both sides vanish simultaneously. This implies $F^c = F^{c'}, F^d = F^{d'}$ and the decomposition is unique. □

Theorem 4.4.3 (*Converse of Theorem 4.4.1*): If F is a function on R such that it has all the properties (i) - (iv) mentioned in Theorem 4.4.1, then there exists a random variable X such that $F(x)$ is the distribution function of X.

Theorem 4.4.4 (*Uniqueness Theorem*): Two random variables have the same distribution *iff* they have the same distribution function.

Remark 4.4.1: If X is a random variable on $(\mathcal{S}, \mathcal{A}, P)$, Theorem 4.4.1 states that $F(x) = P(X \leq x)$ is a distribution function associated with X. Theorem 4.4.3 states that given a distribution function $F(x)$, there exists a random variable X whose distribution function is $F(x)$. Hence we have the following theorem.

Theorem 4.4.5 (*Correspondence Theorem*): Every probability measure P on $(\mathcal{R}, \mathcal{B})$ determines uniquely a d.f. F, through the correspondence

$$P(-\infty, x] = F_X(x). \tag{4.4.6}$$

Evidently, $F(-\infty) = 0, F(+\infty) = 1$. Conversely, every distribution function F of a random variable determines uniquely a probability measure P on $(\mathcal{R}, \mathcal{B})$, through (4.4.6).

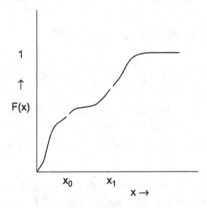

Fig. 4.4.1 $F(x)$ has saltus at $x = x_0$ and $x = x_1$, but is continuous at other points x.

DEFINITION 4.4.1 *General Distribution Function*: If $F(x)$ is monotonically non-decreasing right-continuous with $F(-\infty) = 0, F(+\infty) = c$ (a constant), then $F(x)$ is called a general distribution function. Given a general distribution function, dividing it by $F(+\infty)$, one can get the distribution function of a random variable.

Note 4.4.1: A general distribution function corresponds to a finite Lebesgue-Stieltjes measure (vide Section 3.8). In the case of a continuous random variable, the distribution function is given by an integral and in the case of a discrete random variable, it is given by a sum. (Discrete and continuous random variables have been discussed in the next section.) In stating results equally true for a continuous and discrete random variable, it is advantageous to have a single representation valid for both the cases. Riemann-Stieltjes integral representation (Section 4.8.3) satisfies this property.

DEFINITION 4.4.2 *Mixture of Distribution Functions*: If $F_i(x)$ are distribution functions of random variables, and $\alpha_i \geq 0, \sum_{i=1}^{n} \alpha_i = 1$, then

$$F(x) = \sum_{i=1}^{n} \alpha_i F_i(x) \tag{4.4.7}$$

is also a distribution function of a random variable X. X is said to have a mixture distribution. $F(x)$ is called a mixture of distribution functions $F_i(x)$'s with weights $\alpha_i \geq 0 (i = 1, \ldots, n)$. If $\sum \alpha_i \neq 1, F$ will be a general distribution function.

As a special case, we have $F(x)$ as a mixture of a step function and a continuous distribution function. Let $0 < \alpha < 1$. The function F defined by

$$F(x) = \alpha G(x) + (1 - \alpha)H(x) \tag{4.4.8}$$

where

$$G(x) = \sum_{x_i : x_i \leq x} p(x_i), \quad H(x) = \int_{-\infty}^{x} h(y)dy, \; x \in \mathcal{R} \tag{4.4.9}$$

is also a distribution function of a random variable X. The distribution function F is a mixture of distribution function G of a discrete random variable and H of a continuous random variable. The distribution of X here is of mixed type.

Such a distribution may arise, if one, after making a random experiment with probability of success p and probability of failure $(1 - p)$, chooses distribution function $G(x)$ if there is a success and $H(x)$ if there is a failure.

EXAMPLE 4.4.2: An electric bulb may be fused instantly with probability $(1 - p)$ or may survive up to age x with probability $1 - e^{-\mu x}$ if it is not fused instantly. Therefore

$$F(x) = \begin{cases} 0, & x < 0, \\ 1 - p, & x = 0, \\ 1 - p + p(1 - e^{-\mu x}), & x > 0. \end{cases}$$

The distribution function F is, therefore, a mixture of a step function with jump 1 at $x = 0 (G(x) = 0, x < 0; = 1, x \geq 1)$ and a continuous distribution function $H(x) = 1 - e^{-\mu x}, x > 0, = 0, x \leq 0$ with weights $(1 - p)$ and p respectively.

EXAMPLE 4.4.3: Let X have a distribution function

$$F_X(x) = \begin{cases} 0, & x < 0, \\ \frac{x}{2}, & 0 \leq x < 1, \\ \frac{3}{4}, & 1 \leq x < 2, \\ 1, & 2 \leq x. \end{cases}$$

Here $F_X(x) = \frac{1}{2}(G(x) + H(x))$ where

$$G(x) = \begin{cases} 0, & x < 1, \\ \frac{1}{2}, & 1 \leq x < 2, \\ 1, & 2 \leq x; \end{cases}$$

and

$$H(x) = \begin{cases} 0, & x < 0, \\ x, & 0 \leq x \leq 1, \\ 1, & 1 < x. \end{cases}$$

Thus X has a mixture distribution.

4.5 Discrete and Continuous Random Variables

We shall consider two types of random variables : (i) discrete (ii) continuous.

DEFINITION 4.5.1 *Discrete Random Variables*: Let X be a random variable on $(\mathcal{S}, \mathcal{A}) \to (\mathcal{R}, \mathcal{B})$. If the number of elements in the range of X in $\mathcal{R}$ is finite or countably infinite, then $X(\omega)$ is called a (one dimensional) discrete random variable.

Let $x_1, x_2, \ldots$ be the elements in the range space of X.

Let $p(x_i) = P[X(\omega) = x_i], x_i \in \mathcal{R}$. $p(x_i)$ is called the *probability mass function (pmf)* of X at mass point x_i if

(i) $p(x_i) \geq 0$ for all $x_i \in \mathcal{R}$; (ii)$\sum_{x_i \in R} p(x_i) = 1$.

The distribution function of X is given by

$$F_X(x) = P(X \leq x) = \sum_{x_i \leq x} p(x_i). \tag{4.5.1}$$

Thus F is a step function which remains constant over every interval not containing any mass point, but has at each mass point x_i a step of height p_i (Fig. 4.5.1). The sum in (4.5.1) is a countable sum, since $p_x = 0$ except at jump point $x_i (i = 1, 2, \ldots)$.

Thus, a discrete random variable may be specified by giving a countable set $\{x_i, i = 1, 2, \ldots\}$ of real numbers and a set of probabilities $\{p(x_i), i = 1, 2, \ldots\}$, $p_{x_i} \geq 0$, $\sum_i p(x_i) = 1$.

Fig. 4.5.2 shows the column diagram of pmf of a discrete random variable X.

DEFINITION 4.5.2 *Continuous Random Variable*: Let X be a random variable on $(\mathcal{S}, \mathcal{A}) \to (\mathcal{R}, \mathcal{B})$. The random variable X is said to be of absolutely continuous (or simply of continuous) type *iff* there is a non-negative real-valued Borel measurable function f_X on R such that

$$F_X(x) = \int_{-\infty}^{x} f_X(y)dy. \qquad (4.5.2)$$

In this case, the distribution function $F_X(.)$ is absolutely continuous, $\frac{dF}{dx}$ exists and $\frac{dF}{dx} = f_X(x)$ except for a set of values of x of probability measure zero.

The function $f_X(x)$ is called the *probability density function (pdf)* of the random variable X. Since $F_X(x)$ is continuous everywhere, it has no point of jump. Thus the probability that X takes the value x is $P(X = x) = 0$. For any real numbers $\alpha, \beta (\alpha < \beta)$,

$$P[\alpha < X \leq \beta] = \int_{\alpha+0}^{\beta} f(x)dx = F(\beta) - F(\alpha). \qquad (4.5.3)$$

$P[\alpha < X \leq \beta]$ represents the area under the curve of $f(x)$ between the lines $x = \alpha$ and $x = \beta$ (Fig. 4.5.4).

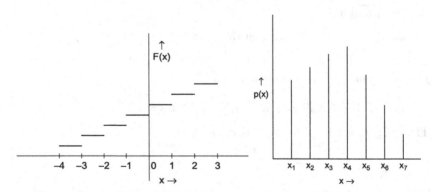

Fig. 4.5.1: df of a discrete r.v.   Fig. 4.5.2: pmf of a discrete r.v.

It follows from the mean value theorem of calculus that

$$P[x < X \leq x + dx] = \int_{x}^{x+dx} f(t)dt = dx f(\xi), \quad x < \xi \leq x + dx. \qquad (4.5.4)$$

If dx is small, (4.5.4) is approximately equal to

$$f(x)dx. \qquad (4.5.5)$$

The properties of the pdf $f(x)$ are:

(i) $f(x) \geq 0 \ \forall \ x$;

(ii) f is integrable (in the Riemann sense) over the domain of x;

(iii) $\int_{-\infty}^{\infty} f(x)dx = 1$.

In fact, any function $f(x)$ satisfying the properties (i), (ii), (iii) above is the probability density function of an absolutely continuous random variable.

Clearly, the range space of a continuous random variable X in $\mathcal{R}$ is an interval or the union of two or more disjoint intervals in $\mathcal{R}$.

Sometimes, as in the case of observing the height and weight of a person, diameter of a product, length of a leaf, X can take any value within a certain range. X in such cases are continuous random variables.

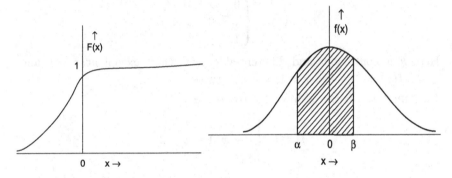

Fig. 4.5.3: df of a continuous r.v. Fig. 4.5.4: Curve of pdf of a continuous r.v.

DEFINITION 4.5.3 *Mixed Type of Random Variable*: Suppose the random variable X assumes certain discrete values $x_1, \ldots, x_n$ with positive probabilities and assumes all values in an interval, say, $[a, b]$. Thus, X is discrete for certain range of values of X and continuous over an interval or some interval not containing $x_1, \ldots, x_n$. Here X is a mixed type random variable.

To each value x_i, we assign a non-negative number $p(x_i) = P(X = x_i)$ such that $\sum_{i=1}^{n} p(x_i) = p(< 1)$. Let f be a function defined over $[a, b]$ such that $f(x) \geq 0 \ \forall \ x \in [a, b], f$ is integrable and $\int_a^b f(x)dx = 1 - p$. The probability (mass or density) function defined in this way satisfies $P(-\infty < X < \infty) = 1$.

EXAMPLE 4.5.1: Let X be a continuous random variable with *pdf* given by

$$f(x) = \begin{cases} ax, & o \leq x < 1, a, \, 1 \leq x < 2, \\ -ax + 3a, & 2 \leq x < 3, \\ 0, & 3 \leq x. \end{cases}$$

Determine a and $F(x)$.

We have $\int_{-\infty}^{\infty} f(x)dx = a\int_0^1 x\,dx + a\int_1^2 dx + \int_2^3 (3a - ax)dx = 1$, when $a = 2$.

$$F(x) = \begin{cases} 0, & x < 0, \\ 2\int_0^x y\,dy = x^2, & 0 \le x < 1, \\ 2\int_0^1 y\,dy + 2\int_1^x dy = 2x - 1, & 1 \le x < 2, \\ 2\int_0^1 y\,dy + 2\int_1^2 dy + \int_2^x (6 - 2y)dy = 6x - x^2 - 5, & 2 \le x < 3, \\ 1, & x \le 3. \end{cases}$$

EXAMPLE 4.5.2: A number is selected at random in the interval $[0,1]$. Suppose X is the square of the number selected. The cumulative distribution function of X is

$$F_X(x) = \begin{cases} 0, & x < 0, \\ \sqrt{x}, & 0 \le x < 1, \\ 1, & \le x. \end{cases}$$

Here F is continuous and differentiable everywhere except at $x = 0$ and $x = 1$. $f(x) = \frac{1}{2\sqrt{x}}, 0 < x < 1; = 0$, otherwise.

EXAMPLE 4.5.3: Let X have the *triangular pdf*,

$$f(x) = \begin{cases} x, & 0 \le x < 1, \\ 2 - x, & 1 \le x < 2, \\ 0, & x \ge 2. \end{cases}$$

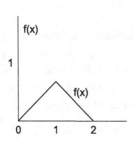

Fig. 4.5.5(a): Graph of f.

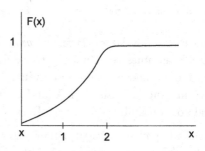

Fig. 4.5.5(b): Graph of F.

It is seen that f is a pdf. Here,

$$F_X(x) = \begin{cases} 0, & x < 0, \\ \int_0^x y\,dy = \frac{x^2}{2}, & 0 \le x < 1, \\ \int_0^1 y\,dy + \int_1^x (2 - y)dy = 2x - \frac{x^2}{2} - 1, & 1 \le x < 2, \\ 1, & x \ge 2. \end{cases}$$

4.6 Independent Random Variables

The notion of stochastic independence introduced in Subsection 2.8.3 for events may be extended to random variables. A real-valued random variable X defines an event $X^{-1}(B_1)$ corresponding to any Borel set B_1 on the real line. Another random variable Y also defines an event $Y^{-1}(B_2)$ for any Borel set B_2 on the real line. Independence of random variables is defined in terms of the independence of such events in the following manner.

DEFINITION 4.6.1: The random variables $X(.)$ and $Y(.)$ are said to be stochastically independent *iff* for each choice of Borel sets B_1 and B_2, the events $X^{-1}(B_1)$ and $Y^{-1}(B_2)$ are independent events.

The definition is equivalent to the statement that the following product rule

$$P(X \in B_1, Y \in B_2) = P(X \in B_1)P(Y \in B_2) \qquad (4.6.1)$$

holds for each choice of Borel sets B_1 and B_2, where $P(X \in M) = P(X(\omega) \in M)$.

EXAMPLE 4.6.1: The indicator function $I_A(.)$ and $I_B(.)$ for events A and B are independent *iff* the events A and B are independent.

Here, $I_A^{-1}(M)$ is one of the sets in the class $\alpha = \{A, A^c, S, \phi\}$ (vide Example 3.3.3), $I_B^{-1}(M)$ is one of the sets in the class $\beta = \{B, B^c, S, \phi\}$ for any Borel set M. Thus, I_A, I_B are independent random variables *iff* any member of the class α is independent of any member of the class β. By the results in Examples 2.6.12 and 2.6.13, this condition holds *iff* A and B are independent.

EXAMPLE 4.6.2: Consider two simple functions $X(.) = \sum_{i=1}^{n} t_i I_{A_i}(.)$ and $Y(.) = \sum_{j=1}^{m} u_j I_{B_j}(.)$, where the class $\alpha = \{A_i, i = 1, \ldots, n\}$ and $\beta = \{B_j, j = 1, \ldots, n\}$ are both partitions of Ω. The simple functions $X(.)$ and $Y(.)$ are independent *iff* $P(A_i B_j) = P(A_i)P(B_j) \; \forall \; i, j$.

Thus, if X and Y are discrete random variables with $P(X = x_i) = p_i(i = 1, 2, \ldots), P(Y = y_j) = q_j(j = 1, 2, \ldots)$, we say that X, Y are stochastically independent *iff* the joint probability $P(X = x_i, Y = y_j) = p_i q_j \; \forall \; i, j$.

If X, Y are continuous random variables with *pdfs* $g(x), h(y)$, respectively, we say X, Y are stochastically independent *iff*

$$P\{x - \tfrac{1}{2}dx \le X \le x + \tfrac{1}{2}dx, \ y - \tfrac{1}{2}dy \le Y \le y + \tfrac{1}{2}dy\}$$

$$= P\{x - \tfrac{1}{2}dx \le X \le X = \tfrac{1}{2}dx\}\{y - \tfrac{1}{2}dy \le Y \le y + \tfrac{1}{2}dy\}$$

$$= g(x)h(y)dxdy$$

for all x, y in the domain of X, Y respectively.

The concept of independence may be applied to any finite or infinite class of random variables as follows.

DEFINITION 4.6.2: A class $\{X_i : i \in J\}$ of random variables is said to be an independent class *iff* for each class $\{M_i : i \in J\}$ of Borel sets, arbitrarily chosen, the class of events $\{X_i^{-1}(M_i) : i \in J\}$ is an independent class of events.

4.7   Integral of a Borel Measurable Function (Random Variable)

4.7.1   *The Lebesgue integral*

Consider the measure space $(\Omega, \mathcal{F}, \mu)$ where μ is a measure on $\mathcal{F}$. If a function $f : \Omega \to \mathcal{R}$ (or $\bar{\mathcal{R}}$) is Borel measurable, we can define the Lebesgue integral of f with respect to μ, written as $\int_\Omega f d\mu, \int_\Omega f(\omega)\mu(d\omega)$ or $\int_\Omega f(\omega)d\mu(\omega)$.

Since a random Variable X is a function on $(\mathcal{S}, \mathcal{A})$ and probability P is a measure, we should be able to define integration of X with respect to P. The Lebesgue integration has been defined explicitly in the Appendix. We state without proof some important results in this regard.

The Lebesgue-integral of a random variable $X(\omega)$ with respect to $(\mathcal{S}, \mathcal{A}, P)$ over $E \in \mathcal{A}$, has the following properties.

(i) If $X(\omega) = k$, a constant, for all $\omega \in \mathcal{S}$, except possibly for a set of probability measure zero (i.e. *a.e.*),

$$\int_E X(\omega)dP(\omega) = kP(E). \tag{4.7.1}$$

(ii) If $\alpha \le X(\omega) \le \beta$ for all $\omega \in \mathcal{S}$ (*a.e.*), where α, β are constants, then

$$\alpha P(E) \le \int_E X(\omega)dP(\omega) \le \beta P(E). \tag{4.7.2}$$

(iii) *Linearity with respect to the integrand:* If $X_1(\omega), X_2(\omega)$ are integrable and α, β are constants, $\alpha X_1(\omega) + \beta X_2(\omega)$ is integrable and

$$\int_E \{\alpha X_1(\omega) + \beta X_2(\omega)\} dP(\omega) = \alpha \int_E X_1(\omega) dP(\omega) + \beta \int_E X_2(\omega) dP(\omega).$$
(4.7.3)

(iv) *Linearity with respect to a measure:* If $\mu(.), \mu'(.), \mu''(.)$ are totally finite measures and a, b are positive constants such that $\mu(.) = a\mu'(.) + b\mu''(.)$, and if $X(.)$ is integrable with respect to each of these measures, then

$$\int_E X d\mu = a \int_E X d\mu' + \int_E X d\mu''.$$
(4.7.4)

(v) *Additivity:* If $\{E_i : i \in I\}$ is a finite or countably infinite partition of E, and if $x(.)$ is an integrable random variable, then

$$\int_E X dP = \sum_{i \in I} X dP.$$
(4.7.5)

(vi) If $X_1(\omega) \le X(\omega) \le X_2(\omega)$ for all $\omega \in S$ (a.e.), where $X_1(\omega), X_2(\omega)$ are integrable,

$$\int_E X_1(\omega) dP(\omega) \le \int_E X(\omega) dP(\omega) \le \int_E X_2(\omega) dP(\omega).$$
(4.7.6)

(vii) $X(\omega)$ is integrable if and only if $|X(\omega)|$ is integrable and further,

$$|\int_E X(\omega) dP(\omega)| \le \int |X(\omega)| dP(\omega).$$
(4.7.7)

(viii) If $\{X_n(\omega)\}$ is a sequence of random variables such that

$$\lim_{n \to \infty} X_n(\omega) = X(\omega) \text{ and } |X_n(\omega)| \le Y(\omega)$$

where $Y(\omega)$ is integrable, then $X(\omega)$ is integrable and

$$\lim_{n \to \infty} \int_E |X_n(\omega) - X(\omega)| dP(\omega) = 0.$$
(4.7.8)

Also,

$$\lim_{n \to \infty} \int_E X_n(\omega) dP(\omega) = \int_E X(\omega) dP(\omega)$$
(4.7.9)

uniformly for all $E \in \mathcal{E}$.

(ix) *Product rule for independent random variables:* If $X(.)$ and $Y(.)$ are independent random variables which are integrable, the product $X(.)Y(.)$ is integrable and

$$\int_\Omega XY dP = \int_\Omega X dP \int_\Omega Y dP.$$
(4.7.10)

4.7.2 *The Lebesgue-Stieltjes integral*

Let $F(.)$ be a distribution function and let $\mu_F(.)$ be the Lebesgue-Stieltjes Measure induced by it (vide Section 3.4). The Lebesgue-Stieltjes integral of a function $g(x)$ defined on the real line $\mathcal{R}$ is given by

$$\int_a^b g(t)dF(t) = \int_{[a,b]} g d\mu_F \tag{4.7.11}$$

where $[a, b]$ is a closed interval $a \leq t \leq b$.

In the special case, where $F(t)$ is given by $F(t) = t$, the measure induced by it is the ordinary Lebesgue measure, which assigns to each interval its length. The Lebesgue-Stieltjes integral reduces in this case to the Lebesgue integral.

4.7.3 *The Riemann-Stieltjes integral*

Let $g(x)$ be a continuous function in $(a, b]$. We divide the interval $(a, b]$ into n subintervals by inserting points $x_1, x_2, \ldots, x_{n-1}$. In order of these intervals, arbitrarily choose points $\xi_1, \xi_2, \ldots, \xi_n$ in these intervals and find the sum

$$S_n = \sum_{i=1}^n g(\xi)[F_X(x_i) - F_X(x_{i-1})], \quad x_0 = a, \quad x_n = b. \tag{4.7.12}$$

If M_i, m_i, respectively, are the upper and lower bound of $g(x)$ in $(x_{i-1}, x_i]$, we define upper and lower sums respectively as

$$S_{nU} = \sum_{i=1}^n M_i[F_X(x_i) - F_X(x_{i-1})], \quad S_{nL} = \sum_{i=1}^n m_i[F_X(x_i) - F_X(x_{i-1})]. \tag{4.7.13}$$

In the same way as for the ordinary Riemann integral, it can be proved that when $n \to \infty$ in such a way that $\max(x_i - x_{i-1}) \to 0$, S_n tends to a definite limit and is called the Riemann-Stieltjes integral of $g(x)$ with respect to $F(x)$ over $(a, b]$ and is given by

$$\lim_{n \to \infty} S_n = \lim_{n \to \infty} S_{nL} = \lim_{\to} S_{nU} = \int_a^b g(x)dF_X(x). \tag{4.7.14}$$

The concepts of a random variable as the limit of a non-decreasing sequence of simple functions, Lebesgue integration of a Borel measurable function including that of a random variable and associated convergence theorems (Monotone convergence Theorem, Fatau's Lemma, Dominated Convergence

Theorems, Fubini's theorem, Radon-Nikodyn Theorem), Lebesgue-Stieltjes integral and Riemann-Stieltjes integral have been explicitly developed in the Appendix.

The reader interested in theory of measure and integration may refer to books by Cramer (1950), Halmos (1958), Loe've (1977), Munroe (1953), Mcshane (1944), Pitt (1963), Ash and Dade (2000), Kingman and Taylor (1966) among others.

4.8 Exercises and Complements

4.1 Prove that a simple function $X(\omega) = \sum_{k=1}^{n} c_k I_{B_k}(\omega)$ (compare with equation (4.2.2)), where the sets B_k are not necessarily disjoint, is a random variable.

4.2 Let X be the number of successes in n repeated trials. Let E_i be the event of a success on the ith trial and A_{rn} the event of exactly r successes in n such trials. Show that $\{E_1, \ldots, E_n\}$ is an independent class and forms a partition of Ω. Also, show that $X(\omega) = \sum_{i=1}^{n} I_{E_i}(\omega)$.

Also, show that $\{A_{rn}, r = 0, 1, \ldots, n\}$ is a partition of S. Hence, $X(\omega) = \sum_{r=0}^{n} r I_{A_{rn}}(\omega)$.

Observe that the set of ω for which $X(\omega) \leq t$ may be expressed as $\{\omega : X(\omega) \leq t\} = \cup_{r=0}^{[t]} A_{rn}$ where $[t]$ is the largest integer which is not greater than t. Therefore, for each real t, the set $\{\omega : X(\omega) \leq t\}$ is an event. Hence $X(.)$ is a random variable.

4.3 Let X be an extended random variable. Then show that $\{\omega : X(\omega) = \infty\}, \{\omega : X(\omega) = -\infty\}, \{\omega : X(\omega) > c\}, \{\omega : X(\omega) = c\}, c \in \mathcal{R}$ are all in S and $-X$ is also an extended random variable.

$[\{\omega : X(\omega) < \infty\} = \cup_{n=1}^{\infty}\{\omega : X(\omega) \leq n\}$; so $\{\omega : X(\omega) < \infty\} \in S$. Therefore, $\{\omega : X(\omega) = \infty\} = \{\omega : X(\omega) < \infty\}^c \in S$.

Also, $\{\omega : X(\omega) = -\infty\} = \cap_{n=-\infty}^{0}\{\omega : X(\omega) \leq n\} \in S$.

$\{\omega : X(\omega) > c\} = \{\omega : X(\omega) \leq c\}^c \in S$.

Again, $\{\omega : X(\omega) = c\} = \cap_{n=1}^{\infty}[\{\omega : X(\omega) \leq c + \frac{1}{n}\} \cap \{\omega : X(\omega) > c - \frac{1}{n}\}]$; so $\{\omega : X(\omega) = c\} \in S$.

$\{\omega : -X(\omega) \leq c\} = \{\omega : X(\omega) > -c\} \cup \{\omega : X(\omega) = -c\} \in S$.

Hence, $-X$ is also an extended random variable.]

4.4 Let $X_1, X_2, \ldots$ be extended random variables. Then show that $\sup_n X_n, \inf_n X_n, \overline{\lim}_n X_n$ and $\underline{\lim}_n X_n$ are also extended random variables.

4.5 Find the distribution functions corresponding to the following probability density functions.

(a)

$$f(x) = \begin{cases} 0, & x < a, \\ \frac{1}{b-a}, & a \le x < b, \\ 0, & x \ge b. \end{cases}$$

(This is called *Uniform* distribution.)

(b)

$$f(x) = \begin{cases} \frac{2}{a}(1 - \frac{x}{a}), & 0 < x < a, \\ 0, & \text{otherwise.} \end{cases}$$

(This is called *Right Triangular* distribution.) Also find $P(a/2 < X < a)$.

(c) $f(x) = \lambda e^{-\lambda x}, x > 0$.

(d) $f(x) = Axe^{-h^2 x^2}, x > 0$. Find the factor A. (This is called *Raleigh* distribution.)

(e) $f(x) = Kx^2 e^{-x/2}, \; x > 0$. Also, find the value of K.

(f)

$$f(x) = \begin{cases} 0, & x < 2, \\ \frac{1}{18(3+2x)}, & 2 \le x \le 4, \\ 0, & x > 4. \end{cases}$$

Find $P[2 \le X \le 3]$.

(g) $f(x) = ae^{-\lambda|x|}, \lambda > 0, -\infty < x < \infty$. (This is known as *Laplace* distribution.) Find the factor $'a'$.

(h)

$$f(x) = \frac{a}{\pi(1 + x^2)}, \quad -\infty < x < \infty.$$

(This is known as *Cauchy* distribution. Find the factor $'a'$. Find $P(-1 < X < 1$.)

(i)

$$f(x) = \begin{cases} ae^{-(x-x_0)}, & x \geq 10, \\ 0, & x < 10. \end{cases}$$

Find the factor $'a'$.

4.6 Does the function $f(x) = \frac{x+1}{\theta(\theta+1)}e^{-x/2}$ $(x > 0)$, where $\theta > 0$, represent a *pdf*? If so, find the corresponding distribution function.

4.7 Do the following functions define distribution functions (d.f.'s)? In case they are d.f.'s, determine the *pdf*, $f(x)$. Draw the graphs of $f(x)$ and $F(x)$.

(a) $F_X(x) = 0, x \leq 1;\ 1 - \frac{1}{x}, x > 1.$
(b) $F_X(x) = 1 - e^{-x}, x \geq 0;\ 0, x < 0.$
(c) $F_X(x) = \frac{2}{\pi}\sin^{-1}(\sqrt{x}),\ 0 \leq x \leq 1.$
(d) $F_X(x) = \frac{x^2}{2} + \frac{1}{2},\ -1 \leq x \leq 1.$
(e) $F_X(x) = e^{3x},\ -\infty < x \leq 0.$
(f) $F_X(x) = \frac{1}{1+\exp\{-\frac{x-\alpha}{\beta}\}},\ \beta > 0, x > 0.$ (This is called the *Logistic* distribution.)

4.8 An angle θ is chosen at random from $[-\frac{\pi}{2}, \frac{\pi}{2}]$ and a line is drawn through the point $(0, a)$ making an angle θ with the y axis. Find the probability density function of the point x at which this line crosses the x-axis.

4.9 Suppose f and g are *pdf*'s on the same interval $\alpha \leq x \leq \beta$. (a) Show that $f + g$ is not a *pdf* on that interval; (b) show that for every number $\alpha, 0 < \alpha < 1, \alpha f(x) + (1 - \alpha)g(x)$ is a *pdf* on that interval.

4.10 A random variable X has a *pdf* $f_1(x)$ with probability p_1 and a *pdf* $f_2(x)$ with probability $p_2, p_1 + p_2 = 1$. Find the *pdf* and distribution function of X.

4.11 Let $X_1, \ldots, X_n$ be discrete random variables. Show that the X_i's are independent *iff* $P[X_1 = x_1, \ldots, X_n = x_n] = \Pi_{i=1}^{n}P(X_i = x_i)\ \forall$ real $x_1, \ldots, x_n$.

4.12 Let $(S, \mathcal{A}, P)$ be the probability space. The classes $\mathcal{C}_i, i \in I$ of sets in $\mathcal{A}$ are said to be independent *iff* for a given class $\mathcal{C}_i$, the events represented by the sets $C_l \in \mathcal{C}_i$ are independent.

If the classes $C_i, i \in I$ are independent, show that if the following sets are added to the C_i, the enlarged classes still remain independent.

(a) Proper difference $A - B$, where $A, B \in C, B \subset A$;
(b) The sets ϕ, S;
(c) Countable disjoint unions of sets in C_i;
(d) Limits of monotone sequences in C_i.

4.13 Let $\Omega = (-6, 6)$, $\mathcal{A}$ the class of all Borel sets in Ω and λ, the Lebesgue measure. Taking $(\Omega, \mathcal{A}, \mu)$ to be the measure space, evaluate $\int_A X d\lambda$ where $A = [-1, 5)$ and

$$X(\omega) = \begin{cases} \frac{1}{2}, & \omega \in (-6, 2] \\ \frac{1}{3}, & \omega = 2, \\ 1, & \omega \in (2, 4], \\ 0, & \omega \in (4, 6]. \end{cases}$$

4.14 Let $\Omega = \{-2, -1, 4, 8\}, A = \{-2, -1, 8\}, \mu(-2) = 2, \mu(-1) = 1, \mu(4) = 4, \mu(8) = 8$. If $X(\omega) = 1$, if $\omega = 4, \omega = 8; = -1$ if $\omega = -2, \omega = -1$, evaluate $\int_A X d\mu$.

4.15 If X_1 and X_2 are independently and identically distributed random variables, prove that

(i) $P\{|X_1 - X_2| > t\} \leq 2P\{|X_1| > \frac{t}{2}\}$.

(ii) if $a \geq 0$ such that $P\{X_1 \geq a\} \leq 1 - p$ and $P\{X_1 \leq -a\} \leq 1 - p$, then $P\{|X_1 - X_2| \geq \epsilon\} \geq pP\{|X_1| > a + \epsilon\}$.

[(i) $\{|X_1| > \frac{t}{2}\} \cup \{|X_2| > \frac{t}{2}\} \supset \{|X_1 - X_2| > t\}$. Hence

$$P\{|X_1 - X_2| > t|\} \leq P\{|X_1| > \frac{t}{2}\} + P\{|X_2| > \frac{t}{2}\} = 2P\{|X_1| > \frac{t}{2}\}.$$

(ii) $\{X_1 \geq a + \epsilon, X_2 \leq a\} \cup \{X_1 \leq -a - \epsilon, X_2 \geq -a\} \subset \{|X_1 - X_2| \geq \epsilon\}$. Hence

$$\begin{aligned} P\{|X_1 - X_2| \geq \epsilon\} &\geq P\{X_1 > a + \epsilon, X_2 \leq a\} \\ &+ P\{X_1 < -a - \epsilon, X_2 \geq -a\} \\ &\geq pP\{X_1 \geq a + \epsilon\} + pP\{x_1 \leq -a - \epsilon\} \\ &= pP\{|X_1| > a + \epsilon\}.] \end{aligned}$$

Chapter 5

Expectation of a Discrete Random Variable

5.1 Introduction

In the previous chapter we have introduced the concepts of discrete random variables and continuous random variables. In the next chapter we shall study the properties of a probability distribution in $\mathcal{R}$ in the general set up. However, since the discrete random variables are sometime easier to deal with, we shall in this chapter consider some simple properties of the probability mass function of a discrete random variable, namely, its expectation, variance and covariance and correlation between two discrete variables.

5.2 Probability Distribution of a Discrete Random Variable

We have seen in Section 4.5 that a discrete random variable takes only a finite or countably finite number of values.

To refresh the idea in brief, suppose the sample space $\mathcal{S}$ consists of a finite number of points $\omega_1, \ldots, \omega_m$. Let X be a point function defined on $\mathcal{S}$ such that its value of ω_i is $X(\omega_i) = x_i, i = 1, \ldots, m$. Here X is a random variable with domain $\mathcal{S}$ and range $\{x_i; i = 1, \ldots, m\}$.

EXAMPLE 5.2.1: A coin is tossed three times. Here $\mathcal{S} = \{\omega_1 = (HHH), \omega_2 = (HHT), \omega_3 = (HTH), \omega_4 = (HTT), \omega_5 = (THH), \omega_6 = (THT), \omega_7 = (TTH), \omega_8 = (TTT)\}$. Let X denote the number of heads. Then $X(\omega_1) = 3, X(\omega_2) = X(\omega_3) = X(\omega_5) = 2; X(\omega_4) = X(\omega_6) = X(\omega_7) = 1; X(\omega_8) = 0$. X is a random variable with domain $\mathcal{S}$ and range $\{0, 1, 2, 3\}$.

EXAMPLE 5.2.2: A coin is tossed until a head appears. Let X be the number of tosses (including the toss in which the head appears) required. Here $S = \{\omega_1 = (H), \omega_2 = (TH), \omega_3 = (TTH), \ldots\}$ and X assumes the countably infinite number of values with $X(\omega_i) = i, i = 1, 2, 3, \ldots$.

We shall in this chapter consider in some detail the properties of a discrete random variable.

DEFINITION 5.2.1: Let X be a discrete random variable taking a countable number of values $x_1, x_2, \ldots$ with probabilities $p_1, p_2, \ldots$; $p_i \geq 0, \sum_{i=1}^{\infty} p_i = 1$. The set of values of X together with their corresponding probabilities $\{x_i, p_i; i = 1, 2, \ldots\}$ is called the *probability distribution* of X.

Note 5.2.1: All the results presented in this chapter will hold if X takes only a finite number of values $x_1, \ldots, x_m$.

For example 5.2.1, writing $P(X = i) = p_i, p_0 = \frac{1}{8}, p_1 = p_2 = \frac{3}{8}, p_3 = \frac{1}{8}; \{i, p_i\}$ is the probability distribution of X. For Example 5.2.2, $p_i = \frac{1}{2^i}; \{i, p_i; i = 1, 2, \ldots\}$ is the probability distribution of X.

5.3 Expectation

DEFINITION 5.3.1 *Expectation*: Let X be a discrete random variable with probability distribution $\{x_i, p_i; i = 1, 2, \ldots\}$. The expected value of X, also called the mean value of X, is defined by

$$E(X) = \sum_{i=1}^{\infty} x_i p_i \tag{5.3.1}$$

provided the sum exists.

As noted before, p_i is defined as the long-run relative frequency with which X takes the value x_i; hence, $E(X)$ may be defined as the long run average value of X.

EXAMPLE 5.3.1: For example 5.2.2, calculate the expected value of the number of trials required (including the last trial in which a head has to appear).

Here, $P(X_i = i) = \frac{1}{2^i} (i = 1, 2, \ldots)$. Thus, $E(X) = \sum_{i=1}^{\infty} \frac{i}{2^i} = \frac{1}{2} + 2 \cdot \frac{1}{2^2} + 3 \cdot \frac{1}{2^3} + \ldots = 2$.

EXAMPLE 5.3.2: Consider the Indicator random variable I_A (vide equation

(3.3.8)). We have

$$E(I_A(\omega)) = 1 \sum_{\omega \in A} p(\omega) + 0. \sum_{\omega \notin A} p(\omega) = \sum_{\omega \in A} p(\omega) = P(A).$$

Thus, the probability of an event A is the expectation of its indicator random variable I_A.

DEFINITION 5.3.2 *Function of a Random Variable and Its Expectation*: Let X be a discrete random variable with domain S and range $R_x = \{x_1, x_2, \ldots\}$. Let $H(X)$ be a function of X defined on R_x with range of values $\{h_1, h_2, \ldots\}$. Clearly, $H(x)$ is also a random variable.

$$P[H(X) = h_i] = \phi_i \text{(say)} = \sum_{\{x_i : H(x_i) = h_i\} = B_i} \text{(say)} \ P(X = x_i) \qquad (5.3.2)$$
$$= \sum_{\omega_k : X(\omega_k) = x_j \in B_i} P(\omega_k) \geq 0.$$

It follows that $\sum_i \phi_i = 1$. Thus, $\{i, \phi_i ; i = 1, 2, \ldots\}$ is the probability distribution of $H(X)$. Its expectation is

$$E[H(X)] = \sum_i h_i P[H(X) = h_i] = \sum_i h_i \phi_i.$$

EXAMPLE 5.3.3: Let X be a random variable with $P(X_k = +_-m) = \frac{1}{2k}, m = 1, \ldots, k$. Find $E(X^2)$.

Here X^2 is a random variable with $P(X^2 = m^2) = P(X = m) + P(X = -m) = \frac{1}{k}, m = 1, \ldots, k$. $E(X^2) = \frac{1}{k} \sum_1^k m^2 = \frac{(k+1)(2k+1)}{6}$.

We now consider some theorems on expectation of a discrete random variable. The properties of expectation of a random variable in the general case has been considered in detail in Section 6.2.

Theorem 5.3.1: Let $X(\omega) = c$, a constant, whatever be the outcome of the experiment. Then $E(X) = c$.

Proof. Follows easily since $x_i = c, i = 1, 2, \ldots$.

Theorem 5.3.2: Let $Y = cX, c$ an arbitrary constant, be a function of X. Then, $E(Y) = cE(X)$.

Proof. $E(Y) = \sum cx_i p_i = c \sum_i x_i p_i = cE(X)$.

Theorem 5.3.3: Random variables X and Y are both defined on $(S, \mathcal{A}, P)$. Let $Z = X + Y$. Then, $E(Z) = E(X) + E(Y)$.

Proof. Let X take the value $x_1, x_2, \ldots$ and Y take the values $y_1, y_2, \ldots$ on $\omega_1, \omega_2, \ldots$ with probabilities $p_1, p_2, \ldots$. Clearly Z is a random variable taking values $x_i + y_i$ on $\omega_i, i = 1, 2, \ldots$. Thus

$$E(Z) = \sum_{i=1}^{\infty}(x_i + y_i)p_i = \sum_i p_i x_i + \sum_i p_i y_i = E(X) + E(Y).$$

Corollary 5.3.3.1: If $Y = a + bX, a, b$, being arbitrary constants $E(Y) = a + bE(X)$.

Theorem 5.3.4: If $X_1, \ldots, X_n$ are all random variables defined on $(\mathcal{S}, \mathcal{A}, P)$ and if $Z = X_1 + \ldots + X_n$, then $E(Z) = E(X_1) + \ldots + E(X_n)$.

Proof. Let X_j take the value x_{ji} with probability p_i on $\omega_i, i = 1, 2, \ldots; j = 1, 2, \ldots, n$. Then Z is a random variable taking value $\sum_{j=1}^{n} x_{ji}$ on $\omega_i, i = 1, 2, \ldots$ Hence

$$E(Z) = \sum_{i=1}^{\infty}(\sum_{j=1}^{n} x_{ji})p_i = \sum_{j=1}^{n}(\sum_{i=1}^{\infty} x_{ji}p_i) = \sum_{j=1}^{n} E(X_j).$$

Corollary 5.3.4.1: If $X_1, \ldots, X_n$ are random variables, all defined on $(\mathcal{S}, \mathcal{A}, P)$ and if $Z = c + a_1 X_1 + \ldots + a_n X_n, c, a_1, \ldots, a_n$ being arbitrary constants, then $E(Z) = c + \sum_{i=1}^{n} a_i E(X_i)$.

EXAMPLE 5.3.4: n numbers are selected from N numbers $1, 2, \ldots, N$ each with equal probability and without replacement $(n < N)$. Find the expectation of the sum of the selected numbers.

Let X_i be the number drawn at the i draw $(i = 1, \ldots, n)$. Since the numbers are drawn without replacement, each of $X_1, \ldots, X_n$ will be different.

$P(X_i = j) = \frac{1}{N}, j = 1, \ldots, N;$

$P(X_2 = j) = \sum_{k(\neq j)=1}^{N} P(X_1 = k)P(X_2 = j | X_i = k) = \sum_{k(\neq j)=1}^{N} \frac{1}{N}\frac{1}{N-1}$

$\quad\quad\quad = \frac{1}{N}, j = 1, \ldots, N.$

It can be similarly shown that $P(X_l = j) = \frac{1}{N}, j = 1, \ldots, N$ and $l = 3, \ldots, n$. Thus each of $X_1, \ldots, X_n$ has the same probability distribution and hence the same expectation. Now $E(X_1) = \sum_{j=1}^{N} j\frac{1}{N} = \frac{N+1}{2}$. Hence $E(X_1 + \ldots + X_n) = \frac{n(N+1)}{2}$.

EXAMPLE 5.3.5: n balls are distributed at random among m cells. Find the expected number of cells that remain empty.

For the cell i we define a random variable X_i such that $X_i = 1(0)$ if the ith cell remains empty (occupied). Then $X = \sum_{i=1}^{m} X_i$ is the total number of empty cells. Now,

$$E(X_i) = P(X_i = 1) = \frac{(m-1)^n}{m^n}.$$

Hence, $E(X) = \frac{m(m-1)^n}{m^n}$.

Lemma 5.3.1: If possible values of X are $0, 1, 2, \ldots$, show that

$$E(X) = P(X > 0) + P(X > 1) + P(X > 2) + \ldots \qquad (5.3.3)$$

The right side of (5.3.3) is

$$\{P(X = 1) + P(X = 2) + \ldots\} + \{P(X = 2) + P(X = 3) + \ldots\}$$
$$\{P(X = 3) + P(X = 4) + \ldots\} + \ldots$$
$$= \sum_{i=1}^{\infty} iP(X = i) = E(X).$$

EXAMPLE 5.3.6: Balls are successively distributed among n cells. Find the expected number of balls that must be distributed before all the cells are occupied.

Let Z be the random variable denoting the number of balls required to achieve a full occupancy. We have seen in Chapter 2, the probability that all the m cells are occupied, when n balls are distributed at random is

$$p_{n,m} = \sum_{k=0}^{m} (-1)^k \binom{m}{k} (\frac{m-k}{m})^n.$$

Hence, $P(Z \leq n) = p_{n,m}$. Hence, by Lemma 5.3.1,

$$E(Z) = \sum_{n=0}^{\infty} P(Z > n) = \sum_{n=0}^{\infty} (1 - p_{n,m})$$

$$= \sum_{n=0}^{\infty} [\binom{m}{1}(\frac{m-1}{m})^n - \binom{m}{2}(\frac{m-2}{m})6n + \ldots + (-1)^m \binom{m}{m-1}(\frac{1}{m})^n]$$

$$= m[\binom{m}{1} - \frac{1}{2}\binom{m}{2} + \frac{1}{3}\binom{m}{3} - \ldots + (-1)^m \frac{1}{m-1}\binom{m}{m-1}].$$

$$(5.3.4)$$

To evaluate (5.3.4), consider the identity

$$\sum_{i=0}^{m-1} (1-t)^i = \frac{1 - (1-t)^m}{t}.$$

Integrating both sides over $0 \leq t \leq 1$,

$$1 + \frac{1}{2} + \frac{1}{3} + \ldots + \frac{1}{m} = \binom{m}{1} - \frac{1}{2}\binom{m}{2} + \frac{1}{3}\binom{m}{3} - \ldots + (-1)^m \frac{1}{m-1}\binom{m}{m-1}.$$

Hence, the required quantity (5.3.4) is

$$m[1 + \frac{1}{2} + \ldots + \frac{1}{m}].$$

5.4   Variance, Covariance, Correlation Coefficient

DEFINITION 5.4.1: *Variance of X*, denoted as $V(X)$ is a measure of dispersion or spread of X (vide Section 6.8) around its mean value $E(X)$ and is given by

$$V(X) = E(X - E(X))^2 = E(X^2) - (E(X))^2.$$

This is also often denoted as σ_X^2. Its positive square root is called *standard deviation (s.d.)* of X.

Theorem 5.4.1: If $X = c$, a constant, whatever be the outcome of the experiment, $V(X) = 0$.

Proof. $X - E(X) = c - 0 = 0$. Thus, $V(X) = 0$.

Theorem 5.4.2: If $Y = cX, c$ a constant, $V(Y) = c^2 V(X)$.

Proof. $V(Y) = E(cX - cE(X))^2 = c^2 V(X)$.

Theorem 5.4.3: If $Y = a + bX, a, b$ arbitrary constant, $V(Y) = b^2 V(X)$.

Proof. $Y - E(Y) = b(X - E(X))$.

EXAMPLE 5.4.1: A number of independent trials are made in each of which an event A may occur with probability p. The trial continues until the event occurs for the first time. If X is the number of trials (including the trial in which A occurs) that must be made, find $E(X)$ and $V(X)$.

Here, $P(X = 1) = p, P(X = 2) = qp, \ldots, P(X = k) = q^{k-1}p, \ldots$, where $q = 1 - p$. In this case X is said to have a *geometric distribution* with parameter p. Thus

$$E(X) = p + 2qp + 3q^2p + \ldots + kq^{k-1}p + \ldots = \tfrac{p}{(1-q)^2} = \tfrac{1}{p};$$
$$E(X^2) = p + 2^2qp + 3^2q^2p + \ldots + k^2q^{k-1}p + \ldots = \tfrac{p(1+q)}{(1-q)^3}.$$

The last equation follows, because

$$\sum_{m=1}^{\infty} mq^m = \frac{q}{(1-q)^2}.$$

Differentiating both sides with respect to q,

$$\sum_{m=1}^{\infty} m^2 q^{m-1} = \frac{(1-q)}{(1-q)^2}.$$

Therefore, $V(X) = \tfrac{q}{p^2}$.

EXAMPLE 5.4.2: Consider a random variable X with *pmf*

$$P(X = r) = c.\frac{\binom{n}{r}}{r+1}, \quad r = 0, 1, \dots, n,$$

c being a constant. Evaluate c and find $E(X), V(X)$.

We have $P(X = r) = c\frac{\binom{n}{r}}{r+1} = c.\frac{\binom{n+1}{r+1}}{n+1}$. Since, $\sum_{r=0}^{\infty} P(X = r) = 1$, $\frac{c}{n+1} \sum_{r=0}^{n} \binom{n+1}{r+1} = 1$, when $c = \frac{n+1}{2^{n+1}-1}$.

$$E(X) = \frac{1}{2^{n+1}-1} \sum_{r=0}^{n} r\binom{n+1}{r+1} = \frac{(n+1)}{2^{n+1}-1} \sum_{r=0}^{n} \binom{n}{r} = 1 - \frac{(n-2)2^n+1}{2^{n+1}-1}.$$

$$E(X^2) = \frac{1}{2^{n+1}-1} \sum_{r=0}^{n} r^2 \binom{n+1}{r+1}$$

$$= \frac{1}{2^{n+1}-1}[n(n+1)\sum_{r=1}^{n} \binom{n-1}{r-1} - (n+1)\sum_{r=0}^{n} \binom{n}{r} + \sum_{r=0}^{n} \binom{n+1}{r+1}]$$

$$= \frac{2^{n-1}(n^2-n+2)-1}{2^{n+1}-1}.$$

$$V(X) = \frac{(n+1)2^{n-1}}{(2^{n+1}-1)^2}[2^{n+1} - (n+2)].$$

DEFINITION 5.4.2 *Joint Distribution of Two Random Variables*: Let X, Y be two random variables, (X, Y) taking the values (x_i, y_j) with probability $p_{ij}, i = 1, \dots, m; j = 1, \dots n$;

$$p_{ij} \geq 0 \; \forall \; i, j; \sum_{i=1}^{m} \sum_{j=1}^{n} p_{ij} = 1. \tag{5.4.1}$$

The set of values $\{(x_i, y_j); p_{ij}\}$ is the joint probability distribution of (X, Y).

In this set up, probability X takes the value x_i is

$$\sum_{j=1}^{n} p_{ij} = p_{io} \; \text{(say)}. \tag{5.4.2}$$

The set of values $\{x_i; p_{io}\}$ is called the *marginal probability distribution* of X.

Similarly, probability that Y takes the value y_j is

$$\sum_{i=1}^{m} p_{ij} = p_{0j} \; \text{(say)}. \tag{5.4.3}$$

The set of values $\{y_j; p_{0j}\}$ is the marginal probability distribution of Y.

The conditional probability that X takes the value x_i, given that Y has taken the value y_j is given by

$$\frac{p_{ij}}{p_{0j}} = p(x_i|y_j) \quad \text{(say)}, \tag{5.4.4}$$

Clearly,

$$p(x_i|y_j) \geq 0 \; \forall i; \; \sum_{i=1}^{n} p_{i|j} = 1.$$

The set of values $\{x_i; p(x_i|y_j)\}$ is called the *conditional probability distribution* of X given $Y = y_j$.

Similarly, the conditional probability that Y takes the value y_j, given that X has taken the value x_j is

$$\frac{p_{ij}}{p_{i0}} = p(y_j|x_i) \quad \text{(say)}. \tag{5.4.5}$$

The set of values $\{y_j; p(y_j|x_i)\}$ is called the conditional probability distribution of Y given that $X = x_i$. Clearly,

$$p(y_j|x_i) \geq 0 \; \forall \; y_j; \; \sum_{j=1}^{n} p(y_j|x_i) = 1.$$

If X, Y are independent, $p(x_i|y_j) = p_{i0}$ and $p(y_j|x_i) = p_{0j}$ for all i, j. Hence

$$p_{ij} = p_{i0}p_{0j} \; \forall \; i, j. \tag{5.4.6}$$

Theorem 5.4.4: Let X, Y be stochastically independent random variables. Then $E(XY) = E(X).E(Y)$.

Proof.

$$\begin{aligned}
E(XY) &= \sum_{i=1}^{m} \sum_{j=1}^{n} x_i y_j p_{ij} \\
&= \sum_{i=1}^{m} \sum_{j=1}^{n} x_i y_j p_{i0} p_{0j} \\
&= \left(\sum_{i=1}^{m} x_i p_{i0}\right) \left(\sum_{j=1}^{n} y_j p_{0j}\right) \\
&= E(X).E(Y).
\end{aligned}$$

Corollary 5.4.4.1: If $X_1, \dots X_k (k \geq 2)$ are stochastically independent random variables, $E(X_1 \dots X_k) = E(X_1) \dots E(X_k)$.

DEFINITION 5.4.3 *Covariance, Correlation Coefficient*: Covariance between two random variables X, Y is defined as

$$\text{Cov}\,(X, Y) = E(X - E(X))(Y - E(Y)) = E(XY) - E(X)E(Y). \tag{5.4.7}$$

The ratio,

$$\rho(X,Y) = \frac{\text{Cov } (X,Y)}{\sqrt{V(X)}\sqrt{V(Y)}} \qquad (5.4.8)$$

is called the correlation coefficient between X and Y.

Both covariance and correlation coefficient measure the association between X and Y. If covariance (correlation coefficient) between X and Y is positive, variables X and Y are said to be positively related. If value of X increases (decreases), value of Y also increases (decreases). If covariance (correlation coefficient) between X and Y is negative, there is a negative association between X and Y. If value of X increases (decreases), value of Y decreases (increases).

If X and Y are stochastically independent, $E(XY) = E(X)E(Y)$ and hence Cov $(X,Y) = 0$ and $\rho(X,Y) = 0$.

However, the converse is not necessarily true. That is, $\text{Cov}(X,Y) = 0$ or $\rho(X,Y) = 0$ does not imply that X and Y are independent. This means that there are situations where X, Y are mutually dependent random variables, yet their correlation coefficient (covariance) is zero. Example 5.4.3 illustrates such a situation.

The following results are easy to verify.

(i)   Cov $(aX, bY) = ab$ Cov (X,Y);
(ii) $\text{Cov}(X + a, Y + b) =$ Cov (X,Y);
(iii) $\rho(aX + b, cY + d) = \rho(aX, cY) =$ (sign of ac).$\rho(X,Y)$,   (5.4.9)

where a, b, c, d are arbitrary constants.

It follows that

$$[\text{Cov } (X,Y)]^2 \le V(X)V(Y). \qquad (5.4.10)$$

(For a proof see Section 6.4.) Hence $\rho^2 \le 1$ or $|\rho| \le 1$. Further details about ρ have been discussed in Section 10.7.

Note 5.4.1: All the definition and the results above readily extend to the case when X, Y have countably infinite number of values.

EXAMPLE 5.4.3: Let X, Y be the results of tosses of two balanced dice. Then X and Y are independently and identically distributed. In particular, $E(X) = E(Y), E(X^2) = E(Y^2), V(X) = V(Y)$. Consider two random variables

$$U = X + Y, \; V = X - Y.$$

Then

$$\text{Cov } (U, V) = E(X^2 - Y^2) - E(X + Y)E(X - Y) = 0.$$

Hence, U, V are uncorrelated.

However, if we consider all the 36 points $\omega = (x_i, y_j); x_i = 1, \ldots, 6; y_j = 1, \ldots, 6$ in the sample space then we notice that if $U(\omega)$ is an odd (even) number, $V(\omega)$ is also an odd (even) number. Therefore, U and V are necessarily dependent on each other.

EXAMPLE 5.4.4: X, Y have joint probability distribution as follows:

$$p_{11} = 1/12, \quad p_{12} = 1/12, \quad p_{13} = 2/12,$$

$$p_{21} = 2/12, \quad p_{22} = 0, \quad p_{23} = 0,$$

$$p_{31} = 1/12, \quad p_{32} = 1/12, \quad p_{33} = 4/12,$$

where $p_{ij} = P(X = i, Y = j)$. Hence, marginal distribution of X is $p_{10} = 4/12, p_{20} = 2/12, p_{30} = 6/12$. Similarly, marginal distribution of Y is $p_{01} = 4/12, p_{02} = 2/12, p_{03} = 6/12$. Thus,

$$E(X) = \frac{13}{6} = E(Y), \quad E(X^2) = \frac{11}{2} = E(Y^2), \quad E(XY) = \frac{29}{6},$$

$$V(X) = V(Y) = \frac{29}{36}, \quad \text{Cov } (X, Y) = \frac{5}{36}, \quad \text{Corr. Coeff. } \rho(X, Y) = \frac{5}{29}.$$

Theorem 5.4.5: If $X_1, \ldots, X_n$ are n random variables with finite variances $\sigma_1^2, \ldots, \sigma_n^2$ respectively, then

$$V(l_1 X_1 + \ldots + l_n X_n) = \sum_{i=1}^{n} l_i^2 \sigma_i^2 + \sum_{i \neq j=1}^{n} \sum l_i l_j \text{ Cov } (X_i, X_j) \quad (5.4.11)$$

where $l_1, \ldots, l_n$ are arbitrary constants.

Proof.

$$V(l_1 X_1 + \ldots l_n X_n) = E(l_1 X_1 + \ldots + l_n X_n - l_1 E(X_1) - \ldots - l_n E(X_n))^2$$
$$= E \left(\sum_{i=1}^{n} l_i (X_i - E(X_i)) \right)^2 = \sum_{i=1}^{n} l_i^2 \sigma_i^2 + \sum \sum_{i \neq j=1}^{n} l_i l_j \text{ Cov } (X_i, X_j)$$
$$= \sum_{i=1}^{n} l_i^2 \sigma_i^2 + \sum \sum_{i \neq j=1}^{n} \rho_{ij} \sigma_i \sigma_j$$

$$(5.4.12)$$

where ρ_{ij} is the correlation coefficient between X_i, X_j.

Corollary 5.4.5.1: If $l_1 = l_2 = \ldots = l_n = \frac{1}{n}$, so that $\sum_{i=1}^{n} l_i X_i = \frac{1}{n} \sum_{i=1}^{n} X_i = \bar{X}$, and if Cov $(X_i, X_j) = 0$, $i \neq j = 1, \ldots, n$, $V(\bar{X}) = \sum_{i=1}^{n} \frac{\sigma_i^2}{n^2}$. If further, $\sigma_i^2 = \sigma^2, i = 1, \ldots, n$, $V(\bar{X}) = \frac{\sigma^2}{n}$.

EXAMPLE 5.4.5: n trials in each of which an event A may or may not occur are performed. If X denotes the number of occurrences of A, find $E(X), V(X)$.

Let $X_i = 1(0)$ if A occurs (does not occur) at the ith trial. Therefore, $X = \sum_{i=1}^{n} X_i$. Let $P(X_i = 1) = p_i, i = 1, \ldots, n$. Then $E(X_i) = p_i, E(X_i^2) = p_i, V(X_i) = p_i(1 - p_i)$, Cov $(X_i, X_j) = E(X_i X_j) - p_i p_j = P(X_i = 1, X_j = 1) = p_{ij} - p_i p_j$ (say) . Hence, $E(X) = \sum_{i=1}^{n} p_i$.

$$V(X) = \sum_{i=1}^{n} V(X_i) + \sum_{i \neq j=1}^{n} \sum (p_{ij} - p_i p_j).$$

If $p_i = p \, \forall \, n$ and $p_{ij} = P \, \forall \, i \neq j, V(X) = np(1 - p) + n(n - 1)(P - p^2)$.

EXAMPLE 5.4.6: Consider the problem of Example 5.4.1. Suppose now the trials are continued until A occurs for the kth time. Find $E(X), V(X)$ where X denotes the required number of trials.

Let X_i be the number of trials starting immediately after the $(i - 1)$th occurrence of A and ending with the ith occurrence of A. Then $X = X_1 + \ldots + X_k$. Clearly, each X_i has a geometric distribution with parameter p. Also, X_i, X_j are independent, $i \neq j = 1, \ldots, k$. Therefore, $E(X) = \frac{k}{p}, V(X) = V(X_1) + \ldots + V(X_n) = \frac{kq}{p^2}$.

EXAMPLE 5.4.7: In the familiar matching problem, where N letters are distributed at random to N addressed envelopes and a match is said to occur if a letter goes to the right envelope, find the expectation and variance of the number of matches.

Let $X_i = 1(0)$ if a match occurs (does not occur) at the ith place. Then $X = X_1 + \ldots + X_n$ is the total number of matches. Now, $P(X_i = 1) = \frac{1}{N}, P(X_i = 1, X_j = 1) = \frac{1}{N(N-1)}, i \neq j = 1, \ldots, N$ (vide Example 2.8.3). Thus, $E(X_i) = \frac{1}{N}, V(X_i) = \frac{N-1}{N^2}$, Cov $(X_i, X_j) = \frac{1}{N^2(N-1)}$. Hence, $E(X) = 1, V(X) = 1$.

EXAMPLE 5.4.8: Units are drawn with replacement from a population of N identifiable units until r distinct units are observed. Find the expectation and variance of the required number of draws.

The first draw contains the first distinct unit. Let X_k be the number of draws after the draw in which the k th distinct unit appears up to the draw in which the $(k+1)$th distinct unit occurs. The required number of draws is $1 + X_1 + \ldots + X_{r-1} = Z$ (say). Now,

$$P(X_1 = 1) = 1 - \frac{1}{N}, P(X_1 = 2) = \frac{1}{N}(1 - \frac{1}{N}), \ldots,$$

$$P(X_1 = k) = (\frac{1}{N})^{k-1}(1 - \frac{1}{N}), \ldots$$

Hence,

$$E(X_1) = 1.\frac{N-1}{N} + 2.\frac{1}{N}.\frac{N-1}{N} + 3.\frac{1}{N^2}.\frac{N-1}{N} + \ldots = \frac{N}{N-1}.$$

In general, $P(X_t = 1) = \frac{N-t}{N}, P(X_t = 2) = \frac{t}{N}.\frac{N-t}{N}$, etc. $E(X_t) = \frac{N}{N-t}, t = 1, 2, \ldots, r-1$. Thus

$$E(Z) = N\{\frac{1}{N} + \frac{1}{N-1} + \frac{1}{N-2} + \ldots + \frac{1}{N-r+1}\} \approx N \int_{N-r+1}^{N} \frac{dx}{x} \text{ for large } N$$

$$= N \log \frac{N}{N-r+1}.$$

$$E(X_t^2) = \frac{N-t}{N}[1 + 2^2(\frac{t}{N}) + 3^2(\frac{t}{N})^2 + 4^2(\frac{t}{N})^3 + \ldots]$$

$$= \frac{N-t}{N}\left[\frac{1}{(1-\frac{t}{N})^2} + \frac{\frac{2t}{N}}{(1-\frac{t}{N})^3}\right] \text{ (vide Example 5.4.1)}$$

$$= \frac{N(N+t)}{(N-t)^2}.$$

Hence, $V(X_t) = \frac{Nt}{(N-t)^2}, t = 1, \ldots, r-1$. Thus,

$$V(Z) = N \sum_{k=1}^{r-1} \frac{t}{(N-t)^2} = N \sum_{t=1}^{r-1}\{\frac{N}{(N-t)^2} - \frac{1}{N-t}\}$$

$$\approx N^2 \int_{N-r+1}^{N} \frac{dx}{x^2} - N \int_{N-r+1}^{N} \frac{dx}{x}$$

$$\text{for large } N = \frac{N(r-1)}{N-r+1} - N \log(\frac{N}{N-r+1}).$$

EXAMPLE 5.4.9: A box contains N objects numbered $1, 2, \ldots, N$ from which n objects are drawn at random with replacement. Find $P(X = k), E(X)$ and $V(X)$ where X denotes the largest number drawn.

We have

$$P(X \le k) = (\frac{k}{N})^n; P(X = k) = P(X \le k) - P(X \le k-1)$$

$$= \frac{k^n - (k-1)^{n-1}}{N^n}.$$

$$E(X) = \sum_{k=1}^{N} k P(X=k) = \frac{1}{N^n} \{ \sum_{k=1}^{N} k^{n+1} - \sum_{k=1}^{N} k(k-1)^n \}$$

$$= \frac{1}{N^n} \sum_{k=1}^{N} \{ k^{n+1} - (k-1)^{n+1} - (k-1)^n \}$$

$$= \frac{1}{N^n} \{ N^{n+1} - \sum_{k=1}^{N} (k-1)^n \}$$

$$\approx \frac{1}{N^n} \{ N^{n+1} - \int_0^N t^n dt \} = \frac{nN}{n+1} \quad \text{for large } N.$$

$$E(X^2) = \frac{1}{N^2} \sum_{k=1}^{N} k^2 P(X=k) = \frac{1}{N^n} \sum_{k=1}^{N} k^{n+2} - \sum_{k=1}^{N} k^2 (k-1)^n$$

$$= \frac{1}{N^n} \sum_{k=1}^{N} \{ k^{n+2} - (k-1)^{n+2} - 2(k-1)^{n+1} - (k-1)^n \}$$

$$\approx \frac{1}{N^n} \{ N^{n+2} - 2\int_0^N t^{n+1} dt - \int_0^N t^n dt \} \approx \frac{nN^2}{(n+2)} \quad \text{for large } N.$$

Thus,

$$V(X) \approx \frac{nN^2}{(n+1)^2(n+2)} \quad \text{for large } N.$$

5.5 Exercises and Complements

5.1 A and B roll a pair of unbiased dice alternatively. A wins if he throws a 6 before B throws a 7; B wins if he throws a 7 before A has a 6. If A begins the game, show that his probability of winning is $\frac{30}{61}$ and the expected number of trials needed for winning is approximately 6.

5.2 A group of m radar units scan a region in which there are n targets $T_1, \ldots, T_n$, for a time period T. During this time interval, each unit, independently of each other, may detect the ith target with probability $p_i (i = 1, \ldots, n)$. Find the expected number of targets that will be detected.

5.3 An urn contains n tickets numbered $1, \ldots, n$. From this m tickets are drawn one by one without replacement. Find expectation and variance of X, where X is the sum of the numbers on the m tickets drawn.

5.4 A number of independent trials are carried out in each of which an event A may occur with probability p. The trials are terminated as soon as A occurs or the number of trials equal N, whichever is earlier. Find $E(X)$ where X is the number of trials.

5.5 In a box containing 2^n tickets, $\binom{n}{r}$ tickets bear the number r each, $r = 0, 1, \ldots, r$. A lot of m tickets are selected at random. Let X be the sum of the numbers on the m selected tickets. Find $E(X), V(X)$.

5.6 If $X_1, \ldots, X_n$ are independently and identically distributed positive random variables, prove that $E(S_k/S_n) = k/n$ where $S_r = \sum_{i=1}^{r} X_i$.

5.7 In a vehicle insurance policy the premium is β dollars in the first year and if no claim is made in any of the first r years, the premium is $\beta\delta^r$ dollars in the $(r+1)$th year $(r = 1, 2, \ldots)$.

If in any year, a claim is made, the premium in that year remains unaffected, but the next year is treated as second year, third year, etc. for calculation of premium. Assuming that the probability that no claim is made in a year is q, calculate the expected amount of premium payable in the nth year.

5.8 In a trial, there are two outcomes S, F, happening with probability p, q, respectively; n such trials are performed. A man who has no real ability to guess decides to allot r outcomes as S's and $(n - r)$ outcomes as F's at random. Given that he scores point $1(0)$ for every correct (incorrect) guess, find the expectation and variance of the total score.

5.9 Consider the problem 5.7 with $p > \frac{1}{2}$. Suppose that the man allots S to a trial with probability λ and F with probability $(1 - \lambda)$. Prove that the probability of a correct guess for any trial is
$$[1 - p + (2p - 1)\lambda].$$
Derive the expectation and variance of his total score.

5.10 In Buffon's needle problem (vide Example 2.9.2), find the expected number of hits.

5.11: A straight line with $6n(n \geq 1)$ sections is divided at random into two parts A and B, A having r sections and B, $(n - r)$ sections, $r \geq 1$. If A is known to be longer than B, find expectation and variance of length of A.

5.12: For n sets $A_1, \ldots, A_n$ show that
$$I_{\cup_{i=1}^{n} A_i} = \sum I_{A_i} - \sum\sum_{i<j} I_{A_i A_j} + \sum\sum\sum_{i<j<k} I_{A_i A_j A_k} + \ldots + (-1)^{n-1} I_{cap_{i=1}^{n} A_i},$$
where I_B is the indicator function of the set B. Hence, prove Theorem 3.6.3 by taking expectation of both sides.

Chapter 6

Some Properties of a Probability Distribution on $\mathcal{R}$

6.1 Introduction

In Chapter 4 we introduced the concepts of discrete probability distributions. Since, discrete distributions are in general easier to deal with we studied the expectation, variance of a discrete random variable, joint distribution of two discrete random variables and association between two discrete random variables in Chapter 5. Here we shall study the properties of a probability distribution under a general set up, covering both discrete and continuous random variables. We shall consider expectations of a random variable in Section 6.2, moments in Sections 6.3 through 6.6, measures of central tendency in Section 6.7, measures of dispersion in Section 6.8 and measures of skewness and kurtosis in Section 6.9. Lastly, some probability inequalities are discussed in Section 6.10.

6.2 Expectation

Let X be a random variable defined on a probability space $(\mathcal{S}, \mathcal{A}, P)$ and let F be its distribution function. The expectation of X is defined by

$$
\begin{aligned}
E(X) &= \int_{-\infty}^{\infty} x \, dF(x) \\
&= \int_{\mathcal{S}} X \, dP \\
&= \int_{-\infty}^{\infty} X \, dP_X
\end{aligned}
\tag{6.2.1}
$$

where the integral on the first line on the right is the Riemann-Stieltjes integral of x with respect to $F(x)$ over $\mathcal{R}$, the integral on the second line is the Lebesgue-Stieltjes integral of the random variable X with respect to $(\mathcal{S}, \mathcal{A}, P)$ over $\mathcal{S}$ and the integral in the last line is the Lebesgue-Stieltjes integral of X with respect to the induced probability space $(\mathcal{R}, \mathcal{B}, P_X)$ over $\mathcal{R}$.

$E(X)$ is said to exist if and only if

$$\int_{-\infty}^{\infty} |x| dF(x) < \infty. \tag{6.2.2}$$

Thus, if X is a discrete random variable with pmf $P(X = x_k) = p(x_k), k = 1, 2, \ldots$, then

$$E(X) = \sum_{k=1}^{\infty} x_k p(x_k) \tag{6.2.3}$$

is the expectation of X, provided the condition

$$\sum_{k=1}^{\infty} |x_k| p(x_k) < \infty \tag{6.2.4}$$

is satisfied.

If X is a continuous random variable with pdf $f(x)$, the integral

$$E(X) = \int_{-\infty}^{\infty} x f(x) dx \tag{6.2.5}$$

is called the expectation of X, provided the condition

$$\int_{-\infty}^{\infty} |x| f(x) dx < \infty \tag{6.2.6}$$

is satisfied.

If we have a mixed distribution $F_X(.) = F_{Xd}(.) + F_{Xc}(.)$, with usual notation

$$E(X) = \sum_{i} x_i P_X(\{x_i\}) + \int x f_{Xc}(t) dx$$

where f_{Xc} denotes the absolutely continuous density function corresponding to F_{Xc}.

Remarks: The expectation of a random variable may not always exist. Before calculating $E(X)$ one must therefore check if (6.2.4) or (6.2.6) is satisfied.

EXAMPLE 6.2.1: The following example shows a situation where (6.2.3) is finite though the condition (6.2.4) is not satisfied.

Let X be a discrete random variable with $P(X = x) = \frac{6}{\pi^2 x^2}(0)$ for $x = 1, -2, 3, -4, \ldots$(otherwise).

$$E(X) = \frac{6}{\pi^2} \sum_{x=1}^{\infty} \frac{(-1)^{x-1}}{x} < \infty.$$

However, $E|X| = \frac{6}{\pi^2} \sum_{x=1}^{\infty} \frac{1}{x} = \infty$. Hence, $E(X)$ does not exist.

EXAMPLE 6.2.2: Let X take the values $0, +_-1, +_-2, \ldots$ with

$$P[X = k] = P[X = -k] = \frac{3}{\pi^2 k^2}, \quad k = 1, 2, \ldots$$

We can write $X(.) = X_+(.) - X_-(.)$ where, for $j > 0 (= 1, 2, \ldots)$, $X_+(j) = j$, $X_-(j) = 0$ and for $j < 0 (= -1, -2, \ldots)$, $X_+(j) = 0$, $X_-(j) = -j$. Also, $X_+(0) = X_-(0) = 0$. Thus,

$$E(X_+) = \sum_{k=1}^{\infty} \frac{3}{\pi^2 k} = \infty = E(X_-).$$

Therefore, $E(X)$ is of the form $\infty - \infty$ and hence does not exist.

If X takes the values $1, 2, \ldots$ with $P[X = k] = \frac{6}{\pi^2 k^2}$, then $E(X_+) = \infty$, $E(X_-) = 0$. Hence, $E(X)$ exists, though it is infinite.

EXAMPLE 6.2.3: The distribution $f(x) = \frac{1}{\pi(1+x^2)}$, $-\infty < x < \infty$ is known as the *Cauchy* distribution. Here $E(X)$ exists if

$$E\{|X|\} = \lim_{a \to -\infty; b \to \infty} \int_a^b \frac{|x|}{\pi(1+x^2)} dx < \infty.$$

However,

$$\int_a^b \frac{|x|}{\pi(1+x^2)} dx = \frac{1}{2\pi} (\log(b^2 + 1)) \to \infty$$

as $b \to \infty$. Hence $E(X)$ does not exist for this distribution. However, the Cauchy principal value

$$\frac{1}{\pi} \int_{-\infty}^{\infty} \frac{x \, dx}{1 + x^2} = \lim_{A \to \infty} \int_{-A}^{A} \frac{x \, dx}{\pi(1 + x^2)}$$

exists and is equal to zero.

This example shows that Lebesgue integral may not exist, while Cauchy principal value may exist.

EXAMPLE 6.2.4: A variate X has *pdf* $f(x) = \frac{1}{x\sqrt{2\pi}} e^{-\frac{1}{2}(\log x)^2}$, $x > 0$.

$$E(X) = \int_0^{\infty} \frac{1}{\sqrt{2\pi}} e^{-\frac{1}{2}(\log x)^2} dx = \frac{1}{\sqrt{2\pi}} \int_{-\infty}^{\infty} e^{-\frac{t^2}{2}} e^t dt \quad \text{(putting } t = \log x\text{)}$$

$$= \frac{\sqrt{e}}{\sqrt{2\pi}} \int_{-\infty}^{\infty} e^{-\frac{1}{2}(t-1)^2} dt = \sqrt{e} \left[\text{as } \frac{1}{\sqrt{2\pi}} \int_{-\infty}^{\infty} e^{-\frac{y^2}{2}} dy = 1 \right].$$

$\square$

We have seen in Theorem 4.2.9 that if X is a random variable on $(\mathcal{S}, \mathcal{A}, P)$, so is any Borel-measurable function $g(X)$ of X. $E(g(X))$, if it exists, is given by

$$E[g(X)] = \int_{-\infty}^{\infty} g(x)dF(x). \qquad (6.2.7)$$

Thus, if X is discrete with *pmf* $p(x_i)$,

$$E[g(X)] = \sum_{i=1}^{\infty} g(x_i)p(x_i). \qquad (6.2.8)$$

If X is continuous with *pdf* $f(x)$,

$$E[g(X)] = \int_{-\infty}^{\infty} g(x)f(x)dx \qquad (6.2.9)$$

provided the expectation exists.

The expression for the mixed case can be written similarly.

EXAMPLE 6.2.5: Suppose X is a mixed type random variable (vide Definition 4.5.3) which probability mass function $P_X(x) = \frac{1}{10}$ for $x = -2, -1, 0, 1, 2$ and has a density function $f(x) = \frac{1}{8}$ for $x \in [-2, 2]$. Find $E(X^3)$.

We check that $\sum_{-2}^{2} P_X(x) + \int_{-2}^{2} f(x)dx = 1$. Hence

$$E(X^3) = \sum_{-2}^{2} x^3 P_X(x) + \int_{-2}^{2} x^3 f(x)dx$$
$$= \frac{1}{10}(4 + 1 + 0 + 1 + 4) + \frac{1}{8}\int_{-2}^{2} x^3 dx = 1 + \frac{2}{3} = \frac{5}{3}. \square$$

Some conditions for the existence of expectation in terms of the distribution function has been given in Theorem 6.3.7.

6.2.1 *Some properties of expectation*

We discuss here some general properties of mathematical expectation which do not depend on the nature of any given probability distribution. These properties are, for the most part, direct consequences of the properties of integrals studied in Section 4.8.

(1) If the random variable X is a constant c, i.e., if $X(\omega) = c$ *a.e.* i.e., if $P[X = c] = 1, E(X) = c$.

Proof.

$$E(X) = \int_{-\infty}^{\infty} x dF(x) = \int_{-\infty}^{\infty} c dF(x) = c \int_{-\infty}^{\infty} dF(x) = c.$$

(2) *Linearity.* If $E(X)$ exists, $E(aX + b)$ exists and is given by

$$E(aX + b) = aE(X) + b$$

where a, b are arbitrary constants.

Proof.

$$\int_{-\infty}^{\infty} |ax+b|dx = \int_{-\infty}^{\infty} |ax|dF(x)+|b| \int_{-\infty}^{\infty} dF(x) < |a| \int_{-\infty}^{\infty} |x|dF(x)+|b| < \infty,$$

since $EX < \infty$. Also,

$$E(aX+b) = \int_{\infty}^{\infty} (ax+b)dF(x) = a \int_{-\infty}^{\infty} xdF(x)+b \int_{-\infty}^{\infty} dF(x) = aE(X)+b.$$

(3) If $g(X)$ is a Borel function of $X, |E(g(X))| \le E|g(X)|$.

Proof.

$$|E(g(X))| = |\int_{-\infty}^{\infty} g(x)dF(x)| \le \int_{-\infty}^{\infty} |g(x)|dF(x) = E|g(X)|.$$

(4) If $g(X) = g_1(X)+g_2(X)$ where $g_1(X), g_2(X)$ are two Borel measurable functions of the random variable X, then

$$E[g(X)] = E[g_1(X)] + E[g_2(X)]$$

provided the expectations exist.

Proof.

$$E[g(X)] = \int_{-\infty}^{\infty}[g_1(x) + g_2(x)]dF(x)$$

$$= \int_{-\infty}^{\infty} g_1(x)dF(x) + \int_{-\infty}^{\infty} g_2(x)dF(x)$$

$$= Eg_1(X) + Eg_2(X).$$

(5) If $g_1(X) \le g_2(X)$ for all possible values of X, then

$$Eg_1(X) \le Eg_2(X).$$

Proof.

$$Eg_1(X) = \int_{-\infty}^{\infty} g_1(X)dF(x) \le \int_{-\infty}^{\infty} g_2(X)dF(x) = Eg_2(X).$$

(6) *Conditional expectation of X given A:* Let X be a random variable defined on (S, A, P) such that $E(X)$ exists. Let A be a measurable subset of the sample space S with $P(A) > 0$. As noted in Section 3.6, $P(.|A)$ is a probability measure on (S, A) so that $(S, A, P(.|A))$ is a probability space.

Clearly, X is also integrable with respect to $P(.|A)$ and we define

$$E(XI_A) = \int_S XdP(.|A) \tag{6.2.10}$$

as the conditional expectation of X given A.

Now,

$$P(.|A) = 0 \text{ on the class } \{B \cap A^c, B \in \mathcal{A}\},$$

and

$$P(.|A) = \frac{P(.)}{P(A)} \text{ on the class } \{B \cap A, B \in \mathcal{A}\}.$$

Hence, we can write

$$E(X|A) = \frac{1}{P(A)} \int_A XdP = \frac{E(XI_A)}{P(A)} \tag{6.2.11}$$

provided $E(XI_A)$ exists and $P(A) > 0$. Here I_A is the indicator function of A. Since $|XI_A| \le |X|, E(XI_A)$ exists if $E(X)$ exists. Clearly,

$$E(XI_A) = \sum_{\omega \in A} X(\omega)P(\omega).$$

In particular, if $X = I_B, B \in \mathcal{A}$,

$$E(I_B|A) = \frac{P(B \cap A)}{P(A)} = P(B|A).$$

Theorem 6.2.1: Let $\{A_i, i = 1, \ldots, k\}$ be a measurable partition of S (i.e. $\cup_{i=1}^k A_i = S, A_i \cap A_j = \phi, i \ne j = 1, \ldots, k$) such that $P(A_i) > 0 \; \forall \; i$ and let $E(X)$ exist. Then

$$E(X) = \sum_{i=1}^k E(X_i|A_i)P(A_i) \left(= E_A[E(X|A)]\right). \tag{6.2.12}$$

Here $E_A(.)$ denotes expectation with respect to A (i.e. for different subsets $A_i, i = 1, \ldots, k$).

Proof.

$$E(X) = \int XdP = \int_{\sum_i A_i} XdP = \sum_i \int XI_{A_i}dP$$

$$\tag{6.2.13}$$

$$= \sum_i E(XI_{A_i}) = \sum_i E(X|A_i)P(A_i)$$

by (6.2.11). $\qquad \qquad \qquad \qquad \qquad \qquad \qquad \qquad \qquad \Box$

Note 6.2.1: The result (6.2.12) holds even if $\{A_n, n \ge 1\}$ is a countable partition of S.

Theorems 5.3.4 and 5.4.4 (Corollary 5.4.1) on expectation and product of discrete random variables respectively hold good for continuous variables also. That is, in the general set up we have the following two theorems.

Theorem 6.2.2: If $X_1, \ldots, X_n$ are random variables, each having range in the real line $\mathcal{R}$, then $E(X_1 + \ldots + X_n) = E(X_1) + \ldots + E(X_n)$.

Proof follows from the linearity property (4.8.26) of integrals of random variables.

Theorem 6.2.3: Let X and Y be two independent random variables. If $E|X| < \infty, E|Y| < \infty$, then $E|XY| < \infty$ and $E(XY) = E(X)E(Y)$.

Conversely, if $E|XY| < \infty$ and neither of X and Y is degenerate at zero, then $E|X| < \infty$ and $E|Y| < \infty$.

Proof. The first part of the theorem follows from the property (4.8.34) of the integration of a general random variable in Subsubsection 4.8.1.1. However, an elaborate proof is given below.

We first consider the case when both X and Y are simple non-negative random variables almost everywhere. Thus, we can write

$$X(.) = \sum_{i=1}^{m} \alpha_i I_{E_i} \text{ and } Y = \sum_{j=1}^{n} \beta_j I_{F_j}$$

where $\{E_i : i = 1, \ldots, m\}$ and $\{F_j : j = 1, \ldots, n\}$ are two partitions of S. Without loss of generality, we may assume that α_i's and β_j's are all distinct, so that

$$E_i = \{\omega \in S : X(\omega) = \alpha_i\}, \ i = 1, \ldots, m,$$

$$F_j = \{\omega \in S : Y(\omega) = \beta_j\}, \ j = 1, \ldots, n.$$

Since X, Y are independent the classes $\{E_1, \ldots, E_m\}$ and $\{F_1, \ldots, F_n\}$ are independent classes of events, so that

$$P(E_i \cap F_j) = P(E_i)P(F_j) \ \forall \ i = 1, \ldots, m; j = 1, \ldots, n.$$

In this case, XY is a simple random variable, and can be written as

$$XY = \sum_{i=1}^{m} \sum_{j=1}^{n} \alpha_i \beta_j I_{E_i \cap F_j}.$$

Hence,

$$E(XY) = \int_S XY \, dP$$

$$= \sum_{i=1}^{m} \sum_{j=1}^{n} \alpha_i \beta_j P(E_i \cap F_j)$$

$$= [\sum_{i=1}^{m} \alpha_i P(E_i)][\sum_{j=1}^{n} \beta_j P(F_j)]$$

$$= E(X)E(Y).$$

We now consider the case, when X and Y are non-negative random variables. For every $n \geq 1$ and every $\omega \in S$, we define

$$X_n(\omega) = \begin{cases} \frac{i-1}{2^n} & \text{if } \frac{i-1}{2^n} \leq X(\omega) \leq \frac{i}{2^n}, \ i = 1, 2, \ldots, 2^n \\ n & \text{if } X(\omega) \geq n, \end{cases}$$

and

$$Y_n(\omega) = \begin{cases} \frac{j-1}{2^n} & \text{if } \frac{j-1}{2^n} \leq Y(\omega) \leq \frac{j}{2^n}, \ j = 1, 2, \ldots, 2^n, \\ n & \text{if } Y(\omega) \geq n. \end{cases}$$

Thus, $\{X_n(\omega)\}, \{Y_n(\omega)\}$ are sequences of non-decreasing, non-negative simple random variables such that $X_n \uparrow X, Y_n \uparrow Y$. Clearly, $0 \leq X_n Y_n \uparrow XY$. Since X, Y are independent, it can be verified that X_n, Y_n are also independent $(n \geq 1)$. It follows that

$$E(X_n Y_n) = E(X_n)E(Y_n) \ \forall \ n.$$

Again, by monotone convergence theorem (Theorem 4.8.2),

$$\lim_{n \to \infty} E(X_n) = E(X), \ \lim_{n \to \infty} E(Y_n) = E(Y),$$

$$\lim_{n \to \infty} E(X_n Y_n) = E(XY).$$

Thus, $E(XY) = E(X)E(Y)$.

Now, suppose that X and Y are arbitrary independent random variables. Then the non-negative random variables X_+ and X_- are independent of Y_+ and Y_-. Also,

$$E(X_+ Y_+) = E(X_+)E(Y_+),$$

$$E(X_+ Y_-) = E(X_+)E(Y_-),$$

$$E(X_- Y_+) = E(X_-)E(Y_+),$$

$$E(X_- Y_-) = E(X_-)E(Y_-).$$

Again, if $E(X), E(Y)$ exist, so do $E(X_+), E(X_-), E(Y_+), E(Y_-)$ and hence $E(X_+Y_+), E(X_+Y_-)$, etc. Therefore, $E|XY| < \infty$ and

$$
\begin{aligned}
E(XY) &= E(X_+ - X_-)(Y_+ - Y_-) \\
&= E(X^+ - X_-)E(Y_+ - Y_-) \\
&= E(X)E(Y).
\end{aligned}
$$

Second part: Suppose that $E|XY| < \infty$. Clearly, $|X| \geq 0$, a.s. and $|Y| \geq 0$, a.s. Also, $|X|$ and $|Y|$ are independent. Therefore, $E|X|E|Y| = E|XY| < \infty$. Since, none of X and Y has a degenerate distribution at zero, it follows that $E|X| < \infty, E|Y| < \infty$. Hence the theorem. □

The above result can be extended as follows.

Theorem 6.2.4: Let $X_1, \ldots, X_n$ be independent random variables and let $h_1, \ldots, h_n$ be Borel measurable functions on R such that $E|h_i(X_i)| < \infty, i = 1, \ldots, n$. Then the following relation holds.

$$
E[\Pi_{i=1}^n h_i(X_i)] = \Pi_{i=1}^n E[h_i(X_i)] \tag{6.2.14}
$$

Also, $E\Pi_{i=1}^n |h_i(X_i)| < \infty$.

Conversely, if $E\Pi_{i=1}^n [h_i(X_i)] < \infty$, then, $E[|h_i(X_i)|] < \infty \ \forall \ i = 1, \ldots, n$, provided the relation (6.2.13) holds and none of the $h_i(X_i)$ has a degenerate distribution at zero $(i = 1, \ldots, n)$.

Proof. We first prove that if $X_1, \ldots, X_n$ are independent random variables, then $h_1(X_1), \ldots, h_n(X_n)$ are independent random variables.

Let $A_1, \ldots, A_n$ be Borel sets in $\mathcal{R}^n$. For the measure space $(\mathcal{S}, \mathcal{A})$,

$$
\{\omega : h_i(X_i(\omega)) \in A_i\} = \{\omega : X_i(\omega) \in h_i^{-1}(A_i)\}.
$$

Hence

$$
P[h_1(X_1) \in A_1, \ldots, h_n(X_n) \in A_n]
$$

$$
= P[X_1 \in h_1^{-1}(A_1), \ldots, X_n \in h_n^{-1}(A_n)]
$$

$$
= P[X_1 \in h_1^{-1}(A_1)] \ldots . P[X_n \in h_n^{-1}(A_n)]
$$

(by independence of $X_1, \ldots, X_n$)

$$
= P[h_1(X_1) \in A_1] \ldots . P[h_n(X_n) \in A_n].
$$

Hence, $h_1(X_1), \ldots, h_n(X_n)$ are independent random variables.

The remaining part of the theorem is an extension of the result in Theorem 6.2.3 and its proof follows similarly.

Corollary 6.2.4.1: If $X_1, \ldots, X_n$ are independent random variables, then

$$E(X_1 \ldots X_n) = E(X_1) \ldots E(X_n).$$

(7) An Inequality for Convex Functions

DEFINITION 6.2.1 *Convex Functions*: A continuous function $g(x)$ is said to be convex in an interval I, if for two values x_1, x_2 in I,

$$\lambda g(x_1) + (1 - \lambda)g(x_2) \geq g(\lambda x_1 + (1 - \lambda)x_2) \qquad (6.2.15)$$

for any $\lambda \in (0, 1)$. That is, any point lying on the straight line joining any two points of $g(x), x \in I$ lies above the point with the same abscissa on the curve (Fig. 6.2.1). For such a function, for every point x_0 in I, there exists a straight line which passes through $(x_0, g(x_0))$ and lies on or below the graph of $g(x)$. Examples of convex functions are:

$$(i)|x|, \ -\infty < x < \infty,$$

$$(ii)x^2, \ -\infty < x < \infty,$$

$$(iii)x^p, p \geq 1, \ x > 0,$$

$$(iv)\frac{1}{x^p}, p > 0, \ x > 0,$$

$$(v)e^x, \ -\infty < x < \infty,$$

$$(vi) - \log x \ 0 < x < \infty.$$

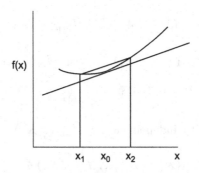

Fig. 6.2.1

All these functions are strictly convex, except (i) and (iii) with $p = 1$. The following inequality holds in the case of expectation of a convex function.

Theorem 6.2.5 *Jensen's Inequality:* Let X be a random variable and $g(X)$ a convex function of X. Then

$$E[g(X)] \geq g[E(X)] \qquad (6.2.16)$$

provided the expectations exist. The equality holds if and only if X has a degenerate distribution $(X(\omega) = k$, a constant a.e.$)$.

Proof. Since $g(x)$ is continuous and convex, there exists a line $l(x) = a + bx$, which passes through the point $(E(X), g(E(X))$ and lies below it, i.e. $l(x) \leq g(x)$ for all x and $l(E(X)) = g(E(X))$. Now, $l[E(X)] = a + bE(X) = E[l(X)]$. Hence, $g[E(X)] = E[l(X)] \leq E[g(X)]$, i.e., $E[g(X)] \geq g[E(X)]$.

EXAMPLE 6.2.6: $g(x) = x^2$ is a convex function. Hence, $E(X^2) \geq [E(X)]^2$ i.e. $E(X - E(X))^2 \geq 0$.

EXAMPLE 6.2.7: Since $|x|^r$ is a convex function of x for $r > 1$,

$$E|X|^r \geq [E|X|]^r. \quad \square$$

An approximate evaluation of expectation of $g(X)$, a function of X, has been given in Theorem 6.8.6.

6.3 Moments

Let X be a random variable with *pmf* $p(x_k)$ or *pdf* $f(x)$. The rth raw (non-central) moment of the distribution of X, if it exists, is defined as

$$\mu'_r = E(X^r) = \sum_{k=1}^{\infty} x_k^r p(x_k) \qquad (6.3.1)$$

$$\text{or} \int_{-\infty}^{\infty} x^r f(x) dx \qquad (6.3.2)$$

according as X is discrete or continuous. The series (6.3.1) or the integral (6.3.2) must be absolutely convergent for μ'_r to exist. The first moment μ'_1, if it exists, is the expectation of X. Clearly, $\mu'_0 = 1$.

The rth moment of X about a point 'a', if it exists, is given by

$$\mu'_r(a) = E(X - a)^r = \sum_{k=1}^{\infty} (x_k - a)^r p(x_k) \qquad (6.3.3)$$

$$\text{or} \int_{-\infty}^{\infty} (x-a)^r f(x)dx \tag{6.3.4}$$

according as X is discrete or continuous.

In particular, if $E(X) = \mu_1' = \mu$ (say), exists, one may take $a = \mu$. The expectation of $(X-\mu)^r$, when it exists, is called the rth central moment of X and is generally denoted as μ_r. Thus

$$\mu_r = \mu_r(\mu)' = E(X-\mu)^r = \sum_{k=1}^{\infty}(x_k-\mu)^r p(x_k) \tag{6.3.5}$$

$$\text{or} \int_{-\infty}^{\infty} (x-\mu)^r f(x)dx \tag{6.3.6}$$

according as X is discrete or continuous. Expanding $(X-\mu)^r$ in the integral (or summation) we get

$$\mu_r = \int_{-\infty}^{\infty} \left[x^r - \binom{r}{1}x^{r-1}\mu + \binom{r}{2}x^{r-2}\mu^2 - \ldots + (-1)^r\mu^r\right] f(x)dx \tag{6.3.7}$$

$$= \mu_r' - \binom{r}{1}\mu_{r-1}'\mu + \binom{r}{2}\mu_{r-2}'\mu^2 - \ldots + (-1)^r\mu^r.$$

The relation (6.3.7) gives the expressions for the central moments in terms of raw moments. Thus,

$$\mu_1 = \mu_1' - \mu_0'\mu = 0; \quad \mu_3 = \mu_3' - 3\mu\mu_2' + 2\mu^3;$$

$$\mu_2(\text{ or } \sigma^2) = \mu_2' - \mu^2; \quad \mu_4 = \mu_4' - 4\mu\mu_3' + 6\mu^2\mu_2' - 3\mu^4.$$

The quantity $\mu_2 = \mu_2' - \mu^2 = E(X^2) - (E(X))^2 = E(X-E(X))^2$, often denoted as σ^2 is called the variance of the distribution of X and is a measure of the spread or dispersion of the values x of the random variable X (Section 6.8).

Similarly, one may express μ_r' in terms of central moments by considering

$$\int_{-\infty}^{\infty} x^r dF(x) = \int_{-\infty}^{\infty} \{(x-\mu) + \mu\}^r dF(x)$$

when

$$\mu_r' = \mu_r + \binom{r}{1}\mu_{r-1}\mu + \binom{r}{2}\mu_{r-2}\mu^2 + \ldots + \mu^r. \tag{6.3.8}$$

In particular,

$$\mu_1' = \mu \quad \mu_3' = \mu_3 + 3\mu_2\mu + \mu^3$$

$$\mu_2' = \mu_2 + \mu^2 \quad \mu_4' = \mu_4 + 4\mu_3\mu + 6\mu_2\mu^2 + \mu^4.$$

We now consider the problems of existence of moments.

Theorem 6.3.1: If for a random variable, μ'_r exists, so does μ'_s for $0 < s < r$.

Proof. For $|X| \geq 1, |X|^s \leq |X|^r$; for $|X| < 1, 1 > |X|^s > |X|^r$. Now,

$$
\begin{aligned}
E|X|^s &= \int_{|x|\leq 1} |x|^s dF(x) + \int_{|x|>1} |x|^s dF(x) \\
&\leq \int_{|x|\leq 1} dF(x) + \int_{|x|>1} |x|^r dF(x) < 1 + E|X|^r < \infty.
\end{aligned}
$$

The following two theorems can be proved similarly.

Theorem 6.3.2: If for a random variable $X, \mu'_r(a)$ exists, so does $\mu'_s(a)$, where a is an arbitrary constant and $0 < s < r$.

Theorem 6.3.3: If for a random variable X, μ_r exists, so does μ_s for $0 < s < r$.

Theorem 6.3.4: If μ'_r exists, $\mu'_r(a)$ exists and is given by

$$
\mu'_r(a) = \mu'_r - \binom{r}{1}\mu'_{r-1}a + \binom{r}{2}\mu'_{r-2}a^2 - \ldots + (-1)^r a^r. \tag{6.3.9}
$$

Proof. The expression (6.3.8) follows as in (6.3.6) by expanding $(x - a)^r$ in $\int_{-\infty}^{\infty} (x - a)^r dF(x)$. Existence follows from the fact

$$
\int_{\infty}^{\infty} |x - a|^r dF(x) \leq \int_{\infty}^{\infty} |x|^r dF(x) + \binom{r}{1}|a| \int_{-\infty}^{\infty} |x|^{r-1} dF(x)
$$

$$
+ \binom{r}{2}|a|^2 \int_{-\infty}^{\infty} |x|^{r-2} dF(x) + \ldots + |a|^r.
$$

Theorem 6.3.5: If $\mu'_r(a)$ exists, μ'_r exists and is given by

$$
\mu'_r = \mu'_r(a) + \binom{r}{1}a\mu'_{r-1}(a) + \binom{r}{2}a^2\mu'_{r-2}(a) + \ldots + a^r. \tag{6.3.10}
$$

Proof. Follows as in (6.3.7) and Theorem 6.3.4.

The quantity $E|X|^r = \nu_r$ (say), when it exists, is called the rth absolute moment of the random variable X.

We now consider some conditions for the existence of moments of a random variable in terms of its distribution function.

Theorem 6.3.6: Let X be a random variable. Then

$$
E|X|^p < \infty \Rightarrow x^p P\{|X| \geq x\} \to 0 \text{ as } x \to \infty. \tag{6.3.11}
$$

Proof. Let F be the distribution function of X. Then

$$E|X|^p = \int_{|t| \geq x} |t|^p dF(x) + \int_{|t| < x} |t|^p dF(x).$$

Since $E|X|^p < \infty$, it follows that $\lim_{x \to \infty} \int_{|t| \geq x} |t|^p dF(x) \to 0$. Again,

$$\int_{|t| \geq x} |t|^p dF(x) \geq x^p P\{|X| \geq x\}.$$

The result follows by taking limits of both sides as $x \to \infty$.

Theorem 6.3.7: Let $X \geq 0$ a.s. and let F be the distribution function of X. Then

$$E|X| < \infty \Rightarrow \int_0^\infty [1 - F(x)] dx < \infty$$

and

$$EX = \int_0^\infty [1 - F(x)] dx. \tag{6.3.12}$$

Proof. Suppose that $EX < \infty$. Then

$$EX = \int_0^\infty x dF(x) = \lim_{n \to \infty} \int_0^n x dF(x).$$

On integration by parts,

$$\begin{aligned}
\int_0^n x dF(x) &= nF(n) - \int_0^n F(x) dx \\
&= -n[1 - F(n)] + \int_0^n [1 - F(x)] dx.
\end{aligned} \tag{6.3.13}$$

By Theorem 6.3.6, since $EX, < \infty, n[1 - F(n)] \to 0$ as $n \to \infty$. Hence the result (6.3.11).

Conversely, suppose $\int_0^\infty [1 - F(x)] dx < \infty$. Now, for every n,

$$\int_0^n x dF(x) \leq \int_0^n [1 - F(x)] dx \text{ (by (6.3.13))} \leq \int_0^\infty [1 - F(x)] dx.$$

Therefore, $EX < \infty$. By virtue of (6.3.13), relation (6.3.12) holds. $\square$

EXAMPLE 6.3.1: Find the rth non-central moment of the rectangular distribution

$$f(x) = \frac{1}{b - a}, \ a < x < b, \ = 0, \text{ otherwise.}$$

Here

$$\mu'_r = \frac{1}{b - a} \int_a^b x^r dx = \frac{b^{r+1} - a^{r+1}}{(b - a)(r + 1)}.$$

Thus, $\mu_1' = \mu = \frac{b+a}{2}, \sigma^2 = \mu_2' - \mu^2 = \frac{(b-a)^2}{12}$.

EXAMPLE 6.3.2: A continuous random variable X has the distribution function $F(x)$ proportional to $\alpha x^\beta - \beta x^\alpha, \alpha > \beta \geq 1, 0 \leq x < 1$. Show that for this distribution $\mu_r = \frac{\alpha\beta}{(\alpha+r)(\beta+r)}$.

Let $F(x) = k(\alpha x^\beta - \beta x^\alpha)$ where k is a suitable constant.

$$f(x) = \frac{dF(x)}{dx} = k\alpha\beta(x^{\beta-1} - x^{\alpha-1}).$$

Since $\int_0^1 f(x)dx = 1, k = \frac{1}{\alpha-\beta}$.

$$\mu_r' = \frac{\alpha\beta}{\alpha-\beta}\int_0^1 x^r(x^{\beta-1} - x^{\alpha-1})dx = \frac{\alpha\beta}{(\alpha+r)(\beta+r)}.$$

EXAMPLE 6.3.3: Find the rth non-central moment of the distribution

$$f(x) = \frac{1}{\Gamma(n)}\alpha^n e^{-\alpha x}x^{n-1}, \quad x > 0, n > 0.$$

(This is known as Gamma distribution (Section 9.4). Here $\Gamma(n) = (n-1)!$ for integral n.)

$$\mu_r' = \frac{\alpha^n}{\Gamma(n)}\int_0^\infty e^{\alpha x}x^{n+r-1}dx = \frac{\Gamma(n+r)}{\alpha^r\Gamma(n)}.$$

Hence, $\mu_1' = \mu = \frac{n}{\alpha}, \mu_2 = \mu_2' - \mu_2^2 = \frac{n}{\alpha^2}$. For $n = 1$, the distribution is known as exponential distribution (Section 9.3), when $\mu = \frac{1}{\alpha}, \sigma^2 = \frac{1}{\alpha^2}$.

EXAMPLE 6.3.4: If $f(x)$ is an odd function of x of period $\frac{1}{2}$, show that $\int_0^\infty x^r x^{-\log x}f(\log x)dx = 0$ for all integral values of r. Hence, show that $dF = x^{-\log x}[1 - \lambda\sin(4\pi\log x)]dx, 0 < x < \infty, 0 \leq \lambda \leq 1$ have the same moments whatever be the value of λ.

Let $I = \int_0^\infty x^{r-\log x}f(\log x)dx$.
Putting $\log_e x = y + \frac{1}{2}$,

$$I = \int_{-\infty}^\infty e^{(y+\frac{1}{2})(r-y-\frac{1}{2})}f(y+\frac{1}{2})e^{y+\frac{1}{2}}dy$$
$$= e^{\frac{1}{2}(r+1)^2}\int_{-\infty}^\infty e^{-(y-\frac{1}{2}r)^2}f(y)dy, \quad \left[\text{since } f(y+\frac{1}{2}) = f(y)\right]. \quad (i)$$

Again, putting $\log_e x = -y + \frac{1}{2}$,

$$I = -e^{\frac{1}{2}(r+1)^2}\int_{-\infty}^\infty e^{-(y+\frac{r}{2})^2}f(y)dy \quad [\text{since } f(-y) = -f(y)]. \quad (ii)$$

Adding (i) and (ii),

$$2I = e^{\frac{1}{2}(r+1)^2}\left[e^{-(y-\frac{r}{2})^2}f(y)dy - \int_{-\infty}^\infty e^{-(y+\frac{r}{2})^2}f(y)dy\right]. \quad (iii)$$

Putting $y - \frac{r}{2} = u$ in the first integral and $y + \frac{r}{2} = v$ in the second integral in (iii) and observing that $f(u + \frac{r}{2}) = f(u), f(v - \frac{r}{2}) = f(v)$,

$$2I = e^{\frac{1}{4}(r+1)^2} \left[\int_{-\infty}^{\infty} e^{-u^2} f(u) du - \int_{-\infty}^{\infty} e^{-v^2} f(v) dv \right] = 0.$$

Hence, $I = 0$. Now, $\sin(4\pi x)$ is an odd function of x of period $\frac{1}{2}$.

Thus $\int_0^{\infty} x^{r - \log x} \sin(4\pi \log x) = 0$.

Hence, the moments of the given distribution do not depend on the value of λ.

6.4   Some Moment Inequalities

Cauchy-Schwarz inequality: Let X, Y be two random variables, both defined on $(\mathcal{S}, \mathcal{A}, P)$ with $E(X^2) < \infty, E(Y^2) < \infty$. Then $E|XY| < \infty$ and the following inequality holds.

$$[E(XY)]^2 \le [EX^2][EY^2]. \tag{6.4.1}$$

Equality in (6.4.1) holds *iff* Y is proportional to X almost surely (*a.s.*)

Proof. Inequality (6.4.1) holds trivially if at least one of X and Y is zero a.s. Thus we assume that $EX^2 > 0$ and $EY^2 > 0$.

Since $|XY| \le \frac{1}{2}(X^2 + Y^2), E|XY| < \infty$, as $E(X^2) < \infty, E(Y^2) < \infty$. Hence $|E(XY)| < E|XY| < \infty$.

For any real number t, let $h(t) = E(tX - Y)^2$. Clearly, $h(t)$ is real and ≥ 0. Hence, the equation

$$h(t) = t^2 E(X^2) - 2t E(XY) + E(Y^2) = 0 \tag{6.4.2}$$

has either no (real) solution or only one solution. It has no (real) solution *iff* $(E(XY))^2 < E(X^2)E(Y^2)$. This proves the inequality in (6.4.1).

The equation (6.4.2) has exactly one solution *iff* $(E(XY))^2 = E(X^2)E(Y^2)$. Therefore, equality in (6.4.1) holds *iff* there exists some (real) t for which $E(tX - Y)^2 = 0$ i.e. $P[tX - Y = 0$ for some real $t] = 1 a.s.$ i.e. Y is proportional to X a.s.

Note 6.4.1: If in (6.4.1) we take $X - E(X), Y - E(Y)$ in place of X, Y respectively, the inequality (6.4.1) reduces to

$$V(X)V(Y) \ge [\text{Cov } (X, Y)]^2 \tag{6.4.3}$$

where $V(X)$, called the variance of X is $E(X - E(X))^2$ and Cov $(X, Y) = E[(X - E(X)(Y - E(Y))$. The positive square root of $V(X)$ is called the standard deviation (s.d) of X. The equality in (6.4.3) holds *iff* Y is a linear function of X a.s.

(6.4.3) implies that the *correlation coefficient* between X and Y, defined by

$$\rho_{X,Y} = \frac{\text{Cov } (X, Y)}{\sqrt{V(X)V(Y)}} \tag{6.4.4}$$

has its absolute value less than unity.

Note 6.4.2: (6.4.1) holds if X, Y are replaced respectively by some Borel measurable functions $g_1(X), g_2(X)$ of random variable X defined on $(\mathcal{S}, \mathcal{A}, P)$, provided $E[g_1(X)]^2 < \infty, E[g_2(X)]^2 < \infty$.

Theorem 6.4.1: If both μ'_{2r}, μ'_{2s} exist, then

(i) $\mu'_{2r}\mu'_{2s} \geq (\mu'_{r+s})^2;$ (6.4.5)

(ii) $\mu_{2r}\mu_{2s} \geq (\mu_{r+s})^2;$ (6.4.6)

In particular,

(iii) $\beta_2 = \frac{\mu_4}{\mu_2^2} \geq 1$, provided $\mu_2 > 0$.   (6.4.7)

Proof. If μ'_{2r}, μ'_{2s} exist, μ_{2r}, μ_{2s} exist; also μ'_{r+s}, μ_{r+s} exist (Theorems 6.3.2 and 6.3.4).

Substituting in (6.4.1) X^r, X^s in place of X, Y respectively, the inequality (6.4.5) is obtained. (6.4.6) follows by putting $(X - E(X))^r, (X - E(X))^s$ in place of X, Y respectively in (6.4.1). (6.4.7) is a special case of (6.4.6) for $r = 2, s = 0$.

Theorem 6.4.2: For any random variable X,

$$\begin{vmatrix} \mu_{2a} & \mu_{a+b} & \mu_{a+c} \\ \mu_{a+b} & \mu_{2b} & \mu_{b+c} \\ \mu_{a+c} & \mu_{b+c} & \mu_{2c} \end{vmatrix} = \delta(a, b, c) \text{ (say) } \geq 0 \tag{6.4.8}$$

where a, b, c are non-negative integers. Hence, provided μ_4 exists,

$$\beta_2 - \beta_1 - 1 \geq 0 \tag{6.4.9}$$

where $\beta_1 = \frac{\mu_3^2}{\mu_2^3}, \beta_2 = \frac{\mu_4}{\mu_2^2}$.

Proof. We have

$$E[AX^a + BX^b + CX^c]^2 \geq 0$$

i.e.

$$E\left[A^2X^{2a} + B^2X^{2b} + C^2X^{2c} + 2ABX^{a+b} + 2ACX^{a+c} + 2BCX^{b+c}\right] \geq 0$$
$$(6.4.10)$$

for all values of A, B, and C. If X is measured from $E(X)$, (6.4.10) reduces to

$$A^2\mu_{2a} + B^2\mu_{2b} + C^2\mu_{2c} + 2AB\mu_{a+b} + 2AC\mu_{a+c} + 2BC\mu_{b+c} \geq 0 \quad (6.4.11)$$

for all values of A, B and C. The left side of (6.4.11) is, therefore, a positive semi-definite form.

A set of necessary and sufficient conditions for this quadratic form to be positive semi-definite is that all principal minors of δ are non-negative, i.e.

$$\mu_{2a} \geq 0, \quad \begin{vmatrix} \mu_{2a} & \mu_{a+b} \\ \mu_{a+b} & \mu_{2b} \end{vmatrix} \geq 0, \ \delta \geq 0.$$

For $a = 0, b = 1, c = 1$, (6.4.8) reduces to

$$\begin{vmatrix} \mu_0 & \mu_1 & \mu_2 \\ \mu_1 & \mu_2 & \mu_3 \\ \mu_2 & \mu_3 & \mu_4 \end{vmatrix} \geq 0 \text{ or } \mu_2\mu_4 - \mu_3^2 - \mu_2^3 \geq 0 \text{ or } \beta_2 - \beta_1 - 1 \geq 0.$$

Since $\beta_1 \geq 0, \beta_2 \geq 1$.

Corollary 6.4.2.1: For any random variable X,

$$\begin{vmatrix} \mu_0 & \mu_1 & \mu_2 & \mu_3 \\ \mu_1 & \mu_2 & \mu_3 & \mu_4 \\ \mu_2 & \mu_3 & \mu_4 & \mu_5 \\ \mu_3 & \mu_4 & \mu_5 & \mu_6 \end{vmatrix} \geq 0.$$

Theorem 6.4.3: For any random variable X, $\nu_r^{\frac{1}{r}}$ is an increasing function of r, ie., $\nu_{r+1}^{\frac{1}{r+1}} \geq \nu_r^{\frac{1}{r}}$ for $r = 1, 2, \ldots$

Proof. We have

$$E\left[|X|^{\frac{r+1}{2}}u + |X|^{\frac{r-1}{2}}v\right]^2 \geq 0 \tag{6.4.12}$$

where u, v are arbitrary constants. Thus (6.4.12) means

$$u^2\nu_{r+1} + v^2\nu_{r-1} + 2uv\nu_r \geq 0$$

is a positive semi-definite quadratic form in u, v. A set of necessary and sufficient conditions for this is

$$\begin{vmatrix} \nu_{r+1} & \nu_r \\ \nu_r & \nu_{r-1} \end{vmatrix} \geq 0, \text{ i.e. } \nu_{r-1}\nu_{r+1} \geq \nu_r^2 \text{ or } \nu_{r-1}^r\nu_{r+1}^r \geq \nu_r^{2r}. \tag{6.4.13}$$

Substituting in (6.4.13), $r = 1, 2, \ldots, r$, we have

$$\nu_0 \nu_2 \geq \nu_1^2 \tag{i}$$

$$\nu_1^2 \nu_3^2 \geq \nu_2^4 \tag{ii}$$

$$\nu_2^3 \nu_4^3 \geq \nu_3^6 \tag{iii}$$

$$\cdots\cdots$$

$$\nu_{r-1}^r \nu_{r+1}^r \geq \nu_r^{2r}. \tag{r}$$

Multiplying (i), ..., (r),

$$\nu_{r+1}^r \geq \nu_r^{r+1} \text{ or } \nu_{r+1}^{\frac{1}{r+1}} \geq \nu_r^{\frac{1}{r}}.$$

Thus, $\frac{1}{r+1} \log \nu_{r+1} \geq \frac{1}{r} \log \nu_r$ or $\frac{1}{r} \log \nu_r$ is an increasing function of r for positive r.

Theorem 6.4.4: For any random variable

$$\nu_{\alpha_1} \nu_{\alpha_2} \ldots \nu_{\alpha_k} \geq \left[\nu_{\frac{\alpha_1 + \ldots + \alpha_k}{k}} \right]^k \tag{6.4.14}$$

where $\alpha_1, \ldots, \alpha_k$ are positive integers.

Proof. Consider the *generalized Cauchy-Schwarz inequality* : For positive random variables $Y_1, \ldots, Y_k$,

$$E(Y_1^k) E(Y_2^k) \ldots E(Y_k^k) \geq [E(Y_1 Y_2 \ldots Y_k)]^k. \tag{6.4.15}$$

Setting in (6.4.15), $Y_1 = |X - \mu|^{\alpha_1/k}, \ldots, Y_k = |X - \mu|^{\alpha_k/k}$, the inequality (6.4.14) follows.

Theorem 6.4.5 (*Liapounoff's Inequality*): For any random variable X,

$$\nu_b^{a-c} \leq \nu_c^{a-b} \nu_a^{b-c} \tag{6.4.16}$$

for $a \geq b \geq c \geq 0$.

Proof. Substitute in (6.4.14),

$$k = a - c; \ \alpha_1 = \ldots = \alpha_{a-b} = c; \ \alpha_{a-b+1} = \alpha_{a-b+2} = \ldots = \alpha_{a-c} = a.$$

Thus, $\frac{\alpha_1 + \ldots + \alpha_k}{k} = b$ and (6.4.16) follows.

Theorem 6.4.6 (*Holder's Inequality*): Let p and q be two positive real numbers satisfying $1 < p, q < \infty$ and $p^{-1} + q^{-1} = 1$. Let X and Y

be two random variables satisfying $E|X|^p < \infty$ and $E|Y|^q < \infty$. Then $E|XY| < \infty$ and the inequality

$$E|XY| \leq [E|X|^p]^{1/p}[E|Y|^q]^{1/q} \tag{6.4.17}$$

holds.

Proof. We first prove the following result. Let p and q be as defined above and let $\alpha \geq 0, \beta \geq 0$. Then the inequality

$$\frac{\alpha^p}{p} + \frac{\beta^q}{q} \geq \alpha\beta \geq 0 \tag{6.4.18}$$

holds. The inequality (6.4.18) is trivially true, if at least one of α, β is zero. We, therefore, assume $\alpha > 0, \beta > 0$. Define a function

$$\psi(t) = \frac{t^p}{p} + \frac{t^{-q}}{q}, \ t \in (0, \infty).$$

Thus, $\psi'(t) = t^{p-1} - t^{-q-1}$. Let $t_0 (> 0)$ be a solution of $\psi'(t) = 0$. Then $t_0^{p+q} = 1$ which implies $t_0 = 1$. Again, $\psi'(t) < 0$ for $0 < t < 1$ and $\psi'(t) > 0$ for $t > 1$. Also,

$$\lim_{t \downarrow 0} \psi(t) = \lim_{t \to \infty} \psi(t) = +\infty.$$

It follows, therefore, that $t = 1$ is the unique minimum of $\psi(t)$ over $(0, \infty)$. Hence,

$$\frac{t^p}{p} + \frac{t^{-q}}{q} \geq 1, \ t \geq 0. \tag{6.4.19}$$

Setting $t = (\alpha^{1/q})(\beta^{1/p})^{-1}$ in (6.4.19), we obtain (6.4.18).

We now come back to the proof of (6.4.17). The inequality is trivially true if at least one of EX^p or $E|X|^q$ is zero. We assume, therefore, $E|X|^p > 0, E|Y|^q > 0$. For each $\omega \in \mathcal{S}$, we set

$$\alpha = \frac{|X(\omega)|}{[E|X|^p]^{1/p}}, \quad \beta = \frac{Y(\omega)}{[E(|Y|^q]^{1/q}}$$

in (6.4.18) to obtain the inequality

$$0 < |XY| \leq [E|X|^p]^{1/p}[E|Y|^q]^{1/q} \left[\frac{1}{p} \frac{|X|^p}{E|X|^p} + \frac{1}{q} \frac{|Y|^q}{E|Y|^q} \right]$$

which holds a.s. Clearly $E|XY| < \infty$. Taking expectation of both sides we obtain (6.4.17). $\qquad \square$

Note 6.4.3: Cauchy-Schwarz inequality (6.4.1) is a special case of (6.4.17) for $p = q = 2$.

Theorem 6.4.7 (*Minkowski's Inequality*): Let $1 \le p < \infty$. Let X and Y be two random variables such that $E|X|^p < \infty, E|Y|^p < \infty$. Then $E|X+Y|^p < \infty$ and the inequality

$$[E|X+Y|^p]^{\frac{1}{p}} \le [E|X|^p]^{1/p} + [E|Y|^p]^{1/p} \qquad (6.4.20)$$

holds.

Proof.

$$\begin{aligned}
E|X+Y|^p &= E[|X+Y|.|X+Y|^{p-1}] \\
&\le E[|X|.|X+Y|^{p-1}] + E[|Y|.|X+Y|^{p-1}] \\
&\le E^{1/p}|X|^p E^{(p-1)/p}|X+Y|^p + E^{1/p}|Y|^p E^{(p-1)/p}|X+Y|^p,
\end{aligned}$$

using Holder's inequality with $s = p/(p-1)$. Dividing both sides by $E^{(p-1)/p}|X+Y|^p$, the result follows.

Theorem 6.4.8 (X_r-*Inequality*):

$$E|X+Y|^r \le C_r[E|X|^r + E|Y|^r] \qquad (6.4.21)$$

where

$$C_r = \begin{cases} 1 & \text{if } r \le 1, \\ 2^{r-1} & \text{if } r \ge 1. \end{cases} \qquad (6.4.22)$$

Proof. We know that if $a > 0, b > 0$,

$$(\frac{a}{a+b})^r + (\frac{b}{a+b})^r \ge 1, \text{ for } r \le 1,$$

i.e.,

$$a^r + b^r \ge (a+b)^r. \qquad (6.4.23)$$

Hence, for all $\omega \in S$ and $r \le 1$,

$$\begin{aligned}
|X(\omega)|^r + |Y(\omega)|^r &\ge [|X(\omega)| + |Y(\omega)|]^r \\
&\ge |X(\omega) + Y(\omega)|^r.
\end{aligned} \qquad (6.4.24)$$

Now, the function

$$\phi(t) = t^r + (1-t)^r \ (r \ge 1; 0 < t < 1)$$

has a minimum at $t = \frac{1}{2}$. Therefore, for $a > 0, b > 0$ and $r \ge 1$,

$$(\frac{a}{a+b})^r + (\frac{b}{a+b})^r \ge \frac{1}{2^{r-1}}$$

or

$$2^{r-1}(a^r + b^r) \ge (a+b)^r. \qquad (6.4.25)$$

Hence, for all $\omega \in S$ and all $r \ge 1$,

$$\begin{aligned}
2^{r-1}[|X(\omega)|^r + |Y(\omega)|^r] &\ge [|X(\omega)| + |Y(\omega)|]^r \\
&\ge |X(\omega) + Y(\omega)|^r.
\end{aligned} \qquad (6.4.26)$$

Taking expectations of (6.4.24) and (6.4.26), the result (6.4.22) is obtained.

The theorem shows that if the rth absolute moments of X and Y exist and are finite, so also is the rth absolute moment of $X + Y$.

6.5   Moments of a Symmetric Probability Distribution

A random variable X (probability distribution of X) is said to be symmetric about a point a if

$$P(X \geq a + x) = P(X \leq a - x) \tag{6.5.1}$$

for all x. In terms of the distribution function of X, this means if

$$F_X(a - x) = 1 - F_X(a + x) + P[X = a + x] \tag{6.5.2}$$

holds for all $x \in \mathcal{R}$, we say that the probability distribution of X is symmetric with a as the center of symmetry. If $a = 0$, then for every x,

$$F_X(-x) = 1 - F_X(x) + P[X = x]. \tag{6.5.3}$$

If X is a continuous random variable, X is symmetric with center a *iff* the *pdf* $f(x)$ of X satisfies,

$$f(x - a) = f(a + x) \text{ for all } x \in \mathcal{R}. \tag{6.5.4}$$

If $a = 0$, we say simply that X is symmetric (or F is symmetric).

It follows immediately that if X is symmetric about $a, E(X)$ exists and is a.

Theorem 6.5.1: If X is symmetric about $a, E(X) = a$ and μ_{2r+1} (if it exists) $= 0, r = 0, 1, 2, \ldots$.

Proof. Suppose X is a continuous random variable with *pdf* $f(x)$.

$$E(X) = \int_{-\infty}^{\infty} x f(x) dx = \int_{-\infty}^{a} x f(x) dx + \int_{a}^{\infty} x f(x) dx = I_1 + I_2 \text{ (say)}.$$

In I_1 put $y = a - x$; then $I_1 = \int_0^{\infty} (a - y) f(a - y) dy$;

In I_2 put $y = x - a$; then $I_2 = \int_0^{\infty} (a + y) f(a + y) dy$.

Hence,

$$E(X) = I_1 + I_2 = 2 \int_0^{\infty} a f(a + y) dy \text{ (by (6.5.4))} = a.$$

Again,

$\mu_{2r+1} = \int_{-\infty}^{\infty} (x - a)^{2r+1} f(x) dx$

$\quad = \int_{-\infty}^{a} (x - a)^{2r+1} f(x) dx + \int_0^{\infty} (x - a)^{2r+1} f(x) dx = I_3 + I_4 \text{ (say)}$.

Now, $I_3 = -\int_0^{\infty} z^{2r+1} f(a - z) dz$(putting $z = a - x$);

$I_4 = \int_0^{\infty} z^{2r+1} f(a + z) dz$(putting $z = x - a$).

Hence, $\mu_{2r+1} = 0$. In case, X is discrete, the result can similarly be proved.

6.6 Factorial Moments

Let X be a discrete random variable with *pmf* $P(X = k) = p_k, k = 0, 1, 2, \ldots$ Then $E[(X)_r]$, where $(X)_r = X(X-1)(X-2)\ldots(X-r+1)$, if it exists, is called the rth *factorial moment* of X about origin and will be denoted as $\mu_{(r)}$.

$$\mu_{(r)} = E[X(X-1)\ldots(X-r+1)] = \sum_{x=r}^{\infty} x(x-1)\ldots(x-r+1)p_r.$$

It readily follows that $\mu_{(r)}$ exists, *iff* μ_r' exists. It is easy to check that
$$\mu_{(1)} = \mu_1'; \quad \mu_{(2)} = \mu_2' - \mu_1';$$
$$\mu_{(3)} = \mu_3' - 3\mu_2' + 2\mu_1';$$
$$\mu_{(4)} = \mu_4' - 6\mu_3' + 11\mu_2' - 6\mu_1'.$$

Again, since $x^2 = (x)_2 + x$, $x^3 = (x)_3 + 3(x)_2 + x$, $x^4 = (x)_4 + 6(x)_3 + 7(x)_2 + x, \ldots$, it follows that
$$\mu_1' = \mu_{(1)}; \quad \mu_2' = \mu_{(2)} + \mu_{(1)};$$
$$\mu_3' = \mu_{(3)} + 3\mu_{(2)} + \mu_{(1)};$$
$$\mu_4' = \mu_{(4)} + 6\mu_{(3)} + 7\mu_{(2)} + \mu_{(1)}.$$

EXAMPLE 6.6.1: Consider the *Poisson* distribution (Section 8.6)
$$P(X = x) = e^{-\lambda}\frac{\lambda^x}{x!}, \quad x = 0, 1, 2, \ldots, \quad x > 0.$$

Here
$$\mu_{(r)} = \sum_{x=0}^{\infty} x_{(r)}e^{-\lambda}\frac{\lambda^x}{x!} = \sum_{x=r}^{\infty} \frac{e^{-\lambda}\lambda^x}{(x-r)!} = \sum_{y=0}^{\infty} e^{-\lambda}\frac{\lambda^{y+r}}{y!} = \lambda^r.$$

Thus, $\mu_{(1)} = \lambda, \sigma^2 = \lambda^2 + \lambda - \lambda^2 = \lambda$.

EXAMPLE 6.6.2: Consider the *discrete uniform* distribution (Section 8.2).
$P(X = x) = \frac{1}{N}, x = 1, 2, \ldots, N$.

Here
$$\mu_{(r)} = \frac{1}{N}\sum_{x=1}^{N}(x)_r.$$

Now,
$$\sum_{j=a}^{b}(j)_n = \frac{(b+1)_{(n+1)} - (a)_{(n+1)}}{n+1}.$$

Hence,
$$\mu_{(r)} = \frac{(N+1)_{(r+1)}}{N(r+1)}, r = 1, 2, \ldots, N-1; \quad \mu_{(N)} = (N-1)!.$$

Thus,
$$\mu = \frac{N+1}{2}, \quad \mu_{(2)} = \frac{N^2-1}{3}, \quad \sigma^2 = \frac{(N+1)(N-1)}{12}.$$

6.7 Different Measures of Central Tendency

The random variable X takes values in the Borel sets B of $\mathcal{R}$ according to some probability law which is called the probability distribution of X. However, the different values of X will generally have a tendency to cluster around some central value. This value may be taken as a representative value of X and measures the *central tendency* (*location parameter*) of the distribution. Different measures of central tendency are given below.

(a) *Mean*: This is the most commonly used measure of central tendency of the distribution of X. This is also called the expectation of X, and if it exists, is given by

$$E(X) = \int_{-\infty}^{\infty} x dF(x).$$

(b) *Median*: If μ_e is a point which divides the whole probability mass of the distribution into two equal parts, each containing probability mass equal to $\frac{1}{2}$, μ_e is called the median of the distribution.

Consider the curve of the probability distribution function $F(x)$ of a continuous random variable X. The abscissa of the point of intersection of this curve with the straight line $y = \frac{1}{2}$ is the unique median μ_e of this distribution. Thus μ_e satisfies

$$F(\mu_e - 0) \le \frac{1}{2} \le F(\mu_e). \tag{6.7.1}$$

For a discrete distribution, it may so happen that $F(x) = \frac{1}{2}$ for all x in the interval $[a, b]$. Then median is taken to be $\frac{1}{2}(a + b)$.

The mean does not always exist, whereas a median always exists (an example is the Cauchy distribution discussed in Example 6.2.1). Even in cases where the mean does exist, the median is sometimes preferred as a measure of central tendency, since the value of the mean may sometimes be largely affected by the presence of some stray probability masses at either end.

(c) *Mode*: Mode is that value of the random variable X which occurs with the highest probability. It, therefore, corresponds to that value x for which the curve of the *pdf* $f(x)$ has the maximum value of the ordinate or in case of a discrete variable that value x for which the probability $P[X = x]$ is the largest. If the *pdf* or *pmf* has a single relative maxima (unique mode) μ_0, then μ_0 may be used as a measure of central tendency of the distribution. For a continuous random variable with *pdf* $f(x)$, if $f(x)$ is twice differentiable, μ_0 can be found out from the condition $f'(\mu_0) =$

$0, f''(\mu_0) < 0$. In case $f(x)$ is not differentiable twice, μ_0 has to be found out from the condition $f(\mu_0) > f(x) \; \forall \; x \neq \mu_0$. For a multimodal distribution X cannot be used as a measure of central tendency.

For a non-negative random variable, one may sometimes use the *geometric mean* (GM) G as a measure of central tendency. This is given by

$$\log G = \int_{-\infty}^{\infty} (\log x) dF(x). \qquad (6.7.2)$$

Sometimes, *harmonic mean* (HM) H may be used as a measure. This is given by

$$H = \frac{1}{\int_{-\infty}^{\infty} (\frac{1}{x}) dF(x)}. \qquad (6.7.3)$$

In a perfectly symmetrical distribution the mean, median and mode coincide. For a moderately asymmetric distribution occurring in practice, it is found empirically that

$$| \text{ Mean - Mode } | = 3(| \text{ Mean - Median } |).$$

EXAMPLE 6.7.1: Find the arithmetic mean, geometric mean and harmonic mean of the distribution

$$f(x) = \frac{1}{B(p.q)} (1-x)^{p-1} x^{q-1} dx, \; 0 \leq x \leq 1, \; p, q > 0 \qquad (6.7.4)$$

where

$$B(p, q) = \int_0^1 (1-x)^{p-1} x^{q-1} dx = \frac{\Gamma(p)\Gamma(q)}{\Gamma(p+q)} \qquad (6.7.5)$$

and $\Gamma(r) = (r-1)!$ for integral r (Section 9.4).

The arithmetic mean is given by

$$E(X) = \frac{p}{p+q}.$$

The GM is given by

$$\log G = E(\log X) = \frac{1}{B(p, q)} \int_0^1 \log(x)(1-x)^{p-1} x^{q-1} dx.$$

Differentiating both sides of (6.7.5) with respect to q (differentiation under the integral sign being permissible as the integral converges uniformly),

$$\int_0^1 (1-x)^{p-1} x^{q-1} \log x dx = \frac{\partial}{\partial q} B(p, q).$$

Hence,

$$\log G = \tfrac{1}{B(p,q)} \tfrac{\partial}{\partial q} B(p,q) = \tfrac{\partial \log B(p,q)}{\partial q}$$

$$= \tfrac{\partial}{\partial q} \log \left[\tfrac{\Gamma(p)\Gamma(q)}{\Gamma(p+q)} \right] = \tfrac{\partial}{\partial q} [\log \Gamma(q) - \log \Gamma(p+q)].$$

The HM is given by

$$\tfrac{1}{H} = E(\tfrac{1}{X}) = \tfrac{1}{B(p,q)} \int_0^1 (1-x)^{p-1} x^{q-2} dx$$

$$= \tfrac{B(p,q-1)}{B(p,q)} = \tfrac{\Gamma(q-1)\Gamma(p+q)}{\Gamma(p+q-1)\Gamma(q)} = \tfrac{p+q-!}{q-1}.$$

Hence,

$$H = \frac{q-1}{p+q-1}.$$

EXAMPLE 6.7.2: Find the mode and median of the distribution with *pdf*

$$f(x) = \frac{\alpha \beta x^{\alpha-1}}{(1+\beta x^\alpha)^2}, \ \beta > 0, \alpha > 1, \ x \geq 0.$$

Let μ_0, μ_e be, respectively, the mode and median of the distribution.

$$\frac{df(x)}{dx} = \frac{\alpha \beta x^{\alpha-2}(1+\beta x^\alpha)[\alpha - 1 - \beta x^\alpha(1+\alpha)]}{(1+\beta x^\alpha)^4}.$$

Hence, $\frac{\partial f(x)}{dx} = 0 \Rightarrow \mu_0 = \left[\frac{\alpha-1}{\beta(1+\alpha)} \right]^{1/\alpha}$. It can be checked that $\left[\frac{d^2 f(x)}{dx^2} \right]_{\mu_0} <$ 0. Again

$$\int_0^{\mu_e} \frac{\alpha \beta x^{\alpha-1}}{(1+\beta x^2)^2} dx = \frac{1}{2}.$$

Setting $1+\beta x^2 = t, \int_1^{1+\beta\mu_e^2} \frac{dt}{t^2} = \frac{1}{2}$ gives $\mu_e = (\frac{1}{\beta})^{1/\alpha}$.

6.8 Measures of Dispersion

Measures of spread or dispersion (sometimes called parameters of concentration) are parameters which give an idea of the scatter (or cluster) of the probability mass of the distribution about the location parameter. These are given below.

(a) *Range*: If X has a distribution such that

$$F(x) = \begin{cases} 0 & \text{if } x < 0 \\ \in (0,1) & \text{if } a \leq x < b \\ = 1 & \text{if } x \geq b, \end{cases}$$

then the difference $b - a$, called the range of X may be taken as a measure of dispersion. This measure does not take into account how the probability mass is distributed over $[a, b]$ and is thus very unsuitable as a good measure of dispersion of the distribution of X.

(b) *Mean Deviation*: It is the expected value of the absolute deviation of X about $A, E|X - A| = MD_A$ (say)., where A is an arbitrary constant. The absolute value of $(X - A)$ is taken, because the dispersion should be indicated by the magnitude of deviation and not by its sign. Its value is large when there is large scattering and small when there is a great concentration about A. Mean deviation is generally taken about the mean.

Theorem 6.8.1: Mean deviation is least when taken about the median, i.e.,

$$E|X - A| \geq E|X - \mu_e| \tag{6.8.1}$$

when A is any number.

Proof. Suppose $A < \mu_e$. Let us write $\mu_e = m$, for simplicity.

$$E|X - A| = \int_{-\infty}^{\infty} |x - A| dF(x) = \int_{-\infty}^{A} (A - x) dF(x) + \int_{A}^{\infty} (x - A) dF(x)$$

$$= \int_{-\infty}^{A} (A - m + m - x) dF(x) + \int_{A}^{\infty} (x - m + m - A) dF(x)$$

$$= (A - m)F(A) + (m - A)(1 - F(A)) + \int_{-\infty}^{A} (m - x) dF(x) + \int_{A}^{\infty} (x - A) dF(x)$$

$$= (m - A)(1 - 2F(A)) + \int_{-\infty}^{\infty} (m - x) dF(x) - \int_{A}^{m} (m - x) dF(x) + \int_{A}^{\infty} (x - m) dF(x) + \int_{m}^{\infty} (x - m) dF(x)$$

$$= (m - A)(1 - 2F(A)) + E|X - m| + \int_{A}^{m} |(x - m) - (m - x)| dF(x)$$

$$= E|X - m| + (m - A)(1 - 2F(A)) + 2 \int_{A}^{m} (x - m) dF(x)$$

$$= E|X - m| + (m - A)(1 - 2F(A)) - 2(m - A) \left[\int_{-\infty}^{m} dF(x) - \int_{-\infty}^{A} dF(x) \right] + 2 \int_{A}^{m} (x - A) dF(x)$$

$$= E|X - A| + (m - A)(1 - 2F(m)) + 2 \int_{A}^{m} (x - A) dF(x).$$

Since $F(m) = \frac{1}{2}$, the middle term vanishes. Since $A < m$ the integral is also positive. Hence the result.

The result for the case $\mu_e < A$ can be proved similarly.

(c) *Variance, Mean Square Error, Standard Deviation*: For measuring the deviation of X around A, one may also consider $E(X - A)^2$ which, like $E|X - A|$, is free from sign and measures the deviation of X around A. This is called the mean square error of X around A, $[MSE_A(X)]$. When A is $E(X) = \mu$, this is called the variance of X, often denoted as $V(X)$. Clearly, this is the second central moment μ_2 of the distribution, $V(X) = E(X-\mu)^2$. We have

$$E(X - A)^2 = E(X - \mu)^2 + (\mu - A)^2$$

so that

$$MSE_A(X) = V(X) + (\mu - A)^2. \tag{6.8.2}$$

Another measure is the positive square root of $V(X)$, called the standard deviation (s.d.) of X, often denoted as $\sigma_X (\sigma_X = +\sqrt{V(X)})$. The positive square root of mean square deviation of X about A, is called the *root mean square deviation of X about A*.

Theorem 6.8.2: $V(X)$ exists *iff* $E(X)$ and $E(X^2)$ exists.

Theorem 6.8.3: For any random variable $X, V(X) \le E(X - A)^2, A \ne \mu$.

Theorem 6.8.4: For any random variable $X, E|X - A| \le \sqrt{E(X - \mu)^2}$, where $\mu = E(X)$.

Proof. In Cauchy-Scwartz inequality, put $|X - \mu|$ in place of X and 1 in place of Y. Taking positive square roots of both sides of the inequality, the result follows.

DEFINITION 6.8.1 *Conditional Variance of X given A*: Let A be a measurable subset of S. The conditional variance of X given A is defined as

$$V(X|A) = E\left[\{X - E(X|A)\}^2 |A\right] = \frac{E\left\{(X - E(X|A))^2 I_A\right\}}{P(A)}. \tag{6.8.3}$$

Theorem 6.8.5: Let $\{A_i\}$ be a measurable partition of the sample space S such that $P(A_i) > 0$ for each $i = 1, \ldots, k$ and let $V(X)$ exist. Then

$$V(X) = \sum_{i=1}^{k} P(A_i)V(X|A_i) + \sum_{i=1}^{k} P(A_i)[E(X|A_i) - E(X)]^2 \tag{6.8.4}$$

$$= E_A[V(X|A)] + V_A[E(X|A)].$$

where $E(X|A_i), V(X|A_i)$ respectively are the conditional expected value of X given A_i (defined in Section 6.2) and conditional variance of X given A_i.

Proof.

$$V(X) = E(X - E(X))^2 = \sum_i P(A_i) E\left\{(X - E(X))^2|A_i\right\}$$

(by Theorem 6.2.1). Now

$$
\begin{aligned}
&P(A_i) E\left\{(X - E(X))^2|A_i\right\} \\
&= E\left[\{(X - E(X))^2\}I_{A_i}\right] \quad \text{(by (6.2.10))} \\
&= \int_{A_i}(X - E(X))^2 dP = \int_{A_i}\{X - E(X|A_i) + E(X|A_i) - E(X)\}^2 dP \\
&= \int_{A_i}(X - E(X|A_i))^2 dP + \int_{A_i}(E(X|A_i) - E(X))^2 dP \\
&= P(A_i)V(X|A_i) + (E(X|A_i) - E(X))^2 P(A_i).
\end{aligned}
$$

Again, $\sum_i P(A_i)V(X|A_i) = E_A[V(X|A)]$. Also

$$\sum_i P(A_i)[E(X|A_i) - E(X)]^2 = E_A[E(X|A) - E_A(E(X|A))]^2 = V_A[E(X|A)]$$

where $V_A(.)$ denotes the variance of $(.)$ with respect to A. Hence the theorem. $\qquad\square$

Theorem 6.8.6 (*Approximate evaluation of expectation and variance of* $g(X)$): Let X be a random variable with $E(X) = \mu$ and variance, $V(X) = \sigma^2$. Let $g(X)$ be a function of X. Then

$$E[g(X)] \approx g(\mu) + \frac{g''(\mu)}{2}\sigma^2 \tag{6.8.5}$$

$$V[g(X)] \approx [g'(\mu)]^2\sigma^2 \tag{6.8.6}$$

provided $g(X)$ is at least twice differentiable at $X = \mu$.

Proof. Expanding $g(X)$ in a Taylor series about $X = \mu$,

$$g(X) = g(\mu) + (X - \mu)g'(\mu) + \frac{(X - \mu)^2}{2!}g''(\mu) + \Delta \tag{6.8.7}$$

where Δ is a remainder. Discarding Δ and taking expectation of both sides (6.8.5) follows.

Ignoring the last two terms in the right hand side of (6.8.7) and taking variance of both sides, (6.8.6) follows. $\qquad\square$

(d) *Quartile Division*: Consider a quantity $p \in (0,1)$. The value $x = \psi_p$ of the random variable X which satisfies

$$F(\psi_p - 0) \leq p \leq F(\psi_p) \tag{6.8.8}$$

is said to be the quantile of X of order p. Comparing with (6.7.1), median is quantile of order 0.5.

If X is a discrete random variable, ψ_p is defined as the smallest value of X (with non-zero probability) satisfying $F(x) \geq p$.

The quantities $\psi_{1/4}, \psi_{1/2}, \psi_{3/4}$ are respectively called the first *quartile* (Q_1), second quartile (Q_2; also median), third quartile (Q_3). The quantiles $\psi_1, \psi_2, \ldots, \psi_9$ are called the first, second, ..., ninth *decile* respectively. The quantiles $\psi_{.01}, \psi_{.02}, \ldots \psi_{.99}$ are called the first, second, ..., 99th *percentile* respectively.

The quantity

$$Q = \frac{Q_3 - Q_1}{2} \qquad (6.8.9)$$

called the *Quartile deviation* (or *semi-interquartile range*) is also taken as a measure of dispersion. The higher the dispersion in values of X, the higher the value of Q.

(e) *Gini's Mean Difference*: The Italian statistician Corrado Gini suggested the quantity Δ_1 based on the expected absolute difference between pairs of units as a measure of dispersion.

If X is a discrete random variable with *pmf* $P(X = x) = p(x)$,

$$\Delta_1 = \sum_{i=1}^{\infty} \sum_{j=1}^{\infty} |x_i - x_j| p(x_i) p(x_j). \qquad (6.8.10)$$

If X is a continuous random variable with *pdf* $f(x)$,

$$\Delta_1 = \int_{-\infty}^{\infty} \int_{-\infty}^{\infty} |x - y| f(x) f(y) dx dy. \qquad (6.8.11)$$

Another measure Δ_2 based on the square of the differences between pairs of observations is obtained by replacing $|x_i - x_j|$ by $(x_i - x_j)^2$ and $|x - y|$ by $(x - y)^2$ in (6.8.10) and (6.8.11) respectively.

Lemma 6.8.1: If the mean $E(X)$ exists, $\Delta_1 = 4 \int_{-\infty}^{\infty} x[F(x) - \frac{1}{2}] dF$.

Proof.

$$\Delta_1 = 2 \int_{-\infty}^{\infty} \int_{-\infty}^{\infty} (x-y) dF(x) dF(y) = 2 \int_{-\infty}^{\infty} dF(x) \int_{-\infty}^{x} (x-y) dF(y)$$

$$= 2 \int_{-\infty}^{\infty} dF(x) \int_{-\infty}^{x} F(y) dy$$
(on integration by parts taking $(x-y)$ as the first factor)

$$= 2 \int_{-\infty}^{\infty} dF(x) \{ [y F(y)]_{-\infty}^{x} - \int_{-\infty}^{x} y dF(y) \}$$

$$= 2 \int_{-\infty}^{\infty} x F(x) [dF(x)] - 2 \int_{-\infty}^{\infty} dF(x) \int_{-\infty}^{x} y dF(y).$$
(6.8.12)

The last term in the right hand side of (6.8.12) is

$$2 \int_{-\infty}^{\infty} y dF(y) \int_{y}^{\infty} dF(x) = 2 \int_{-\infty}^{\infty} y dF(y)(1 - F(y)).$$

Hence

$$\Delta_1 = 2 \int_{-\infty}^{\infty} x dF(x) - 2 \int_{-\infty}^{\infty} x(1 - F(x)) dF(x) = 4 \int_{-\infty}^{\infty} x [F(x) - \frac{1}{2}] dF(x).$$

Lemma 6.8.2: $\Delta_2 = 2\sigma_X^2$.

Proof. Let X be continuous with $E(X) = \mu$.

$$\Delta_2 = \int_{-\infty}^{\infty} \int_{-\infty}^{\infty} [(x-\mu) - (y-\mu)]^2 f(x) f(y) dx dy$$

$$= \int_{-\infty}^{\infty} (x-\mu)^2 f(x) dx + \int_{-\infty}^{\infty} (y-\mu)^2 f(y) dy +$$
$$2 \int_{-\infty}^{\infty} (x-\mu) f(x) dx \int_{-\infty}^{\infty} (y-\mu) f(y) dy = 2\sigma_X^2.$$

Similarly, when X is discrete.

Lemma 6.8.3: If X is a continuous random variable with distribution function $F(x)$,

$$\Delta_1 = 2 \int_{-\infty}^{\infty} F(x)\{1 - F(x)\} dx.$$
(6.8.13)

Proof.

$$\Delta_1 = \int_{-\infty}^{\infty} \int_{-\infty}^{\infty} |x - y| dF(x) dF(y)$$

$$= 2 \int_{-\infty}^{\infty} \int_{y=-\infty}^{x} (x-y) dF(x) dF(y)$$

$$= 2 \int_{-\infty}^{\infty} dF(x) \int_{-\infty}^{x} (x-y) dF(y).$$

Now, on integration by parts, taking $(x - y)$ as the first function,

$$\int_{-\infty}^{x} (x - y)dF(y) = \int_{-\infty}^{x} F(y)dy.$$

Hence,

$$\Delta_1 = 2 \int_{-\infty}^{\infty} dF(x) \int_{-\infty}^{x} F(y)dy = 2 \int_{-\infty}^{\infty} dF(x)\phi(x)$$
$$\text{where } \phi(x) = \int_{-\infty}^{x} F(y)dy$$

$$= 2 \, F(x)\phi(x)]_{-\infty}^{\infty} - 2 \int_{-\infty}^{\infty} F(x)F(x)dx \text{ (integration by parts)}$$

$$= 2 \int_{-\infty}^{\infty} F(x)[1 - F(x)]dx$$

since

$$F(x)\phi(x)]_{-\infty}^{\infty} = \int_{-\infty}^{\infty} F(x)dx.$$

(f) *Coefficient of Variation*: If $E(X)$ and σ_X exist, coefficient of variation (CV) of X is defined as $\frac{\sigma_X}{E(X)} = C_X$ (say) and is taken as a a measure of dispersion. Unlike the previous measures, it has no unit and is often expressed in percentages ($100C_x\%$).

(g) *Gini's Coefficient of Concentration* defined as $G = \frac{\Delta_1}{2E(X)}$ is another measure, often used by the scientists in measuring social inequality in terms of concentration of money, resources, etc.

EXAMPLE 6.8.1: X is a continuous random variable with *pdf* $f(x) = Ce^{-x/\sigma}, x \geq 0, \sigma > 0$. Find its mean, median, variance and quartile deviation.

Since $\int_0^{\infty} f(x)dx = 1, C = \frac{1}{\sigma}$,

$$\mu = \frac{1}{\sigma} \int_0^{\infty} xe^{-x\sigma} dx = -xe^{-x/\sigma}]_0^{\infty} + \int_0^{\infty} e^{-x/\sigma} dx = \sigma.$$

If m be the median

$$\int_0^m f(x)dx = \frac{1}{2}, \text{ i.e. } \frac{1}{\sigma} \int_0^m e^{-x/\sigma} dx = \frac{1}{2}.$$

This gives $e^{m/\sigma} = 2$ or $m = \sigma \log_e 2$. Again,

$$\int_0^{\psi_{1/4}} \frac{1}{\sigma} e^{-x/\sigma} dx = \frac{1}{4}.$$

This gives $\psi_{1/4} = \sigma(\log 4 - \log 3)$. Similarly,

$$\int_0^{\psi_{3/4}} \frac{1}{\sigma} e^{-x\sigma} dx = \frac{3}{4} \text{ gives } \psi_{3/4} = \sigma \log 4.$$

Hence, quartile deviation is $\frac{1}{2}(\psi_{3/4} - \psi_{1/4}) = \frac{\sigma \log 3}{2}$, $\mu_2' = \frac{1}{\sigma} \int_0^{\infty} x^2 e^{-x/\sigma} dx = 2\sigma^2$. Thus, $\mu_2 = \sigma^2$.

6.9 Measures of Skewness and Kurtosis

6.9.1 *Measure of skewness*

By *skewness* we mean the asymmetry of the distribution. A probability density curve (or probability mass histogram) is positively skewed if it has more probability mass to the right of the mode (Fig. 6.9.1); is negatively skewed if more mass is distributed to the left of the mode (Fig. 6.9.2). A measure of skewness should be unit-free and should take positive (negative) values for a positively (negatively) skewed distribution. For a symmetrical distribution, $\mu_3 = 0$; μ_3 is positive (negative) for a positively (negatively) skewed distribution. To make it dimensionless, we divide it by σ^3. Thus

$$\sqrt{\beta_1} = \frac{\mu_3}{\sigma^3} = \frac{\mu_3}{\mu_2^{3/2}} \qquad (6.9.1)$$

or β_1 is often taken as a measure of skewness (if μ_3 exists). For a symmetrical distribution $\beta_1 = 0$. The value of β_1 gives a measure of departure from symmetry, that is, of skewness. It can take any value between $-\infty$ and ∞.

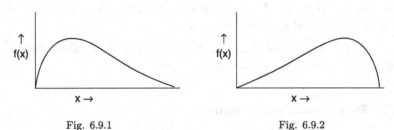

Fig. 6.9.1   Fig. 6.9.2

Other measures are:

(i) $\frac{E(X) - \mu_0}{\sigma}$ (6.9.2)

(ii) $\frac{3(E(X) - \mu_e)}{\sigma}$ (6.9.3)

    (since, mean − mode ≈ 3(mean − median) empirically);

(iii) For a symmetrical distribution $Q_3 - \mu_e = \mu_e - Q_1$. Hence, $(Q_3 - \mu_3) - (\mu_e - Q_1)$ can be taken as a measure of skewness. To make it dimensionless, we divide it by $Q_3 - Q_1$ to obtain

$$\frac{Q_3 - 2Q_2 + Q_1}{Q_3 - Q_1} \qquad (6.9.4)$$

    another measure.

For a symmetrical distribution each measure has value zero. For a positively skewed distribution, $\mu_3 > 0, E(X) > \mu_e > \mu_0; Q_3 - Q_2 > Q_2 - Q_1$ and each measure has positive value. For a negatively skewed distribution, the inequalities are reversed and each measure has a negative value.

Since $|E(X) - \mu_e| \leq E|X - \mu_e| \leq E|X - \mu| \leq \sigma$, (6.9.3) varies between -3 and +3, (6.9.2) and (6.9.4) being approximately equal, have the same limits. For an extremely positively (negatively) skewed distribution, $Q_2 - Q_1 \to 0$ $(Q_3 - Q_2 \to 0)$ and hence (6.9.4) $\to 1(-1)$.

6.9.2 *Measures of kurtosis*

By *kurtosis* we mean the flatness or peakedness of the curve of probability distribution. The quantity

$$\beta_2 - 3 = \frac{\mu_4}{\mu_2^2} - 3 = \gamma \text{ (say)} \tag{6.9.5}$$

is taken as a measure of kurtosis.

For a Normal distribution (Section 9.7), $\beta_2 = 3$ and $\gamma = 0$. If for a distribution $\beta_2 > 3$, the peak of the distribution is sharper than that of a normal curve and it is called a *Leptokurtic* distribution.

If $\beta_2 < 3$, i.e., $\gamma < 0$, the distribution is said to be *Platykurtik*. Here, the peak is flatter than that of the normal curve.

6.10 Some Probability Inequalities

Theorem 6.10.1: Let $g(X)$ be a non-negative Borel measurable function of the random variable X. If $E[g(X)]$ exists, then for every $\epsilon > 0$,

$$P\{g(X) \geq \epsilon\} \leq \frac{E[g(X)]}{\epsilon}. \tag{6.10.1}$$

Proof.

$$E[g(X)] = \int_A g(x)dF(x) + \int_{A^c} g(x)dF(x)$$

where $A = \{x : g(x) \geq \epsilon\}$. Hence,

$$E[g(X)] \geq \int_A g(x)dF(x) \geq \epsilon \int_A dF(x) = \epsilon P\{g(X) \geq \epsilon\}.$$

Hence the theorem.

Corollary 6.10.1.1: If $g(X) = |X|^r$ and $x = t^r, t > 0, r > 0$ and ν_r exists, then

$$P\{|X| \geq t\} \leq \frac{E|X|^r}{t^r}. \qquad (6.10.2)$$

For $r = 1$, we have

$$P\{|X| \geq t\} = \frac{E|X|}{t} \qquad (6.10.3)$$

which is known as *Markov's inequality*.

Taking $g(X) = (X - \mu)^2$, $t = \epsilon^2 \sigma^2$ where $E(X) = \mu, V(X) = \sigma^2$, we have

$$P\{|X - \mu| \geq \epsilon\sigma\} \leq \frac{1}{\epsilon^2} \qquad (6.10.4)$$

which is known as Chebychev's Inequality.

Theorem 6.10.2 (*Chebychev's Inequality*): Let X be a random variable with finite variance $\sigma^2(> 0)$ and $E(X) = \mu$. Then

$$P\{|X - \mu| \geq \epsilon\} \leq \frac{\sigma^2}{\epsilon^2} \qquad (6.10.5)$$

for any $\epsilon > 0$.

Chebychev's inequality gives an upper bound for the probability that X deviates from its mean by more than ϵ units. No assumption about the distribution of X is required except that it has a finite variance. Many probability inequalities have been developed starting from Chebychev's inequality and many sharper bounds for $P(|X - \mu| \geq \epsilon)$ than the one given by (6.10.5) exist. However, in general, the bound given by the Chebychev's inequality can hardly be improved upon.

EXAMPLE 6.10.1: Suppose $X_1, \ldots, X_{n+1}$ are the results of $(n+1)$ Bernoulli trials with $P(X_i = 1) = p, P(X_i = 0) = 1-p, i = 1, \ldots, n+1$. Suppose also that Y_n is equal to the number of i such that $X_i = X_{i+1} = 1, i = 1, \ldots, n$. Using Chebychev's inequality, find an upper bound to

$$P\{|\frac{Y_n}{n} - p^2| \geq \epsilon\}, \ \epsilon > 0.$$

Let $Z_i = 1$ if $X_i = 1, X_{i+1} = 1$ and 0 otherwise. Then $Y_n = Z_1 + \ldots + Z_n$, $E(Z_i) = p^2, V(Z_i) = p^2(1 - p^2)$, Cov $(Z_i, Z_{i+1}) = E(Z_i Z_{i+1}) - p^4 = p^3(1 - p)$, because $E(Z_i Z_{i+1}) = P(Z_i Z_{i+1} = 1) = P(Z_i = 1)P(Z_{i+1} = $

$1|Z_i = 1) = p^2(\frac{p^3}{p^2}) = p^3$. Hence, $E(Y_n) = np^2, V(Y_n) = p^2[n(1 + 3p)(1 - p) + 2p(p - 1)]$. By (6.10.5),

$$P\left(|\frac{Y_n}{n} - p^2| \geq \epsilon\right) \leq \frac{V(Y_n)}{n^2\epsilon^2} = \frac{p^2(1 + 3p)(1 - p)}{n\epsilon^2} + \frac{2p^3(p - 1)}{n^2\epsilon^2}$$

$$< \frac{p^2(1 + 3p)(1 - p)}{n\epsilon^2}.$$

Theorem 6.10.3 (*One-sided Chebychev's Inequality*): If X is a random variable with finite variance $\sigma^2(> 0)$ and $E(X) = \mu$, then

$$P|X \geq \mu + t\sigma] \leq E(\frac{X - \mu - a\sigma}{t\sigma - a\sigma})^2 \tag{6.10.6}$$

for any $t > 0$ and a.

Proof. Define a random variable Y such that

$$Y = \begin{cases} 1 & \text{if } X \geq \mu + t\sigma, \\ 0 & \text{otherwise.} \end{cases}$$

Hence,

$$E(Y) = P(X \geq \mu + t\sigma).$$

Now $(\frac{X-\mu-a\sigma}{t\sigma-a\sigma})^2 \geq Y$ for all values of X a.s. Taking expectation of both sides the result follows.

It can be shown that the equality in (6.10.6) holds when X takes only two values, $\mu + t\sigma$ and $\mu + a\sigma$ with probabilities p and $1 - p$, respectively. The conditions $E(X) = \mu, V(X) = \sigma^2$ give,

$$a = -\frac{1}{t}, p = \frac{1}{1 + t^2}$$

in which case

$$E(\frac{X - \mu - a\sigma}{t\sigma - a\sigma})^2 = p.$$

The inequality (6.10.6) can thus be put in the alternative form

$$P\{X \geq \mu + t\sigma\} \leq \frac{1}{1 + t^2}. \tag{6.10.7}$$

Corollary 6.10.3.1:

$$P[X \leq \mu - t\sigma] = P[-X \geq -\mu + t\sigma] \leq \frac{1}{1 + t^2}. \tag{6.10.8}$$

Corollary 6.10.3.2: The distribution function $F(x)$ of a random variable X with mean μ and variance σ^2 satisfies the inequalities:

$$(i) F(x) \leq \frac{\sigma^2}{\sigma^2 + (x - \mu)^2}, \quad \text{for } x \leq \mu.$$

$$(ii) F(x) \geq \frac{(x - \mu)^2}{\sigma^2 + (x - \mu)^2} \quad \text{for } x \geq \mu.$$

Proof. (i) Put $x = \mu - t\sigma$ in (6.10.8).

(ii) Put $x = \mu + t\sigma$ in (6.10.7). $\qquad\qquad\square$

Both Markov's inequality and Chebychev's inequality are special cases of the following basic inequality.

Theorem 6.10.4: Let X be an arbitrary random variable on the probability space (S, A, P) and g an non-negative Borel function which maps $\mathcal{R} \to \mathcal{R}$. If g is even and is non-decreasing on $[0, \infty)$, then for every $\epsilon > 0$,

$$\frac{E[g(X)] - g(\epsilon)}{a.s. \sup g(X)} \leq P[|X| > \epsilon] \leq \frac{E[g(X)]}{g(\epsilon)}. \qquad (6.10.9)$$

Proof. Since g is a Borel function on $\mathcal{R}$, it follows that $g(X)$ is A-measurable and hence is a random variable. Since g is non-negative, its integral over S exists and

$$E[g(X)] = \int_A g(X)dP + \int_{A^c} g(X)dP \qquad (6.10.10)$$

where $A = \{\omega : |X(\omega)| \geq \epsilon\}$. On A, since g is non-decreasing and even,

$$g(\epsilon) \leq g(X) \leq a.s. \sup g(X). \qquad (6.10.11)$$

Hence

$$g(\epsilon).P(A) \leq \int_A g(X)dP \leq (a.s. \sup g(X)).P(A). \qquad (6.10.12)$$

Since on $A^c, 0 \leq g(X) \leq g(\epsilon)$,

$$0 \leq \int_{A^c} g(X)dP \leq g(\epsilon)P(A^c) \leq g(\epsilon). \qquad (6.10.13)$$

Adding (6.10.12) and (6.10.13),

$$g(\epsilon).P(A) \leq E[g(X)] \leq a.s. \sup g(X).P(A) + g(\epsilon).$$

Hence the result (6.10.9).

Corollary 6.10.4.1:

$$\frac{E|X|^r - \epsilon^r}{a.s.\sup|X|^r} \leq P(|X| \geq \epsilon) \leq \frac{E|X|^r}{\epsilon^r}. \qquad (6.10.14)$$

Proof follows by putting in (6.10.9) $g(x) = |x|^r, r > 0$. The right hand side of (6.10.14) is called the Markov's inequality for $r = 1$ and Chebyshev's inequality for $r = 2$.

The left hand side of (6.10.9) is non-trivial only if $g(X)$ is a bounded random variable. If $g(X)$ is unbounded, a.s. sup $g(X)$ is infinite and the left hand side of (6.10.9) reduces to zero.

Taking

$$g(X) = \frac{|X|^r}{1 + |X|^r}, \quad r > 0 \qquad (6.10.15)$$

when $g(x)$ is even, monotonically increasing with a.s. sup $g(x) = 1$. In this case, the inequality (6.10.9) becomes

$$E[\frac{|X|^r}{1 + |X|^r}] - \frac{\epsilon^r}{1 + \epsilon^r} \leq P[|X| \geq \epsilon] \leq \frac{1 + \epsilon^r}{\epsilon^r} E[\frac{|X|^r}{1 + |X|^r}]. \qquad (6.10.16)$$

6.11   Exercises and Complements

6.1 Examine if for the distribution $P(X = k) = \frac{1}{2^k}, k = 1, 2, \ldots, E(X)$ exists. Find $E(X)$ if it exists.

6.2   Let X be a discrete random variable with $P[X = n] = \frac{1}{2^n}, n = 0, 1, 2, \ldots$. Let $Y = g(X)$ where $g(n) = (-1)^{n+1}\frac{2^n}{n}$. Show that $E(Y)$ does not exist, although the series $\sum_{n=1}^{\infty} g(n)P[X = n]$ is conditionally convergent.

6.3   Show that for the *triangular distribution*

$$dF = \frac{1}{\alpha}[1 - \frac{|x - \beta|}{\alpha}]dx, \quad |x - \beta| > \alpha$$

the mean is β and variance is $\frac{\alpha^2}{6}$.

6.4 Find the mean, variance, mean deviation about mean of the random variable X with *pdf*

(a) $f(x) = \frac{1}{2}e^{-|x|}, \; (-\infty < x < \infty)$;

(b)

$$f(x) = \begin{cases} \frac{1}{2b} & \text{if } |x - a| \le b \\ 0 & \text{otherwise} . \end{cases}$$

$[(a)\mu = 0, \mu_2 = 2; 1(b)\mu = a, \mu_2 = \frac{b^2}{3}; \frac{b}{2}]$

6.5 For the continuous distribution

$$dF = \frac{3}{4}x(2 - x)dx, \ 0 \le x \le 2$$

find the first four moments, mean deviation about mean, skewness and kurtosis.

$[\mu = 1, \mu_2 = \frac{1}{5}, \mu_3 = 0, \mu_4 = \frac{3}{35}, \text{m.d.} = \frac{3}{8}, \beta_1 = 0, \beta_2 = -\frac{6}{7}.]$

6.6 X has a continuous distribution with

$$dF = k(x - x^2)dx, \ 0 \le x \le 1,$$

k being a constant. Find the arithmetic mean, harmonic mean, mode and median of the distribution.

$[k = 6, \text{a.m.} = \frac{1}{2}, \text{h.m.} = \frac{1}{3}, \text{mode} = \frac{1}{2}, \text{median} = \frac{1}{2}.]$

6.7 A continuous random variable X has the distribution function

$$F(x) = 1 - e^{\alpha \tan x}, \ 0 \le x \le \frac{\pi}{2}, \ 0 < \alpha < 1.$$

Find the *pdf* of X and show that mode of the distribution is $\mu_0 = \frac{1}{2}(\pi - \sin^{-1}\alpha)$.

6.8 Show that for the *rectangular distribution*

$$dF = dx, \ 0 \le x \le 1,$$

$\mu'_1 = \frac{1}{2}, \mu_2 = \frac{1}{12}$, mean deviation $= \frac{1}{4}, \Delta_1 = \frac{1}{3}$. Show also that $E(\frac{1}{X})$ does not exist.

6.9 A continuous random variable X has the *pdf* proportional to

$$(\alpha + |x|)e^{-\beta|x|}, \ -\infty < x < \infty$$

where α, β are constants. Find β_2, the coefficient of kurtosis of X and show that $\frac{1}{3} < \beta_2 < 3$, irrespective of values of α and β.

6.10 For a random variable X with *pdf*

$$f(x) = \begin{cases} x/2 & 0 \le x < 1, \\ 1/2, & 1 \le x < 2, \\ (3-x)/2, & 2 \le x < 3, \\ 0 & x \ge 3 \end{cases}$$

show that moments of all orders exist. Find the mean and variance of X.

6.11 Obtain the ratio, (quartile deviation)/(standard deviation) for the distribution with *pdf*

$$f(x) = \begin{cases} 2(b+x)/b(a+b), & -b \le x < 0, \\ 2(a-x)/a(a+b), & 0 \le x < a, \\ 0 & x \ge a. \end{cases}$$

6.12 Let X be a random variable with *Pareto distribution*

$$f(x) = \frac{ak^a}{x^{a+1}}, \quad a > 0, x \ge k > 0.$$

Show that the distribution function is

$$F(x) = 1 - \left(\frac{k}{x}\right)^a;$$

$E|X|^r < \infty$ for $r < a$ and $\mu_r' = \frac{ak^r}{a-r}; \mu = \frac{ak}{a-1}$ (if $a > 1$) and $V(X) = \frac{ak^2}{(a-1)(a-2)}$ (if $a > 2$), the mean deviation is $\frac{2k}{a-1}(1-a^{-1})^{a-1}$; the geometric mean is $G = ke^{1/a}$; the harmonic mean is $H = k(1 + \frac{1}{a})$; median is $m = k2^{1/a}$.

6.13 Show that for the *Raleigh density*

$$f(x) = \frac{x}{\alpha^2}e^{-\frac{x^2}{2\alpha^2}}, \quad x > 0,$$

$$E[X^n] = \begin{cases} \sqrt{\frac{\pi}{2}}1.3.\ldots n\alpha^n & \text{for } n \text{ odd}, \\ 2^k k! \alpha^{2k} & \text{for } n = 2k. \end{cases}$$

In particular,

$$E[X] = \alpha\sqrt{\frac{\pi}{2}}, \quad E(X^2) = 2\alpha^2, \quad \sigma_X^2 = (2 - \frac{\pi}{2})\alpha^2.$$

6.14 Show that for the *Maxwell density*

$$f(x) = \frac{\sqrt{2}}{\alpha^3\sqrt{\pi}}x^2 e^{-\frac{x^2}{2\alpha^2}}, \quad x > 0,$$

$$E[X^n] = \begin{cases} 1.3.\dots(n+1)\alpha^n & \text{for } n \text{ even} \\ \sqrt{\frac{2}{\pi}} 2^k k! \alpha^{2k-1} & \text{for } n = 2k-1. \end{cases}$$

In particular,

$$E[X] = 2\alpha\sqrt{\frac{2}{\pi}}, \quad E[X^2] = 3\alpha^2.$$

6.15 For the distribution in Example 6.3.2, show that when $\alpha = 2\beta$, the median is $(1 - \frac{1}{\sqrt{2}})^{2/\alpha}$.

6.16 Find the skewness of the following distributions:

(i) $f(x) = \alpha x^{\alpha-1}$, $0 \le x \le 1; = 0$ otherwise $(\alpha > 1)$;
(ii) $f(x) = x(1-x)^{\alpha-1}$; $0 \le x \le 1; = 0$ otherwise $(\alpha > 1)$;
(iii) $f(x) = 1, 0 \le x \le 1; = 0$ otherwise.

6.17 For a continuous random variable X the distribution function is

$$F(x) = 1 - e^{-\beta x^2}, \quad \beta > 0, \ 0 < x < \infty.$$

Find the median of the distribution. Also, if m, m_0, σ denote, respectively, the mean, mode and the standard deviation of the distribution, show that

$$2m_0^2 - m^2 = \sigma^2 \text{ and } m_0 = m\sqrt{\frac{2}{\pi}}.$$

$$\left[\text{median} = \sqrt{\frac{\log_e 2}{\beta}}, \ m = \frac{1}{2}\sqrt{\frac{2}{\beta}}, \ \sigma^2 = \frac{4-\pi}{4\beta}, \ m_0 = \frac{1}{\sqrt{2\beta}}.\right]$$

6.18 For the *Poisson* random variable with *pmf*

$$P(X = x) = e^{-\lambda}\frac{\lambda^x}{x!}, \quad x = 0, 1, 2, \dots$$

show that $E(X) = \lambda, E(X^2) = \lambda + \lambda^2, E(X^3) = \lambda + 3\lambda^2 + \lambda^3, E(X^4) = \lambda + 7\lambda^2 + 6\lambda^3 + \lambda^4, \mu_2 = \lambda = \mu_3, \mu_4 = \lambda + 3\lambda^2$.

Also find expectation of kXe^{-kX} where k is a constant.

$$\left[E(.) = \sum_0^\infty (kxe^{-kx})(e^{-\lambda}\frac{\lambda^x}{x!}) = \lambda k \exp\{\lambda(e^{-k}-1) - k\}.\right]$$

6.19 For the distribution

$$F(x) = \begin{cases} 0 & \text{if } x < -1 \\ a + b \text{ arc sin } x & \text{if } -1 \le x \le 1, \\ 1 & \text{if } x > 1, \end{cases}$$

find $E(X), V(X)$.

$[E(X) = 0, V(X) = \frac{1}{2}.]$

6.20 A random variable X has the *pdf*

$$f(x) = k.e^{-b(x-a)}, \ a \le x < \infty,$$

k, a, b being constants. Show that $k = b = \frac{1}{\sigma}, a = m - \sigma$ where m, σ, respectively are the mean and standard deviation of the distribution. Find β_1, β_2 of the distribution.

$[\beta_1 = 4, \beta_2 = 9.]$

6.21 For the distribution

$$dF = k(1 + \frac{x}{\alpha})^{m-1}e^{-\frac{mx}{\alpha}} dx, \ -\alpha \le x < \infty,$$

find the constant k and first four moments. Show that $8\beta_2 = 4 + 9\beta_1$.

$\left[k = m^m e^{-m}/\alpha\Gamma(m), \ \mu_1' = \alpha, \ \mu_2 = \frac{\alpha^2}{2m}, \ \mu_3 = \frac{\alpha^3}{3m^2}, \ \mu_4 = \frac{\alpha^4(m+2)}{8m^2}. \right]$

6.22 For the distribution

$$dF = k(1 - \frac{x^2}{a^2})^{-h}dx, \ 0 < h < 1, \ |x| < a,$$

k, a constant, show that

$$\mu_{r+1}' = \frac{ra^2}{r+2-2h}\mu_{r-1}'.$$

[Integrate by parts.]

6.23 For the curve

$$y = y_0 x^r (1 - x)^s, \ o \le x \le 1,$$

find skewness Sk = (mean - median)/ s.d. and $\sqrt{\beta_1}$.

$\left[S_k = \frac{s-r}{s+r}\sqrt{\frac{r+s+3}{(r+1)(s+1)}}, \ \sqrt{\beta_1} = \frac{2(s-r)}{r+s+!}\sqrt{\frac{r+s+3}{(r+1)(s+1)}}. \right]$

6.24 For the distribution

$$dF = \frac{x}{\sigma^2}e^{-\frac{x^2}{2\sigma^2}} dx, \ x > 0$$

find $Q_3 - Q_1$ and $(Q_3 - Q_1)/s.d.$

$\left[\sigma\sqrt{2}[\sqrt{\log 4} - \sqrt{\log \frac{4}{3}}], \ (\sqrt{\log 4} - \sqrt{\log \frac{4}{3}})/\sqrt{1 - \frac{\pi}{2}} \right].$

6.25 Prove that if $y = g(x)$ is a monotone function and μ_x is the median of X, then $g(\mu_e)$ is median of $Y = g(X)$.

6.26 For a continuous random variable X defined in the range of $a \le x \le b$ and having *pdf* proportional to $g(x)$, prove that μ_e, the median of the distribution satisfies the relation,

$$2 \int_0^{\mu_e} g(x)dx = \int_a^b g(x)dx.$$

6.27 Show that if m be the mode of X, then $a + bm (b \ne 0)$ is the mode of $Y = a + bX$.

6.28 If for a random variable X, μ_{2r+2} exists, show that

$$(\mu_{2r+1})^2 \ge \mu_{2r}\mu_{2r+2}.$$

[In Cauchy-Schwartz inequality $E(Y^2)E(Z^2) \ge (E(YZ))^2$, put $Y = (X - \mu)^r, Z = (X - \mu)^{r+1}$ where $\mu = E(X)$.]

6.29 Let X be a random variable satisfying

$$\frac{P(|X| > \alpha k)}{P(|X| > k)} \to 0 \text{ as } k \to \infty \text{ for } \alpha > 1.$$

Show that

$$E|X|^r < \infty \text{ for } r > 0.$$

6.30 Suppose X is a non-negative random variable whose mean (μ) exists. Prove that

(a) $\lim_{t \to \infty} t[1 - F(t)] = 0$

and hence in case X is absolutely continuous

(b) $\mu = \int_0^\infty [1 - F(x)]dx$.

[See Theorem 6.3.7.]

6.31 Show that a random variable X possesses moments of all orders, *iff*

$$\lim_{n \to \infty} \sup[P\{|X| > \alpha^n\}]^{1/n} = 0 \text{ for } \alpha > 1.$$

(Laha and Rohatgi, 1979)

6.32 Let X be a random variable and $0 < r < 1$. Suppose that $nP[|X| > n^{1/r}] \to 0$ as $n \to \infty$. Show that

$$n^{1-\frac{1}{r}} \int_{|x|>n^{1/r}} x\, dP\{X \le x\} \to 0 \text{ as } n \to \infty.$$

(Laha and Rohatgi, 1979)

6.33 Show that for any two random variables $X(.)$ and $Y(.)$ whose squares are integrable, $|E|XY|| \le \max\{E[|X|^2], E[|Y|^2]\}$.

6.34 Let X have a Poisson distribution with parameter λ (vide Exercise 6.18). Prove that

(i) $P(X \le \frac{\lambda}{2}) \le \frac{4}{\lambda}$;

(ii) $P(X \ge 2\lambda) \le \frac{1}{\lambda}$.

6.35 For the *geometric distribution* $P(X = n) = p_n = 2^{-n}, n = 1, 2, \ldots,$ show that Chebychev's inequality gives $P[|X - 2| > 2] < \frac{1}{2}$ while the actual probability is $\frac{1}{16}$.

6.36 If $X_1, \ldots, X_n$ are independently and identically distributed random variables having zero means and unit variances, show that

$$P(\sum_{i=1}^{n} X_i^2 \ge n\lambda) \le \frac{1}{n} \text{ for } \lambda > 0.$$

[Apply Markov's inequality.]

6.37 Let X be a random variable with zero mean and finite variance σ^2. Let also $E|X|^4 < \infty$. Then for $k > 1$, show that

$$P\{|X| \ge k\sigma\} \ge \frac{\mu_4 - \sigma^4}{\mu_4 + \sigma^4 k^4 - 2k^2\sigma^4}$$

where $\mu_4 = EX^4$.

6.38 Show that for any distribution $\int_{-\infty}^{\infty}(1 - \frac{x^2}{t^2})dF(X) \le \int_{-t}^{t} dF(x)$ and hence deduce that $P[|X - E(X)| > k\sigma] \le \frac{1}{k^2}$ when $k > 0$ and $\sigma^2 = V(X)$.

$$[\int_{-\infty}^{\infty}(1 - \frac{x^2}{t^2})dF(x) = \int_{-t}^{t}(1 - \frac{x^2}{t^2})dF(x) + \int_{-t}^{t}(1 - \frac{x^2}{t^2})dF(x)$$
$$+ \int_{t}^{\infty}(1 - \frac{x^2}{t^2})dF(x)$$
$$\le \int_{-t}^{t}(1 - \frac{x^2}{t^2})dF(x) \le \int_{-t}^{t} dF(x).$$

Now,

$$P(|X| > k\sigma) = 1 - \int_{-k\sigma}^{k\sigma} dF(x) \leq 1 - \int_{-\infty}^{\infty}(1 - \tfrac{x^2}{k^2\sigma^2})dF(x)$$
$$= \int_{-\infty}^{\infty} \tfrac{x^2}{k^2\sigma^2} dF(x) = \tfrac{1}{k^2}$$

when the origin is at the mean.]

6.39 Show that if $M(X)$ is an increasing function of $|X|$, then $P[|X|. > a] \leq \frac{E[M(X)]}{M(a)}$, provided $E[M(X)]$ exists.

[Define a random variable Y such that $Y = 1(0)$ if $|X| > a$ (otherwise). Hence, $E(Y) = P(|X| > a)$. Now, $M(X) \geq YM(a)$. The result follows by taking expectation.]

6.40 Let X be such that $E(e^{aX})$ exists, $a > 0$. Show that for any t,

$$P(X > t) < \frac{E(e^{aX})}{e^{at}}.$$

[Proceed as in exercise 6.39 since e^{ax} is a non-negative function of x.]

6.41 Suppose $E(X^2) < \infty$.

(i) Show that for any $a > 0$ and $x > 0, P(X \geq x) \leq \frac{E(X+a)^2}{(x+a)^2}$.

(ii) Show that if $E(X) = 0$, then $P(X \geq x) \leq \frac{E(X^2)}{E(X^2)+x^2}$.

[Hints: To prove (ii), minimize the last expression in (i) with respect to a.]

6.42 Let X, Y be independent random variables and assume that $E(X) = 0$. Show that $E|X - Y| \geq E|Y|$.

6.43 Let g be a non-negative non-decreasing function. Prove that if $E[g(|X - \mu|)]$, where $\mu = E(X)$, exists then

$$P(|X - \mu| > t) < \frac{E[g(|X - \mu|)]}{g(t)}.$$

[Proceed as in Exercise 6.39.]

6.44 (a) Let X be a non-negative random variable. Considering the discontinuous *Heaviside function* $H(X - a)$ where

$$H(Y) = \begin{cases} 1 & \text{for } Y > 0, \\ 0 & \text{for } Y \leq 0, \end{cases}$$

$'a'$ being a positive constant, prove Markov's inequality $P(X > a) < \frac{E(X)}{a}$.

(b) Let X be a random variable with finite $E(X^2)$. Considering the inequality

$$H(|X - b| - a) \leq (\frac{X - b}{a})^2$$

where $a(> 0), b$ are arbitrary constants, show that

$$P(|X - b| > a) \leq \frac{E(X - b)^2}{a^2}.$$

[(a) $H(X - a) = 1(0)$ if $\frac{X}{a} > 1$ (otherwise). Hence $H(X - a) \leq \frac{X}{a}(X \geq 0)$. Thus, $E[H(X - a)] = P(X > a) \leq \frac{E(X)}{a}$.

(b) $H(|X - b| - a) = 1(0)$ if $\frac{|X - b|}{a} > 1$ (otherwise). Hence, $H(|X - b| - a) \leq \frac{(X - b)^2}{a^2}$. The result follows by taking expectation.]

6.45 (*Bernstein's Inequality*): Let $M(r) = E(e^{r(X - \mu)}), \mu = EX, r > 0$. Then show that

$$P[X \geq \mu + \frac{t + \log M(r)}{r}] \leq e^{-t}, \ t > 0.$$

6.46 For any real random variable X with finite expectation, show that

$$\sum_{n=1}^{\infty} P(|X| \geq n) \leq E(|X|) < 1 + \sum_{n=1}^{\infty} P(|X| \geq n).$$

[Let $Y = |X|$ and suppose that G is the distribution function of Y.

$$E(Y) = \int_0^{\infty} y dG(y) = \sum_{i=0}^{\infty} \int_i^{(i+1)-} y dG(y).$$

Now,

$$i \int_i^{(i+1)-} dG(y) \leq \int_i^{(i+1)-} y dG(y) \leq (i + 1) \int_i^{(i+1)-} dG(y),$$

i.e. $iP[i \leq Y < i + 1] \leq \int_i^{(i+1)-} y dG(y) \leq (i + 1)P[i \leq Y < i + 1]$.

Summing over all i,

$$\sum_{i=0}^{\infty} iP[i \leq Y < i + 1] \leq E(Y) \leq \sum_{i=0}^{\infty} (i + 1)P[i \leq Y < i + 1].$$

Now,

$$\sum_{i=0}^{\infty} iP[i \leq Y < i + 1] = \sum_{n=1}^{\infty} P(|X| \geq n),$$

$$\sum_{i=0}^{\infty} (i + 1)P[i \leq Y < i + 1] = 1 + \sum_{n=1}^{\infty} P(|X| \geq n)].$$

Appendix 6.A

Some Inequalities

1. *Cauchy-Schwarz Inequality:* Let $(x_1, \ldots, x_n), (y_1, \ldots, y_n)$ be two sets of real numbers. Then

$$\sum_{i=1}^{n} x_i^2 \sum_{i=1}^{n} y_i^2 \geq \left(\sum_{i=1}^{n} x_i y_i \right)^2, \qquad (6.A.1)$$

equality holds when and only when $x_i \propto y_i, i = 1, \ldots, n$.

Proof. Consider the quadratic form in two real variables a, b,

$$\sum_{i=1}^{n} (ax_i + by_i)^2 = a^2 \sum_{i=1}^{n} x_i^2 + 2ab \sum_{i=1}^{n} x_i y_i + b^2 \sum_{i=1}^{n} y_i^2$$

$$= [a \ b] \begin{bmatrix} \sum_{i=1}^{n} x_i^2 & \sum_{i=1}^{n} x_i y_i \\ \sum_{i=1}^{n} x_i y_i & \sum_{i=1}^{n} y_i^2 \end{bmatrix} \begin{bmatrix} a \\ b \end{bmatrix}$$

$$= [a \ b] \Delta \begin{bmatrix} a \\ b \end{bmatrix} \text{ (say)}$$

which is always ≥ 0 for all values of a, b and hence has a non-negative determinant Δ. The equality is attained when $\Delta = 0$, which means there exist real numbers a, b such that $ax_i + by_i = 0 \ \forall \ i = 1, 2, \ldots, n$, that is $x_i \propto y_i, i = 1, \ldots, n$.

(6.A.1) also applies to infinite sums, i.e., $\sum x_i^2 \sum y_i^2 \geq (\sum x_i y_i)^2$ and also to complex numbers, i.e., $\sum |x_i|^2 \sum |y_i|^2 \geq \sum |x_i y_i|^2$ where x_i, y_i are arbitrary complex numbers and $|t|$ denotes modulus of complex number t.

Inequality (6.A.1) also applies to integrals. Let $f(x)$ and $g(y)$ be two arbitrary real functions of x and y respectively. Then

$$\int f^2(x)dx \int g^2(y)dy \geq \{\int fgdxdy\}^2. \tag{6.A.2}$$

2. *Holder's Inequality:* Let $(x_1, \ldots, x_n), (y_1, \ldots, y_n)$ be two sets of non-negative real numbers. Then

$$(\sum_{i=1}^{n} x_i^p)^{\frac{1}{p}}(\sum_{i=1}^{n} y_i^2)^{\frac{1}{q}} \geq \sum_{i=1}^{n} x_iy_i \tag{6.A.3}$$

where $\frac{1}{p} + \frac{1}{q} = 1, p > 1$. The equality holds *iff* $y_i \propto x_i^{p-1}$.

(6.A.3) also applies to integrals. Let $f(x), g(y)$ be two arbitrary real non-negative functions of x, y respectively. Then

$$(\int f^p dx)^{\frac{1}{p}}(\int g^q dy)^{\frac{1}{q}} \geq \int fgdxdy \tag{6.A.4}$$

where $\frac{1}{p} + \frac{1}{q} = 1, p > 1$.

More generally, for sets of non-negative real numbers $(x_1, \ldots, x_n), (y_1, \ldots, y_n), (z_1, \ldots, z_n), \ldots,$

$$(\sum x_i^p)^{\frac{1}{p}}(\sum y_i^q)^{\frac{1}{q}}(\sum z_i^r)^{\frac{1}{r}} \cdots \geq \sum x_iy_iz_i \cdots \tag{6.A.5}$$

if $\frac{1}{p} + \frac{1}{q} + \frac{1}{r} + \ldots = 1$.

3. *Minkowski's Inequality:* For two sets of non-negative real numbers $(x_1, \ldots, x_n), (y_1, \ldots, y_n)$ and $p \geq 1$,

$$(\sum x_i^p)^{\frac{1}{p}} + (\sum y_i^p)^{\frac{1}{p}} \geq [\sum (x_i + y_i)^p]^{\frac{1}{p}}. \tag{6.A.6}$$

Chapter 7

Generating Functions

7.1 Introduction

We have seen that moments, factorial moments, cumulants are important properties of probability distributions. We consider different generating functions of these quantities in this chapter. Section 7.2 addresses the probability generating function of a discrete random variable. The moment generating function is considered in Section 7.3, factorial moment generating function in Section 7.4, cumulant generating function in Section 7.5 and characteristic functions in Section 7.6. Standard discrete and continuous distributions will be considered in the next two chapters.

7.2 Probability Generating Function

Let X be an integer-valued discrete random variable with $P(X = k) = p_k, k = 0, 1, 2, \ldots$ and $\sum_{k=0}^{\infty} p_k = 1$.

DEFINITION 7.2.1: The function defined by

$$P_X(t) = E(t^X) = \sum_{k=0}^{\infty} p_k t^k, \ |t| < 1 \tag{7.2.1}$$

is called the *probability generating function* (*pgf*) of X. Since $P_X(1) = 1$, series (7.2.1) is uniformly and absolutely convergent in $|t| \leq 1$ and $P_X(t)$ is a continuous function of t.

Every *pgf* determines a unique probability distribution, that is, a unique set of probabilities $\{p_k\}$. Coefficients of t^k in the expansion of $P_X(t)$ gives $P(X = k)$. The *pgf* is used for discrete variables only.

EXAMPLE 7.2.1: X is uniformly distributed over $0, 1, \ldots, n$ with $P(X =$

$k) = \frac{1}{n+1}, k = 0, 1, \ldots, n.$

$$P_X(t) = \frac{1}{n+1} \sum_{k=0}^{n} t^k = \frac{1}{(1-t)(n+1)}[1 - t^{n+1}], \ |t| < 1.$$

EXAMPLE 7.2.2: X has a Binomial distribution, $P(X = k) = \binom{n}{k}p^k(1 - p)^{n-k}$, $0 < p < 1, k = 0, 1, \ldots, n.$ Here

$$P_X(t) = \sum_{k=0}^{n} t^k \binom{n}{k} p^k (1 - p)^{n-k} = (1 + p(t - 1))^n.$$

EXAMPLE 7.2.3: X has a Poisson distribution, $P(X = k) = e^{-\lambda}\frac{\lambda^k}{k!}$, $k = 0, 1, 2, \ldots$

$$P_X(t) = \sum_{k=0}^{\infty} (\lambda t)^k \frac{e^{-\lambda}}{k!} = e^{-\lambda(1-t)}, \ |t| < 1.$$

EXAMPLE 7.2.4: X has a Geometric distribution, $P(X = k) = pq^k, k = 0, 1, \ldots; 0 < p < 1; q = 1 - p.$

$$P_X(t) = \sum_{k=0}^{\infty} t^k pq^k = \frac{p}{1 - tq}, |t| < 1.$$

EXAMPLE 7.2.5: X has a Negative Binomial distribution with parameters $n, p, : P(X = x) = \binom{-n}{x}p^n(p - 1)^x, x = 0, 1, 2, \ldots$

$$P_X(t) = \sum_{x=0}^{\infty} t^x \binom{-n}{x} p^n (p - 1)^x = p^n \sum_{x=0}^{\infty} \binom{-n}{x} [t(p - 1)]^x$$

$$= p^n[1 + t(p - 1)]^{-n}, \quad \text{provided } |t(p - 1)| < 1.$$

Note that when $n = 1$, the *pgf* becomes identical with the *pgf* of a Geometric distribution with parameter p (Example 7.2.4). Hence, if $X_1, X_2, \ldots, X_n$ are independently distributed random variables, each having a Geometric distribution with parameter p, their sum $X_1 + \ldots + X_n$ follows a negative binomial distribution with parameters n, p (follows by Theorem 7.2.2. Also, see Example 7.2.6).

Theorem 7.2.1: Let X be an integer-valued random variable. Consider $Y = a + bX, a, b$ being arbitrary constants. The *pgf* of Y is

$$P_Y(t) = E(t^{aX+b}) = t^b P_{aX}(t). \tag{7.2.2}$$

Theorem 7.2.2: Let $X_1, X_2, \ldots, X_n$ be n independent random variables with *pgf*s $P_1(t), P_2(t), \ldots, P_n(t)$, respectively. The *pgf* of the sum $S_n = X_1 + \ldots + X_n$ is

$$P_{S_n}(t) = P_1(t) \ldots P_n(t). \qquad (7.2.3)$$

Proof. $P_{S_n}(t) = E(t^{S_n}) = E(t^{X_1}) \ldots E(t^{X_n}) = P_1(t) \ldots P_n(t), |t| < 1$, since $t^{X_1}, \ldots, t^{X_n}$ are independent random variables, $X_1, \ldots, X_n$ being independent random variables.

EXAMPLE 7.2.6: Independent Bernoullian trials with probability of success p are performed $0 < p < 1$. Let X_k denote the number of failures following the $(k-1)$th success and preceding the kth success. Then $S_n = X_1 + \ldots + X_n$ is the number of failures preceding the nth success. Here, $X_1, \ldots, X_n$ are independently and identically distributed random variables, each X_i having the Geometric distribution mentioned in Example 7.2.4. Hence, $P_{S_n}(t) = (\frac{p}{1-tq})^n, |t| < 1$. Coefficient of t^k is $\binom{-n}{k} p^k (-q)^k, k = 0, 1, \ldots$ which is the probability of k failures preceding the nth success (vide Section 8.8). $\square$

We now consider the *pgf* of a sum of indicator random variables.

Theorem 7.2.3: Suppose $I = I_1 + \ldots + I_n$ where each I_i is an indicator random variable. Let A_i be the event for which I_i is an indicator random variable, i.e., $P(I_i = 1) = P(A_i)$. The *pgf* of I is

$$P_I(t) = 1 + (t-1)B_1 + (t-1)^2 B_2 + \ldots + (t-1)^n B_n, \quad |t| < 1 \qquad (7.2.4)$$

where

$$B_k = \sum \ldots \sum_{i_1 < \ldots < i_k = 1}^{n} P(A_{i_1} \ldots A_{i_k}).$$

Proof. Consider the random function

$$Y = [1 + (t-1)I_1] \ldots [1 + (t-1)I_n]. \qquad (7.2.5)$$

Suppose for a sample point $\omega, I(\omega) = k$. It is easy to verify that for this point $Y(\omega) = t^k$. Hence, $P[I = k] = P[Y = t^k], k = 0, 1, \ldots, n$. Thus

$$\begin{aligned} E(Y) &= t^0 P(Y = t^0) + tP(Y = t) + t^2 P(Y = t^2) + \ldots + t^n P(Y = t^n) \\ &= t^0 P(I = 0) + tP(I = 1) + t^2 P(I = 2) + \ldots + t^n P(I = n) = P_I(t). \end{aligned} \qquad (7.2.6)$$

Again, from (7.2.5),

$$Y = 1 + (t-1) \sum_i I_i + (t-1)^2 \sum \sum_{i<j} I_i I_j + \ldots (t-1)^n I_1 \ldots I_n.$$

Hence,

$$E(Y) = 1 + (t-1)B_1 + (t-1)^2 B_2 + \ldots + (t-1)^n B_n. \qquad (7.2.7)$$

Combining (7.2.6) and (7.2.7), the result follows.

EXAMPLE 7.2.7: A unbiased coin is tossed n times. Find the *pgf* of the number of heads.

Let $X_i = 1(0)$ if the ith toss results in a head (tail). Then $X = X_1 + \ldots + X_n$ = total number of heads. Here

$$B_k = \text{ Probability of getting k head } = \binom{n}{k}\frac{1}{2^n}.$$

$$P_X(t) = \sum_{k=0}^{n} \binom{n}{k}\frac{(t-1)^k}{2^n} = (\frac{1+t}{2})^n, \ |t| < 1.$$

EXAMPLE 7.2.8: Find the *pgf* of the total score obtained when n balanced dice are rolled.

Let X_i denote the score on the ith dice $(i = 1, \ldots, n)$, X_i has a uniform distribution over $1, \ldots, 6$ and its *pgf* is

$$P_i(t) = \frac{t(1-t^6)}{6(1-t)}.$$

Hence, *pgf* of $S_n = \sum_{i=1}^{n} X_i$ is

$$P(t) = [\frac{t(1-t^6)}{6(1-t)}]^n = (\frac{t}{6})^n (1-t^6)^n (1-t)^{-n}$$

$$= (\frac{t}{6})^n \{\sum_{i=0}^{n} \binom{n}{i}(-1)^i t^{6i}\}\{\sum_{j=0}^{\infty} \binom{-n}{j}(-1)^j t^j\}$$

$$= (\frac{t}{6})^n \sum_{i=0}^{n} \sum_{j=0}^{n} \binom{n}{i}\binom{-n}{j}(-1)^{i+j} t^{6i+j}$$

$$= (\frac{t}{6})^n \sum_{k=0}^{\infty} t^k \sum_{i} \binom{n}{i}\binom{-n}{k-6i}(-1)^{i+k}$$

where $k = j + 6i$. Prob. $(S_n = n + k) = $ coefficient of t^{n+k} in $P(t)$

$$= \frac{1}{6^n} \sum_{i} \binom{n}{i}\binom{-n}{k-6i}(-1)^{i+k} \qquad (7.2.8)$$

where the ith term in the summand in (7.2.8) is zero whenever $i > n$ or $6i > k$.

EXAMPLE 7.2.9: Let X be a random variable with $P(X = k) = p_k, k = 1, 2, \ldots$ Let $P(X > k) = q_k, k = 0, 1, \ldots$ and $Q(t) = \sum_{j=0}^{\infty} q_j t^j$. Show that the *pgf* of the tail probabilities q_k is

$$Q(t) = \frac{1 - P(t)}{1 - t}, \ |t| < 1 \tag{7.2.9}$$

where $P(t)$ is the *pgf* of X. Show also that

$$\lim_{t \uparrow 1} \frac{1 - P(t)}{1 - t} = E(X) \tag{7.2.10}$$

where $t \uparrow 1$ means t approaches 1 from left.

Here

$$\begin{aligned}
Q(t) &= 1 + q_1 t + q_2 t^2 + \ldots = 1.t^0 + (1 - p_1)t + (1 - p_1 - p_2)t^2 + \ldots \\
&= (1 + t + t^2 + \ldots) - p_1 t(1 + t + t^2 + \ldots) - p_2 t^2(1 + t + t^2 + \ldots) - \\
&= \frac{1 - P(t)}{1 - t}.
\end{aligned}$$

Hence, $\sum_{j=0}^{\infty} t^j P(X > j) = \frac{1 - P(t)}{1 - t}$. Taking limit of both sides as $t \uparrow 1$, (7.2.9) follows from Lemma 5.3.1. □

Since the power series $\sum_{k=0}^{\infty} p_k t^k$ converges absolutely when $|t| = 1$, its radius of convergence is ≤ 1, and inside the unit disk (i.e., for $|t| < 1$), one can differentiate term by term, at any order. Thus, if a distribution is specified by its *pgf*, the factorial moments (about origin) can be obtained as follows:

$$\frac{\partial^r P_X(t)}{\partial t^r} = \sum_{x} x(x - 1) \ldots (x - r + 1)t^{x-r} p_x. \tag{7.2.11}$$

Hence, $\left. \frac{\partial^r P_X(t)}{\partial t^r} \right]_{t=1} = \mu_{(r)}$.

EXAMPLE 7.2.10: In Example 7.2.4, $P'(t) = \frac{pq}{(1-tq)^2}, P''(t) = \frac{2pq^2}{(1-tq)^3}$. Hence,

$$E(X) = \frac{q}{p}, E(X^2) = \frac{q}{p} + \frac{2pq^2}{p^3}$$

when $V(X) = \frac{q}{p^2}$.

In Example 7.2.3, $P'(t) = \lambda e^{-\lambda(1-t)}, P''(t) = \lambda^2 e^{-\lambda(1-t)}$. Hence, $E(X) = \lambda, E\{X(X - 1)\} = \lambda^2$ when $V(X) = \lambda$.

Theorem 7.2.4: Suppose X is a non-negative, integer-valued variable and its moments of all orders exist. Then

$$P_X(t) = \sum_{k=0}^{\infty} [E\{(X)_k\}] \frac{(t - 1)^k}{k!}, \ |t| \leq 1. \tag{7.2.12}$$

Proof. We have $P_X(t) = p_0 + p_1 t + p_2 t^2 + \ldots$ Now

$$t^i = (t - 1 + 1)^i = \sum_{k=0}^{i} \binom{i}{k}(t-1)^k.$$

Hence,

$$P_X(t) = \sum_{i=0}^{\infty} p_i \sum_{k=0}^{i} \binom{i}{k}(t-1)^k = \sum_{k=0}^{\infty}(t-1)^k \sum_{i=k}^{\infty} \binom{i}{k}p_i.$$

The rearrangement in the order of summation is valid, because (7.2.12) is absolutely convergent for $|t| \leq 1$. Since, $\sum_{i=k}^{\infty} \binom{i}{k}p_i = \frac{1}{k!}E\{(X)_k\}$, the result follows. $\qquad\square$

The multivariate *pgf* will be considered in Section 10.4.

7.3 Moment Generating Function

DEFINITION 7.3.1: Let X be a random variable defined on $(\mathcal{S}, \mathcal{A}, P)$. The function

$$M_X(t) = E(e^{tX}) \qquad (7.3.1)$$

defined for all real values of t is known as the *moment generating function* (*mgf*) of X, if the expectation of e^{tX} exists for some $|t| < T$. For a discrete random variable with *pmf* $P(X = x_k) = p_k, k = 0, 1, 2, \ldots$

$$M_X(t) = \sum_{k=0}^{\infty} e^{tx_k} p_k. \qquad (7.3.2)$$

In the case of a continuous probability distribution, specified by a *pdf* f, the *mgf* is

$$M_X(t) = \int_{-\infty}^{\infty} e^{tx} f(x) dx. \qquad (7.3.3)$$

Since for every fixed t, e^{tx} is a positive function of x, it follows that $M_X(t)$ is either finite or infinite. We say that the probability law possesses a *mgf* if there exists a positive number T such that $M_X(t)$ is finite for $|t| < T$.

If a *mgf* exists for some t in $[-T, T]$, then one may form the successive derivatives by successively differentiating under the integral or summation sign. Thus,

$$M'(t) = \frac{d}{dt}M(t) = E(\frac{\partial}{\partial t}e^{tX}) = E(Xe^{tX})$$

$$M''(t) = \frac{d^2}{dt^2} M(t) = E(\frac{\partial}{\partial t} X e^{tX}) = E(X^2 e^{tX}).$$

Setting $t = 0$, we have

$$M^{(k)}(0) = E(X^k), \text{ for all positive integral } k. \qquad (7.3.4)$$

Thus, $M'(0) = E(X), M''(0) = E(X^2)$, etc. In this case, all moments of the distribution of X exist and are given by the successive derivatives of $M_X(t)$ at $t = 0$.

If the *mgf* $M(t)$ is finite for some $|t| < T$ (for some $T > 0$), one can express $M(t)$ uniquely in a Maclaurin series expansion,

$$M(t) = \sum_{j=0}^{\infty} \frac{M^{(j)}(0)}{j!} t^j = \sum_{j=0}^{\infty} \frac{\mu'_j}{j!} t^j \qquad (7.3.5)$$

so that μ'_j is the coefficient of $\frac{t^j}{j!}$ in the expression (7.3.5).

The validity of the result (7.3.5) depends crucially on the fact whether or not the formal differentiation under the integral sign is permissible. The answer is affirmative if the corresponding derivatives are absolutely integrable for $t = 0$, which is ensured by the assumption of existence of moments.

We have seen that there exists probability laws without finite moments (e.g. Cauchy distribution). Consequently, there also exist probability laws which do not possess *mgf*s. The following results hold for a *mgf*.

Theorem 7.3.1: (i) A *mgf*, if it exists, uniquely determines the distribution function of a random variable X and conversely, if the *mgf* exists, it is unique.

Thus, if X, Y be two random variables with *mgf*s $M_X(t), M_Y(t)$ respectively, such that $M_X(t) = M_Y(t)$ for all values of t, then X, Y have the same probability distribution.

(ii) Even if the moments of all order exist for a random variable, it does not follow that the *mgf* exists.

(iii) If not all the moments of a random variable exist, one can not find the distribution function of X from the moments, even from the moments that do exist.

Since there exist probability laws for which *mgf* does not exist, utility of *mgf* in determining the distribution function is somewhat limited. It is more convenient to use the characteristic function (discussed in Section 7.6),

$\psi_X(t) = E(e^{itX})$, where $i = \sqrt{-1}$, the imaginary quantity and t is any real number. $\psi_X(t)$ always exists and uniquely determines the distribution of X. $\psi_X(t)$ can also be used, like *mgf*, to determine those moments which do exist.

The relation between the convergence of a sequence of *mgf*s to the convergence of a sequence of distribution functions, the so-called *Continuity Theorem for moments* has been stated in Subsection 12.2.4.

EXAMPLE 7.3.1: Consider the Binomial distribution, $P(X = k) = \binom{n}{k}p^k q^{n-k}, 0 < p < 1, q = 1 - p$.

$$M(t) = \sum_{k=0}^{n} e^{tk}\binom{n}{k}p^k q^{n-k} = (q + pe^t)^n \quad \text{exists for } -\infty < t < \infty.$$

$$M'(t) = npe^t(pe^t + q)^{n-1},$$
$$M''(t) = npe^t(pe^t + q)^{n-1} + n(n-1)p^2 e^{2t}(pe^t + q)^{n-2}.$$

Thus,

$$M'(0) = E(X) = np$$
$$M''(0) = E(X^2) = npq + n^2 p^2$$
$$V(X) = npq.$$

EXAMPLE 7.3.2: Poisson distribution with parameter λ:

$$P(X = k) = e^{-\lambda}\frac{\lambda^k}{k!}, \ k = 0, 1, 2, \ldots \quad M(t) = \sum_{k=0}^{\infty} e^{-\lambda}\frac{\lambda^k}{k!}e^{tk} = e^{\lambda(e^t - 1)}.$$

$$M'(t) = e^{\lambda(e^t - 1)}\lambda e^t, \ M''(t) = \lambda^2 e^{\lambda}(e^t - 1)e^{2t} + \lambda e^t e^{\lambda(e^t - 1)}.$$

Thus, $E(X) = M'(0) = \lambda, E(X^2) = M''(0) = \lambda^2 + \lambda, V(X) = \lambda$. For truncated Poisson distribution, truncated at $X = 1$,

$$P(X = k) = \frac{e^{-\lambda}\lambda^k}{k!(1 - e^{-\lambda})}, \ k = 1, 2, \ldots$$
$$M_X(t) = \frac{(e^{\lambda e^t} - 1)}{(e^{\lambda} - 1)}.$$

EXAMPLE 7.3.3: Geometric distribution with parameter $p, 0 < p < 1, P(X = k) = pq^k, q = 1 - p, k = 0, 1, 2, \ldots; M(t) = \frac{P}{1-qe^t}$, exists for t such that $qe^t < 1$. it follows that $E(X) = \frac{q}{p}, V(X) = \frac{q}{p^2}$.

EXAMPLE 7.3.4:

$$f(x) = \frac{6}{\pi^2 x^2}, \ x = 1, 2, \ldots; \ = 0 \text{ otherwise.}$$

$$M_X(t) = \frac{1}{\pi^2} \sum_{x=0}^{\infty} \frac{6e^{tx}}{\pi^2 x^2} \text{ is infinite for every } t > 0.$$

The *mgf*, therefore, does not exist.

EXAMPLE 7.3.5: The *mgf* of the continuous uniform distribution on $[a, b]$ (vide Section 9.2) is

$$M_X(t) = \int_a^b \frac{e^{tx}}{b - a} dx = \frac{e^{bt} - e^{at}}{t(b - a)}. \tag{7.3.6}$$

Determination of values of the derivatives of $M_X(t)$ at $t = 0$ requires repeated application of L' Hospital rule. We can, however, expand exponentials into power series and after some algebra, obtain

$$M_X(t) = 1 + \frac{1}{2!} \frac{b^2 - a^2}{b - a} t + \frac{1}{3!} \frac{b^3 - a^3}{b - a} t^3 + \ldots \tag{7.3.7}$$

This is a Maclaurin series expansion of $M_X(t)$ about $t = 0$, so that we must have

$$\mu_r' = M_X^{(r)}(0) = \frac{r!}{(r+1)!} \frac{b^{r+1} - a^{r+1}}{b - a}$$

$$= \frac{1}{r+1} [b^r + b^{r-1}a + b^{r-2}a^2 + \ldots + a^r]. \quad \square$$

The *mgf* of $Z = \frac{X-a}{h}$, where a, h are arbitrary constants, is

$$M_Z(t) = E(e^{tZ}) = e^{-\frac{at}{h}} M_X\left(\frac{t}{h}\right). \tag{7.3.8}$$

The *mgf* of X about its mean μ, $M_{X-\mu}(t)$ gives the central moments of X, μ_r. $M_{X-\mu}(t) = \sum_{j=0}^{\infty} \frac{\mu_j}{j!} t^j$, provided the *mgf* exists for some $|t| < T$.

Theorem 7.3.2: The *mgf* of the sum $S_n = X_1 + \ldots + X_n$ of n independent random variables $X_1, \ldots, X_n$ is

$$M_{S_n}(t) = M_{X_1}(t) \ldots M_{X_n}(t) \tag{7.3.9}$$

where $M_{X_i}(t)$ is the *mgf* of X_i, provided all the *mgf*s exist.

Proof.

$$M_{S_n}(t) = E(e^{t \sum_{i=1}^{n} X_i}) = (\Pi_{i=1}^{n} e^{tX_i})$$
$$= \Pi_{i=1}^{n} E(e^{tX_i})$$

because of independence.

EXAMPLE 7.3.6: X is a normal variable with mean μ and s.d. σ,

$$f(x) = \frac{1}{\sigma\sqrt{2\pi}} e^{-\frac{(x-\mu)^2}{2\sigma^2}}, \quad -\infty < x < \infty.$$

The *mgf* of X about μ is

$$M_{X-\mu}(t) = E[e^{t(X-\mu)}] = \frac{1}{\sigma\sqrt{2\pi}} \int_{-\infty}^{\infty} e^{t(x-\mu)} e^{-\frac{(x-\mu)^2}{2\sigma^2}} dx$$

$$= \frac{e^{t^2\sigma^2/2}}{\sigma\sqrt{2\pi}} \int_{-\infty}^{\infty} e^{-\frac{1}{2\sigma^2}\{(x-\mu)-\sigma^2 t\}^2} dx \qquad (7.3.10)$$

$$= e^{\frac{t^2\sigma^2}{2}} = \sum_{j=0}^{\infty} \frac{(t^2\sigma^2)^j}{2^j j!}.$$

Hence

$$E(X - \mu) = 0 \quad \text{or} \quad E(X) = \mu,$$

$$E(X - \mu)^{2r+1} = 0, \quad E(X - \mu)^{2r} = \mu_{2r} = \frac{\sigma^{2r}(2r)!}{2^r r!}$$

$$= (2r - 1)(2r - 3)\ldots 3.1.\sigma^{2r} \text{ for } r = 1, 2, \ldots$$

The *mgf* of X is $e^{\mu t + \frac{t^2\sigma^2}{2}}$.

EXAMPLE 7.3.7: Consider Example 7.3.6. Let $Y = \alpha X + \beta$ where α, β are constants. The *mgf* of Y is

$$M_Y(t) = s^{\beta t} M_X(\alpha t) = e^{(\beta+\alpha\mu)t + (\alpha\sigma)^2 \frac{t^2}{2}}.$$

Thus, Y follows a Normal distribution with mean $\beta + \alpha\mu$ and variance $\alpha^2\sigma^2$ (by Theorem 7.3.1).

EXAMPLE 7.3.8: Consider n independent Poisson variables $X_1, \ldots, X_n$. By Theorem 7.3.3, the *mgf* of $S_n = X_1 + \ldots + X_n$ is $\exp\{\sum_{i=1}^{n} \lambda_i(e^t - 1)\}$. Thus, by Theorem 7.3.1, S_n has a Poisson distribution with parameter $\sum_{i=1}^{n} \lambda_i$. This is called the reproductive property of Poisson distribution.

EXAMPLE 7.3.9: Find the *mgf*s of the random variables U and V, having continuous Uniform distributions on $[0, 1]$ and $[-1, 1]$ respectively.

If X follows uniform distribution on $[a, b]$, $Y = \frac{X-a}{b-a}$ follows Uniform over $[0, 1]$. Now $Y = \alpha X + \beta$ where $\alpha = \frac{1}{b-a}, \beta = -\frac{a}{b-a}$. It is known $M_Y(t) = e^{\beta t} M_{\alpha X}(t)$. Hence, from (7.3.6), $Y = U$ has the *mgf*

$$M_U(t) = e^{-\frac{at}{b-a}} M_X\left(\frac{t}{b-a}\right) = \frac{e^t - 1}{t}.$$

Again, let $V = 2U - 1$. Then V will be Uniform on $[-1, 1]$. Here $V = \alpha U + \beta$ where $\alpha = 2, \beta = -1$. Hence

$$M_V(t) = e^{-t} M_U(2t) = \frac{e^t - e^{-t}}{2t} = \frac{\sin ht}{2}.$$

7.4 Factorial Moment Generating Function

DEFINITION 7.4.1: The *factorial moment generating function* (*fmgf*) of a discrete random variable X about the origin is defined as

$$\phi_X(t) = E[(1+t)^X]$$
$$= E[e^{X \log_e(1+t)}] = M_X[\log_e(1+t)]. \qquad (7.4.1)$$

Since, $(1+t)^x = \sum_{t=0}^{\infty} \frac{(x)_r}{r!} t^r$, we get $\mu_{(r)} =$ coefficient of $\frac{t^r}{r!}$ in the expansion of $\phi_X(t)$ in ascending powers of t. Also,

$$\frac{d^r}{dt^r} \phi_X(t)]_{t=0} = \mu_{(r)}. \qquad (7.4.2)$$

EXAMPLE 7.4.1: In the familiar matching problem with n cards (Example 2.8.3), the probability distribution of the number of matches X is

$$P[X = m] = \frac{1}{m!} \sum_{k=0}^{n-m} (-1)^k \frac{1}{k!}, \quad m = 0, 1, \ldots, n.$$

The *mgf* of X is

$$M_X(t) = E(e^{tX}) = \frac{1}{m!} \sum_{k=0}^{n-m} (-1)^k \frac{e^{tX}}{k!} = \sum_{r=0}^{n} \frac{1}{r!} (e^t - 1)^r.$$

Hence, the *fmgf* is

$$\phi_X(t) = \sum_{r=0}^{\infty} \frac{t^r}{r!}.$$

7.5 Cumulant Generating Function

DEFINITION 7.5.1: If the logarithm of the *mgf* of a random variable X can be expanded in a convergent series in powers of t, namely,

$$K_X(t) = \log_e M_X(t) = \sum_{r=1}^{\infty} k_r(X) \frac{t^r}{r!} \qquad (7.5.1)$$

then the coefficients $k_1(X), k_2(X), \ldots$ are called the first cumulant, second cumulant, etc. of the distribution of X and $K_X(t)$ is called the *cumulant generating function* (cgf) of the distribution of X.

We have

$$k_r(X) = \frac{d^r}{dt^r} \log_e M_X(t)]_{t=0}. \qquad (7.5.2)$$

If the *mgf* exists, so does the *cgf*. Evidently, $k_r(X)$ is a function of the moments of X. The first few cumulants are as follows.

$$
\begin{aligned}
k_1 &= & \mu_1' &= \mu \\
k_2 &= & \mu_2' - (\mu_1')^2 &= \sigma^2 \\
k_3 &= & \mu_3' - 3\mu_2'\mu_1' + 2(\mu_1')^3 &= \mu_3
\end{aligned}
\tag{7.5.3}
$$

$\cdots\cdots\cdots$

and conversely,

$$
\begin{aligned}
\mu_1' &= k_1 \\
\mu_2' &= k_2 + k_1^2 \\
\mu_3' &= k_3 + 3k_1 k_2 + k_1^3
\end{aligned}
\tag{7.5.4}
$$

$\cdots \quad \cdots\cdots$

Now, for any constant a and h, the *mgf* of $Z = \frac{X-a}{h}$ is

$$
M_Z(t) = E[e^{t(\frac{X-a}{h})}] = e^{-\frac{at}{h}} M_X\left(\frac{t}{h}\right).
$$

Hence,

$$
K_Z(t) = \log_e M_{\frac{X-a}{h}}(t) = -\frac{at}{h} + \log_e M_X\left(\frac{t}{h}\right)
$$

$$
= -\frac{at}{h} + \frac{t}{h} k_1(X) + \frac{t^2}{2!h^2} k_2(X) + \frac{t^3}{3!h^3} k_3(X) + \cdots
$$

$$
= (-a + k_1(X))\frac{t}{h} + \frac{t^2}{2!h^2} k_2(X) + \frac{t^3}{3!h^3} k_3(X) + \cdots
$$

Thus,

$$
k_1(Z) = \frac{1}{h}[-a + k_1(X)], \quad k_r(Z) = \frac{k_r(X)}{h^r}, \quad r \geq 2.
\tag{7.5.5}
$$

Therefore, only the first cumulant is affected by the change of origin. All the cumulants are affected by change of scale. For this reason, the cumulants are sometimes called *semi-invariants*. The cumulants were originally defined and studied by Thiele (1903).

Putting $a = \mu$ and $h = 1$, we get the central cumulants (cumulants about the mean μ) which are the same as the cumulants about zero except for the first cumulant which becomes zero.

Theorem 7.5.1: If $X_1, \ldots, X_n$ are independent random variables and if the *mgf* $M_{X_i}(t)(i = 1, \ldots, n)$ exist, the *cgf* of $S_n = X_1 + \ldots + X_n$ is

$$
K_{S_n}(t) = \sum_{i=1}^{n} K_{X_i}(t).
\tag{7.5.6}
$$

Proof. $K_{S_n}(t) = \log_e\{M_{S_n}(t)\} = \log_e\{\Pi_{i=1}^n M_{X_i}(t)\}$. Hence the proof. □

It follows that

$$k_r\left(\sum_{i=1}^n X_i\right) = \sum_{i=1}^n k_r(X_i) \ \forall \ r. \tag{7.5.7}$$

Thus, the cumulant of the sum of independent random variables is the sum of the cumulants. This makes the name 'cumulant' appropriate.

EXAMPLE 7.5.1: Poisson distribution: $P(X = k) = e^{-\lambda}\frac{\lambda^k}{k!}$. The *cgf* is $\lambda(e^\lambda - 1)$. Hence, $k_r = \lambda, r = 1, 2, \ldots$

Normal distribution (vide Example 7.3.6). The *cgf* about μ is $\frac{t^2\sigma^2}{2}$. Hence, $k_1 = 0, k_2 = \sigma^2, k_r = 0, r \geq 2$.

7.6 Characteristic Function

As stated in Section 7.3, the *mgf* of $X, E(e^{tX})$, where t is real and belongs to a certain interval of $\mathcal{R}$ may not always exist. But the characteristic function, defined below, always exists.

DEFINITION 7.6.1: The *characteristic function* (c.f.) of a random variable X is defined as

$$\psi_X(t) = E(e^{itX}) \tag{7.6.1}$$

where t is any real number and $i = \sqrt{-1}$.

Hence, for a discrete random variable with *pmf* $P(X = x_k) = p_k, k = 0, 1, 2, \ldots,$

$$\psi_X(t) = \sum_{k=0}^\infty e^{itx_k} p_k \tag{7.6.2}$$

and for a continuous random variable with *pdf* $f(x)$,

$$\psi_X(t) = \int_{-\infty}^\infty e^{itx} f(x)dx. \tag{7.6.3}$$

Since $e^{itx} = \cos tx + i \sin tx$ and $\cos tx$ and $\sin tx$ are both integrable over $\mathcal{R}$ for any real $t, \psi(t)$ is a complex number whose real and imaginary parts are finite for every value of t.

The rth non-central moment of X, when it exists, can be determined from the c.f. by the formula,

$$\frac{d^r}{i^r dt^r}\bigg]_{t=0} = \mu'_r. \tag{7.6.4}$$

Note 7.6.1: The characteristic function $\psi_X(t) = \int e^{itx} dF(x)$ is the Fourier-Stieltjes transformation of F, while, the *mgf* $E(e^{tX}) = \int e^{tX} dF(x)$ is the Laplace-Stieltjes transformation of F.

Note 7.6.2: For real values of t, $M(t)$ may not always exist; however, since $|e^{itx}|$ is a bounded and continuous function for all finite real t and x, the characteristic function always exists. Similarly, $M(t)$ always exists when t is purely imaginary, and if either exists, we have $\psi(t) = M(it)$.   □

The properties of c.f. are:

(i) $\psi(0) = 1$.

(ii) $|\psi(t)| \leq 1$ for all values of t, because $|\psi(t)| = |E(e^{itX})| \leq E|e^{itX}| = E(1) = 1$.

(iii) If $\psi(t)$ is the characteristic function of $X, \bar{\psi}(t)$, complex conjugate of $\psi(t)$, is the characteristic function of $(-X)$.

Proof. $E(e^{-itX}) = E(\cos tX - i \sin tX) = E(\cos tX) - iE(\sin tX) = \bar{\psi}(t)$.

(iv) $\psi(t)$ is uniformly continuous on $\mathcal{R}$.

Proof.

$$\psi(t) = \int_{-\infty}^{\infty} e^{itx} dF(x).$$

$$|\psi(t+h) - \psi(t)| = \int_{-\infty}^{\infty} e^{itx}(e^{ihx} - 1) dF(x)$$

$$\leq \int_{-\infty}^{\infty} |e^{ihx} - 1| dF(x) = 2 \int_{-\infty}^{\infty} |\sin \frac{hx}{2}| dF(x).$$

Hence,

$$|\psi(t+h) - \psi(t)| \leq 2 \int_{-\infty}^{-A} |\sin \frac{hx}{2}| dF(x) + 2 \int_{-A}^{B} |\sin \frac{hx}{2}| dF(x)$$
$$+ 2 \int_{B}^{\infty} |\sin \frac{hx}{2}| dF(x) \quad \text{for } A > 0, B > 0.$$

(7.6.5)

The right hand side of (7.6.5) is independent of t. By choosing A and B sufficiently large and h sufficiently small, the first and the third term on the right hand side of (7.6.5) can be made sufficiently small and the proof follows.

Theorem 7.6.1: Let F be a distribution function with finite moments up to order n. Then the characteristic function $\psi(t)$ of F has continuous derivatives up to order n and the relation

$$\frac{d^k}{i^k d^k} \psi(t) \bigg]_{t=0} = \mu'_k \qquad (7.6.6)$$

holds for $k = 1, 2, \ldots, n$. Moreover, ψ admits of the expansion

$$\psi(t) = 1 + \sum_{k=1}^{n} \mu'_k \frac{(it)^k}{k!} + o(t^n), \quad \text{as } t \to 0. \qquad (7.6.7)$$

Conversely, suppose that the characteristic function ψ of a distribution function F has an expansion of the form

$$\psi(t) = 1 + \sum_{k=1}^{n} a_k \frac{(it)^k}{k!} + o(t^n), \quad \text{as } t \to 0. \qquad (7.6.8)$$

Then F has finite moments up to order n if n is even and up to order $n-1$ if n is odd. Moreover, in this case, $a_k = \mu'_k \ \forall \ k$.

Proof. Suppose that $|\mu'_k| < \infty$ for $k = 1, 2, \ldots, n$. Clearly, $\int_{\infty}^{\infty} e^{itx} x^k dF(x)$ converges uniformly and absolutely for all $t \in \mathcal{R}$ and $k = 1, 2, \ldots, n$. Note that

$$\frac{\psi(t+h) - \psi(t)}{h} = \int_{-\infty}^{\infty} e^{itx} \frac{e^{ihx} - 1}{h} dF(x),$$

and $|e^{ihx} - 1| \leq |hx|$. Hence,

$$\left| \frac{\psi(t+h) - \psi(t)}{h} \right| \leq \int_{-\infty}^{\infty} |x| dF(x) < \infty.$$

Using the Lebesgue dominated convergence theorem (Theorem A.2.3), we have

$$\frac{d\psi(t)}{dt} = \lim_{h \to 0} \frac{\psi(t+h) - \psi(t)}{h} = i \int_{-\infty}^{\infty} e^{itx} x dF(x).$$

Thus, the first derivative $\psi^{(1)}$ exists and (7.6.6) holds for $n = 1$. Proceeding successively, it is concluded that ψ has all the derivatives up to order n and the relation (7.6.6) holds. It also follows immediately that $\psi^{(k)}$ is continuous on $\mathcal{R}$ for $1 \leq k \leq n$.

In the neighborhood of $t = 0$, we expand $\psi(t)$ in Maclaurin's series to get

$$\psi(t) = 1 + \sum_{k=1}^{n} \psi^{(k)}(0) \frac{t^k}{k!} + R_n(t),$$

where

$$R_n(t) = \frac{t^n}{n!} [\psi^{(n)}(\theta t) - \psi^{(n)}(\theta)], \quad 0 < \theta < 1.$$

Clearly,

$$|R_n(t)| \leq \frac{|t|^n}{n!} \int_{-\infty}^{\infty} |x|^n |e^{i\theta tx} - 1| dF(x).$$

We conclude from the Lebesgue dominated convergence theorem (Theorem A.2.3) that

$$R_n(t) = o(t^n), \quad \text{as } t \to 0.$$

This completes the proof of the first part.

Conversely, suppose that ψ can be expanded in the form (7.6.7) and n is even, say, $n = 2m$. Then ψ has a finite derivative of order $2m$ at $t = 0$, which is given by

$$\psi^{(2m)}(0) = \lim_{h \to 0} \int_{-\infty}^{\infty} \left(\frac{e^{ihx} - e^{-ihx}}{2h} \right)^{2m} dF(x)$$

$$= (-1)^m \lim_{h \to 0} \int_{-\infty}^{\infty} \left(\frac{\sin hx}{h} \right)^{2m} dF(x).$$

By Fatou's Lemma (Lemma A.2.1), we conclude that

$$\int_{-\infty}^{\infty} x^{2m} dF(x) \leq \lim_{h \to 0} \int_{-\infty}^{\infty} \left(\frac{\sin hx}{h} \right)^{2m} dF(x) < \infty$$

which proves that μ'_{2m} exists. If n is odd, we repeat the procedure with $n - 1$ and conclude that $\mu'_{n-1} < \infty$. Clearly, $a_k = \mu'_k \ \forall \ k$.

Corollary 7.6.1.1: ψ has continuous derivatives of all orders *iff* F has finite moments of all orders.

Corollary 7.6.1.2: Let ψ be of the form $\psi(t) = 1 + o(t^{2+\delta})$ where $\delta > 0$ as $t \to 0$. Then ψ is characteristic function of the distribution degenerate at zero.

Proof. Clearly, $\psi(t) = 1 + o(t^2)$ as $t \to 0$. It therefore, follows from Theorem 7.6.1, that $\mu'_1 = \mu'_2 = 0$. Hence the result.

Corollary 7.6.1.3: Let $a(t) = o(t)$ as $t \to 0$ and $a(-t) = -a(t)$. Then the only characteristic function of the form $\psi(t) = 1 + a(t) + o(t)$ as $t \to 0$, is the function $\psi(t) = 1$.

Proof. We have

$$|\psi(t)|^2 = [1 + a(t) + o(t)][1 - a(t) + o(t)]$$
$$= 1 + o(t^2) \text{ as } t \to 0.$$

Since, $\psi(t)$ is a characteristic function, it follows that $|\psi(t)|^2 = \psi(t)\psi(-t) = 1$. This implies that $\psi(t) = e^{i\alpha t}, \alpha \in \mathcal{R}$. Therefore,

$$\psi(t) = 1 + i\alpha t - \frac{1}{2}\alpha^2 t^2 + o(t^2).$$

Hence, ψ is of the form $1 + a(t) + o(t)$ iff $\alpha = 0$. Hence, $\psi(t) = 1$.

EXAMPLE 7.6.1: The functions $e^{-t^2}, (1+t^2)^{-1}$ are not characteristic functions.

Theorem 7.6.2 (*Inversion Theorem*): Let F be a distribution function and let $\psi(t)$ be its characteristic function. Then for any continuity points a and $a + h$ $(h > 0)$ of $F(x)$,

$$F(a + h) - F(a - h) = \lim_{T \to \infty} \frac{1}{\pi} \int_{-T}^{T} \frac{\sin ht}{t} e^{-ita} \psi(t) dt. \qquad (7.6.9)$$

Proof. We note that the integral $\int_0^a (\frac{\sin t}{t}) dt$ exists, is bounded for all $a > 0$ and tends to the limit $\frac{\pi}{2}$ as $a \to +\infty$. We let

$$\theta(h, T) = \frac{2}{\pi} \int_0^T \frac{\sin ht}{t} dt.$$

Clearly, $\theta(h, T)$ is uniformly bounded for all $h \in \mathcal{R}$ and $T > 0$. and $\theta(-h, T) = -\theta(h, T)$. Moreover,

$$\lim_{T \to \infty} \theta(h, T) = \begin{cases} 1, & h > 0, \\ 0, & h = 0, \\ -1, & h < 0. \end{cases}$$

For $T > 0$, we set

$$\begin{aligned} I_T &= \frac{1}{\pi} \int_{-T}^{T} \frac{\sin ht}{t} e^{-iat} \psi(t) dt \\ &= \frac{1}{\pi} \int_{-T}^{T} \frac{\sin ht}{t} e^{-ita} [\int_{-\infty}^{\infty} e^{itx} dF(x)] dt \end{aligned}$$

and noting that for $h > 0$,

$$|\frac{\sin ht}{t} e^{it(x-a)}| \le h$$

we conclude that I_T exists and is finite. Using Fubini's Theorem (Theorem A.2.4) to change the order of integration we obtain,

$$\begin{aligned} I_T &= \frac{1}{\pi} \int_{-\infty}^{\infty} dF(x) \int_{-T}^{T} \frac{\sin ht}{t} e^{it(x-a)} dt \\ &= \frac{2}{\pi} \int_{-\infty}^{\infty} dF(x) \int_0^T \frac{\sin ht}{t} \cos[(x - a)t] dt \\ &= \int_{-\infty}^{\infty} g(a, T) dF(x) \end{aligned}$$

where

$$\begin{aligned} g(a, T) &= \frac{2}{\pi} \int_0^T \frac{\sin ht}{t} \cos[(x - a)t] dt \\ &= \frac{1}{\pi} \int_0^T \frac{\sin[(x-a+h)t]}{t} dt - \frac{1}{\pi} \int_0^T \frac{\sin[(x-a-h)t]}{t} dt \\ &= \frac{1}{2} \theta(x - a + h, T) - \frac{1}{2} \theta(x - a - h, T). \end{aligned}$$

Also, $g(x, T)$ is uniformly bounded for all $x \in \mathcal{R}$ and all $T > 0$. Again,

$$\lim_{T \to \infty} g(x, T) = \begin{cases} 0 & \text{if } x < a - h, \\ \frac{1}{2} & \text{if } x = a - h, \\ 1 & \text{if } a - h < x < a + h, \\ \frac{1}{2} & \text{if } x = a + h, \\ 0 & \text{if } x > a + h. \end{cases}$$

Using Lebesgue dominated convergence theorem (Theorem A.2.3), we can take the limit as $T \to \infty$ under the integral sign and obtain

$$\lim_{T \to \infty} I_T = \int_{-\infty}^{\infty} [\lim_{T \to \infty} g(x, T)] dF(x)$$
$$= \int_{a-h}^{a+h} dF(x) = F(a + h) - F(a - h)$$

since, $a - h, a + h$ are continuity points of F. This completes the proof.

Corollary 7.6.2.1: Let $a, b(a < b)$ be continuity points of F. Then

$$F(b) - F(a) = \lim_{T \to \infty} \int_{-T}^{T} \frac{e^{-ita} - e^{-itb}}{it} \psi(t) dt. \qquad (7.6.10)$$

Theorem 7.6.3 (*Uniqueness Theorem*): Two distribution functions $F_1(x)$ and $F_2(x)$ are identical *iff* their characteristic functions $\psi_1(t)$ and $\psi_2(t)$ are identical.

Proof. Let F_1 and F_2 be two distribution functions with characteristic functions $\psi_1(t)$ and $\psi_2(t)$ respectively. Suppose that $\psi_1(t) = \psi_2(t)$ for every $t \in \mathcal{R}$. Let $a, b(a < b)$ be continuity points of both F_1 and F_2. Then (7.6.10) gives

$$F_1(b) - F_1(a) = F_2(b) - F_2(a).$$

Letting $a \to -\infty$, we obtain $F_1(b) = F_2(b)$ so that $F_1(x) = F_2(x)$ at all continuity points $x \in \mathcal{R}$. Hence, $F_1 \equiv F_2$. The converse follows from the definition of Characteristic function.

Note 7.6.3: We note that $\psi_1(t) = \psi_2(t)$ for all t in some real interval on $\mathcal{R}$ does not imply that $F_1 = F_2$ on $\mathcal{R}$.

Theorem 7.6.4 (*Fourier Inversion Theorem*): Suppose ψ is absolutely integrable on $\mathcal{R}$. Then the corresponding distribution function F is absolutely continuous. Moreover, the *pdf* $f = F'$ exists, is bounded and uniformly continuous on $\mathcal{R}$ and is given by

$$f(x) = \frac{1}{2\pi} \int_{-\infty}^{\infty} e^{-itx} \psi(t) dt, \ x \in \mathcal{R}. \qquad (7.6.11)$$

Proof. Since $\int_{-\infty}^{\infty} |\psi(t)| dt < \infty$, the integrand on the right side of (7.6.9) is dominated by an absolutely integrable function. Hence, we can write

$$F(x+h) - F(x-h) = \frac{1}{\pi} \int_{-\infty}^{\infty} \frac{\sin ht}{t} e^{-itx} \psi(t) dt \qquad (7.6.12)$$

where $x-h, x+h$ are continuity points of F. Taking limits of both sides as $h \downarrow 0$, we obtain

$$F(x) - F(x-0) = 0.$$

Therefore, F is continuous to the left. It follows that F is continuous on the right. Using a similar argument on

$$\frac{F(x+h) - F(x-h)}{2h} = \frac{1}{2\pi} \int_{-\infty}^{\infty} \frac{\sin ht}{ht} e^{-itx} \psi(t) dt$$

we conclude that F is differentiable everywhere on $\mathcal{R}$ and for $x \in \mathcal{R}$, we have

$$\begin{aligned} F'(x) &= \frac{1}{2\pi} \int_{-\infty}^{\infty} \lim_{h \downarrow 0} \frac{\sin ht}{ht} e^{-itx} \psi(t) dt \\ &= \frac{1}{2\pi} \int_{-\infty}^{\infty} e^{-itx} \psi(t) dt. \end{aligned}$$

Hence, F is absolutely continuous and $f = F'$ exists and is bounded on $\mathcal{R}$. We note that

$$|f(x+h) - f(x)| \le \frac{1}{2\pi} \int_{-\infty}^{\infty} |\sin \frac{ht}{2}| |\psi(t)| dt$$

so that uniform continuity of f follows from the Lebesgue dominated convergence theorem (Theorem A.2.3).

Note 7.6.4: There exist absolutely continuous distribution functions whose characteristic functions are not absolutely integrable on $\mathcal{R}$.

Theorem 7.6.5: $\psi(t)$ is real and even *iff* $F(x)$ is symmetric.

Proof.

$$\psi_X(t) = \int_{-\infty}^{\infty} (\cos tx + i \sin tx) dF(x).$$

If X is symmetric, the contribution of the sin term vanishes, because

$$\int_{\infty}^{0-} \sin tx \, dF(x) + \int_{0+}^{\infty} \sin tx \, dF(x) + (\sin 0)(P(X=0))$$

$$= \int_{0+}^{\infty} -\sin tx \, dF(x) + \int_{0+}^{\infty} \sin tx \, dF(x) = 0.$$

Conversely, if $\psi_X(t)$ is real, the random variable $Y = -X$ has the characteristic function,

$$\psi_Y(t) = \int_{-\infty}^{\infty} e^{-itx} dF(x) = \psi_X(-t) = \bar{\psi}_X(t).$$

Since, $\psi_X(t)$ is real, $\psi_X(t) = \bar{\psi}_X(t)$. Thus, X and Y have the same characteristic function.

Again, $P(X \le -x) = P(-X \ge x) = P(Y \ge x)$, i.e., $F_X(-x) = 1 - P(Y < x) = 1 - F_Y(x - 0)$ (by Theorem 7.6.2). Hence, F_X is symmetric about zero.

EXAMPLE 7.6.2: For the Normal distribution

$$f(x) = \frac{1}{\sqrt{2\pi}\sigma} e^{-\frac{(x-\mu)^2}{\sigma^2}} \quad -\infty < x < \infty,$$

$$\psi(t) = \int_{-\infty}^{\infty} \frac{e^{itx}}{\sqrt{2\pi}\sigma} e^{-\frac{(x-\mu)^2}{2\sigma^2}} dx$$

$$= \int_{-\infty}^{\infty} \frac{1}{\sqrt{2\pi}} e^{it(\mu+\sigma y)} e^{-\frac{y^2}{2}} dy \quad \text{where } y = \frac{x-\mu}{y} \qquad (7.6.13)$$

$$= \frac{e^{it\mu}}{\sqrt{2\pi}} \int_{-\infty}^{\infty} e^{-\frac{1}{2}(y^2 - 2it\sigma y)} dy$$

$$= e^{it\mu - \frac{t^2\sigma^2}{2}} \int_{-\infty}^{\infty} \frac{1}{\sqrt{2\pi}} e^{-\frac{1}{2}(y-it\sigma)^2} dy = e^{it\mu - \frac{t^2\sigma^2}{2}},$$

the integral being unity, since it is the integral over a Normal distribution. Hence,

$$\psi'(t) = e^{it\mu - \frac{t^2\sigma^2}{2}} (i\mu - t\sigma^2)$$

$$\psi''(t) = e^{it\mu - \frac{t^2\sigma^2}{2}} [(i\mu - t\sigma^2)^2 - \sigma^2].$$

Thus, $\psi'(0) = i\mu$, $\psi''(0) = i^2(\mu^2 + \sigma^2)$.

EXAMPLE 7.6.3: For the Gamma distribution,

$$f(x) = \frac{\lambda^n}{\Gamma(n)} e^{-\lambda x} x^{n-1}, \quad 0 < x < \infty, \quad \lambda > 0$$

$$\psi(t) = \int_0^{\infty} \frac{\lambda^n}{\Gamma(n)} e^{-(\lambda - it)x} x^{n-1} dx$$

$$= \frac{\lambda^n}{(\lambda - it)^n} \int_0^{\infty} \frac{(\lambda - it)^n}{\Gamma(n)} e^{-(\lambda - it)x} x^{n-1} dx$$

$$= \left(\frac{\lambda}{\lambda - it}\right)^n = \left(1 - \frac{it}{\lambda}\right)^{-n}.$$

For the Exponential distribution, which is a special case of Gamma distribution with $n = 1$,

$$\psi(t) = (1 - \tfrac{it}{\lambda})^{-1}, \quad \psi'(t) = (-n)(1 - \tfrac{it}{\lambda})^{-n-1}(-\tfrac{i}{\lambda})$$

$$\psi''(t) = (-n)(-n-1)(1 - \tfrac{it}{\lambda})^{-n-2}(\tfrac{t^2}{\lambda^2})$$

$$\psi'(0) = \tfrac{i\lambda}{n}, \quad \psi''(0) = \tfrac{n(n+1)i^2}{\lambda^2}.$$

EXAMPLE 7.6.4: Laplace distribution $f(x) = \tfrac{1}{2}e^{-|x|}$, $-\infty < x < \infty$.

$$\psi(t) = \tfrac{1}{2}\int_{-\infty}^{\infty} e^{itx}e^{-|x|}dx = \tfrac{1}{2}\int_{-\infty}^{\infty}(\cos tx + i\sin tx)e^{-|x|}dx$$
$$= \int_0^{\infty} \cos txe^{-x}dx.$$

Now, $\int_0^{\infty} e^{-x}\cos txdx = 1 - i\int_0^{\infty} e^{-x}\sin txdx$ (on integration by parts, taking $\cos tx$ as the first function).

Similarly, $\int_0^{\infty} e^{-x}\sin txdx = i\int_0^{\infty} e^{-x}\cos txdx$.

Hence $\psi(t) = \tfrac{1}{1+t^2}$.

Note that $\psi(t)$ is absolutely integrable and by (7.6.11),

$$\int_{-\infty}^{\infty} \frac{e^{-itx}}{1+t^2}dt = \pi e^{-|t|}. \tag{7.6.14}$$

EXAMPLE 7.6.5: For the Cauchy distribution

$$f(x) = \frac{1}{\pi(1+x^2)}, \quad -\infty < x < \infty, \quad \psi(t) = \int_{-\infty}^{\infty} \frac{e^{itx}}{\pi(1+x^2)}.$$

To evaluate this integral, consider the result (7.6.14). Replacing t by $-t$ and interchanging t and x,

$$\int_{-\infty}^{\infty} \frac{e^{itx}}{1+x^2}dx = \pi e^{-|t|}.$$

Hence, $\psi(t) = e^{-|t|}$. $\psi(t)$ is not differentiable at the point $t = 0$. This shows that for the Cauchy distribution, no moment exists.

EXAMPLE 7.6.6: Consider the characteristic function $\psi(t) = (\cos ht)^{-1}$ for $t \in \mathcal{R}$. Since, ψ is real and even, it corresponds to a symmetric distribution function (Theorem 7.6.4). Now,

$$\int_{-\infty}^{\infty} |\psi(t)|dt = \int_{-\infty}^{\infty} \sec htdt = \tan^{-1}(\sin ht)]_{-\infty}^{\infty} < \infty.$$

Therefore, by Fourier Inversion theorem, the *pdf* can be written as

$$f(x) = \frac{1}{2\pi} \int_{-\infty}^{\infty} e^{-itx} \psi(t) dt$$
$$= \frac{1}{2} \cos h(\frac{\pi x}{2}).$$

Theorem 7.6.6: If a and b are real constants, then the characteristic function of $Y = a + bX$ is

$$\psi_Y(t) = e^{itb} \psi_X(at). \tag{7.6.15}$$

In particular,

$$\psi_{\frac{X-\mu}{\sigma}} = e^{-\frac{it\mu}{\sigma}} \psi_X(\frac{t}{\sigma}). \tag{7.6.16}$$

EXAMPLE 7.6.7: Consider the Normal distribution in Example 7.6.2. The characteristic function of

$$Y = \frac{X - \mu}{\sigma} \text{ is } e^{-\frac{t^2}{2}} = \sum_{j=0}^{\infty} \frac{1}{j!} (-\frac{t^2}{2})^j.$$

Since the coefficient of $\frac{(it)^j}{j!}$ gives the jth moment μ_j of the distribution of Y is

$$\frac{j!}{2^{\frac{j}{2}}(\frac{j}{2})!} \quad \text{if j is even },$$
$$0 \qquad \text{if j is odd }.$$

Theorem 7.6.7: The characteristic function of the sum $S_n = X_1 + \ldots + X_n$ of n independent variables $X_1, \ldots, X_n$ is

$$\psi_{S_n}(t) = \psi_{X_1}(t) \ldots \psi_{X_n}(t) \tag{7.6.17}$$

where $\psi_{X_i}(t)$ is the characteristic function of $X_i, i = 1, \ldots, n$.

Proof.

$$\psi_{S_n}(t) = E(e^{it \sum_{i=1}^{n} X_i}) = E(\Pi_{i=1}^{n} e^{itX_i})$$
$$= \Pi_{i=1}^{n} E(e^{itX_i}),$$

because of independence.

EXAMPLE 7.6.8: Let $X_1, \ldots, X_n$ be n independent variables such that X_i is Normal with mean μ_i and variance $\sigma_i^2, i = 1, \ldots, n$. The characteristic function of $X = \sum_{i=1}^{n} X_i$ is

$$\psi_X(t) = E^{itX} = \Pi_{i=1}^{n}(Ee^{itX_i})$$
$$= \exp\{t \sum_{i=1}^{n} \mu_i - \frac{t^2}{2} \sum_{i=1}^{n} \sigma_i^2\}.$$

It follows from the uniqueness theorem that X has a Normal distribution with mean $\sum_{i=1}^{n} \mu$ and variance $\sum_{i=1}^{n} \sigma_i^2$. □

We state without proof another important result.

Theorem 7.6.8: The necessary and sufficient conditions for a bounded and continuous function to be the characteristic function of a distribution are that $\psi(0) = 1$ and the function

$$\Psi(x, A) = \int_0^A \int_0^A \psi(t - u)e^{i(t-u)x}\,dt\,du$$

is real and non-negative for all x and $A > 0$. □

In Subsection 12.2.4 we shall state two important theorems on characteristic function, the Continuity Theorems which give a necessary and sufficient condition for the weak convergence of a sequence of distribution functions to a distribution function.

For further detail on characteristic functions, the reader may refer to Lukacs (1970, 1982), Moran (1968), Laha (1982), among others.

7.7 Exercises and Complements

7.1 Find the *pgf* of the random variable X when

(a) $P(X = k) = [e^{-\mu}/(1 - e^{-\mu})]\frac{\mu^k}{k!}, k = 1, 2, \dots; \lambda > 0$

(b) $P(X = k) = pq^k/(1 - q^{N+1})), k = 0, 1, \dots, N; 0 < p < 1, q = 1 - p$

(c) X is the number of tosses required to obtain a head for the first time in tossing a biased coin, with $P(\text{head}) = p$.

7.2 In the familiar matching problem (Example 2.6.3) with two decks of n cards each, let X be the number of matches. Show that the *pgf* of X is $\sum_{k=0}^{n} \frac{(t-1)^k}{k!}$.

7.3 (a) If $P(t)$ is a *pgf*, show that $\frac{1}{2-P(t)}$ is a *pgf*.

(b) Show that $P(t) = \frac{2}{1+t}$ is not a *ogf*, even though $P(1) = 1$.

[(a)

$$Q(t) = \frac{1}{2-P(t)} = \frac{1}{2}(1 - \frac{P(t)}{2})^{-1}$$
$$= \frac{1}{2}[1 + \frac{1}{2}P(t) + (\frac{P(t)}{2})^2 + (\frac{P(t)}{2})^3 + \dots]$$

$Q(t) = 1$, since $P(1) = 1$; also coefficient of $(P(t))^i$ is non-negative, $i = 1, 2, \ldots$

(b) Coefficient of $t^i, i = 1, 2, \ldots$ are not all non-negative.]

7.4 Balls are distributed among m cells. Let Z be the number of balls that must be distributed to occupy all the m cells. Find the *pgf* of Z.

$[Z = m + X_1 + \ldots + X_{m-1}$ where X_i is the number of balls following the occupancy of the ith cell for the first time and preceding the occupancy of the $(i+1)$th cell for the first time.

X_1 follows a Geometric distribution with probability of success $p = 1 - \frac{1}{m}$. The *pgf* of X_1 is $\frac{(m-1)/m}{1-t/m}$ (Example 7.2.4).

Similarly, X_r follows a Geometric distribution with probability of success $p = 1 - \frac{r}{m}, r = 2, 3, \ldots, m-1$. Also, $X_1, \ldots, X_{m-1}$ are independent. Hence *pgf* of Z is

$$t^m \Pi_{r=1}^{m-1} \frac{(m-r)}{(m-tr)}.]$$

7.5 n balls are randomly distributed among m cells. Find the *pgf* of X, the number of unoccupied cells.

[Let $X_i = 1(0)$ if the ith cell is empty (occupied); $X = \sum_{i=1}^{m} X_i$ is the number of empty cells. By the notation of Theorem 7.2.3, $B_i = $ probability that i cells remain empty is $\binom{m}{i}(\frac{m-i}{m})^n$. Hence, by (7.2.4),

$$P_X(t) = \sum_{i=0}^{m} \binom{m}{i}(\frac{m-i}{m})^n (t-1)^i.]$$

7.6 Show that for any integer-valued random variable X,

$$\sum_{n=0}^{\infty} t^n P(X \leq n) = (1-t)^{-1} P(t),$$

where $P(t)$ is the *pgf* of X.

7.7 Let X, Y be independent random variables each assuming values $1, 2, \ldots, a$ with probability $\frac{1}{a}$. Find the *pgf* of $X - Y$ and hence determine the probability that X exceeds Y.

$\left[P_{X-Y}(t) = E(t^{X-Y}) = E(t^X)E(t^{-Y}), X, Y \text{ being independent. } E(t^X) = \frac{t(1-t^a)}{a(1-t)}, E(t^{-Y}) = \frac{1-t^a}{at^a(1-t)}. \text{ Hence, } P_{X-Y}(t) = \frac{1}{a^2 t^{a-1}}\left(\frac{1-t^a}{1-t}\right)^2.\right.$

$$P(X > Y) = \sum_{r=1}^{a-1} \text{Coefficient of } t^r \text{ in } P_{X-Y}(t)$$

$$= \sum_{r=1}^{a-1} \text{Coefficient of } t^{r+a-1} \text{ in } \frac{1}{a^2}\left(\frac{1-t^a}{1-t}\right)^2$$

$$= \frac{1}{a^2}\sum_{r=1}^{a-1}(a-r) = \frac{1}{2} - \frac{1}{2a}.\Big]$$

7.8 Let X, Y be independent random variables with *pgf* $U(t), V(t)$ respectively. Show that $P(X - Y) = j$ is the coefficient of t^j in $U(t)V(\frac{1}{t})$, when $j = 0, +_-1, +-2, \ldots$

7.9 A random variable X has *pmf*

$$P(X = x) = \frac{1}{x!}[1 - \frac{1}{1!} + \frac{1}{2!} - \ldots +_- \frac{1}{(n-x)!}], \quad x = 0, 1, \ldots, n.$$

Show that the *pgf* is $P(t) = \sum_{i=0}^{n} \frac{(t-1)^i}{i!}$. Hence, find the mean, variance, factorial moments of the distribution of X.

7.10 Let $X_1, \ldots, X_n$ be *iid* random variables with $P(X_i = k) = \frac{1}{a}, k = 0, 1, \ldots, a-1; i = 1, 2, \ldots, n$. Let $X = X_1 + \ldots + X_n$. Show that *pgf* of X is $P_X(t) = \{\frac{1-t^a}{a(1-t)}\}^n$ and hence

$$P(X = j) = \frac{1}{a^n}\sum_{n=0}^{\infty}(-1)^{u+j+au}\binom{n}{u}\binom{-n}{j-au}.$$

7.11 Suppose X has *pgf* $P(t)$ and that $E(X^3)$ exists.

Show that (i) $E(X^3) = P^{(3)}(1) + 3P^{(2)}(1) + P^{(1)}(1)$; (ii) $E(X^2) = P^{(2)}(1) + P^{(i)}(1)$, where $P^{(i)}(1)$ denotes the *i*th derivative of $P(t)$ at $t = 1$.

7.12 Show that the mgf of a discrete random variable X with $P(X = k) = \frac{1}{n}$ for $k = 0, 1, \ldots, n-1$ is

$$M_X(t) = \begin{cases} \frac{e^{nt}-1}{n(e^t-1)} & \text{for } t \neq 0 \\ 1 & \text{for } t = 0. \end{cases}$$

7.13 Let X be the number of tosses of a biased coin, up to and including the first head. Assume that probability of getting a head is p. Fine the *mgf* of X. Hence determine the *mgf* of Y = number of tails preceding the first head.

7.14 Show that the function $E(\frac{1}{1-tX})$ can be used to generate the moments of the random variable X.

7.15 Find the *mgf*, expectation and variance of X having *pdf* $f(x) = \frac{1}{2}e^{-\frac{x}{2}}, x > 0$. [*mgf* $= \frac{1}{1-2t}$]

7.16: Show that

$$e^{itx} = 1 + (e^{it} - 1)x_{(1)} + (e^{it} - 1)^2 \frac{x_{(2)}}{2!} + \ldots + (e^{it} - 1)^r \frac{x_{(r)}}{r!} + \ldots$$

where $x_{(r)} = x(x-1)\ldots(x-r+1)$. Hence, show that $\mu_{(r)} = \Delta^r \phi(t)]_{t=0}$ where $\Delta = \frac{d}{d(e^t)}, \phi(t) = E(e^{itX})$.

[Let $e^{itx} = A_0 + A_1 x_{(1)} + A_2 \frac{x_{(2)}}{2!} + \ldots + A_r \frac{x_{(r)}}{r!} + \ldots$, an identity in $\frac{x_{(r)}}{r!}$. Putting $x = 0, 1, 2, \ldots$ and solving for $A_0, A_1, A_2, \ldots, A_0 = 1, A_1 = e^{it} - 1, A_2 = (e^{it} - 1)^2, \ldots$. Hence,

$$E(e^{itX}) = A_0 + A_1 E(X_{(1)}) + \frac{A_2}{2!} E(X_{(2)}) + \ldots + \frac{A_r}{r!} E(X_{(r)}) + \ldots$$
$$= A_0 + A_1 \mu_{(1)} + \frac{A_2}{2!} \mu_{(2)} + \ldots + \frac{A_r}{r!} \mu_{(r)} + \ldots$$

Hence $\Delta^r \phi(t)]_{t=0} = \mu_{(r)}$.]

7.17 Define a factorial *mgf* and show that if the factorial *mgf* of a function $f(x)$ is $G(t)$, then the factorial *mgf* of $(-\nabla)^r f(x)$ is $t^r G(t)$ where $\nabla f(x) = f(x) - f(x-1)$.

[Factorial *mgf*, $G(t) = E[(1+t)^X] = \sum_x (1+t)^x f(x)$.

fmgf of $f(X-1) - f(X) = \sum_x (1+t)^x \{f(x-1) - f(x)\}$.
$$= (1+t) \sum_x (1+t)^{x-1} f(x-1) - \sum_x (1+t)^x f(x) = tG(t).$$

Now, $\nabla^2 f(x) = \nabla(\nabla f(x)) = f(x) - 2f(x-1) + f(x-2)$. *fmgf* of $\nabla^2 f(x) = t^2 G(t)$.

Thus the result follows by induction.]

7.18 With usual notations, prove that

$$\frac{\partial \mu'_r}{\partial k_j} = \binom{r}{j} \mu'_{r-j}.$$

[We have

$$\exp\{k_1 t + k_2 \frac{t^2}{2!} + \ldots + k_r \frac{t^r}{r!} + \ldots\} = 1 + \mu'_1 t + \mu'_2 \frac{t^2}{2!} + \ldots$$

Differentiating with respect to k_j,

$$\frac{t^j}{j!}\exp\{k_1 t + k_2\frac{t^2}{2!} + \ldots + k_r\frac{t^r}{r!} + \ldots\} = \sum_{r=0}^{\infty}\frac{\partial\mu'_r}{\partial k_j}\cdot\frac{t^r}{r!}$$

or

$$\frac{t^j}{j!}\{1 + \mu'_1 t + \mu'_2\frac{t^2}{2!} + \ldots\} = \sum_{r=0}^{\infty}\frac{\partial\mu'_r}{\partial k_j}\frac{t^r}{r!}.$$

The result follows by equating the coefficient of t^r from both sides.]

7.19 For the Logistic distribution of Exercise 4.4(f) in Section 4.9, find the *mgf*. Hence, show that $E(X) = \alpha, V(X) = \frac{\beta^2\pi^2}{3}$.

[Substituting $Y = \frac{X-\alpha}{\beta}$, the *mgf* of Y is $M_Y(t) = B(1 - t, 1 + t)$, where $B(p, q)$ is the Beta function with parameter p, q. Beta function has been discussed in detail in Section 9.5.2.]

7.20 For the Power Series distribution discussed in Section 8.9,

$$P\{X = j\} = a_j\frac{\beta^j}{f(\theta)}, \quad j = 0, 1, 2, \ldots$$

$\theta > 0$, where $a_j \geq 0, f(\theta) = \sum_{j=0}^{\infty} a_j\theta^j$, find the *pgf* and *mgf*.

Show that, for this distribution

$$\mu'_{r+1} = \theta\frac{d\mu'_r}{d\theta} + \mu'_r\mu'_1$$

$$\theta\frac{d\mu_r}{d\theta} = \mu_{r+1} - r\theta\frac{d\mu'_1}{d\theta}\mu_{r-1}$$

$$\mu'_1 = \theta\frac{d\log f(\theta)}{d\theta}, \quad \mu_2 = \theta^2\frac{d^2\log f(\theta)}{d\theta^2} + \mu'_1$$

$$= \theta^2\frac{f''(\theta)}{f(\theta)} - \theta^2(\frac{f'(\theta)}{f(\theta)})^2 + \theta\frac{f'(\theta)}{f(\theta)}.$$

7.21 Show that for a symmetric distribution all the cumulants of odd order except the first vanish.

[If the distribution of X is symmetric about a, we transform from X to $X - a$. The *mgf* about a is

$$M_{X-a}(t) = E(e^{t(X-a)}) = [1 + \mu_2\frac{t^2}{2!} + \mu_4\frac{t^4}{4!} + \ldots + \mu_{2r}\frac{t^{2r}}{2r!}],$$

since all the existent moments of odd order about 'a' vanish.

$$K_{X-a}(t) = \log[1 + \mu_2\frac{t^2}{2!} + \ldots].$$

Hence, $k_1(X) = a + k_1(X - a), k_{2r-1}(X) = k_{2r-1}(X - a) =$ coefficient of $\frac{t^{2r-1}}{(2r-1)!} = 0$.]

7.22 A random variable X has moments $\mu'_r = r!$. Find its characteristic function and deduce its distribution.

$$\left[\psi(t) = \frac{1}{1-it}; f(x) = e^{-x}, 0 < x < \infty.\right]$$

7.23 Show that the distribution for which the characteristic function is $e^{-|t|}$ has the *pdf* $f(x) = \frac{1}{\pi(1+x^2)}, -\infty < x < \infty.$

7.24 Find the *pdf*s corresponding to the following characteristic functions.

   (a) $\psi(t) = 1 - |t|$ if $|t| \le 1$; and $= 0$ if $|t| > 1$.
   (b) $\psi(t) = (\cos ht)^{-1}$.
   (c) $\psi(t) = \cos at$.
   (d) $\psi(t) = t(\sin t)^{-1}$.
   (e) $\psi(t) = (\cos ht)^{-2}$.

7.25 Let $\psi_n(t)(n \ge 1)$ and $\psi(t)$ be absolutely integrable characteristic functions such that $\int_{-\infty}^{\infty} |\psi_n(t) - \psi(t)|dt \to 0$ as $n \to \infty$. Show that if f_n and f are the corresponding probaility density functions, then $f_n \to f$ as $n \to \infty$.

Chapter 8

Some Discrete Distributions on $\mathcal{R}^1$

8.1 Introduction

We consider in this chapter some special univariate discrete distributions and study their properties. The discrete uniform distribution is considered in Section 8.2, Bernoulli distribution in Section 8.3, binomial distribution in Section 8.4, hypergeometric distribution in Section 8.5. Sections 8.6 through 8.8 consider respectively Poisson distribution, geometric distribution and negative Binomial distribution. Along with each distribution, its truncated form has been considered. We conclude this chapter by a brief discussion on the power series distribution of which all the above distributions come as special cases.

8.2 The Discrete Uniform Distribution

A random variable X is said to have a discrete uniform distribution over the range $[1, N]$ if its *pmf* is

$$P(X = x) = \begin{cases} \frac{1}{N} & \text{for } x = 1, 2, \ldots, N, \\ 0, & \text{otherwise .} \end{cases} \tag{8.2.1}$$

The quantity N is the parameter of the distribution and lies in the set of all positive integers. (8.2.1) is also called the *discrete rectangular distribution*.

$$E(X) = \frac{1}{N} \sum_{i=1}^{N} i = \frac{N+1}{2}, \quad E(X^2) = \frac{1}{N} \sum_{i=1}^{N} i^2 = \frac{(N+1)(2N+1)}{6}$$

$$V(X) = E(X^2) - (E(X))^2 = \frac{(N+1)(N-1)}{12}.$$

The *pgf* is

$$P_X(t) = E(t^X) = \frac{1}{N} \sum_{x=1}^{N} t^x = \frac{t(1-t^N)}{N(1-t)}, \quad t \neq 1; = 1 \text{ if } t = 1.$$

The *mgf* is

$$M_X(t) = E(e^{tX}) = \frac{1}{N} \sum_{x=1}^{N} e^{tx} = \frac{e^t(1-e^{Nt})}{N(1-e^t)}$$
$$= \frac{e^{\frac{(N+1)t}{2}} \sin h(\frac{Nt}{2})}{N \sin h(\frac{t}{2})}.$$

The first 10 moments and some other results are given in Pierce (1940).

When X is distributed uniformly over the range $[0, N]$, its *pmf* is

$$P[X = x] = \begin{cases} \frac{1}{N+1}, & x = 0, 1, \ldots, N \\ 0, & \text{otherwise.} \end{cases}$$

Here the *pgf* is

$$P_X(t) = \frac{1}{N+1} \frac{1-t^{N+1}}{1-t}, \quad t \neq 1; = 1 \text{ if } t = 1.$$

The *mgf* is

$$M_X(t) = \frac{1}{N+1} \frac{e^{(N+1)t} - 1}{e^t - 1}.$$

EXAMPLE 8.2.1: Let $X_i(i = 1, \ldots, n)$ be independent random variables, each having rectangular distribution $P(X_i = x) = \frac{1}{N}, x = 1, \ldots, N; i = 1, \ldots, n$. Find the distribution of M_n and N_n where $M_n = \min_{i=1,\ldots,n} X_i$, and $N_n = \max_{i=1,\ldots,N} X_i$.

$$P(M_n = r) = P(X_i \geq r, i = 1, \ldots, n) - P(X_i \geq r+1, i = 1, \ldots, n)$$

$$= (\tfrac{N-r+1}{N})^n - (\tfrac{N-r}{N})^n = \frac{(N-r+1)^n - (N-r)^n}{N^n}.$$

Similarly,

$$P(N_n = r) = P(X_i \leq r, i = 1, \ldots, n) - P(X_i \leq r-1, i = 1, \ldots, n)$$

$$= (\tfrac{r}{N})^n - (\tfrac{r-1}{N})^n = \frac{r^n - (r-1)^n}{N^n}.$$

8.3 The Bernoulli Distribution

A random variable X is said to have a Bernoulli distribution with parameter p if its *pmf* is given by

$$P(X = x) = \begin{cases} p^x(1-p)^{1-x} & \text{for } x = 0, 1, \\ 0 & \text{otherwise.} \end{cases} \tag{8.3.1}$$

The parameter p satisfies $0 \leq p \leq 1$. Often $(1 - p)$ is denoted as q.

$$E(X) = 1.p + o.q = p, \ V(X) = pq.$$

The *pgf* is

$$P_X(t) = pt + q.$$

The *mgf* is

$$M_X(t) = q + pe^t.$$

A random variable whose outcomes are of two types, success S and failure F, occurring with probabilities p and q respectively is called a Bernoulii trial. If for this experiment, a random variable X is defined such that it takes value 1 when S occurs and 0 when F occurs, then X follows a Bernoulli distribution.

EXAMPLE 8.3.1: For an arbitrary probability space $(\mathcal{S}, \mathcal{A}, P)$ and for an event $A \in \mathcal{A}$ consider the indicator random variable $I_A(\omega) = 1(0)$ if $\omega \in A (\notin A)$. Then $P(I_A = 1) = P(\omega \in A) = P(A)$. The variable I_A follows a Bernoulli distribution with parameter $p = P(A)$.

8.4 The Binomial Distribution

A random variable X is said to have a binomial distribution with parameters (n, p) if its *pmf* is given by

$$P(X = x) = b(x : n, p) = \begin{cases} \binom{n}{x} p^x q^{n-x} & \text{for } x = 0, 1, \ldots, n, \\ 0 & \text{otherwise,} \end{cases} \qquad (8.4.1)$$

where n ranges over the set of positive integers and $p \in [0, 1]$. The probabilities in (8.4.1) are the terms of binomial expansion of $(q + p)^n$. For this reason, the distribution is called a binomial distribution.

For $n = 1$, (8.4.1) reduces to the Bernoulli distribution. For this reason, sometimes Bernoulli distribution is called a point binomial distribution.

The binomial distribution arises in the following circumstances. Consider n independent Bernoulli trials each having probability of success $'p'$. Let X be the random variable denoting the number of successes in these trials. The sample space for such an experiment is given by

$$S = \{(t_1, \ldots, t_n), t_i = S \text{ or } F, i = 1, \ldots, n\},$$

t_i denoting the result of the ith trial. Since the trials are independent, the probability of any specified outcome, say, $(S, S, F, F, \ldots, S)$ is $p.p.q.q \ldots p$.

Hence $P(X = x) = P[\text{there are } x \text{ successes and } (n - x) \text{ failures}] = \binom{n}{x}p^x q^{n-x}$, since any outcome with x successes and $(n - x)$ failures has probability $p^x q^{n-x}$ and there are $\binom{n}{x}$ such mutually exclusive outcomes. Hence X follows a binomial distribution.

EXAMPLE 8.4.1: An urn contains a white balls and b black balls. From this, n balls are drawn *with replacement*. Find the probability that the sample contains x white balls and $n - x$ black balls.

Since sampling is with replacement, probability that a white ball is selected at any particular draw is $\frac{a}{a+b}$. We have therefore a series of n independent Bernoulli trials with probability of success $\frac{a}{a+b}$ at each trial. Hence, X, the number of white balls in the sample follows a binomial distribution with parameters n and $\left(\frac{a}{a+b}\right)$, when

$$P(X = x) = \binom{n}{x}\left(\frac{a}{a+b}\right)^x \left(\frac{b}{a+b}\right)^{n-x}, \quad x = 0, 1, \ldots, n. \quad \square$$

Considering (8.4.1),

$$\sum_{x=0}^{n} P(X = x) = \sum_{x=0}^{n} \binom{n}{x}p^x q^{n-x} = (p + q)^n = 1.$$

The rth factorial moment of X is

$$\mu_{(r)} = E[(X)_r] = \sum_{x=0}^{n}(x)_r \binom{n}{x}p^x q^{n-x} = (n)_r p^r \sum_{y=0}^{n-r}\binom{n-r}{y}p^y q^{n-r-y}$$
$$\text{where } y = x - r$$
$$= (n)_r p^r (p + q)^{n-r} = (n)_r p^r.$$

Thus,

$$\mu_1' = np$$
$$\mu_2' = \mu_{(2)} + \mu_{(1)} = np + n(n - 1)p^2$$
$$\mu_3' = \mu_{(3)} + 3\mu_{(2)} + \mu_{(1)} = np + 3n(n - 1)p^2 + n(n - 1)(n - 2)p^3$$
$$\mu_4' = \mu_{(4)} + 6\mu_{(3)} + 7\mu_{(2)} + \mu_{(1)}$$
$$= np + 7n(n - 1)p^2 + 6n(n - 1)(n - 2)p^3 + n(n - 1)(n - 2)(n - 3)p^4.$$

The central moments are

$$\mu_2 = npq(< \mu_1'), \quad \mu_3 = npq(q - p), \quad \mu_4 = 3(npq)^2 + npq(1 - 6pq),$$

$$\sqrt{\beta_1} = \frac{q - p}{\sqrt{npq}}, \quad \beta_2 = 3 + \frac{1 - 6pq}{npq}, \quad \beta_2 - \beta_1 - 3 = -\frac{2}{n}.$$

Thus the skewness of the distribution is positive (negative) if $p < (>)\frac{1}{2}$. The distribution is symmetrical *iff* $p = \frac{1}{2}$. For a fixed value of p, the point (β_1, β_2) falls on the straight line

$$\frac{\beta_2 - 3}{\beta_1} = 1 - \frac{2pq}{(q-p)^2}.$$

As $n \to \infty$, p remaining finite, the point (β_1, β_2) approaches the point $(0, 3)$ as a limit, that is, the distribution tends to be symmetric and mesokurtic $(\beta_2 = 3)$. Figures 8.4.1a and 8.4.1b represent binomial distributions with parameters $(10, 0.35)$ and $(10, 0.45)$ respectively.

The following recurrence relation holds for the moments.

Lemma 8.4.1: For the binomial distribution

$$\mu_{r+1} = pq(nr\mu_{r-1} + \frac{d\mu_r}{dp}). \tag{8.4.2}$$

Proof.

$$\mu_r = \sum_{j=0}^{n} \binom{n}{j} p^j q^{n-j} (j - np)^r$$
$$\frac{d\mu_r}{dp} = \sum_{j=0}^{n} \binom{n}{j} jp^{j-1} q^{n-j} (j - np)^r - \sum_{j=0}^{n} (n-j) \binom{n}{j} p^j q^{n-j-1} (j - np)^r$$
$$\quad -nr \sum_{j=0}^{n} \binom{n}{j} p^j q^{n-j} (j - np)^{r-1}$$
$$= \sum_{j=0}^{n} \binom{n}{j} p^{j-1} q^{n-j-1} (j - np)^r (jq - np + jp) - nr\mu_{r-1}.$$

Hence

$$pq\frac{d\mu_r}{dp} = \sum_{j=0}^{n} \binom{n}{j} p^j q^{n-j} (j - np)^{r+1} - npqr\mu_{r-1}$$

when $\mu_{r+1} = pq(\frac{d\mu_r}{dp} + nr\mu_{r-1})$.

A similar relation holds for the non-central moments:

$$\mu'_{r+1} = pq(\frac{n}{q}\mu'_r + \frac{d\mu'_r}{dp}). \tag{8.4.3}$$

The mean deviation about mean is given by

$$E[|X - np|] = n\binom{n-1}{[np]} p^{[np]+1} q^{n-[np]} \tag{8.4.4}$$

where $[\theta]$ denotes the highest integer contained in θ.

The ratio

$$\frac{b(k+1; n, p)}{b(k; n, p)} = \frac{n-k}{k+1}\frac{p}{q}, \quad k = 0, 1, \ldots, n-1$$

shows that $b(k+1; n, p) > (<)b(k; n, p)$ if $k < (>)(n+1)p - 1$. Hence, as k increases, the quantities $P(X = k)$ increase to a maximum value at the integer k^* satisfying

$$(n+1)p - 1 < k^* < (n+1)p, \qquad (8.4.5)$$

provided $(n+1)p$ is not an integer and thereafter they decrease. The integer k^* is the mode of the distribution.

If $(n+1)p$ is an integer,

$$P[X = (n+1)p - 1] = P[X = (n+1)p]$$

and this common value is the maximum among the values of $P[x = k]$. Here, mode is at $X = (n+1)p - 1$ and $X = (n+1)p$.

The *pgf* is

$$P_X(t) = E(t^X) = (q + pt)^n, \ |t| \le 1 \quad \text{(vide Example 7.2.2)}.$$

The *mgf* about origin is

$$M_X(t) = E(e^{tX}) = (q + pe^t)^n, \ |t| < \infty \quad \text{(vide Example 7.3.1)}.$$

The *mgf* about mean is

$$M_{X-\mu}(t) = (qe^{-pt} + pe^{tq})^n.$$

The *cgf* about mean is

$$K_{X-\mu}(t) = n\log_e(q + pe^t)^n$$
$$= n\log_e[1 + pq\frac{t^2}{2!} + pq(q-p)\frac{t^3}{3!} + pq(q^2 - pq + p^2)\frac{t^4}{4!} + \ldots].$$

Equating the coefficients of $\frac{t^2}{2!}, \frac{t^3}{3!}, \ldots$ on either side we obtain, after some simplification,

$$k_2 = npq, \ k_3 = npq(q-p), \ k_4 = npq(1 - 6pq).$$

The factorial *mgf* is

$$[(1+t)^X] = \sum_{x=0}^{n}(1+t)^x \binom{n}{x}p^x q^{n-x} = (1 + pt)^n.$$

Theorem 8.4.1: Let $X_i(i = 1, \ldots, k)$ be independent binomial variables with parameters (n, p_i). Then $S_n = \sum_{i=1}^{n} X_i$ has a binomial distribution with parameters (n, p) where $n = \sum_{i=1}^{k} n_i$.

Proof. The *mgf* of S_n is

$$P_{S_n}(t) = \Pi_{i=1}^{k}P_{X_i}(t) = \Pi_{i=1}^{k}(q + pe^t)^{n_i} = (q + pe^t)^n$$

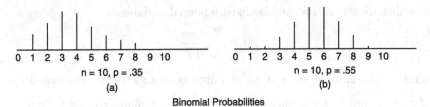

Binomial Probabilities

Fig. 8.4.1

which is the *mgf* of a binomial random variable with parameters (n, p). The result follows from the uniqueness property of *mgf* (Theorem 7.3.1). □

Consider two independent variables X_1, X_2 distributed as $b(x : n_1, p)$ and $b(x; n_2, p)$ respectively. The conditional probability $X_1 = k$ given that $X_1 + X_2 = x$ is

$$P[X_1 = k | X_1 + X_2 = x] = \frac{P(X_1=k, X_2=x-k)}{P(X_1+X_2=x)}$$

$$= \frac{\binom{n_1}{k} p^k q^{n_1-k} \binom{n_2}{x-k} p^{x-k} q^{n_2-x+k}}{\binom{n_1+n_2}{x} p^x q^{n_1+n_2-x}} \tag{8.4.6}$$

$$= \frac{\binom{n_1}{k}\binom{n_2}{x-k}}{\binom{n_1+n_2}{x}},$$

$\max(0, x - n_2) \le k \le \min(n_1, x)$. (8.4.6) is known as a hypergeometric distribution (discussed in Section 8.5).

The distribution of the difference $(X_1 - X_2)$ is

$$P(X_1 - X_2 = k) = \sum_{x_1} \binom{n_1}{x_1}\binom{n_2}{x_1 - k} p^{2x_1-k} q^{n_1+n_2-(2x_1-k)} \tag{8.4.7}$$

where the summation is over the limits, $\max(0, k) \le x_1 \le \min(n_1, n_2 + k)$. When $p = q = \frac{1}{2}$, (8.4.7) reduces to

$$P(X_1 - X_2 = k) = \frac{\binom{n_1+n_2}{n_2+k}}{2^{n_1+n_2}} \quad (-n_2 \le k \le n_1).$$

The distribution of the standardized binomial variable

$$X' = \frac{X - np}{\sqrt{npq}}$$

tends in the limit to the unit normal distribution as $n \to \infty$ (Section 9.7). If $n \to \infty$ and $p \to 0$ in such a way that $np = \lambda$ remains constant, then

$$b(x; n, p) \to e^{-\lambda} \frac{\lambda^x}{x!}.$$

This approximation has been discussed in Lemma 8.5.1.

8.4.1 *The truncated binomial distribution*

A singly truncated binomial distribution is formed if the binomial distribution is truncated at only one end, that is, if only the values $0, 1, \ldots, K_1 - 1(K_1 \geq 1)$ or the values $n - K_2 + 1, \ldots, n(K_2 \geq 1)$ are omitted from the values of a binomial variable. The distribution obtained by omitting the value 0 only, is

$$P(X = r) = \frac{\binom{n}{r} p^r q^{n-r}}{1 - q^n}, \quad r = 1, \ldots, n. \tag{8.4.8}$$

The jth raw moment of the truncated distribution (8.4.8) is obtained by dividing the corresponding moment of the complete distribution (8.4.1) by $1 - q^n$. In particular, its mean and variance are $\frac{np}{1-q^n}$ and $\frac{npq}{1-q^n} - \frac{n^2 p^2 q^n}{(1-q^n)^2}$ respecively.

A binomial distribution truncated on the left at $x = K$ is

$$\frac{\binom{n}{x} p^x q^{n-x}}{\sum_{x=K}^{n} \binom{n}{x} p^x q^{n-x}} = \frac{1}{H_K(z)} \binom{n}{x} z^x, \quad x = K, K+1, \ldots, n \tag{8.4.9}$$

where $z = \frac{p}{q}$, $H_K(z) = \sum_{K}^{n} \binom{n}{x} z^x$.

Mean of the distribution is

$$\mu_1' = \frac{1}{H_K(z)} \sum_{K}^{n} x \binom{n}{x} z^x = \frac{z H_K'(z)}{H_K(z)}. \tag{8.4.10}$$

For the distribution (8.4.9) the recuurence relation (8.4.2) holds, i.e.,

$$\mu_{r+1} = pq \left(nr\mu_{r-1} + \frac{d\mu_r}{dp} \right). \tag{8.4.11}$$

Proof.

$$\mu_r = \frac{1}{H_K(z)} \sum_K^n (x - \mu_1')^r \binom{n}{x} z^x$$

$$z\frac{d\mu_r}{dz} = \frac{d}{dz}\{\sum_K^n (x - \mu_1')^r \binom{n}{x} z^x\} \frac{z}{H_K(z)} - \{\sum_K^n (x - \mu_1')^r \binom{n}{x} z^{x+1}\} \frac{H_K'(z)}{(H_K(z))^2}$$

$$= \frac{1}{H_K(z)} \sum_K^n \binom{n}{x}(x - \mu_1')^r x z^x - z r \frac{d\mu_1'}{dz} \frac{1}{H_K(z)} \sum_K^n \binom{n}{x} z^x (x - \mu_1')^{r-1}$$

$$- \frac{zH_K'(z)}{H_K(z)} \frac{1}{H_K(z)} \sum_K^n (x - \mu_1')^r \binom{n}{x} z^x.$$

$$(8.4.12)$$

The first and the last term together on the right hand side of (8.4.12) give

$$\frac{1}{H_K(z)} \sum_K^n \binom{n}{x} z^x (x - \mu_1')^{r+1} = \mu_{r+1}, \quad \text{(using (8.4.9))}.$$

Hence, (8.4.12) reduces to

$$\mu_{r+1} = z[\frac{d\mu_r}{dz} + r\frac{d\mu_1'}{dz}\mu_{r-1}].$$

(8.4.11) follows, using $\frac{dz}{dp} = \frac{1}{q^2}$.

The relation (8.4.3) holds similarly for the distribution (8.4.9).

A doubly truncated binomial variable is formed from (8.4.1) by omitting the values of x such that $0 \le x < K_1$ and $x - K_2 < x \le n($ with $0 \le K_1 \le n - K_2 \le n)$. For the resulting distribution

$$P(X = j) = \frac{\binom{n}{j}p^j q^{n-j}}{\sum_{r=K_1}^{n-K_2} \binom{n}{r}p^r q^{n-r}}, \quad j = K_1, K_1 + 1, \ldots, n - K_2.$$

8.5 The Hypergeometric Distribution

A random variable X is said to have a hypergeometric distribution with parameters $N, M,$ and n if its *pmf* is given by

$$P(X = x) = H(x; N, M, n) \text{ (say)} = \begin{cases} \frac{\binom{M}{x}\binom{N-M}{n-x}}{\binom{N}{n}} & \text{for } \max(0, n - N + M), \\ & \le x(\text{integer}) \le \min(M, n) \\ 0 & \text{(otherwise)}, \end{cases}$$

$$(8.5.1)$$

where N is a positive integer, M is a positive integer not exceeding N and n is a positive integer that is at most N.

The formula in (8.5.1) is valid for integer values of x such that

$$0 \le x \le M \quad \text{and} \quad 0 \le n - x \le N - M;$$

that is,

$$\max(0, n - N + M) \leq x \leq \min(M, n).$$

Note that the same distribution is obtained if n and M are interchanged.

The quantities in the right side of (8.5.1) are the successive terms in the expansion of

$$\frac{(N-n)!(N-M)!}{N!(N-M-n)!} F(-n, -M, N-M-n+1, 1)$$

where

$$F(\alpha, \beta, \gamma, t) = 1 + \frac{\alpha\beta}{\gamma}\frac{t}{1} + \frac{\alpha(\alpha+1)\beta(\beta+1)}{\gamma(\gamma+1)}\frac{t^2}{2!} + \cdots$$

is a hypergeometric function.

The distribution arises in the following circumstances. Suppose an urn contains M white balls and $N - M$ black balls. From this, n balls are selected without replacement. The probability that the sample contains x white balls and $(n-x)$ black balls is given by (8.5.1). The hypergeometric distribution, therefore, arises in sampling from a dichotomous population when sampling is without replacement whereas a binomial distribution arises when sampling is with replacement.

We have

$$H(x; N, M, n) = \binom{n}{x}\frac{(M)_x(N-M)_{(n-x)}}{(N)_n} \quad \text{for } x = 0, 1, \ldots, n.$$

The rth factorial moment about origin is

$$\mu_{(r)} = \sum_{x=0}^{n}(x)_r \frac{\binom{n}{x}(M)_x(N-M)_{(n-r)}}{(N)_n}$$

$$= \frac{1}{(N)_n}\sum_{y=0}^{n-r}(n-y)_r\binom{n}{y}(M)_{(n-y)}(N-M)_y \quad \text{writing } n-x=y$$

$$= \frac{(M)_r(n)_r}{(N)_n}\sum_{y=0}^{n-r}\binom{n-r}{y}(M-r)_{(n-r-y)}(N-M)_y$$

$$= \frac{(M)_r(n)_r}{(N)_n}(N-r)_{(n-r)} \quad [\text{ as } \sum_{x=0}^{n}H(x; N, M, n) = 1.]$$

$$= \frac{(M)_r(n)_r}{(N)_r}.$$

Hence

$$E(X) = \frac{Mn}{N} = np \ (\text{ writing } p = \frac{M}{N}, q = 1 - p)$$

$$V(X) = \frac{N-n}{N-1}npq,$$

$$\sqrt{\beta_1} = \frac{q-p}{[(\frac{N-n}{N-1})npq]^{1/2}} \frac{N-2n}{N-2},$$

$$\beta_2 = \frac{3(N-1)(N+6)}{(N-2)(N-3)} + \frac{N(N^2-1)}{(N-n)(N^2-5N+6)}$$
$$[1 - \frac{6N}{N+1}(pq + \frac{n(N-n)}{N^2})]\frac{1}{npq}.$$

The following recurrence relations hold:

$$H(x+1; N, M, n) = \frac{(M-x)(n-x)}{(x+1)(N-M-n+x+1)} H(x; N, M, n)$$

$$H(x; N, M+1, n) = \frac{(M+1)(N-M-n+x)}{(N-M)(M+1-x)} H(x; N, M, n)$$

(8.5.2)

$$H(x; N, M, n+1) = \frac{(N-M-n+x)(n+1)}{(n+1-x)(N-n)} H(x; N, M, n)$$

$$H(x; N+1, M, n) = \frac{(N+1-n)(N+1-M)}{(N+1-n-M+x)(N+1)} H(x; N, M, n).$$

From the first equation of (8.5.2) it follows that $H(x+1; N, M, n) \geq (\leq)$ $H(x; N, M, n)$ according as

$$x \leq (\geq) \frac{(n+1)(M+1)}{N+2} - 1.$$

Hence the distribution has mode at k^2 where k^2 is the greatest integer which does not exceed $\frac{(n+1)(M+1)}{N+2}$. If $\frac{(n+1)(M+1)}{N+2}$ is an integer, the distribution is binomial having modes at

$$x = \frac{(n+1)(M+1)}{N+2} - 1 \text{ and } x = \frac{(n+1)(M+1)}{N+2}.$$

The following binomial approximation to a hypergeometric distribution holds:

$$H(x; N, M, n) = \binom{n}{x} p^x (1-p)^{n-x}$$

(8.5.3)

where $p = \frac{M}{N}$, when $N \to \infty$, provided $\frac{n}{N}$ is negligible.

If $\frac{M}{N}$ is small, but n is large, the Poisson approximation,

$$H(x; N, M, n) \approx \frac{(np)^x}{x!} e^{-np}$$

(8.5.4)

where $p = \frac{M}{N}$, holds. The Poisson distribution has been discussed in Section 8.6.

8.5.1 *The positive hypergeometric distribution*

This is a truncated hypergeometric distribution where the zero observed value is omitted. However, if $n > (N - M)$, there is no zero observed value. If $n \leq (N - M)$, the distribution is

$$P(X = x) = \frac{\binom{M}{x}\binom{N-M}{n-x}}{[\binom{N}{n} - \binom{N-M}{n}]}, \quad x = 1, \ldots, \min(n, M).$$

8.5.2 *The negative hypergeometric distribution*

This distribution arises in the following circumstances. Suppose we have k black balls and w white balls in an urn. Balls are drawn one at a time, without replacement. We are interested in the probability distribution of the number R of draws needed to obtain a specified number k of black balls.

$$\begin{aligned} P(R = r) &= P((k - 1) \text{ black balls are drawn in the first } (r - 1) \text{ draws })\\ &\quad P(\text{ a black ball is drawn from an urn containing } (b - k + 1)\\ &\quad \text{black balls and } (w - r + k) \text{ white balls }) \end{aligned}$$

$$= \binom{r-1}{k-1} \frac{b(b-1)\ldots(b-k+2)w(w-1)\ldots(w-r+k+1)}{(b+w)(b+w-1)\ldots(b+w-r+2)} \frac{b-k+1}{b+w-r+1}$$

$$= \binom{r-1}{k-1} \frac{(b)_k(w)_{r-k}}{(b+w)_r}, \quad r = k, k + 1, \ldots$$

The distribution of R is a negative hypergeometric distribution with parameters b, w, k.

The jth factorial moment of R is

$$E[R(R + 1) \ldots (R + j - 1)] = \frac{k^{(j)}(b + w + 1)^{(j)}}{(b + 1)^{(j)}}$$

where $n^{(t)} = n(n - 1) \ldots (n + t - 1)$.

In particular,

$$E(R) = \frac{k(b + w + 1)}{b + 1}, \quad V(R) = \frac{k(b + w + 1)w(b + 1 - k)}{(b + 1)^2(b + 2)}.$$

8.6 The Poisson Distribution

A random variable X is said to have a Poisson distribution with parameter $\lambda(> 0)$ if its *pmf* can be written as

$$P(X = x) = P(x; \lambda) = \begin{cases} e^{-\lambda}\frac{\lambda^x}{x!}, & x = 0, 1, 2, \ldots \\ 0 & \text{otherwise.} \end{cases} \qquad (8.6.1)$$

The *pgf* is $P_X(t) = e^{-\lambda(1-t)}, |t| \leq 1$ (vide Example 7.2.3).

$$\mu_1' = \sum_{x=0}^{\infty} xe^{-\lambda}\frac{\lambda^x}{x!} = \lambda e^{-\lambda}(1 + \frac{\lambda}{1} + \frac{\lambda^2}{2!} + \ldots) = \lambda e^{-\lambda}e^{\lambda} = \lambda,$$
$$\mu_2' = \mu_1' + \mu_{(2)} = \lambda + \sum_{x=0}^{\infty} x(x-1)e^{-\lambda}\frac{\lambda^x}{x!} = \lambda + \lambda^2.$$

Hence, $\mu_2 = \lambda$.

The *mgf* about zero is

$$M_X(t) = e^{\lambda(e^\lambda - 1)} \quad \text{(vide Example 7.3.2)}.$$

Hence

$$\mu_3' = \lambda + 3\lambda^2 + \lambda^3, \ \mu_4' = \lambda + 7\lambda^2 + 6\lambda^3 + \lambda^4, \ \mu_3 = \lambda, \ \mu_4 = \lambda + 3\lambda^2,$$

$$\sqrt{\beta_1} = \frac{1}{\sqrt{\lambda}}, \ \beta_2 = 3 + \frac{1}{\lambda}.$$

Thus, for any Poisson distribution,

$$\mu_1' = \mu_2, \quad \text{and} \quad \beta_2 - \beta_1 - 3 = 0.$$

Lemma 8.6.1: The central moments of a Poisson distribution satisfy the following recurrence relation

$$\mu_{r+1} = \lambda(r\mu_{r-1} + \frac{d\mu_r}{d\lambda}). \tag{8.6.2}$$

Proof. We have

$$\mu_r = \sum_{x=0}^{\infty}(x - \lambda)^r e^{-\lambda}\frac{x^r}{r!}$$
$$\frac{d\mu_r}{d\lambda} = \sum_{x=0}^{\infty}\left[-r(x-\lambda)^{r-1} + (x-\lambda)^r(\frac{x}{\lambda} - 1)\right]e^{-\lambda}\frac{\lambda^x}{x!}$$
$$= -r\mu_{r-1} + \frac{1}{\lambda}\lambda_{r+1}.$$

Hence the proof. $\square$

The ratio $\frac{P(x;\lambda)}{P(x-1;\lambda)} = \frac{\lambda}{x}$ shows that $P(x;\lambda) > (<)P(\lambda - 1; x)$ according as $x < (>)\lambda$. Hence, if λ is not an integer, the distribution has mode at the integral value x^* satisfying

$$\lambda - 1 < x^* < \lambda.$$

If λ is an integer, $P[X = \lambda - 1] = P[X = \lambda]$ and the distribution has modes at $x = \lambda - 1$ and $x = \lambda$.

The mean deviation of the Poisson distribution is

$$E[|X - \lambda|] = \frac{2e^{-\lambda}\lambda^{[\lambda]+1}}{[\lambda]}.$$

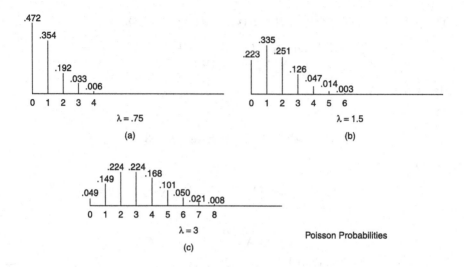

Fig. 8.6.1

The figures 8.6.1(a), (b) and (c) represent Poisson distribution with $\lambda = 0.75, 1.5$ and 3 respectively.

We have seen in Example 7.3.8 that if $X_1, \ldots, X_n$ are n independent Poisson variates, X_i having parameter λ_i, then the sum $S_n = \sum_{i=1}^{n} X_i$ has *mgf* $M_{S_n}(t) = e^{\sum_{i=1}^{n} \lambda_i (e^t - 1)}$. It, therefore, follows by the uniqueness property of *mgf* (Theorem 7.3.1) that S_n follows a Poisson distribution with parameter $\sum_{i=1}^{n} \lambda_i$. This is called the *reproductive property* of the Poisson distribution.

The Poisson distribution provides a reliable model for many random phenomena. Since the values of a Poisson variable are non-negative integers, the distribution may be used to describe experiments in which the observed variable is a count. The Poisson distribution may also be used to describe events occurring randomly and independently in time (space), e.g., number of telephone calls coming into the telephone exchange in a period of time of length t, number of fatal traffic accidents occurring per week in a given region, number of cars passing through a street in time t, number of α particles emitted by a radio-active substance reaching a given portion of space during a period of time of length t, number of flying-bomb hits over a region divided into several small strips. In all such cases, we may assume that the conditions of the experiment remain constant in time (space) and the non-overlapping intervals of time (space) are stochastically independent in the sense that the information concerning the number of events in one interval does not give any information about the other. These two assumptions provide the basis for a random variable to obey Poisson law. More specifically, let us assume the following:

(i) The probability that exactly one event occurs during a small time interval of length dt is $\lambda dt + o(dt)$ [$o(dt)$ denotes an unspecified function which is of smaller order than dt; $\lim_{dt\to 0} \frac{o(dt)}{dt} = 0$.]

(ii) The probability that more than one event occur during a small interval of length dt is $o(dt)$.

(iii) Number of occurrences in non-overlapping time intervals are independent.

The quantity λ may be called the mean rate of occurrence.

Theorem 8.6.1: Let X be a random variable denoting the number of occurrences of an event in a time interval of length t. If the assumptions (i) - (iii) above are satisfied, X follows a Poisson distribution with parameter λt.

Proof. Let $P_x(t)$ be the probability that x events occur during the time interval $(0, t)$, $x = 0, 1, 2, \ldots$

Hence,

$$
\begin{aligned}
P_x(t + dt) &= P(\text{no event in } (0, t + dt]) \\
&= P(\text{no event in } (0, t] \text{ and no event in } (t, t + dt]) \qquad (8.6.3) \\
&= P(\text{no event in } (0, t]) P((\text{no event in}(t, t + dt])
\end{aligned}
$$

Now

$$
\begin{aligned}
&P(\text{no event in } (t, t + dt]) \\
&= 1 - P(\text{one event in}(t, t + dt]) - P(\text{two or more events in } (t, t + dt]) \\
&= 1 - \lambda dt - o(dt).
\end{aligned}
$$

Hence from (8.6.3)

$$
\frac{P_x(t + dt) - P_x(t)}{dt} = -\lambda P_x(t) + \frac{o(dt)}{dt} P_0(t).
$$

Taking limit as $dt \to 0$

$$P_0'(t) = -\lambda P_0(t).$$

The solution of this differential equation is

$$P_0(t) = e^{-\lambda t} \text{ since } P_0(0) = 1.$$

For $x = 1, 2, \ldots$

$$
\begin{aligned}
P_x(t + dt) &= P(x \text{ events in } (0, t] \text{ and no event in } (t, t + dt]) \\
&\quad + P(x - 1 \text{ events in } (0, t] \text{ and one event in } (t, t + dt]) \\
&\quad + \sum_{h=2}^{x} P(x - h \text{ events in } (0, t] \text{ and h events in } (t, t + dt]) \\
&= P_x(t)(1 - \lambda dt - 0(dt)) + P_{x-1}(\lambda dt + o(dt)) \\
&\quad + \sum_{h=2}^{x} P_{x-h}(t)(o(dt)),
\end{aligned}
$$

i.e.,

$$P_x(t + dt) = P_x(t)(1 - \lambda dt) + P_{x-1}(t)\lambda dt + o(dt)$$

or

$$\frac{P_x(t + dt) - P_x(t)}{dt} = -\lambda[P_x(t) - P_{x-1}(t)] + \frac{o(dt)}{dt}.$$

Taking limit as $dt \to 0$,

$$P_x'(t) = \lambda[P_{x-1}(t) - P_x(t)]. \tag{8.6.4}$$

Hence, for $x = 1$,

$$P_1'(t) + \lambda P_1(t) = \lambda e^{-\lambda t}.$$

Multiplying both sides by $e^{\lambda t}$ and integrating, we obtain

$$e^{\lambda t} P_1(t)\lambda = \lambda + K$$

where K is a constant.

The initial condition $P_1(0) = 0$ implies

$$P_1(t) = \lambda t e^{-\lambda t}. \tag{8.6.5}$$

Similarly, $P_2(t)$ is found from (8.6.4) by using (8.6.5). Induction on n shows that

$$P_n(t) = e^{-\lambda t} \frac{(\lambda t)^n}{n!}.$$

EXAMPLE 8.6.1: A telephone switch-board receives 20 calls on an average during an hour. Find the probability that during a period of 5 minutes (a) no call is received, (b) exactly 3 calls are received (c) more than 5 calls are received. Assume that the time is measured in minutes.

We assume that the number of incoming calls during any time period obeys a Poisson law (which follows on the basis of assumptions (i) - (iii) above). The mean rate of occurrence of calls is 0.33 calls per minute. Hence, the number of calls in a 5-minute period follows a Poisson distribution with parameter $\lambda = 1.65$.

(a) P(no call in a 5-minute period) $= e^{-1.65} = 0.192$.

(b) P(3 calls in a 5-minute period) $= e^{-1.65}\frac{(1.65)^3}{3!} = 0.144$.

(c) P(more than 5 calls in a 5-minute period) $= \sum_{x=0}^{\infty} e^{-1.65}\frac{(1.65)^x}{x!} = 0.007$.

EXAMPLE 8.6.2: A traffic light has a constant probability λdt of changing light-signals, - from red to green or the reverse,- in an infinitesimal interval of length dt. Prove that a car arriving at a random instant has a probability $\frac{1}{2}$ of passing through without waiting and a probability element $\frac{1}{2}\lambda e^{-\lambda w}$ of waiting for a time period of $w(> 0)$ units before passing.

A car would pass at any moment with or without waiting. Since both the cases are mutually exclusive and equally likely, the probability of each case is $\frac{1}{2}$.

The car waits for a time period w if (i) the traffic light is red when it arrives and (ii) does not change to green within the next time period w and (iii) therefater (within time dt) changes. The three events are mutuay independent and probability of (i) and (iii) are $\frac{1}{2}$ and λdt respectively.

Let $\frac{w}{dt} = n$. The chance that the traffic light does not change in time w is

$$(1 - \lambda dt)^n = (1 - \lambda dt)^{\frac{w}{dt}} = (1 - x)^{\frac{w\lambda}{x}}.$$

which on taking limit as $dt \to 0$(i.e., $x \to 0$) reduces to $e^{-w\lambda}$.

The required probability is, therefore, $\frac{1}{2}e^{-w\lambda}\lambda dt$.

EXAMPLE 8.6.3: Suppose that the number of seeds produced by a tree follows a Poisson distribution with parameter λ and the probability of a seed developing into a sapling is p. Assuming mutual independence of the events, show that the number of saplings follows a Poisson distribution with parameter λp.

Let X, Y be the random variables denoting respectively the number of seeds and the number of saplings.

$$P(X = n, Y = k) = (e^{-\lambda}\frac{\lambda^n}{n!})\left(\binom{n}{k}p^k q^{n-k}\right), \quad q = 1 - p, k \leq n$$

(by independence).

$$P(Y = k) = \sum_{n=k}^{\infty} P(X = n, Y = k) = e^{-\lambda}(\tfrac{p}{q})^k \sum_{n=k}^{\infty} \frac{(q\lambda)^n}{k!(n-k)!}$$
$$= e^{-\lambda}(\tfrac{p}{q})^k (q\lambda)^k \sum_{n-k=0}^{\infty} \frac{(q\lambda)^{n-k}}{k!(n-k)!} = e^{-\lambda}(\tfrac{p}{q})^k \frac{(q\lambda)^k}{k!} e^{q\lambda}$$
$$= e^{-\lambda p}\frac{(\lambda p)^k}{k!}, \quad k = 0, 1, \ldots$$

EXAMPLE 8.6.4: Let X_1, X_2 be two independent Poisson variables with parameters λ_1 and λ_2 respectively. Show that the conditional distribution of X_1 given $X_1 + X_2$ is binomial.

For positive integer $k, n(k < n)$,

$$P(X_1 = k | X_1 + X_2 = n) = \frac{P(X_1 = k, X_2 = n-k)}{P(X_1 + X_2 = n)}$$
$$= \frac{(e^{-\lambda_1}\frac{\lambda_1^k}{k!})(e^{-\lambda_2}\frac{\lambda_2^{n-k}}{(n-k)!})}{e^{-(\lambda_1+\lambda_2)}\frac{(\lambda_1+\lambda_2)^n}{n!}}$$

(since X_1, X_2 are independent and
$X_1 + X_2$ is a Poisson $(\lambda_1 + \lambda_2)$ variable)

$$= \binom{n}{k}p^k(1 - p)^{n-k} \text{ where } p = \frac{\lambda_1}{\lambda_1+\lambda_2},$$
$$k = 0, 1, \ldots, n.$$

Hence the conditional distribution of X_1 given $X_1 + X_2 = n$, is binomial with parameters n and $p = \frac{\lambda_1}{\lambda_1+\lambda_2}$.

EXAMPLE 8.6.5: The number of female insects in a given region follows a Poisson distribution with parameter λ. The number of eggs laid by each female insect follows a Poisson distribution with mean μ. Find the probability distribution of number of eggs in the region.

Let X denote the number of female insects and Y the number of eggs laid down in the region.

$$P(X = T) = e^{-\lambda}\frac{\lambda^t}{t!}.$$

Let Y_i denote the number of eggs laid down by the ith insect. Hence

$$P(Y = s | X = t) = P(Y_1 = s_1, Y_2 = s_2, \ldots, Y_t = s_t; \sum_{i=1}^{t} s_i = s)$$
$$= \sum_{s_1,\ldots,s_i:\sum s_i=s}(e^{-\mu}\frac{\mu^{s_1}}{s_1!}) \ldots (e^{-\mu}\frac{\mu^{s_t}}{s_t!})$$
$$= \sum{}' e^{-\mu t}\frac{\mu^s}{s_1! \ldots s_t!} \text{ (say)}$$
$$= \frac{1}{s!}e^{-\mu t}(\mu t)^s.$$

Hence,

$$P(X = t; Y = s) = (e^{-\lambda}\tfrac{\lambda^t}{t!})(e^{-\mu t}\tfrac{(\mu t)^s}{s!})$$
$$P(Y = s) \quad = \sum_{t=0}^{\infty} P(X = t; Y = s) = \frac{e^{-\lambda}\mu^s}{s!}\sum_{t=0}^{\infty}\tfrac{\lambda^t t^s}{t!}e^{-\mu t}$$

which is the distribution of Y. The *mgf* of Y is

$$M_Y(h) = E(e^{hY}) = e^{-\lambda}\sum_{s=0}^{\infty}\frac{e^{hs}e^{-\lambda}\mu^s}{s!}\sum_{t=0}^{\infty}\tfrac{\lambda^t t^s}{t!}e^{-\mu t}$$
$$= e^{-\lambda}\sum_{t=0}^{\infty}\sum_{s=0}^{\infty}e^{-\mu t}\frac{(\mu t e^h)^s \lambda^t}{s! t!}$$

$$= e^{-\lambda}\sum_{t=0}^{\infty}\frac{\{\lambda e^{\mu(e^h-1)}\}^t}{t!}$$
$$= e^{-\lambda}e^{\lambda e^{\mu(e^h-1)}} = e^{\lambda(e^{\mu(e^h-1)}-1)}$$

$$= \exp\left[\lambda\{e^{\mu(h+\frac{h^2}{2!}+\cdots)} - 1\}\right]$$

$$= \exp[\lambda\{\mu(h + \tfrac{h^2}{2!} + \ldots) + \tfrac{\mu^2}{2!}(h + \tfrac{h^2}{2!} + \ldots)^2 + \ldots\}]$$

$$= 1 + \lambda\left[\mu(h + \tfrac{h^2}{2!} + \ldots) + \tfrac{\mu^2}{2!}(h + \tfrac{h^2}{2!} + \ldots)^2 + \ldots\right]$$
$$+ \tfrac{\lambda^2}{2!}\left[\mu(h + \tfrac{h^2}{2!} + \ldots) + \tfrac{\mu^2}{2!}(h + \tfrac{h^2}{2!} + \ldots)^2 + \ldots\right]^2$$
$$+ \ldots$$

Hence,

$$\mu_1' = \text{coefficient of } h = \mu\lambda, \quad \mu_2' = \text{coefficient of } \tfrac{h^2}{2!} = \lambda(\mu^2 + \mu + \lambda\mu^2),$$
$$\mu_2 = \lambda\mu(\mu + \lambda).$$

The following lemma shows that a binomial distribution tends to a Poisson distribution under certain conditions.

Lemma 8.6.2: Let X be a binomial random variable with parameters n and p. If $n \to \infty$ and $p \to 0$ in such a way that $np = \lambda$ remains constant then

$$b(x; n, p) \to e^{-\lambda}\frac{\lambda^x}{x!}$$

for a fixed integer x.

Proof.

$$b(x; n, p) = \binom{n}{x}p^x q^{n-x} = \frac{n!}{x!(n-x)!}(\tfrac{\lambda}{n})^x(1 - \tfrac{\lambda}{n})^{n-x}$$

$$= \{\tfrac{\lambda^x}{x!}(1 - \tfrac{\lambda}{n})^n(1 - \tfrac{\lambda}{n})^{-x}\}\{\frac{n!}{n^x(n-x)!}\}.$$

Hence

$$\lim_{n\to\infty} b(x;n,p) = e^{-\lambda}\frac{\lambda^x}{x!}\left[\lim_{n\to\infty}\frac{n!}{n^x(n-x)!}\right]. \qquad (8.6.6)$$

Using Stirling's approximation for $n!$,

$$n! \approx \sqrt{2\pi}e^{-n}n^{n+\frac{1}{2}}$$

the quantity in [ ] in (8.6.6) is

$$\lim_{n\to\infty}\left[\frac{e^{-x}n^{n+\frac{1}{2}-x}}{(n-x)^{n-x+\frac{1}{2}}}\right] = \lim_{n\to\infty}\left[e^x(1-\frac{x}{n})^n(1-\frac{x}{n})^{-x+\frac{1}{2}}\right]^{-1} = e^x.e^{-x} = 1.$$

Hence the proof. □

8.6.1 *The truncated Poisson distribution*

Consider a truncated Poisson variable, truncated at $x = 0$, whose *pmf* is

$$P(X = x) = e^{-\lambda}\frac{\lambda^x}{x!}(1-e^{-\lambda})^{-1} = \frac{\lambda^x}{x!(e^\lambda - 1)}, \quad x = 1, 2, \ldots$$

$$E(X) = \frac{\lambda}{1-e^{-\lambda}}, V(X) = (1-e^{-\lambda})^{-1}[1-\lambda e^{-\lambda}(1-e^{-\lambda})^{-1}].$$

The rth factorial moment is

$$\mu_{(r)} = \frac{\lambda^r}{(1-e^{-\lambda})}.$$

8.7 The Geometric Distribution

A random variable X is said to have a geometric distribution with parameter p if its *pmf* is

$$P(X = x) = g(x;p) = \begin{cases} p(1-p)^x, & x = 0, 1, \ldots \\ 0, & \text{(otherwise)}. \end{cases} \qquad (8.7.1)$$

The parameter p satisfies $0 < p < 1$.

The geometric distribution arises in the following circumstances. Bernoulli trials with constant probability p of success are performed until the first success occurs. The probability that exactly x failures precede the first success is given by (8.7.1).

The *pgf* is

$$P_X(t) = p(1-tq)^{-1} \quad \text{(vide Example 7.2.4)}.$$

The *mgf* about origin is

$$M_X(t) = p(1 - qe^t)^{-1} \quad \text{(vide Example 7.3.3)} .$$

Thus

$$\mu_1' = \frac{q}{p}, \mu_2' = \frac{q}{p} + 2\frac{q^2}{p^2}, \ \mu_2 = \frac{q}{p^2}.$$

$$\sqrt{\beta_1} = \frac{1+q}{\sqrt{q}}, \ \beta_2 = 9 + \frac{p^2}{q}, \ \beta_2 - \beta_1 = 5.$$

The factorial moment of order r is

$$\mu_{(r)} = r!(\frac{q}{p})^r.$$

A very interesting property of geometric distribution is that the conditional probability that a geometric random variable X is greater than or equal to $j + k$ given that X is greater than j is equal to the conditional probability that it will be greater than or equal to k. Thus if a sequence of repeated independent Bernoulli trials, each with probability p of success, is performed until the first success occurs, the probability that the number of failures X preceding the first success will be $\geq j + k$ given that $X \geq j$, is equal to the unconditional probability that $X \geq k$. The event that the waiting time for the first success has already been at least j, does not affect the probability of future waiting time. As a crude example if the arrival of a bus of a particular route at a specified bus-stop in each minute (unit of time) is considered as a Bernoulli trial with constant probability p and if these trials are assumed independent, the probability that a person has to wait, say, at least for 5 minutes, remains independent of whether he has actually waited for 5 minutes or is just coming. This is called the "lack of memory" property of the geometric distribution. This is proved in the following theorem.

Theorem 8.7.1: If X has a geometric distribution, then for any two positive integers j and k,

$$P[X \geq j + k | X \geq j] = P[X \geq k].$$

Proof.

$$P[X \geq j + k | X \geq j] = \frac{P\{X \geq j+k\}}{P\{X \geq j\}} = \frac{\sum_{x=j+k}^{\infty} pq^x}{\sum_{y=j}^{\infty} pq^y} = q^k.$$

Again, $P\{X \geq k\} = p\sum_{y=k}^{\infty} q^y = q^k.$ $\qquad\qquad \square$

Some authors define a geometric random variable as the waiting time till the first success, including the trial on which the success occurs. Here

$$P(X = x) = pq^{x-1}, \ x = 1, 2, \ldots$$

Here $E(X) = \frac{1}{p}, V(X) = \frac{q}{p^2}, M_X(t) = \frac{pe^t}{1-qe^t}.$

EXAMPLE 8.7.1: Let X_1, X_2 be independent random variables each having geometric distribution $q^x p, x = 0, 1, \ldots$ Show that the conditional distribution of X_1 given $X_1 + X_2$ is rectangular.

$$P(X_1 = r | X_1 + X_2 = n) = \frac{P(X_1=r, X_2=n-r)}{P(X_1+X_2=n)}$$

$$= \frac{P(X_1=r)P(X_2=n-r)}{P(X_1+X_2=n)} \quad \text{(by independence).}$$
$$\tag{8.7.2}$$

Now

$$P(X_1+X_2 = n) = \sum_{x=0}^{\infty} P(X_1=x)P(X_2=n-x) = \sum_{x=0}^{n} q^x pq^{n-x} p = p^2 q^n(q+1).$$

Hence, value of the expression in (8.7.2) is $\frac{1}{n+1}$.

EXAMPLE 8.7.2: Let X_1, X_2 be two independent random variables, each having geometric distribution $q^k p, k = 0, 1, \ldots$ Find the probability distribution of $Z = \max(X_1, X_2)$.

$$\begin{aligned}
P(Z = j) &= P(Z = j, X_1 = j, X_2 < j) + P(Z = j, X_2 = j, X_1 < j) \\
&\quad + P(Z = j, X_1 = j, X_2 = j) \\
&= 2q^j p \sum_{k=0}^{j-1} q^k p + (q^j p)^2 = 2q^j p^2 \left(\frac{1-q^j}{1-q}\right) + q^{2j} p^2 \\
&= 2q^j p(1 - q^j) + q^{2j} p^2 = 2q^j p - q^{2j} p - q^{2j+1} p.
\end{aligned}$$

8.8   The Negative Binomial Distribution

A random variable X is said to have a negative binomial distribution with parameters k and p if its *pmf* is given by

$$P(X = x) = b'(x; k, p) = \begin{cases} \binom{k+x-1}{x} p^k q^x = \binom{-k}{x} p^k (-q)^x, \ x = 0, 1, 2, \ldots \\ 0, \qquad\qquad\qquad\qquad\qquad\qquad \text{(otherwise)} \end{cases}$$
$$\tag{8.8.1}$$

where the parameters satisfy $0 < p < 1, k = 1, 2, 3, \ldots$

A negative binomial distribution arises in the following circumstances. An urn contains np white balls and nq black balls. Balls are drawn at random with replacement. The probability that exactly $x + k$ trials will be required to produce k white balls is

$$\binom{x+k-1}{k-1}q^x p^k = \frac{(x+k-1)(x+k-2)\ldots k}{x!}q^x p^k$$
$$= (-1)^x \frac{(-k)(-k-1)\ldots(-k-x+1)}{x!}q^x p^k = (-1)^x \binom{-k}{x}p^k q^x.$$

Clearly,

$$\sum_{x=0}^{\infty} b'(x; k, p) = p^k \sum_{x=0}^{\infty} \binom{-k}{x}(-q)^x = p^k (1 - q)^{-k} = 1.$$

The distribution (8.8.1) remains meaningful even when k is not an integer. When k is an integer the distribution is sometimes called a Pascal distribution. In this case (8.8.1) is also called a discrete waiting time distribution as it gives the probability that the waiting time is x (in terms of failures) to attain the kth success in an unending sequence of Bernoulli trials. For $k = 1$, the distribution reduces to the geometric distribution.

It is convenient to express the distribution in terms of parameters $Q = \frac{1}{p}, P = \frac{1-p}{p}$ so that $Q - P = 1$. The expression (8.8.1) reduces to

$$\binom{k+x-1}{x}\left(\frac{P}{1+P}\right)^x (1 + P)^{-k} = Q^{-k}\binom{k+x-1}{x}\left(\frac{P}{Q}\right)^x$$
$$= \binom{k+x-1}{x}\left(\frac{P}{Q}\right)^x \left(1 - \frac{P}{Q}\right)^k, \quad x = 0, 1, 2, \ldots$$
$$(8.8.2)$$

which is the $(x + 1)$th term in the expansion of $(Q - P)^{-k}$.

The *pgf* is

$$E(t^X) = (Q - P)^{-k} = p^k (1 - tq)^{-k}, |t| \leq 1 \quad \text{(vide Example 7.1)}.$$

The *mgf* about origin is

$$M_X(t) = E(t^X) = \sum_{x=0}^{\infty} \binom{k+x-1}{x}\left(\frac{Pe^t}{Q}\right)^x \left(1 - \frac{P}{Q}\right)^k$$
$$= Q^{-k} \sum_{x=0}^{\infty} \binom{-k}{x}\left(-\frac{Pe^t}{Q}\right)^x = (Q - Pe^t)^{-k}$$
$$= p^k (1 - qe^t)^{-k} \quad \text{for } qe^t < 1.$$

Thus,

$$\mu_1' = M_X'(t)\big|_{t=0} = \left|kPe^t(Q - Pe^t)^{-(k+1)}\right|_{t=0} = kP$$
$$\mu_2' = M_X(t)''\big|_{t=0}$$
$$= \left|kPe^t(Q - Pe^t)^{-(k+1)} + k(k+1)P^2 e^{2t}(Q - Pe^t)^{-(k+2)}\right|_{t=0}$$
$$= kP + k(k+1)P^2.$$

Hence,

$$\mu_2 = kPQ = k\frac{q}{p^2}(> \mu_1').$$

Note that the variance of the negative binomial distribution is greater than its mean (for binomial distribution, the variance is less than its mean).

The jth factorial moment about zero is

$$\mu_{(j)} = \sum_{x=j}^{\infty} (x)_j \binom{k+x-1}{x} (\tfrac{P}{Q})^x (1 - \tfrac{P}{Q})^k$$
$$= (k+j-1)_{(j)} (\tfrac{P}{Q})^j Q^{-k} \sum_{x=j}^{\infty} \binom{k+j+x-j-1}{k+j-1} (\tfrac{P}{Q})^{x-j}$$
$$= (k+j-1)_{(j)} P^j.$$

From these it follows that

$$\sqrt{\beta_1} = \frac{Q+P}{\sqrt{kPQ}}, \quad \beta_2 = 3 + \frac{1+6PQ}{kPQ}.$$

The point (β_1, β_2) lies on the straight line

$$\beta_2 - \beta_1 - 3 = \frac{2}{k}.$$

The cumulants satisfy the following recurrence relation

$$k_{r+1} = PQ \frac{dk_r}{dQ}, \quad r \geq 1.$$

The ratio

$$\frac{b'(x+1; k, p)}{b'(x; k, p)} = \frac{(k+x)P}{(x+1)Q}$$

shows that $b'(x+1; k, p) \geq (\leq) b'(x; k, p)$ according as $x \leq (\geq) kP - Q$. Hence if $kP > Q$ there is a mode at x^* where x^* satisfies $kP - Q - 1 < x^* < kP - Q$. If $kP < Q$, the mode is at $x = 0$.

Lemma 8.8.1: As $k \to \infty$ and $P \to 0$ with kP remaining fixed at λ, $b'(x; k, p) \to e^{-\lambda} \frac{\lambda^x}{x!}$.

Proof.

$$\lim_{k \to \infty, P \to 0} b'(x; k, p) = \lim_{k \to \infty, P \to 0} \frac{(k+x-1)\ldots k}{x!} \left(\frac{P}{1+P}\right)^x (1+P)^{-k}$$

$$= \lim_{k \to \infty, P \to 0} \frac{1}{x!} (1 + \tfrac{x-1}{k})$$
$$(1 + \tfrac{x-2}{k}) \ldots 1 . \lambda^x (1 + \tfrac{\lambda}{k})^{-x} (1 + \tfrac{\lambda}{k})^{-k}$$
$$= e^{-\lambda} \frac{\lambda^x}{x!}.$$

Theorem 8.8.1: Let $X_1, \ldots, X_n$ be independent negative binomial variables, X_i having parameters $(k_i, p), i = 1, \ldots, n$. Then $S_n = X_1 + \ldots + X_n$ has a negative binomial distribution with parameters (k, p) where $k = \sum_{i=1}^{n} k_i$.

Proof. *mgf* of S_n is

$$M_{S_n}(t) = \Pi_{i=1}^n M_{X_i}(t) = \Pi_{i=1}^n M_{X_i}(t) = \Pi_{i=1}^n p^{k_i}(1-qe^t)^{-k_i} = p^k(1-qe^t)^{-k}$$

which is the *mgf* of a negative binomial distribution with parameters k, p. Hence the result follows by the uniqueness property of *mgf* (Theorem 7.3.1).

Corollary 8.8.1.1: Let $X_1, \ldots, X_n$ be independent geometric random variables, each with parameter p. Then S_n has a negative binomial distribution with parameter (n, p).

8.8.1   *The truncated negative binomial distribution*

Consider a negative binomial distribution truncated at $x = 0$. The *pmf* of the truncated distribution is

$$b_t'(x; k, p) = \binom{k+x-1}{x} \frac{p^k q^x}{1-p^k}$$
$$= \binom{k+x-1}{x}(1 - Q^{-k})^{-1}(\tfrac{P}{Q})^x(1 - \tfrac{P}{Q})^k, \quad x = 1, 2, \ldots$$

The moments about zero of this distribution are obtained by multiplying the raw moments of the complete distribution by $(1 - Q^{-k})^{-1}$. Thus, for $b_t'(x; k, p)$

$$E(X) = \frac{kP}{1-Q^{-k}}, \quad E(X^2) = \frac{kPQ+(kP)^2}{1-Q^{-k}},$$
$$V(X) = \frac{kPQ}{1-Q^{-k}}\left[1 - k(\tfrac{P}{Q})\{(1 - Q^{-k})^{-1} - 1\}\right].$$

8.9   The Power Series Distribution

Consider the power series

$$f(z) = \sum_{x=0}^{\infty} a_x z^x$$

where a_x are real non-negative constants, $0 < z < r$ (radius of convergence of $f(z)$) so that $f(z)$ is finite. Define a random variable X with *pmf*

$$P[X = x] = P_x = \frac{a_x z^x}{f(z)}, \quad x = 0, 1, 2, \ldots \tag{8.9.1}$$

Clearly, $P_x \geq 0, \sum_{x=0}^{\infty} P_x = 1$ so that X has a probability distribution P_x. The distribution of X is called the *power series distribution* (*psd*). Its mean is

$$\mu_1' = E(X) = \frac{1}{f(z)} \sum_{x=0}^{\infty} x a_x z^x.$$

Now

$$f'(z) = \frac{1}{z} \sum_{x=1}^{\infty} x a_x z^x.$$

Hence

$$\mu_1' = \frac{z f'(z)}{f(z)} = z \frac{d \log f(z)}{dz}, \quad \mu_r' = \frac{1}{f(z)} \sum_{x=0}^{\infty} x^r a_z z^x, \quad r = 1, 2, \ldots \quad (8.9.2)$$

Now

$$\frac{d\mu_r'}{dz} = \frac{1}{f(z)} \sum_{x=0}^{\infty} a_x x^{r+1} z^{x-1} = \left(\sum_x x^r a_x z^x \right) \frac{f'(z)}{(f(z))^2}.$$

Thus

$$z \frac{d\mu_r'}{dz} = \frac{1}{f(z)} \sum_x a_x x^{r+1} z^x = \left(\sum_x x^r a_x z^x \right) \frac{z f'(z)}{(f(z))^2},$$

i.e., $z \frac{d\mu_r'}{dz} = \mu_{r+1}' = \mu_r' \mu_1'$. Thus, we have the recurrence relation

$$\mu_{r+1}' = z \frac{d\mu_r'}{dz} + \mu_r' \mu_1'. \quad (8.9.3)$$

The central moments are

$$\mu_r = \frac{1}{f(z)} \sum_{x=0}^{\infty} (x - \mu_1')^r a_z z^x.$$

Now,

$$z \frac{d\mu_r}{dz} = \frac{d}{dz} \left\{ \sum_{x=0}^{\infty} (x - \mu_1')^r a_x z^x \right\} \frac{z}{f(z)} - \left\{ \sum_{x=0}^{\infty} (x - \mu_1')^r a_x z^{x+1} \right\} \frac{f'(z)}{(f(z))^2}.$$

Again,

$$\frac{d}{dz} \left\{ \sum_x (x - \mu_1')^r a_x z^x \right\} = \sum_x a_x \left[z^x r (x - \mu_1')^{r-1} \left(-\frac{d\mu_1'}{dz} \right) + (x - \mu_1')^r x z^{x-1} \right].$$

Thus

$$z \frac{d\mu_r}{dz} = \frac{1}{f(z)} \sum_x a_x \cdot x (x - \mu_1')^r z^x - x r \frac{d\mu_1'}{dz} \frac{1}{f(z)} \sum_x a_x z^x \\ (x - \mu_1')^{r-1} - \frac{z f'(z)}{(f(z))^2} \sum_x (x - \mu_1')^r a_x z^x. \quad (8.9.4)$$

The first and the last term in the right hand side of (8.9.4) together simplify to, by virtue of (8.9.1),

$$\frac{1}{f(z)} \sum_x a_x z^x (x - \mu_1')^{r+1} = \mu_{r+1}.$$

Hence, we can write (8.9.4) as

$$z\frac{d\mu_r}{dz} = \mu_{r+1} - rz\frac{d\mu_1'}{dz}\mu_{r-1}$$

when we have the recurrence relation between the central moments,

$$\mu_{r+1} = z\left[\frac{d\mu_r}{dz} + r\frac{d\mu_1'}{dz}\mu_{r-1}\right]. \tag{8.9.5}$$

Putting $r = 1, \mu_0 = 1, \mu_1 = 0$, we get the variance of X as

$$\mu_2 = z\frac{d\mu_1'}{dz} = z\frac{d}{dz}\{z\frac{d}{dz}\log f(z)\} = z^2\frac{d^2\log f(z)}{dz^2} + \mu_1'. \tag{8.9.6}$$

Now,

$$\frac{d^2\log f(z)}{dz^2} = \frac{f''(z)f(z) - (f'(z))^2}{(f(z))^2}.$$

Hence,

$$\mu_2 = z^2\frac{f''(z)}{f(z)} - z^2(\frac{f'(z)}{f(z)})^2 + z\frac{f'(z)}{f(z)}. \tag{8.9.7}$$

From (8.9.7), (8.9.5) can be written as

$$\mu_{r+1} = z\frac{d\mu_r}{dz} + r\mu_2\mu_{r-1} \tag{8.9.8}$$

The characteristic function of X is

$$\psi_X(t) = \sum_x e^{itx}P_x = \frac{1}{f(z)}\sum_x e^{itx}a_x z^x = \frac{f(e^{it}z)}{f(z)}. \tag{8.9.9}$$

8.9.1 *Some special cases*

(a) Poisson distribution: Let $f(z) = 1 + z + \frac{z^2}{2!} + \frac{z^3}{3!} + \dots$ Here $a_x = \frac{1}{x!}$. Thus, $P_x = \frac{a_x z^x}{f(z)} = e^{-z}\frac{z^x}{x!}, x = 0, 1, 2, \dots$, which is the Poisson distribution.

From (8.9.2) and (8.9.6) respectively, $\mu_1' = z, \mu_2 = z$.

The recurrence relation (8.9.5) assumes the form

$$\mu_{r+1} = z(\frac{d\mu_r}{dz} + r\mu_{r-1}).$$

(b) Binomial distribution: Let $f(z) = (1+z)^n = 1 + \binom{n}{1}z + \binom{n}{2}z^2 + \dots + z^n$. Here $a_x = \binom{n}{x}, x = 0, 1, \dots, n$.

$$P_x = \binom{n}{x}(\frac{z}{1+z})^x(\frac{1}{1+z})^{n-x}.$$

which follows a binomial distribution with parameters n and $p = \frac{z}{1+z}$.

From (8.9.2), $\mu_1' = z\frac{f'(z)}{f(z)} = np$.

From (8.9.6) $\mu_z = z\frac{d\mu_1'}{dz} = \frac{nz}{(1+z)^2} = np(1-p)$.

The recurrence relations (8.9.3) and (8.9.5), respectively, assume the form

$$\mu_{r+1}' = z\frac{d\mu_r'}{dz} + n(\frac{z}{1+z})\mu_r' = pq(\frac{d\mu_r'}{dp} + \frac{n}{q}\mu_r') \quad \text{(relation 8.4.3)}$$

and

$$\mu_{r+1}' = \frac{z}{(1+z)^2}[(1+z)^2\frac{d\mu_r}{dz} + nr\mu_{r-1}] = pq[\frac{d\mu_r}{dp} + nr\mu_{r-1}] \quad \text{(relation 8.4.2)}$$

(d) Negative Binomial distribution: Let $f(z) = (1-z)^{-k}, k > 0, 0 < z < 1$.
Hence

$$f(z) = 1 + \binom{k}{1}z + \binom{k+1}{2}z^2 + \binom{k+2}{3}z^3 + \ldots + \binom{k+x-1}{x}z^x + \ldots$$

Here $a_x = \binom{k+x-1}{x}$.

Hence

$$P_x = \frac{a_x z^x}{f(z)} = \binom{k+x-1}{x}z^x(1-z)^k$$

which is the negative binomial distribution. By (8.9.2)

$$E(X) = z\frac{d}{dz}\log\{(1-z)^k\} = \frac{zk}{1-z}.$$

The recurrence relation for the central moments is

$$\mu_{r+1} = z[\frac{d\mu_r}{dz} + r\mu_{r-1}\frac{k}{(1-z)^2}].$$

Hence

$$\mu_2 = \frac{kz}{(1-z)^2}, \quad \mu_3 = \frac{k(1+z)z}{(1-z)^3},$$
$$\mu_4 = z[\frac{d\mu_3}{dz} + 3\mu_2\frac{k}{(1-z)^2}] = \frac{kz(1+4z+3kz+z^2)}{(1-z)^3}.$$

The characteristic function for the distribution is

$$\psi(t) = \frac{f(e^{it}z)}{f(z)} = (\frac{1-e^{it}z}{1-z})^{-k}.$$

(d) Geometric distribution: Let $f(z) = (1-z)^{-1} = 1 + z + z^2 + \ldots, |z| < 1$.
Here $P_x = z^x(1-z)$ which is the geometric distribution with probability of success $(1-z)$.

$$\mu_1' = z\frac{f'(z)}{f(z)} = \frac{z}{1-z}, \quad \mu_2 = z\frac{d\mu_1'}{dz} = \frac{z(1-2z)}{(1-z)^2}.$$

(e) Logarithmic series distribution: Let $f(z) = -\log(1-z) = z + \frac{z^2}{2} + \frac{z^3}{3} + \ldots, 0 < z < 1$.

Hence $P_x = -\frac{z^x}{x\log(1-z)}$, which is the logarithmic series distribution.

$$\mu'_1 = z\frac{f'(z)}{f(z)} = -\frac{z}{(1-z)\log(1-z)}$$
$$\frac{d\mu'_1}{dz} = -\frac{\log(1-z)+z}{(1-z)^2\{\log(1-z)\}^2}.$$

Hence

$$\mu_2 = -\frac{z^2 + z\log(1-z)}{(1-z)^2\{\log(1-z)\}^2}.$$

Thus,

$$\mu_{r+1} = z\left[\frac{d\mu_r}{dz} - r\mu_{r-1}\frac{z+\log(1-z)}{(1-z)^2\{\log(1-z)\}^2}\right].$$

The characteristic function is

$$\psi(t) = \frac{f(e^{it}z)}{f(z)} = \frac{\log(1 - e^{it}z)}{\log(1-z)}.$$

We now consider an interesting theorem characterizing the property of a Poisson distribution.

Theorem 8.9.1: The equality of mean and variance is a necessary and sufficient condition for a *psd* to be a Poisson distribution.

For any *psd* given by (8.9.1), $\mu_2 = z\frac{d\mu'_1}{dz}$.

When $f(z) = e^z$, the *psd* gives Poisson distribution for which $\mu'_1 = z\frac{f'(z)}{f(z)} = z$ and hence $\mu_2 = z$.

To prove the sufficiency part, let $\mu_2 = \mu'_1$ which implies $\frac{d\mu'_1}{dz} = \frac{\mu'_1}{dz}$.

Integrating both sides,

$$\log\mu'_1 = \log(cz)$$

where c is a positive constant. Hence

$$\mu'_1 = cz.$$

Again, by (8.9.2),

$$z\frac{d\log f(z)}{dz} = cz \quad \text{or} \quad \frac{d\log f(z)}{dz} = c.$$

Integrating both sides,

$$\log f(z) = cz + \log k, \quad k \text{ a positive constant.}$$

Thus

$$f(z) = ke^{cz}$$

from which it follows that

$$P_x = e^{-cz}\frac{(cz)^x}{x!}$$

so that X follows a Poisson distribution with parameter cz.

8.10   Exercises and Complements

8.1 Show that the *pgf* of the sum S_K of K independent discrete uniform variables, each over $[1, N]$ is

$$\left\{\frac{t(1 - t^N)}{N(1 - t)}\right\}^K .$$

By expanding the numerator and the reciprocal of the denominator by using the binomial theorem, and collecting the coefficients of t^n in the expansion, find the probability $P(S_K = n)$.

[Hints: Consult Example 7.2.7.]

8.2 A dice is cast repeatedly until an ace turns up. Find the expected number of tosses required.

8.3 Find the expected number of aces drawn in a bridge hand of 13 cards. Also find the expected number of face cards drawn.

8.4 Let X have a probability distribution defined by the function $f(x) = \frac{1}{\pi(1+x^2)}$. Let $Y = 1(0)$ if $X \geq (\leq)1$. Find the probability distribution of Y.

$$\left[\text{Bernoulli}, p = \tfrac{1}{4}\right]$$

8.5   Let X, Y be independent geometric random variables. Show that $\min(X, Y)$ and $X - Y$ are independent random variables.

8.6 Let $X_1, \ldots, X_n$ be independent geometric random variables with parameters $p_1, \ldots, p_n$ respectively. Show that $Z = \min(X_1, \ldots, X_n)$ is a geometric random variable with parameter $p = 1 - \prod_{i=1}^{n}(1 - p_i)$.

8.7 Let X and Y be two independently and identically distributed random variables with $P[X = x] = p_x, x = 0, 1, 2, \ldots$

Suppose

$$P[X = x | X + Y = s] = P[X = s - 1 | X + Y = s] = \frac{1}{s + 1}, \ s \geq 0.$$

Show that X and Y are geometric random variables.

8.8 Let X be a non-negative integer-valued random variable satisfying

$$P[X > j + k | X > j] = P[X \geq k]$$

for any $j(> 0), k(> 0)$. Show that then X must have a geometric distribution.

8.9 Show that for a binomial distribution $b(x; n, p) = \binom{n}{x} p^x q^{n-x}$,

$$P\{X \geq j\} \leq b(j; n, p) \frac{q(j+1)}{j+1-(n+1)p} \quad \text{for } j \geq np$$
$$P\{X \leq j\} \leq b(j; n, p) \frac{(n-j+1)}{(n+1)p-j} \quad \text{for } j \leq np.$$

8.10 Let $X_1, \ldots, X_n$ be n independently and identically distributed Bernoulli variables with parameter p. Show that the sum $S_n = \sum_{i=1}^{n} X_i$ is distributed as a binomial variable with parameters (n, p).

8.11 Let $X_1, \ldots, X_n$ be independent Bernoulli variables, each with parameter p. Set $S_j = X_1 + \ldots + X_j$. Show that $P[S_m = r | S_n = s], m < n$ does not depend on p. Identify the probability distribution and obtain the first two moments of it.

[Hypergeometric]

8.12 Let X_1, X_2 be independent binomial variables with parameters $(n_i, \frac{1}{2}), i = 1, 2$. Find the *pmf* of $X_1 - X_2 + n_2$.

[$(n_2 - X_2)$ is binomial with parameters $(n_2, \frac{1}{2})$.

$$P(X_1 - X_2 + n_2 = k) = \binom{n_1 + n_2}{n_2 + k} / 2^{n_1 + n_2}, (-n_2 \leq k \leq n_1).]$$

8.13 In a population of n elements, n_1 are red and $n_2 (= n - n_1)$ are black. A group of r elements is chosen at random without replacement. Find out the probability $q_k (k = 0, 1, \ldots, \min(n_1, r))$ that the group contains exactly k red elements; hence show that if $np = n_1, q = 1 - p$,

$$\binom{r}{k} (p - \frac{k}{n})^k (q - \frac{r-k}{n})^{r-k} < q_k < \binom{r}{k} p^k q^{r-k} (1 - \frac{r}{n})^{-r}.$$

Justify how you can approximate this q_k by a binomial mass function.

8.14 Show that

$$\frac{\binom{Np}{x} \binom{Nq}{n-x}}{\binom{N}{n}} \to \binom{n}{x} p^x q^{n-x}, \quad q = 1 - p, 0 < p < 1$$

as $N \to \infty$ for any fixed n, provided $\frac{n}{N}$ is not negligible.

8.15 An urn contains a white and b black balls. A sample of two balls is chosen without replacement. Under what conditions will the probability that both the chosen balls are of the same color would be greater than $\frac{1}{2}$?

[$(a - b)^2 > a + b$]

8.16 Show that

$$\binom{k+x-1}{x}p^k q^x \to e^{-\lambda}\frac{\lambda^x}{x!}$$

as $p \to 1$ and $k \to \infty$ in such a way that $k(1-p) = \lambda$ remains constant.

8.17 Obtain by two methods, the probability that in a series of Bernoulli trials (with probability p of success) $'a'$ successes occur before at most $'b'$ failures. Hence prove the identity

$$\sum_{k=0}^{b-1}\binom{a+b-1}{a+b}p^k q^{b-k-1} = \sum_{k=0}^{b-1}\binom{a+k-1}{k}q^k.$$

[Probability that in $(a+b-1)$ trials, $(a+b)$ successes and $b-k-1$ failures occur is

$$\binom{a+b-1}{a+k}p^{a+k}q^{a-k-1}.$$

Hence, probability of getting $'a'$ successes before $'b'$ failures is

$$\sum_{k=0}^{b-1}\binom{a+b-1}{a+k}p^{a+k}q^{b-k-1}. \qquad (i)$$

Alternatively, probability that in $(a+k)$ trials the last trial is a success and the preceding $(n+k-1)$ trials contain $(a-1)$ successes is

$$\binom{a+k-1}{a-1}p^a q^k.$$

The required probability is therefore

$$p^a\sum_{k=0}^{b-1}\binom{a+k-1}{k}q^k.$$

Hence the identity.]

8.18 Let X_1, X_2 be two independent negative binomial variables, X_i having parameters $(k_i, p), i = 1, 2$. Find the conditional probability that $X_1 = l_1$, given that $X_1 + X_2 = r$. Show that when $k_1 = k_2 = 1$, than $P[X_1 = l|X_1 + X_2 = r] = \frac{1}{r+1}, l = 0, 1, \ldots, r; r = 0, 1, 2, \ldots$

[Hints:

$$P(X_1 = l|X_1 + X_2 = r) = \frac{\binom{l+k_1-1}{t}\binom{r+k_2-t-1}{r-t}}{\binom{r+k_1+k_2-1}{r}}, r = 0, 1, 2, \ldots]$$

8.19 Consider a random variable X distributed as $H(x; N, M, n)$. Show that for fixed n and fixed $p = \frac{M}{N}$, the rth factorial moment of this distribution tends to that of a binomial distribution with parameters (n, p) as $N \to \infty$.

8.20 An urn contains N balls numbered $1, \ldots, N$. From these balls n balls are drawn at random without replacement. Let X_i denote the number drawn on the ith ball ($i = 1, \ldots, n$) and $S_n = \sum_{i=1}^{n} X_i$. Find $E(S_n), V(S_n)$.

8.21 Telephone calls arrive at a certain exchange at an average rate of 3 calls per minute. Assuming that the number of incoming calls follow a Poisson distribution, find the probability that 3 or more calls arrive in a 15-second time interval.

$$[0.0406]$$

8.22 A Geiger Muller counter registrars on an average 50 counts per minute in the neighborhood of a certain radioactive substance. Find the probability there will be (i) 3 counts in a 6-second period (ii) x counts in a T-second period.

$$[(i)\ 0.140;\ (ii)\ e^{-.833T} \frac{(0.833T)^x}{x!}]$$

8.23 Find the probability generating function of the sum S of the numbers obtained when one ball is drawn from each of m urns, each of which contains n balls numbered $1, 2, \ldots, n$. Hence find the probability distribution of S.

[Hints: Let X_i be the number on the ball drawn from the ith urn ($i = 1, \ldots, m$). Each X_i has a uniform distribution $P(X_i = k) = \frac{1}{n}, k = 1, \ldots, n$. Hence proceed as in Example 7.2.7.]

Chapter 9

Some Continuous Distributions on $\mathcal{R}^1$

9.1 Introduction

We consider some univariate continuous distributions in this chapter. The uniform distribution is discussed in Section 9.2, Sections 9.3 through 9.6 discuss respectively exponential, gamma (chi-square), beta and Cauchy distribution. Subsequently, the celebrated normal distribution is discussed. We consider lognormal, Laplace, Pareto, Weibull, extreme-value and logistic distributions in Sections 9.8 to 9.13 respectively.

9.2 The Continuous Uniform Distribution

A random variable X is said to have a continuous uniform (rectangular) distribution on the interval $[a, b](-\infty < a < b < \infty)$ if its *pdf* is given by

$$f(x; a, b) = \begin{cases} \frac{1}{b-a} & \text{if } a \leq x \leq b \\ 0, & \text{otherwise,} \end{cases} \tag{9.2.1}$$

a, b are parameters of the distribution. The *cdf* is

$$F_X(x) = \begin{cases} 0, & x \leq a \\ \frac{x-a}{b-a}, & a < x < b \\ 1, & x \geq b. \end{cases} \tag{9.2.2}$$

$$E(X) = \int_a^b \frac{x}{b-a} dx = \frac{a+b}{2},$$

$$E(X^2) = \frac{b^2+ab+a^2}{3},$$

$$V(X) = \frac{(b-a)^2}{12}.$$

The *mgf* is

$$M_X(t) = \int_a^b \frac{e^{tx}}{b-a}\,dx = \frac{e^{bt} - e^{at}}{(b-a)t}, \quad t \neq 0.$$

The distribution can also be defined over the open interval (a, b) or semi-closed intervals $[a, b), (a, b]$. The distribution function remains the same as (9.2.2) for all these distributions (Fig. 9.2.1).

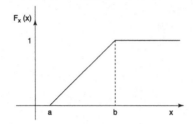

Fig. 9.2.1   cdf of a rectangular distribution

If the variable X is considered to be distributed uniformly over $(0, 1)$, then the nth order raw moment is

$$\mu'_n = \frac{1}{n+1}.$$

The characteristic function is

$$\Psi(t) = \int_0^1 e^{itx}\,dx = \frac{e^{it} - 1}{it}.$$

It is sometimes convenient to shift the origin of this distribution to the point $\frac{1}{2}$ so that X is uniformly distributed over the interval $(-\frac{1}{2}, \frac{1}{2})$. In this case,

$$\mu'_{2r+1} = \mu_{2r+1} = 0$$
$$\mu'_{2r} = \mu_{2r} = \frac{1}{(n+1)2^n} \quad \text{where } n = 2r.$$

The characteristic function is

$$\Psi(t) = \int_{-\frac{1}{2}}^{\frac{1}{2}} e^{itx}\,dx = 2\frac{\sin\frac{it}{2}}{it}.$$

9.3   The Exponential Distribution

A random variable X is defined to have an (negative) exponential distribution with parameters $\lambda(>0)$ if its *pdf* is given by

$$f(x; \lambda) = \begin{cases} \lambda e^{-\lambda x}, & 0 < x < \infty, \lambda > 0 \\ 0, & \text{otherwise} . \end{cases} \tag{9.3.1}$$

The cumulative distribution function is

$$F_X(x) = \begin{cases} 0, & x < 0 \\ 1 - e^{-\lambda x}, & 0 \le x. \end{cases}$$

The *mgf* of the distribution is

$$M_X(t) = \frac{1}{1-(\frac{t}{\lambda})}$$
$$\mu_1' = \tfrac{1}{\lambda}, \; \mu_r' = \tfrac{r!}{\lambda^r}, \; r = 2, 3, \dots$$
$$\mu_2 = \tfrac{1}{\lambda^2}, \; \mu_3 = \tfrac{2}{\lambda^3}, \; \mu_4 = \tfrac{9}{\lambda^4},$$
$$\sqrt{\beta_1} = 2, \; \beta_2 = 9.$$

The mean deviation is $\frac{2}{e\lambda}$. The ration (mean deviation)/(s.d.) $= \frac{2}{e} = 0.736$. The median of the distribution is $\frac{1}{\lambda} \log_e 2$. The mode is at $x = 0$.

The cumulant generating function is

$$\log(1 - \frac{t}{\lambda})$$

with cumulants,

$$k_1 = \frac{1}{\lambda}, k_r = (r-1)!\lambda^{-r}, r > 1.$$

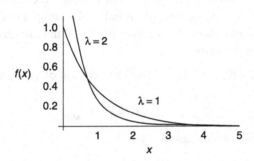

Fig. 9.3.1   Exponential density

Theorem 9.3.1 For an exponential random variable X with parameter λ,

$$P[X > a + b | X > a] = P[X > b] \quad \text{for } a > 0, b > 0.$$

Proof.

$$P[X > a + b | X > a] = \frac{P[X > a + b]}{P[X > a]} = \frac{e^{-\lambda(a+b)}}{e^{-\lambda a}} = e^{-\lambda b} = P[X > b]. \quad \Box$$

If X represents the life-time of a given component, say, an electric bulb, Theorem 9.3.1 states that the probability that a bulb will be operative for at least $(a+b)$ units given that it has already run for $'a'$ units is the same as its initial probability for surviving for at least $'b'$ units. This means that the future lifetime of an individual has the same distribution, no matter how old it is at present. This is the 'lack of memory' property of the exponential distribution. A similar property was noted for the geometric distribution (Theorem 8.7.1).

Suppose that the events occur randomly in time in such a way that the number of occurrences on a fixed time interval $[0, t]$ has a Poisson distribution with parameter λt. Suppose one of the events has just taken place. Probability that the next event will occur not befor time T is $P[$ waiting time $\geq T] = P[$ no occurrence in time interval of length $T] = e^{-\lambda T}$. Hence $P[X \leq T] = 1 - e^{-\lambda T}(T \geq 0)$ so that X, the waiting time has an exponential distribution.

Thus, exponential distribution and Poisson distribution are related (vide also Example 9.4.1).

Because of its mathematical simplicity, an exponential distribution is often used as approximate representation of complex models suitable for a particular situation. The lifetime of an industrial component can often be represented by an exponential random variable. This distribution is often used in life-testing problems.

An exponential distribution truncated by exclusion of values exceeding $'A'$ has the *pdf*

$$f_t(x; \lambda) = \lambda(1 - e^{-\lambda A})^{-1}e^{-\lambda x}, \ 0 < x < A. \tag{9.3.2}$$

The expected value of this distribution is

$$\frac{1}{\lambda}[1 - A\lambda(e^{\lambda A} - 1)^{-1}].$$

9.4 The Gamma Distribution and the Chi-Square Distribution

9.4.1 *The gamma function*

For $\alpha > 0$, the gamma function is defined as

$$\Gamma(\alpha) = \int_0^\infty t^{\alpha - 1}e^{-t}dt \ (\alpha > 0). \tag{9.4.1}$$

$\Gamma(1) = 1$. For $\alpha > 1$, integration by parts yields

$$\Gamma(\alpha) = (\alpha - 1)\Gamma(\alpha - 1)). \tag{9.4.2}$$

If $\alpha = n$, a positive integer,

$$\Gamma(n) = (n - 1)!. \tag{9.4.3}$$

Also,

$$\Gamma(\frac{1}{2}) = \sqrt{\pi}. \tag{9.4.4}$$

Gauss' multiplication formula is

$$\Gamma(n\alpha) = (2\pi)^{-\frac{n-1}{2}} n^{n\alpha - \frac{1}{2}} \Pi_{j=0}^{n-1}(\alpha + jn^{-1}). \tag{9.4.5}$$

Duplication formula ($n = 2$):

$$\Gamma(2\alpha) = (2\pi)^{-\frac{1}{2}} 2^{2\alpha - \frac{1}{2}} \Gamma(\alpha)\Gamma(\alpha + \frac{1}{2}). \tag{9.4.6}$$

The incomplete gamma function is

$$\Gamma_y(\alpha) = \int_0^y e^{-t} t^{\alpha - 1} dt. \tag{9.4.7}$$

The incomplete gamma function ratio is

$$\frac{\Gamma_y(\alpha)}{\Gamma(\alpha)}. \tag{9.4.8}$$

Stirling's approximation formula: There is a useful approximation to $\Gamma(\alpha)$ known as Stirling's formula. This states that

$$\Gamma(\alpha) \approx \sqrt{2\pi} e^{-\alpha} \alpha^{\alpha - \frac{1}{2}}, \quad \alpha \to \infty.$$

For α a positive integer n, we obtain the following approximation to $n! = \Gamma(n + 1)$,

$$n! \approx \sqrt{2\pi} e^{-(n+1)} (n + 1)^{n + \frac{1}{2}}.$$

This is usually written as

$$n! \approx \sqrt{2\pi} e^{-n} n^{n + \frac{1}{2}} \tag{9.4.9}$$

by using the fact that

$$\{\frac{n+1}{n}\}^{n + \frac{1}{2}} = (1 + \frac{1}{n})^n (1 + \frac{1}{n})^{\frac{1}{2}} \approx e \tag{9.4.10}$$

as $n \to \infty$.

9.4.2 *The gamma distribution*

A random variable X is defined to have a gamma distribution with parameters $(\lambda, p)(\lambda > 0, p > 0)$ if its *pdf* is given by

$$f_G(x; \lambda, p) = \begin{cases} \frac{\lambda}{\Gamma(p)} e^{-\lambda x} (\lambda x)^{p-1}, & 0 < x < \infty, \\ 0, & \text{otherwise.} \end{cases} \qquad (9.4.11)$$

The *mgf* is

$$\begin{aligned} M_X(t) = E(e^{tX}) &= \int_0^\infty \frac{\lambda^p}{\Gamma(p)} e^{tx} e^{-\lambda x} x^{p-1} dx \\ &= (\tfrac{\lambda}{\lambda - t})^p \int_0^\infty \frac{(\lambda - t)^p}{\Gamma(p)} x^{p-1} e^{-(\lambda - t)x} dx = (\tfrac{\lambda}{\lambda - t})^p. \end{aligned} \qquad (9.4.12)$$

Hence,

$$\begin{aligned} \mu_1' &= \tfrac{p}{\lambda}, \ \mu_2' = \tfrac{p(p+1)}{\lambda^2}, \ \mu_r' = \tfrac{\Gamma(p+r)}{\Gamma(p)\lambda^r} \\ \mu_2 &= \tfrac{p}{\lambda^2}, \ \mu_3 = \tfrac{2p}{\lambda^3}, \ \mu_4 = \tfrac{3p^2 + 6p}{\lambda^4}, \\ \sqrt{\beta_1} &= \tfrac{2}{\sqrt{p}}, \ \beta_2 = 3 + \tfrac{6}{p}. \end{aligned}$$

The cumulative distribution function is given by the incomplete gamma function ratio

$$F(x) = P(X \le x) = \frac{\Gamma_x(p)}{\Gamma(p)}.$$

The gamma distributions for $p = 1$ are known as the exponential distributions. These have values λ at $x = 0$ and are convex, decreasing with value 0 as $x \to \infty$.

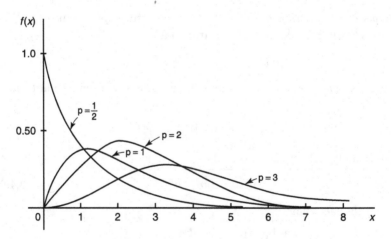

Fig. 9.4.1 Standard gamma densities

For $p < 1$ the gamma densities are also convex decreasing functions having limiting value ∞ at $x = 0$ and 0 at $x = \infty$. For $p > 1$, the gamma densities have a unique maximum at $\frac{p-1}{\lambda}$ and possible points of inflexion at

$$\frac{p-1}{\lambda}\{1 +_- \sqrt{1 - \frac{p-2}{p-1}}\}.$$

Consequently, if $1 < p < 2$, there is only a point of inflexion at

$$\frac{p-1}{\lambda}(1 + \sqrt{1 - \frac{p-2}{p-1}})$$

while for $p > 2$, both are points of inflexion for the density.

The standard form of the gamma distribution is obtained by putting $\lambda = 1$, giving

$$f_G(x; p) = \frac{e^{-x} x^{p-1}}{\Gamma(p)} \quad (x \geq 0). \tag{9.4.13}$$

The standard distribution (9.4.13) has a single mode at $x = p - 1$ if $p \geq 1$. If $p < 1$, $f_G(x; p) \to \infty$ as $x \to 0$.

Some typical standard gamma distributions are shown in Fig. 9.4.1. It is seen that as p increases, the shape of the curve becomes similar to that of the normal probability distribution (discussed in Section 9.7).

9.4.3 *The chi-square distribution*

A random variable X is said to have a chi-square distribution with n degrees of freedom, if its *pdf* is given by

$$f(x) = \frac{1}{2^{\frac{n}{2}} \Gamma(\frac{n}{2})} e^{-\frac{x}{2}} x^{\frac{n}{2}-1}, \ 0 < x < \infty. \tag{9.4.14}$$

Clearly, $\frac{x}{2}$ has a standard gamma distribution (9.4.13) with parameter $p = \frac{n}{2}$.

Theorem 9.4.1: If $X_1, \ldots, X_n$ are independent random variables such that X_j is distributed as gamma with parameters (λ, p_j), then their sum $S_n = \sum_{j=1}^{n} X_j$ is distributed as gamma with parameters (λ, p) where $p = \sum_{j=1}^{n} p_j$.

Proof. The *mgf* of S_n is

$$M_{S_n}(t) = \Pi_{j=1}^{n} M_{X_j}(t)$$

$$= \Pi_{j=1}^{n} (\frac{\lambda}{\lambda - t})^{p_j} = (\frac{\lambda}{\lambda - t})^{p}.$$

The result follows by the uniqueness property of the *mgf*.

Corollary 9.4.1.1: If $X_1, \ldots, X_n$ are independent exponential random variables, each with parameter λ, their sum $S_n = \sum_{j=1}^{n} X_j$ is a gamma variable with parameter (λ, n).

Corollary 9.4.1.2: If $X_1, \ldots, X_k$ are independent chi-square distribution with degree of freedom n_j, then their sum $S_k = \sum_{j=1}^{k} X_j$ has a chi-square distribution with degree of freedom $n = \sum_{j=1}^{k} n_j$.

Theorem 9.4.2: Let X be a Poisson random variable with parameter λ. Then

$$P\{X \leq t\} = \frac{1}{\Gamma(T+1)} \int_{\lambda}^{\infty} e^{-x} x^t dx. \qquad (9.4.15)$$

(9.4.15) expresses the distribution function of X in terms of an incomplete gamma function.

Proof. Integrating by parts

$$\frac{1}{t!} \int_{\lambda}^{\infty} e^{-x} x^t dx = \frac{1}{t!} \lambda^t e^{-\lambda} + \int_{\lambda}^{\infty} \frac{1}{(t-1)!} x^{t-1} e^{-x} dx = \ldots$$
$$= e^{-\lambda} (1 + \lambda + \frac{\lambda^2}{2!} + \ldots + \frac{\lambda^t}{t!}).$$

EXAMPLE 9.4.1: The life-time T of a component has an exponential distribution

$$f_T(t) = \lambda e^{-\lambda t} \quad (t \geq 0, \lambda > 0).$$

Assume that as soon as a component fails it is replaced by another component with exactly the same distribution of life-time. Show that the probability distribution of number of failures X in a period of fixed length τ is Poisson with parameter $\lambda \tau$.

Let T_i denote the life-time of the ith component.

$$P(X = k) = P(T_1 + \ldots + T_{k+1} > \tau) - P(T_1 + \ldots + T_k > \tau). \qquad (9.4.16)$$

For a fixed m, $\sum_{j=1}^{m} T_j$ follows a gamma distribution with parameter (λ, m). Hence

$$P(T_1 + \ldots + T_{k+1} > \tau) = \int_{\tau}^{\infty} \frac{\lambda}{\Gamma(k+1)} e^{-\lambda y} (\lambda y)^k dy$$
$$= \frac{1}{\Gamma(k+1)} \int_{\lambda+}^{\infty} e^{-z} z^k dz \quad \text{where } z = \lambda y$$
$$= \sum_{j=0}^{\infty} e^{-\lambda \tau} \frac{(\lambda \tau)^j}{j!} \quad \text{(by Theorem 9.4.1)}.$$

Hence from (9.4.15),

$$P(X = k) = \sum_{j=0}^{k} e^{-\lambda \tau} \frac{(\lambda \tau)^j}{j!} - \sum_{j=0}^{k-1} e^{-\lambda \tau} \frac{(\lambda \tau)^j}{j!} = e^{-\lambda \tau} \frac{(\lambda \tau)^k}{k!}.$$

9.5 The Beta Distribution

9.5.1 *The beta function*

The beta function is defined as

$$B(p,q) = \int_0^1 t^{p-1}(1-t)^{q-1}dt, \quad (p, q > 0) \qquad (9.5.1)$$

$$= 2\int_0^{\frac{\pi}{2}} \sin^{2p-1}\theta \cos^{2q-1}\theta \, d\theta. \qquad (9.5.2)$$

In terms of gamma function

$$B(p,q) = \frac{\Gamma(p)\Gamma(q)}{\Gamma(p+q)} = B(q,p). \qquad (9.5.3)$$

The incomplete beta function is

$$B_y(p,q) = \int_0^y t^{p-1}(1-t)^{q-1}dt. \qquad (9.5.4)$$

The incomplete beta function ratio is

$$I_y(p,q) = \frac{B_y(p,q)}{B(p,q)}. \qquad (9.5.5)$$

9.5.2 *The beta distribution*

A random variable X is said to have a beta distribution with parameters $a, b(a > 0, b > 0)$ if its *pdf* is given by

$$f(x; a, b) = \begin{cases} \frac{1}{B(a,b)}x^{a-1}(1-x)^{b-1}, & 0 < x < 1, a, b > 0 \\ 0, & \text{otherwise.} \end{cases} \qquad (9.5.6)$$

Its rth raw moment is

$$\mu_r' = \frac{1}{B(p,q)}\int_0^1 x^{r+a-1}(1-x)^{b-1}dx = \frac{B(r+a,b)}{B(a,b)} = \frac{\Gamma(r+a)\Gamma(a+b)}{\Gamma(a)\Gamma(r+a+b)}.$$

In particular,

$$\mu_1' = \frac{a}{a+b}, \quad \mu_2 = \frac{ab}{(a+b)^2(a+b+1)},$$

$$\mu_3 = \frac{2ab(b-a)}{(a+b)^3(a+b+1)(a+b+2)}, \quad \mu_4 = \frac{3ab\{ab(a+b-6)+2(a+b)^2\}}{(a+b)^4(a+b+1)(a+b+2)(a+b+3)}$$

$$\sqrt{\beta_1} = \frac{2(b-a)\sqrt{a+b+1}}{\sqrt{ab}(a+b+2)}, \quad \beta_2 = \frac{3(a+b+1)\{ab(a+b-6)+2(a+b)^2\}}{ab(a+b+2)(a+b+3)}.$$

The *mgf* of the distribution is

$$M_X(t) = E(e^{tX}) = \frac{1}{B(p,q)} \int_0^1 e^{tx} x^{q-1} (1-x)^{b-1} dx.$$

The *mgf* does not have a simple form. However, the moments of all orders exist and hence one can write

$$M_X(t) = \sum_{j=0}^{\infty} \mu'_j \frac{t^j}{j!} = \sum_{j=0}^{\infty} \frac{t^j}{j!} \frac{\Gamma(a+j)\Gamma(a+b)}{\Gamma(a)\Gamma(a+b+j)}.$$

The cumulative distribution function is

$$F_X(u) = P(X \le u) = \begin{cases} 0, & u \le 0 \\ \frac{1}{B(a,b)} \int_0^u x^{a-1}(1-x)^{b-1} dx, & 0 < u < 1 \\ \quad = \frac{B_u(a,b)}{B(a,b)} = I_u(a,b) & \\ 1, & u \ge 1. \end{cases}$$

The incomplete beta function $I_u(a,b)$ has been tabulated extensively.

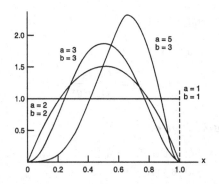

Fig. 9.5.1   Beta densities

The behavior of the beta distribution depends on the values of a and b. If $a = b = 1$, the density is the uniform density on $(0, 1)$.

If $a < 1$, the density approaches ∞ at $x = 0$ and if $b < 1$, also at 1. In this case, there is a unique maximum at $x = \frac{a-1}{a+b-2}$ and the function is a convex function.

If $a < 1$ and $b = 1$, the density is 1 at $x = 1$, is decreasing and convex.

If $a < 1$ and $b > 1$, the density is 0 at 1, convex and decreasing. If $b < 1$, then the behavior is as described above with 0 and 1 reversed.

If $a > 1$ and $b > 1$, the density is 0 at 1 and has a unique maximum at $\frac{a-1}{a+b-2}$.

The shape of the densities thus depend on a, b and may have 0, 1 or 2 points of inflexion.

Since the density can assume a variety of shapes depending on different values of the parameters, this distribution has an wide application to model experiments of different types.

The distribution (9.5.6) is also called the beta distribution of the first type. When nothing has been said about the type of Beta distribution, it will be understood as the Beta distribution of the first type.

Figure 9.5.1 shows some beta density curves for different values of (a, b).

Let X be Beta-distributed with parameters (a, b) and let $U = (d - c)X + c$ where $-\infty < c < d < \infty$. Then the distribution of U is called a Beta distribution on (c, d). Its *pdf* is

$$f_U(u; a, b; c, d) = \frac{1}{B(a,b)}(u - c)^{a-1}(d - u)^{b-1}(d - c)^{-(a+b+1)}, \ c < u < d.$$
(9.5.7)

If X is Beta-distributed, the distribution of $\frac{1}{X}$ is called the inverse Beta distribution. Its *pdf* is

$$f_Y(y; a, b) = B(a,b)^{-1}y^{-(a+b)}(y - 1)^{b-1}, 1 < y < \infty.$$
(9.5.8)

The distribution of $Z = \frac{1}{(X-1)}$ is the inverse Beta distribution on $(0, \infty)$. Its *pdf* is

$$f_Z(z; a, b) = \frac{1}{B(a,b)}(1 + z)^{(a+b)}z^{b-1}, \ z > 0.$$
(9.5.9)

9.5.2.1 *The beta distribution of the second type*

A random variable X is said to have a beta distribution of the second type with parameters $(a, b)(a > 0, b > 0)$ if its *pdf* is given by

$$f(x; a, b) = \begin{cases} \frac{1}{B(a,b)} \frac{x^{a-1}}{(1+x)^{a+b}}, & x \geq 0, \ a > 0, b > 0 \\ 0, & \text{otherwise.} \end{cases}$$
(9.5.10)

The rth raw moment is

$$\mu_r' = \frac{1}{B(a,b)} \int_0^\infty \frac{x^{a+r-1}}{(1+x)^{a+b}}dx.$$

Letting $(1 + x) = \frac{1}{y}$,

$$\mu_r' = \frac{1}{B(a,b)} \int_0^1 y^{b-r-1}(1-y)^{a+r-1}dy = \frac{B(b-r, a+r)}{B(a,b)} = \frac{\Gamma(b-r)\Gamma(a+r)}{\Gamma(a)\Gamma(b)}.$$

In particular,

$$\mu_1' = \frac{a}{b-1}, \ \mu_2 = \frac{a(a+b-1)}{(b-1)^2(b-2)}.$$

9.6   The Cauchy Distribution

A random variable X is said to have a Cauchy distribution with parameter θ and $\lambda(\lambda > 0)$ if its *pdf* is given by

$$f(x; \theta, \lambda) = \frac{\lambda}{\pi} \cdot \frac{1}{\lambda^2 + (x - \theta)^2}, \quad -\infty < x < \infty, \ \lambda > 0. \tag{9.6.1}$$

We check that

$$\int_{-\infty}^{\infty} f(x)dx = \frac{1}{\pi} \int_{-\infty}^{\infty} \frac{dz}{1+z^2} \text{ putting } z = \frac{x-\theta}{\lambda}$$

$$= \frac{2}{\pi} \tan^{-1} z \Big|_0^{\infty} = 1.$$

The cumulative distribution of X is

$$F_X(x) = \frac{1}{2} + \frac{1}{\pi} \tan^{-1}(\frac{x - \theta}{\lambda}).$$

The parameters θ and λ are location and scale parameters respectively. The distribution is symmetrical about $x = \theta$ and hence the median is at $x = \theta$; the upper and the lower quantiles are at $x = \theta +_- \lambda$. The *pdf* has points of inflexion at $\theta +_- \frac{\lambda}{\sqrt{3}}$. The mode of the distribution is at $x = \theta$.

The distribution does not possess finite moments of order greater than or equal to 1. Hence the distribution does not possess finite expected value or standard deviation (Example 6.2.3).

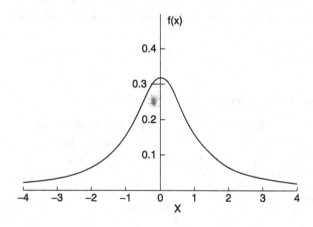

Fig. 9.6.1   Cauchy density

Now,

$$\mu_1' = \frac{\lambda}{\pi} \int_{-\infty}^{\infty} \frac{xdx}{\lambda^2+(x-\theta)^2} dx = \frac{1}{\pi} \int_{-\infty}^{\infty} \frac{(\theta+\lambda z)}{1+z^2} dz \ (\text{ putting } \frac{x-\theta}{\lambda} = z)$$

$$\tag{9.6.2}$$

$$= \theta + \frac{\lambda}{\pi} \int_{-\infty}^{\infty} \frac{zdz}{1+z^2} = \theta + \frac{\lambda}{\pi} \lim_{A\to\infty; B\to\infty} \int_{-A}^{B} \frac{zdz}{1+z^2}.$$

The integral in (9.6.2) does not exist if A, B approach infinity independently. Consequently, the mean of the distribution does not exist. However, when $A = B$, the limit of this integral exists and the limit is zero. Hence if we take the principal value of this integral, the value of this expression in (9.6.2) is θ.

The characteristic function of this distribution is

$$\psi_X(t) = E(e^{itX}) = e^{it\theta - |t|\lambda} \quad \text{(vide Example 7.6.5)}.$$

The standard form of the distribution, represented in Fig.9.6.1 is obtained by putting $\theta = 0, \lambda = 1$.

Theorem 9.6.1: If $X_1, \ldots, X_n$ are independent Cauchy variables, X_j having parameters $(\lambda_j, \theta_j), j = 1, \ldots, n$, then the sum $S_n = \sum_{j=1}^n X_j$ is a Cauchy variable with parameters $\lambda' = \sum_{j=1}^n \lambda_j, \theta' = \sum_{j=1}^n \theta_j$.

Proof. The characteristic function of S_n is

$$\psi_{S_n}(t) = \Pi_{j=1}^n \Psi_{X_j}(t) = \exp\left[it\sum_{j=1}^n \theta_j - |t|\sum_{j=1}^n \lambda_j\right]$$

which is the characteristic function of a Cauchy variable with parameters λ', θ'. Hence the theorem follows by the uniqueness property of the characteristic function (Theorem 7.6.3).

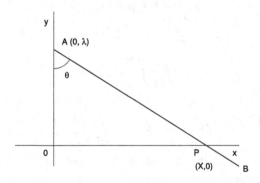

Fig. 9.6.2

Corollary 9.6.1.1: The arithmetic mean $\bar{X} = \frac{1}{n}\sum_{i=1}^n X_i$ of n independently and identically distributed Cauchy variables with parameters (λ, θ) is a Cauchy variable with the same set of parameters (λ, θ).

EXAMPLE 9.6.1: A straight line AB is allowed to move through a fixed point A with coordinates $(0, \lambda)$ and the length of the intersection X it makes with the x-axis is noted. AB makes an angle $\theta(= \angle OAP)$ with the y-axis. Assuming that θ has an uniform distribution between $-\frac{\pi}{2}$ (corresponding to $OP = -\infty$) and $\frac{\pi}{2}$ (corresponding to $OP = \infty$), find the probability density function of X (Fig. 9.6.2).

We note that $\theta = \tan^{-1}(\frac{X}{\lambda})$. Now, the probability that the random variable θ takes any value between $\theta_1 - \frac{1}{2}d\theta$ and $\theta_1 + \frac{1}{2}d\theta$ where θ_1 is an arbitrary, but fixed value is $\frac{d\theta}{\pi} = \frac{\lambda dx}{\pi(\lambda^2 + x_1^2)}$ where x_1 is the value of X corresponding to $\theta = \theta_1$. But the latter is also the probability that X takes any value between $x_1 - \frac{1}{2}dx$ and $x_1 + \frac{1}{2}dx$. Hence, X is a Cauchy variable with parameters (λ, θ).

9.7   The Normal Distribution

The *pdf* of a normal random variable is given by

$$f(x; \mu, \sigma) = \frac{1}{\sqrt{2\pi}\sigma} e^{-\frac{1}{2\sigma^2}(x-\mu)^2}, \quad -\infty < x < \infty \qquad (9.7.1)$$

where the parameters μ and σ satisfy $-\infty < \mu < \infty$, $\sigma > 0$.

It is easy to check that

$$I = \int_{-\infty}^{\infty} f(x; \mu, \sigma) dx = 1. \qquad (9.7.2)$$

For this we set $\frac{x-\mu}{\sigma} = z$. Then

$$I = \frac{1}{\sqrt{2\pi}} \int_{-\infty}^{\infty} e^{-\frac{z^2}{2}} dz.$$

Then

$$I^2 = \frac{1}{2\pi} \int_{-\infty}^{\infty} e^{-\frac{z^2}{2}} dz \int_{-\infty}^{\infty} e^{-\frac{s^2}{2}} ds = \frac{1}{2\pi} \int_{-\infty}^{\infty} \int_{-\infty}^{\infty} e^{-\frac{z^2+s^2}{2}} ds dz.$$

Consider now the polar transformation $s = r\cos\theta, z = r\sin\theta, 0 < r < \infty, 0 < \theta < 2\pi$. Then

$$I^2 = \frac{1}{2\pi} \int_0^{\infty} \int_0^{2\pi} r e^{-\frac{1}{2}r^2} dr d\theta = \int_0^{\infty} r e^{-\frac{1}{2}r^2} dr = 1, \quad \text{when } I = 1. \square$$

We note that $f(\mu + y) = f(\mu - y)$ for all values of y, so that the probability curve (9.7.1) is symmetric about $x = \mu$. The distribution has the unique mode at $x = \mu$ and the modal value of the probability density function is $(\sigma\sqrt{2\pi})^{-1} = 0.3979\sigma^{-1}$. The ordinate of the *pdf* curve decreases rapidly as $|x - \mu|$ increases numerically.

Since the ordinate at $x = \mu$ divides the area under the *pdf* curve into two equal parts, the median of the curve is also at μ. For this distribution, the mean ($= \mu$, shown below), median and the mode coincide.

The density function has two points of inflexion at $x = \mu +_- \sigma$. (The points of inflexion of a curve are the points at which the curve changes from concave upwards to convex downwards and are obtained by solving the equation $\frac{d^2 f(x)}{dx^2} = 0$ and verifying that at these points $\frac{d^3 f(x)}{dx^3} \neq 0$.)

Fig. 9.7.1 shows the *pdf* of a $N(\mu, \sigma^2)$-distribution with $\mu = 50$ and $\sigma = 5$. Fig. 9.7.2 gives examples of some normal pdf's and cdf's.

The mean deviation about mean is

$$
\begin{aligned}
E|X - \mu| &= \tfrac{1}{\sigma\sqrt{2\pi}} \int_{-\infty}^{\infty} |x - \mu| e^{-\frac{1}{2\sigma^2}(x-\mu)^2} dx \\
&= \tfrac{\sigma}{\sqrt{2\pi}} \int_{-\infty}^{\infty} |z| e^{-\frac{z^2}{2}} dz \quad \text{where } z = \tfrac{x-\mu}{\sigma} \\
&= \tfrac{\sigma}{\sqrt{2\pi}} \left[-\int_{-\infty}^{0} z e^{-\frac{z^2}{2}} dz + \int_{0}^{\infty} z e^{-\frac{z^2}{2}} dz \right] \\
&= \sigma\sqrt{\tfrac{2}{\pi}} \int_{0}^{\infty} z e^{-\frac{z^2}{2}} dz = \sigma\sqrt{\tfrac{2}{\pi}} = 0.7979\sigma.
\end{aligned}
\tag{9.7.3}
$$

The odd order central moment is

$$
\begin{aligned}
\mu_{2r+1} &= \tfrac{1}{\sqrt{2\pi}\sigma} \int_{-\infty}^{\infty} (x - \mu)^{2r+1} e^{-\frac{1}{2\sigma^2}(x-\mu)^2} dx \\
&= \tfrac{1}{\sqrt{2\pi}\sigma} \left[\int_{-\infty}^{\mu} (x - \mu)^{2r+1} e^{-\frac{1}{2\sigma^2}(x-\mu)^2} dx + \right. \\
&\qquad \left. \int_{\mu}^{\infty} (x - \mu)^{2r+1} e^{-\frac{1}{2\sigma^2}(x-\mu)^2} dx \right] \\
&= \tfrac{1}{\sqrt{2\pi}\sigma} [I_1 + I_2] \quad \text{(say)}.
\end{aligned}
$$

In I_1 we put $(\mu - x) = y$ and in I_2 we put $x - \mu = y$, when $I_1 = -I_2$. Therefore,

$$
\mu_{2r+1} = 0, r = 1, 2, \ldots
\tag{9.7.4}
$$

Similarly,

$$
\begin{aligned}
\mu_{2r} &= \tfrac{1}{\sigma\sqrt{2\pi}} \int_{-\infty}^{\infty} (x - \mu)^{2r} e^{-\frac{1}{2\sigma^2}(x-\mu)^2} dx \\
&= \sqrt{\tfrac{2}{\sigma}} \int_{0}^{\infty} z^{2r} e^{\frac{z^2}{2}} dz \quad \text{where } z = \tfrac{x-\mu}{\sigma} \\
&= \tfrac{2^r}{\sqrt{\pi}} \sigma^{2r} \Gamma(r + \tfrac{1}{2}) \quad \text{(using gamma function in (9.4.1))} \\
&= \tfrac{\sigma^{2r}}{\sqrt{\pi}} 2^r (r - \tfrac{1}{2})(r - \tfrac{3}{2}) \ldots \tfrac{1}{2} \Gamma(\tfrac{1}{2}) \quad \text{(using (9.4.3) and (9.4.5))} \\
&= \sigma^{2r} (2r - 1)(2r - 3) \ldots 3.1.
\end{aligned}
$$

In particular,

$$
\mu_1' = 0, \quad \mu_2 = \sigma^2, \quad \mu_4 = 3\sigma^4
$$
$$
\beta_1 = 0, \quad \beta_2 = 3.
$$

The moment generating function about $x = 0$ is

$$M_X(t) = E(e^{tX}) = e^{\mu t + \frac{1}{2}t^2\sigma^2} \quad \text{(vide Example 7.3.6).} \tag{9.7.5}$$

The *mgf* about mean μ is

$$M_{X-\mu}(t) = e^{\frac{1}{2}t^2\sigma^2}. \tag{9.7.6}$$

The cumulant generating function about $x = 0$ is

$$K_X(t) = \log_e M_X(t) = \mu t + \frac{1}{2}t^2\sigma^2.$$

Consequently,

$$k_1 = \mu, \ k_2 = \sigma^2, \ k_r = 0, r \geq 3.$$

The characteristic function about $x = 0$ is

$$\psi_X(t) = e^{i\mu t - \frac{1}{2}t^2\sigma^2}. \tag{9.7.7}$$

The characteristic function about mean is

$$\psi_{X-\mu}(t) = e^{-\frac{1}{2}t^2\sigma^2}. \tag{9.7.8}$$

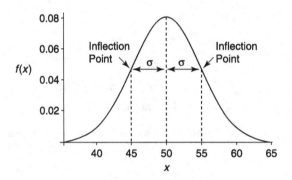

Fig. 9.7.1   The *pdf* of $N(50, 25)$ distribution

Theorem 9.7.1: If $X_1, \ldots, X_n$ are independent normal variables, X_i having mean μ_i and variance σ_i^2, then $Y = \sum_{i=1}^n a_i X_i$ is a normal variable with mean $\mu_* = \sum_{i=1}^n a_i \mu_i$ and variance $\sigma_*^2 = \sum_{i=1}^n a_i^2 \sigma_i^2$.

Proof. The characteristic function of Y is

$$\begin{aligned}
\psi_Y(t) &= E(e^{itY}) = E\{e^{it\sum_{i=1}^n a_i X_i}) \\
&= E\left\{\Pi_{i=1}^n e^{ita_i X_i}\right\} = \Pi_{i=1}^n E(e^{ita_i X_i}) \quad \text{(by virtue of independence)} \\
&= \Pi_{i=1}^n \psi_{a_i X_i}(t) = \Pi_{i=1}^n (e^{a_i \mu_i t - \frac{1}{2}a_i^2 \sigma_i^2 t^2}) = e^{t\mu_* - \frac{1}{2}t^2\sigma_*^2}
\end{aligned}$$

which is the characteristic function of a $N(\mu_*, \sigma_*^2)$-random variable. Hence the result follows by the uniqueness property of the characteristic function (Theorem 7.6.3).

Corollary 9.7.1.1: Let $X_1, \ldots, X_n$ be independent normal variables, X_i being $N(\mu_i, \sigma_i^2)$. Then the sum $S_n = \sum_{i=1}^{n} X_i$ is an $N(\mu_*, \sigma_*^2)$-variable where $\mu_* = \sum_{i=1}^{n} \mu_i$ and $\sigma_*^2 = \sum_{i=1}^{n} \sigma_i^2$.

Corollary 9.7.1.2: If $X_1, \ldots, X_n$ are independently and identically distributed $N(\mu, \sigma^2)$-variables, then the arithmetic mean $\bar{X} = \frac{S_n}{n}$ is an $N(\mu, \frac{\sigma^2}{n})$-random variable.

The variable $\frac{X-\mu}{\sigma} = \tau$ (say) is called a standard or unit normal variable. Its *pdf* is, therefore,

$$\phi(\tau) = \frac{1}{\sqrt{2\pi}} e^{-\frac{\tau^2}{2}}. \tag{9.7.9}$$

Clearly, $E(\tau) = 0, V(\tau) = 1$.

Also, its *mgf* and characteristic functions are, respectively,

$$M_\tau(t) = e^{\frac{\tau^2}{2}}, \ \psi_\tau(t) = e^{-\frac{\tau^2}{2}}. \tag{9.7.10}$$

The ordinate of the unit normal distribution (9.7.9) is denoted as $\phi(t)$. Clearly, $\phi(-t) = \phi(t)$, for any $t > 0$. Cumulative distribution function of τ,

$$F_\tau(u) = P[\tau \le u] = \frac{1}{2\pi} \int_{-\infty}^{u} e^{-\frac{\tau^2}{2}} d\tau \tag{9.7.11}$$

is denoted by $\Phi(u)$. $\Phi(u)$ represents the area of a standard normal curve to the left of the ordinate at $\tau = u$. Clearly, $\Phi(0) = 0.5, \Phi(-\infty) = 0, \Phi(\infty) = 1, \Phi(-t) = 1 - \Phi(t), P[|\tau| \le t] = 2\Phi(t) - 1$ for $t > 0$.

The ordinates $\phi(t)$ and the area $\Phi(t)$ under the curve (9.7.9) up to the left of the ordinate at t have been extensively tabulated and are reproduced in Tables A.1 and A.2 respectively at the end of this book. It is seen that

$$P\{|\tau| > 1\} = 0.318, \ P\{|\tau| > 2\} = 0.046, \ P\{|\tau| > 3\} = 0.0027.$$

Hence for a $N(\mu, \sigma)$-variable X, $P\{|X - \mu| > 3\sigma\} = 0.002$. Therefore, for all practical purposes, a normal distribution is concentrated within three standard deviation units from the mean. It is also seen that

$$\Phi(1.96) = 0.9750, \ \Phi(2.58) = 0.9951.$$

If X is a $N(\mu, \sigma^2)$-variable, for any two integers $a < b$,

$$P(a \le X \le b) = \frac{1}{\sigma\sqrt{2\pi}} \int_a^b e^{-\frac{(x-\mu)^2}{2\sigma^2}} \, dx = \frac{1}{\sqrt{2\pi}} \int_{\frac{a-\mu}{\sigma}}^{\frac{b-\mu}{\sigma}} e^{-\frac{\tau^2}{2}} \, d\tau, \quad \text{where } \tau = \frac{x-\mu}{\sigma}$$

$$= \Phi\left(\frac{b-\mu}{\sigma}\right) - \Phi\left(\frac{a-\mu}{\sigma}\right).$$

(9.7.12)

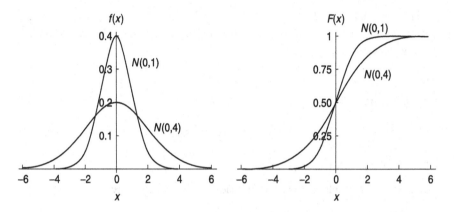

Fig. 9.7.2   Some normal *pdf*s and *cdf*s

Normal distribution arises as a reasonable approximation to distributions of many biological, physical and psychological variables. Distributions such as those relating to measurements on plants and animals, scores of students in a test are approximately of the normal form. Many theoretical reasons for the use of normal distribution are based on the *central limit theorems* (discussed in Chapter 12). These theorems state the conditions under which the distribution of the standardized sums of random variables tends to a unit normal distribution as the number of variables in the sum increases. Many theoretical distributions, often of a complex nature, are approximated by a normal distribution with the same expected value and the same standard deviation.

EXAMPLE 9.7.1: The results of a particular examination are as given in the following table. It is known that a candidate fails if he obtains less than 40% marks while he must obtain at least 75% marks in order to pass with distinction. Hence, determine the mean and the standard deviation of the distribution of marks, assuming it to be normal.

Let μ and σ be the mean and standard deviation of the distribution of X, the marks of a student. Hence

$$F_X(40) = \Phi(\frac{40 - \mu}{\sigma}) = 0.43 = \Phi(-0.18),$$

$$F_X(75) = \Phi(\frac{75 - \mu}{\sigma}) = 0.85 = \Phi(1.04),$$

because, 43% of the students secured less than 40 marks, 85% of the students secured less than 75 marks. Hence,

$$\frac{40 - \mu}{\sigma} = -0.18, \quad \frac{75 - \mu}{\sigma} = 1.04, \quad \text{when } \mu = 45.16, \sigma = 28.68.$$

Result	Percentage of Students
Passed with distinction	15
Passed without distinction	42
Failed	43

We shall now consider the normal approximation to the binomial distribution. This is given by the following theorem.

Theorem 9.7.2 *De Moivre-Laplace Theorem:* Let X be a binomial variable with mean np and variance npq. If $n \to \infty$, the probability distribution of the standardized binomial variable $\frac{X - np}{\sqrt{npq}}$ tends to that of a standard normal variable.

Proof. We have

$$P(X = r) = p(n, r) = \frac{n!}{r!(n - r)!} p^r q^{n-r}.$$

Applying Stirling's approximation (9.4.10) to factorials,

$$\lim_{n \to \infty} P(n, r) = \frac{e^{-n} n^{n+\frac{1}{2}} p^r q^{n-r}}{\sqrt{2\pi} e^{-r} r^{r+\frac{1}{2}} e^{-(n-r)} (n-r)^{n-r+\frac{1}{2}}}$$
$$= \frac{A}{\sqrt{2\pi npq}} \quad \text{(say)}$$

where

$$A = (\frac{np}{r})^{r+\frac{1}{2}} (\frac{nq}{n-r})^{n-r+\frac{1}{2}} = (1 + z_r\sqrt{\frac{q}{np}})^{-(np+z_r\sqrt{npq}+\frac{1}{2})}$$
$$(1 - z_r\sqrt{\frac{p}{nq}})^{-(nq-z_r\sqrt{npq}+\frac{1}{2})}$$

where $z_r = \frac{(r-np)}{\sqrt{npq}}$. Therefore,

$$\log_e A = -(np + z_r\sqrt{npq} + \tfrac{1}{2})\log_e(1 + z_r\sqrt{\tfrac{q}{np}})$$
$$-(nq - z_r\sqrt{npq} + \tfrac{1}{2})\log_e(1 - z_r\sqrt{\tfrac{p}{nq}}).$$

Assuming $z_r = \min.(\sqrt{\tfrac{np}{q}}, \sqrt{\tfrac{nq}{p}})$,

$$\log_e A = -(np + z_r\sqrt{npq} + \tfrac{1}{2})\left[z_r\sqrt{\tfrac{q}{np}} - \tfrac{z_r^2 q}{2np} + \tfrac{z_r^2 q^{\frac{3}{2}}}{3(np)^{\frac{3}{2}}} + 0(n^{-2})\right]$$
$$+ (nq - z_r\sqrt{npq} + \tfrac{1}{2})\left[z_r\sqrt{\tfrac{p}{nq}} + \tfrac{z_r^2 p}{2nq} + \tfrac{z_r^3 p^{\frac{3}{2}}}{3(nq)^{\frac{3}{2}}} + 0(n^{-2})\right] \qquad (9.7.13)$$
$$= -\tfrac{z_r^2}{2} - \tfrac{z_r(q-p)}{2\sqrt{npq}} + \tfrac{z_r^2}{4npq}(p^2 + q^2) + \tfrac{z_r^3}{6(npq)^{\frac{3}{2}}}(q^3 - p^3) + 0(n^{-2}).$$

As $n \to \infty$, (9.7.13) $\to -\tfrac{z_r^2}{2}$.

As r takes values $0, 1, \ldots, n$ and $n \to \infty$, z_r takes values between $-\infty$ and ∞. Also, as r increases by value 1, z_r increases by the amount $\frac{1}{\sqrt{npq}}$, which for large n, would be taken as dz. Hence z_r can be taken as a continuous variable taking values between $-\infty$ and ∞ and

$$\lim_{n\to\infty} P(n, r) = \frac{1}{\sqrt{2\pi}} e^{-\frac{z_r^2}{2}} dz, \quad -\infty < z_r < \infty. \qquad (9.7.14)$$

(For an alternative proof, see Section 12.2.)

Corollary 9.7.2.1: Let X be a binomial variable with parameters (n, p). For a fixed a and $b(a < b)$,

$$P[a \le X \le b] \to \Phi(z_b + \frac{1}{2\sqrt{npq}}) - \Phi(z_a - \frac{1}{2\sqrt{npq}}). \qquad (9.7.15)$$

Proof. By (9.7.14),

$$P[a \le X \le b] \approx \sum_{r=a}^{b} \phi(z_r) \qquad (9.7.16)$$

as $n \to \infty$.

By the mean value theorem of integral calculus $\left[\int_a^b f(x)dx \approx (b-a)f(x_0)\right.$ for an $x_0 \in [a, b]\Big]$,

$$\Phi(z_r + \frac{1}{2\sqrt{npq}}) - \Phi(z_r - \frac{1}{2\sqrt{npq}}) \approx \frac{1}{\sqrt{npq}}\phi(z_r') \quad \text{for some}$$

$$z_r - \frac{1}{2\sqrt{npq}} < z_r' < z_r + \frac{1}{2\sqrt{npq}}.$$

Hence,

$$\phi(z_r)dz = \phi(z_r')dze^{\frac{1}{2}(z_r'^2 - z_r^2)} = \left[\Phi(z_r + \frac{dz}{2}) - \Phi(z_r - \frac{dz}{2})\right]e^{\frac{1}{2}(z_r'^2 - z_r^2)}$$

where $dz = \frac{1}{\sqrt{npq}}$. Now,

$$\frac{1}{2}|z_r'^2 - z_r^2| = \frac{1}{2}|z_r' - z_r||z_r' + z_r| < dz\left[|z_r| + \frac{dz}{4}\right] < \epsilon$$

for sufficiently large n (since $dz = \frac{1}{\sqrt{npq}}$). Hence,

$$\phi(z_r)dz \approx \Phi(z_r + \frac{dz}{2}) - \Phi(z_r - \frac{dz}{2}).$$

when the result follows from (9.7.14). ☐

The equation (9.7.15) shows that for large n, an approximate value of the probability that a binomial random variable falls in an interval can be obtained from the standard normal probability distribution.

The following approximations hold for $\Phi(x)$:

(a) $\frac{1}{2}\left[1 + \sqrt{1 - e^{-\frac{x^2}{2}}}\right] \leq \Phi(x) \leq \frac{1}{2}\left[1 + \sqrt{1 - e^{-x^2}}\right]$

(b) $\Phi(x) \approx \frac{1}{2}\left[1 + \left\{1 - \exp(-\frac{2x^2}{\pi})\right\}^{\frac{1}{2}}\right].$

(c) $\phi(x)\{\frac{1}{x} - \frac{1}{x^3}\} < 1 - \Phi(x) < \frac{\phi(x)}{x}.$

(d) For large x, $1 - \Phi(x) \approx \frac{\phi(x)}{x}.$

9.7.1 *Truncated normal distribution*

A random variable X has a doubly truncated normal distribution if its *pdf* is given by

$$f_t(x; \mu, \sigma) = \frac{\frac{1}{\sigma\sqrt{2\pi}}e^{-\frac{1}{2\sigma^2}(x-\mu)^2}}{\int_A^B \frac{1}{\sigma\sqrt{2\pi}}e^{-\frac{1}{2}(x-\mu)^2}dx}, \quad (A \leq x \leq B)$$

$$= \left[\frac{1}{\sigma}(\phi(\frac{x-\mu}{\sigma}))\right]/T$$

(9.7.17)

where

$$T = \Phi(\frac{B-\mu}{\sigma}) - \Phi(\frac{A-\mu}{\sigma}).$$

The distribution is singly truncated from below or above if B is replaced by ∞ or A is replaced by $-\infty$ respectively. The expected value and the variance of the distribution (9.7.17) is

$$E(X) = \mu + \tfrac{\sigma}{T}[\phi_1 - \phi_2]$$
$$V(X) = \sigma^2 \left[1 + T^{-1}\left\{(\tfrac{A-\mu}{\sigma})\phi_1 - (\tfrac{B-\mu}{\sigma})\phi_2\right\} - \left\{\tfrac{\phi_1-\phi_2}{T}\right\}^2\right]$$

where $\phi_1 = \phi(\tfrac{A-\mu}{\sigma}), \phi_2 = \phi(\tfrac{B-\mu}{\sigma})$.

EXAMPLE 9.7.2: A normal distribution $f(x) = \frac{1}{\sigma\sqrt{2\pi}}e^{-\frac{1}{2\sigma^2}(x-\mu)^2}$, $-\infty < x < \infty$ is truncated at $x = A$, that is, values $x \leq A$ are discarded. Find the mean and variance of the truncated distribution.

Here

$$\mu'_1 = \left[\int_A^\infty xf(x)dx\right]\left[\int_A^\infty f(x)dx\right]^{-1}$$
$$= \left[\int_{A_0}^\infty (\mu + z\sigma)e^{-\frac{z^2}{2}}dz\right]\left[\int_{A_0}^\infty e^{-\frac{z^2}{2}}dz\right]^{-1}$$
$$(\text{ where } z = \tfrac{x-\mu}{\sigma}, \ A_0 = \tfrac{A-\mu}{\sigma})$$
$$= \mu + \tfrac{\sigma}{M\sqrt{2\pi}}e^{-\frac{z^2}{2}} \text{ where } M = 1 - \Phi(A_0).$$

$$\mu'_2 = \left[\int_A^\infty x^2 f(x)dx\right]\left[\int_A^\infty f(x)dx\right]^{-1} = \tfrac{1}{M\sqrt{2\pi}}\int_{A_0}^\infty (\mu + z\sigma)^2 dz$$

$$= \mu^2 + \tfrac{2\mu\sigma}{M\sqrt{2\pi}}e^{-\frac{A^2}{2}} + \tfrac{\sigma^2}{M\sqrt{2\pi}}\int_{A_0}^\infty z^2 e^{-\frac{z^2}{2}}dz.$$

Now,

$$\int_{A_0}^\infty z^2 e^{\frac{z^2}{2}}dz = z\int e^{-\frac{z^2}{2}}d(\tfrac{z^2}{2})\ |_{A_0}^\infty - \int_{A_0}^\infty \left[\int e^{-\frac{z^2}{2}}d(\tfrac{z^2}{2})\right]dz$$

$$= -ze^{-\frac{z^2}{2}}\ |_{A_0}^\infty + \int_{A_0}^\infty e^{-\frac{z^2}{2}}dz = A_0 e^{-\frac{A_0^2}{2}} + M\sqrt{2\pi}.$$

Hence,

$$\mu_2 = \mu'_2 - \mu'^2_1 = \sigma^2\left[1 + \frac{A_0 e^{-\frac{A_0^2}{2}}}{M\sqrt{2\pi}} - \frac{1}{2\pi M^2}e^{-A_0^2}\right].$$

9.8   The Log-Normal Distribution

A random variable X is said to have a log-normal distribution if $\log_e X$ has a normal distribution $(-\infty < \log_e X < \infty)$. If $\log_e X$ has mean μ and standard deviation σ, the probability distribution of $\log_e X$ is given by

$$f(\log x)d(\log x) = \frac{1}{\sigma\sqrt{2\pi}}e^{-\frac{1}{2\sigma^2}(\log x-\mu)^2}d(\log x), \quad -\infty < \log_e x < \infty$$

(9.8.1)

and hence

$$f(x)dx = \frac{1}{\sigma^x\sqrt{2\pi}}e^{-\frac{1}{2\sigma^2}(\log x-\mu)^2}dx, \quad 0 < x < \infty.$$
(9.8.2)

The rth raw moment of the distribution is

$$\mu'_r = \int_0^\infty \frac{x^r}{\sigma^x\sqrt{2\pi}}e^{-\frac{1}{2\sigma^2}(\log x-\mu)^2}dx.$$
(9.8.3)

Substituting $\frac{\log x-\mu}{\sigma} = z$, we have

$$\mu'_r = \frac{1}{\sqrt{2\pi}}\int_{-\infty}^\infty e^{r\sigma z+\mu r-\frac{z^2}{2}}dz = e^{\mu r+\frac{r^2\sigma^2}{2}}\cdot\frac{1}{\sqrt{2\pi}}\int_0^\infty e^{-\frac{1}{2}(x-r\sigma)^2}dz$$
$$= e^{\mu r+\frac{r^2\sigma^2}{2}}.$$

In particular,

$$\mu'_1 = e^{\mu+\frac{\sigma^2}{2}}, \ \mu'_2 = e^{2\mu+2\sigma^2}, \ \mu'_3 = e^{3\mu+\frac{9}{2}\sigma^2}, \ \mu'_4 = e^{4\mu+8\sigma^2}.$$

Denoting $e^{\sigma^2} = \alpha$,

$$\mu_2 = \alpha(\alpha-1)e^{2\mu}, \ \mu_3 = \alpha^{\frac{3}{2}}(\alpha-1)^2(\alpha+2)e^{3\mu},$$

$$\mu_4 = \alpha^2(\alpha-1)^2(\alpha^4+2\alpha^3+3\alpha^2-3)e^{4\mu}.$$

In general,

$$\mu_r = \frac{\mu_2^{\frac{r}{2}}}{(\alpha-1)^{\frac{r}{2}}}\sum_{j=0}^r(-1)^j\binom{r}{j}\alpha^{\frac{(r-j)(r-j-1)}{2}}$$
$$\sqrt{\beta_1} = \sqrt{(\alpha-1)}(\alpha+2), \ \beta_2 = \alpha^4+2\alpha^3+3\alpha^2-3.$$

The curve is therefore positively skewed. Note that neither $\sqrt{\beta_1}$, nor β_2 depend on μ.

The coefficient of variation is $\sqrt{(\alpha-1)}$ and does not depend on μ.

The distribution is unimodal because $\frac{df(x)}{dx} = 0$ gives

$$-\frac{1}{\sigma x^2\sqrt{2\pi}}e^{-\frac{1}{2\sigma^2}(\log x-\mu)^2}\left[\frac{1}{\sigma^2}(\log x - \mu) + 1\right] = 0$$

and hence the mode is $x_m = e^{(\mu-\sigma^2)}$.

The value X_α such that $P[X \le X_\alpha] = \alpha$ is related to the corresponding percentile z_α of the unit normal variable $Z = \frac{\log X-\mu}{\sigma}$ as follows.

$$P[X \le X_\alpha] = \alpha \Rightarrow P[\log X \le \log X_\alpha] = \alpha$$
$$\Rightarrow P\left[\frac{(\log X-\mu)}{\sigma} \le \frac{(\log X_\alpha-\mu)}{\sigma}\right] = \alpha$$

when

$$z_\alpha = \frac{\log X_\alpha - \mu}{\sigma} \text{ or } X_\alpha = e^{(\mu + \sigma z_\alpha)}.$$

In particular, for $\alpha = 0.50, z_\alpha = 0$, when the median of the distribution is e^α. For this distribution, therefore,

$$\mu_1' > \text{median} > \text{mode} \quad \text{and} \quad \frac{\text{mode}}{\mu_1'} = (\frac{\text{median}}{\mu_1'})^2.$$

The standardized log-normal distribution is

$$\xi = \frac{X - \mu_1'}{\sqrt{\mu_2}} = \frac{e^{Z\sigma} - e^{\frac{\sigma^2}{2}}}{[e^{\sigma^2}(e^{\sigma^2} - 1)]^{\frac{1}{2}}} \tag{9.8.4}$$

where z is as defined in (9.8.3). As $\sigma \to 0$, the distribution of ξ tends to a unit normal distribution.

Fig. 9.8.1 shows some standardized log-normal distributions.

A log-normal distribution often gives a good approximation to a normal distribution that has a small absolute value of coefficient of variation. Many biological measurements of quantities which are necessarily non-negative have log-normal distributions.

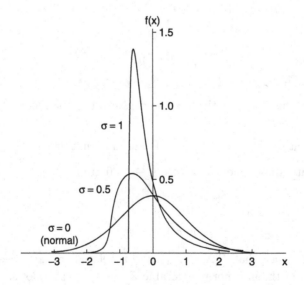

Fig. 9.8.1   Standardized log-normal distributions

9.9   The Double Exponential (Laplace) Distribution

A random variable X is defined to have a double exponential distribution (or Laplace distribution) if its *pdf* is given by

$$f(x; \mu, \sigma) = \frac{1}{2\sigma} e^{-\frac{|x-\mu|}{\sigma}} \quad (\sigma > 0) \tag{9.9.1}$$

where $-\infty < \mu < \infty$ and $\sigma > 0$. Here $E(X) = \mu, V(X) = 2\sigma^2, \sqrt{\beta_1} = 0, \beta_2 = 6$. The distribution is symmetric about $x = 0$. The mean deviation is $E(|X|) = 1$.

The mode is at $x = \mu$. The cumulative distribution function is

$$F_X(x) = \frac{1}{2} \exp\left[-\frac{(\mu - x)}{\sigma}\right] \text{ for } x \le \mu; \; 1 - \frac{1}{2} \exp\left[-\frac{(x - \mu)}{\sigma}\right] \text{ for } x \ge \mu.$$

The lower and the upper quartiles are $\mu +_- \sigma \log_e 2$. Fig. 9.9.1 shows the form of the distribution.

The standardized double exponential distribution is obtained by setting $\mu = 0$ and $\sigma = 1$. Its *pdf* is

$$f(x) = \frac{1}{2} e^{-|x|}. \tag{9.9.2}$$

The characteristic function corresponding to (9.9.2) is

$$\Psi_X(t) = E(e^{tX}) = \frac{1}{1 + t^2} \quad \text{(Example 7.6.4)}.$$

The *mgf* is

$$M_X(t) = \frac{1}{1 - t^2}.$$

The central moments are

$$\mu_r = 0, \quad \text{if r is odd}; \; \mu_r = r! \text{ if r is even}.$$

The cumulant generating function is $K_X(t) = -\log(1 - t^2)$. Thus
$k_r = 0$ if r is odd; $k_r = 2[(r - 1)!]$ if r is even.

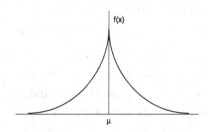

Fig. 9.9.1   Laplace density

9.10   The Pareto Distribution

A random variable X is said to have a Pareto distribution with parameters α, β if its *pdf* is given by

$$f(x) = \frac{\alpha \beta^\alpha}{x^{\alpha+1}}, \quad \alpha > 0, x \ge \beta > 0. \tag{9.10.1}$$

The distribution function is

$$F_X(x) = 1 - \left(\frac{\beta}{x}\right)^\alpha.$$

Provided $r < \alpha$,

$$\mu'_r = \frac{\alpha \beta^r}{\alpha - r}.$$

In particular,

$$\mu'_1 = \frac{\alpha\beta}{\alpha-1}, \ (\alpha > 1), \ \mu_2 = \frac{\alpha\beta^2}{(\alpha-1)^2(\alpha-2)} \ (\alpha > 2),$$

$$\sqrt{\beta_1} = 2\frac{\alpha+1}{\alpha-3}\sqrt{\frac{\alpha-2}{\alpha}}, \ \alpha > 2,$$

$$\beta_2 = \frac{3(\alpha-2)(3\alpha^2+\alpha+2)}{\alpha(\alpha-3)(\alpha-2)}, \ \alpha > 4.$$

As $\alpha \to \infty$, $\sqrt{\beta_1} \to 2$ and $\beta_2 \to 9$. The mode of the distribution is at $x = \beta$. The medain is at $x_m = \beta(2^{\frac{1}{\alpha}})$. The population geometric mean is $\beta.e^{\frac{1}{\alpha}}$. The population harmonic mean is $\beta(1 + \frac{1}{\alpha})$.

The mean deviation is $2\beta(\alpha - 1)^{-1}(1 - \alpha^{-1})^{\alpha-1}$ for $\alpha > 1$.

$$\frac{\text{Mean Deviation}}{\text{Standard Deviation}} = 2(1 - 2\alpha^{-1})^{\frac{1}{2}}(1 - \alpha^{-1})^{(\alpha-1)}.$$

As $\alpha \to \infty$, this ratio tends to $2e^{-1} = 0.735$.

Two other forms of the distribution were proposed by Pareto. One known as Pareto's distribution of the second kind, sometimes also known as *Lomax distribution* is given by

$$F_X(x) = 1 - \frac{C^\alpha}{(x+C)^\alpha}, \quad x > 0, \alpha > 0. \tag{9.10.2}$$

Its standardized form is obtained by putting $C = 1$, when its *pdf* is

$$f_X(x) = \frac{\alpha}{(1+x)^{\alpha+1}}, \quad x > 0, \ a > 0.$$

In many socio-economic investigations, like distribution of population sizes, occurrences of natural resources, size of firms, distribution of income over populations, Pareto curve gives a good fit to the data towats the extremes of the range, while log-normal distribution fits well over the other parts of the data. Lomax (1954) used his distribution in the analysis of business failure data.

9.11 The Weibull Distribution

A random variable X has an Weibull distribution with parameters $c(> 0), a(> 0), \mu$ if its *pdf* is

$$f(x; c, a, \mu) = \frac{c}{a}(\frac{x - \mu}{a})^{c-1} \exp\left[-\left(\frac{x - \mu}{a}\right)^c\right], \ x > \mu. \tag{9.11.1}$$

The cumulative distribution function is

$$F_X(x) = 1 - e^{\{\frac{x-\mu}{a}\}^2}, \ x > \mu.$$

For $c > 1$, the *pdf* (9.11.1) tends to 0 as $x \to \mu$ and the curve is unimodal with mode at

$$x_m = a(\frac{c - 1}{c})^{\frac{1}{c}} + \mu. \tag{9.11.2}$$

As $c \to \infty$, this value tends to $a + \mu$ very rapidly.

For $0 < c \le 1$, the mode is at μ and the density is an increasing function of $x \ \forall \ x > \mu$.

The median of the distribution is at

$$x = a(\log 2)^{\frac{1}{c}} + \mu.$$

The standard form of the distribution is obtained by putting $\mu = 0, \alpha = 1$ when the density is

$$f(x) = cx^{c-1}e^{-x^c} \ (x > 0), (c > 0). \tag{9.11.3}$$

The corresponding cumulative distribution function is

$$F_X(x) = 1 - e^{-x^c}, \ x > 0, c > 0.$$

Here

$$\mu'_r \quad = \Gamma(\tfrac{r}{c} + 1),$$

$$E(X) = \Gamma(\tfrac{1}{c} + 1), \ V(X) = \Gamma(\tfrac{2}{c} + 1) - (\Gamma(\tfrac{1}{c} + 1))^2.$$

For $c = 1$, (9.11.3) reduces to the exponential density.

The distribution is frequently used in the theory of reliability. Figures 9.11.1(a),(b) give standardized Weibull distributions with $c = 2$ and $c = 5$ respectively.

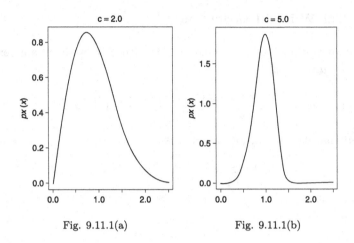

Fig. 9.11.1(a)   Fig. 9.11.1(b)

9.12   The Extreme Value Distribution

The *pdf* of this distribution with parameter $\theta(> 0)$ and μ is
$$f(x; \theta, \mu) = \theta^{-1} e^{-\frac{(x-\mu)}{\sigma}} \exp[-e^{-\frac{(x-\mu)}{\sigma}}];$$
$$E(X) \quad = \mu - 0.57721\sigma, \ V(X) = \pi^2 \frac{\theta^2}{6}.$$
Mode $= \mu$, median $= \mu + \sigma \log \log 2$.

9.13   The Logistic Distribution

The cumulative distribution function for this distribution is
$$F_X(x) = \left[1 + \exp\{-\frac{(x-\mu)}{\sigma}\}\right]^{-1}, \ \sigma > 0. \tag{9.13.1}$$
The corresponding *pdf* is
$$f_X(x) = \frac{e^{-(x-\mu)/\sigma}}{\sigma[1 + e^{-\frac{(x-\mu)}{\sigma}}]^2}, \ -\infty < x < \infty, \ \sigma > 0;$$

$$E(X) = \mu, \ V(X) = \frac{\sigma^2 \pi^2}{3}.$$
The function (9.13.1) is widely used to represent growth functions with x representing time.

9.14   Exercises and Complements

9.1 Show that if X is any random variable having a cumulative distribution function $F_X(x)$ which is absolutely continuous, then the random variable $Y = F_X(x)$ has a uniform distribution over $(0, 1)$.

9.2 Suppose $X_1, \ldots, X_n$ are *iid* variables, each having a rectangular distribution over $(0, 1)$. Show that the sum $Z = X_1 + \ldots + X_n$ has the probability density

$$\frac{1}{(n-1)!}\{x^{n-1} - \binom{n}{1}(x-1)^{n-1} + \binom{n}{2}(x-2)^{n-1} - \ldots\},$$

the series ending with a term in $(x - [x])^{n-1}$, where $[x]$ is the integral part of x.

[Hints: Proceed by using the characteristic function of Z and then by using the inversion theorem.]

9.3 Suppose $X_1, \ldots, X_n$ are *iid* rectangular variables over $(0, 1)$. Show that the distribution of $Y_i = -\log_e X_i$ is given by

$$e^{-y_i} dy_i \ (0 < y_i < \infty);$$

hence, the distribution of $S_n = Y_1 + \ldots + Y_n = -\log(X_1 \ldots X_n)$ is given by

$$\frac{1}{\Gamma(n)} S_n^{n-1} e^{-S_n} dS_n$$

and that of $T_n = \Pi_{i=1}^n X_i$ has the distribution

$$\frac{1}{\Gamma(n)} (-\log T_n)^{n-1} dT_n.$$

9.4 Show that for the rectangular distribution

$$f(x) = \frac{1}{b}, \ a \le x \le a + b,$$

the *mgf* is $e^{at} \frac{[e^{bt} - 1]}{bt}$, $\mu_1' = a + \frac{b}{2} (= \text{median})$, $\mu_2 = \frac{b^2}{12}$, $\mu_r = 0$ for r odd, $= (\frac{b}{2})^r / (r + 1)$ for r even.

9.5 Suppose X has a negative exponential distribution with parameter λ such that $P[X \le 1] = P[X > 1]$. Find $\text{var}(X)$.

9.6 Show that for the *triangular distribution*

$$f(x) = \begin{cases} \frac{2x}{A}, & 0 \le x < A \\ \frac{2(1-x)}{(1-A)}, & A \le x < 1, \end{cases}$$

$\mu_1' = \frac{A+1}{3}$, $\mu_2 = \frac{1}{18}(1 - A + A^2)$.

9.7 Show that for the *erlang distribution*

$$f(x) = \frac{(\frac{x}{b})^{c-1} e^{-\frac{x}{b}}}{b[(c-1)!]}, \ 0 \le x, \infty, \ b > 0, \ c(\text{integer}) > 0,$$

the *mgf* is $(1 - bt)^{-c}, t > \frac{1}{b}$; $\mu_1' = bc$, $\mu_2 = b^2c$, $\sqrt{\beta_1} = \frac{2}{\sqrt{c}}$, $\beta_2 = 3 + \frac{6}{c}$, mode is $b(c - 1)$.

9.8 Suppose X has a Gamma distribution with parameters (λ, p) with p an integer. Show that the distribution function of X is

$$P(X \geq x) = \sum_{j=0}^{p-1} e^{-\lambda x} \frac{(x\lambda)^j}{j!}.$$

9.9 Let X have a Gamma distribution with parameters $\lambda = 1$ and p. Show that the *pdf* of $Z = X^{-1}$ is

$$f(z; 1, p) = \frac{1}{\Gamma(p)} e^{-z^{-1}} z^{-1-p}.$$

Hence, show that

$$E(Z) = (p - 1), \ p \neq 1,$$
$$V(Z) = (p - 1)^{-2}(p - 2), \ p \neq 1, 2.$$

9.10 For a beta distribution with *pdf*

$$f(x) = \frac{1}{B(p, q)}(1 - x)^{p-1}x^{q-1}, \ 0 < x < 1, \ p > 0, q > 0,$$

verify that the harmonic mean is less than the arithmetic mean.

9.11 Show that for the *power function* distribution,

$$f(x) = cx^{c-1}, \ 0 \leq x \leq 1,$$
$$\mu_r' = \frac{c}{c+r}, \ \mu_2 = \frac{c}{(c+2)(c+1)^2},$$

the mode is at $x = 1$ for $c > 1$, at $x = 0$ for $c < 1$ and the median is at $x = \left(\frac{1}{2}\right)^{\frac{1}{c}}$.

9.12 Let N be a non-negative random variable with geometric distribution, $P(N = n) = (1 - e^{-u})e^{-nu}, n = 0, 1, 2, \ldots$ Let M be a positive integer with *pmf*

$$P(M = m) = \frac{e^{-u}\mu^m}{m!(1 - e^{-\mu})}, \ m = 1, 2, \ldots$$

Let $X_1, \ldots, X_M$ be M independent variables, each uniform over $(0, 1)$. Show that

$$Y = \mu[N + \min(X_1, \ldots, X_N)]$$

has the standard exponential distribution.

9.13 Show that the constant K such that $\frac{K}{(1+x^2)^m}$, $x \in \mathcal{R}$ is a *pdf* is

$$K = \frac{\Gamma(m)}{\Gamma(m - \frac{1}{2})\Gamma(\frac{1}{2})}.$$

Show also that this variable has no integral moment of order greater than the integral part of $2m - 2$, but moments of lower orders can be obtained in terms of gamma function. (For $m = 1$, this is called the Cauchy distribution.)

9.14 Let θ be a random angle which is uniformly distributed in the range $(-\frac{\pi}{2}, \frac{\pi}{2})$. Show that the *pdf* of $X = \tan \theta$ is $\frac{1}{\pi(1+x^2)}$.

9.15 Suppose $X_1, \ldots, X_n$ is a sequence of n independent random variables, each having Cauchy distribution $\frac{1}{\pi(1+x^2)}$. Show that $\bar{X} = \frac{1}{n}\sum_{i=1}^{n} X_i$ has the characteristic function $e^{-|t|}$. Hence, show that $\bar{X}$ has the same distribution as $X_i (i = 1, \ldots, n)$. (This astounding property is due to non-existence of the second moment of the Cauchy distribution).

9.16 Let $\bar{X} = \frac{1}{n}(X_1 + \ldots + X_n)$ where X_i's are *iid* $N(\mu, \sigma^2)$-variables. Show that for any $\epsilon > 0$,

$$\lim_{n \to \infty} P(|\bar{X} - \mu| \geq \epsilon) = 0.$$

[Since $\bar{X} \cap N(\mu, \frac{\sigma^2}{n})$,

$$P(|\bar{X} - \mu| \geq \epsilon) = 1 - \int_{-\epsilon\sqrt{n}/\sigma}^{\epsilon\sqrt{n}/\sigma} e^{-t^2 2} dt \quad \text{where } t = \frac{\sqrt{n}(x - \mu)}{\sigma}.$$

Hence,

$$\lim_{n \to \infty} P(|\bar{X} - \mu| \geq \epsilon) = 1 - \frac{1}{\sqrt{2\pi}} \int_{-\infty}^{\infty} e^{-\frac{t^2}{2}} dt = 0.]$$

9.17 Show that for $x > 0$,

$$1 - \Phi(x) = \frac{1}{\sqrt{2\pi}} e^{-\frac{x^2}{2}} \{\frac{1}{x} - \frac{1}{x^3} + \frac{1.3}{x^5} - \frac{1.3.5}{x^7} + \ldots + (-1)^k \frac{1.3.\ldots.(2k-1)}{x^{2k+1}} + \ldots\}.$$

$$(i)$$

Hence, show that

$$1 - \Phi(x) < \frac{\phi(x)}{x} \quad \text{for } x > 0.$$

$$(ii)$$

[

$$1 - \Phi(x) = \frac{1}{\sqrt{2\pi}} \int_x^\infty e^{-\frac{u^2}{2}} \, du.$$

Now,

$$\int_x^\infty e^{-\frac{u^2}{2}} \, du = \int_x^\infty \frac{1}{u}(e^{-\frac{u^2}{2}}) \, du = \frac{1}{x}e^{-\frac{x^2}{2}} - \int_x^\infty \frac{1}{u^3}(ue^{-\frac{u^2}{2}}) \, du, \qquad (iii)$$

by integrating by parts with $\frac{1}{u}$ as the first factor.

Similarly, evaluating $\int_x^\infty \frac{1}{u^3}(ue^{\frac{u^2}{2}}) \, du$ by integration by parts, with $\frac{1}{u^3}$ as the firs factor, and proceeding similarly, the first part follows.

From (iii), $\int_x^\infty e^{-\frac{u^2}{2}} \, du < \frac{1}{x}e^{-\frac{x^2}{2}}$ for $x > 0$. Hence the second part.]

9.18 Suppose X, Y are independent random variables. Show that if $Z = XY$ has a log-normal distribution, then X, Y each has a log-normal distribution and conversely.

Chapter 10

Probability Distributions on $\mathcal{R}^n$

10.1 Introduction

In Chapter 4 we considered a single random variable X and its probability distribution on $\mathcal{R}^1$. In many experiments, however, observations are expressed as a number of several numerical quantities. In measuring the height, weight and the intelligence quotient of a number of school students, we have three observations for each individual. In such cases, we have, therefore, a vector of random variables $\mathbf{X} = (X_1, \ldots, X_n)$ where each X_i is a random variable. In this chapter we shall study the probability distributions of n-dimensional random vectors $\mathbf{X}$ on $\mathcal{R}^n$.

Section 10.2 considers joint probability distributions, marginal probability distributions, conditional probability distributions and the respective distribution functions. The concept of statistical independence is also considered. Joint expectation and moments are the subject matter of the next section. Multivariate probability generating functions, moment generating functions and characteristic functions are considered subsequently. Conditional expectation, conditional variance and regression are discussed in Section 10.7. The remaining two sections deal with two important multivariate distributions, namely, multinomial distribution and bivariate normal distribution. We have refrained from discussing other multivariate distributions, because, they are too many and are beyond the scope of this book.

10.2 Probability Distribution of a Random Vector

Consider the probability space $(\mathcal{S}, \mathcal{A}, P)$ arising out of a random experiment. A vector of functions $\mathbf{X} = (X_1, X_2, \ldots, X_n)$ which maps $\mathcal{S}$ into $\mathcal{R}^n$

is said to be a random vector if, for every Borel set in $\mathcal{B}_n$,

$$\mathbf{X}^{-1}(B) = \{\omega \in \mathcal{S} : \mathbf{X}(\omega) \in B\} \in \mathcal{A}, \ B \in \mathcal{B}_n, \qquad (10.2.1)$$

that is, $\mathbf{X}$ is a measurable transformation of $(\mathcal{S}, \mathcal{A})$ into $(\mathcal{R}^n, \mathcal{B}_n)$.

As in the univariate case, to verify if a certain vector of functions is a random vector, it is not necessary to verify if $\mathbf{X}^{-1}(B) \in \mathcal{A}$ for each $B \in \mathcal{B}_n$. It is sufficient if we verify that $\mathbf{X}^{-1}(\mathcal{C}) \in \mathcal{B}_n$, where $\mathcal{C}$ is any class of subsets of $\mathcal{R}^n$ which generate $\mathcal{B}_n$.

We therefore define a random vector as follows.

DEFINITION 10.2.1 (*Random Vector*): Let $(\mathcal{S}, \mathcal{A}, P)$ be a probability space. A vector of functions $\mathbf{X} = (X_1, X_2, \ldots, X_n)$ which maps $\mathcal{S}$ into $\mathcal{R}^n$ is called a random vector *iff* for each $a_i \in \mathcal{R}^1 (i = 1, 2, \ldots, n)$

$$\{\omega : X_1(\omega) \le a_1, \ldots, X_n(\omega) \le a_n\} \subset \mathcal{A} \qquad (10.2.2)$$

i.e.

$$\mathbf{X}^{-1}[I = \{(x_1, \ldots, x_n) : -\infty < x_i \le a_i, \ i = 1, \ldots, n\}] \subset \mathcal{A} \qquad (10.2.3)$$

i.e. if $\mathbf{X}$ is a $\mathcal{A}$-measurable function. Here I is an n-dimensional interval.

EXAMPLE 10.2.1: A fair coin is tossed twice. Let us assign the score $1(2)$ for $H(T)$. Let X_1, X_2 denote respectively the number at the first toss, sum of scores at the two tosses. Here $\mathbf{X} = (X_1, X_2)$ and

$$\mathbf{X}^{-1}\{(-\infty, a_1], (-\infty, a_2]\} = \begin{cases} \phi, & a_1 < 0, a_2 < 1, \\ (HH), & 0 < a_1 \le 1, 1 < a_2 \le 2, \\ (HT), & 0 < a_1 \le 1, 2 < a_2 \le 3, \\ (TH), & 1 < a_1 \le 2, 2 < a_2 \le 3, \\ (TT), & 1 < a_1 \le 2, 3 < a_2 \le 4, \\ \mathcal{S}, & a_1 > 2, a_2 > 4. \end{cases}$$

Thus $\mathbf{X}$ is a random vector.

Theorem 10.2.1: $\mathbf{X} = (X_1, \ldots, X_n)$ is a random vector *iff* each $X_j, j = 1, \ldots, n$ is a random variable.

Proof. Let $B \in \mathcal{R}^n$ be such that

$$B = \Pi_{j=1}^n B_j, \ B_j \in \mathcal{R}^1, \ j = 1, \ldots, n.$$

Then,

$$\begin{aligned} \mathbf{X}^{-1}(B) &= \{\omega \in \mathcal{S} : X_1(\omega) \in B_1, \ldots, X_n(\omega) \in B_n\} \\ &= \{X_1^{-1}(B_1) \cap \ldots \cap X_n^{-1}(B_n)\} \\ &= \cap_{j=1}^n X_j^{-1}(B_j). \end{aligned} \qquad (10.2.4)$$

Suppose now $\mathbf{X}$ is a random vector. Then for every $B \in \mathcal{B}_n, \mathbf{X}^{-1}(B) \in \mathcal{A}$. In particular, let $B_j \in \mathcal{B}_1, j = 1, \ldots, n$. Then it follows from (10.2.4) that $\cap_{j=1}^n X_j^{-1}(B_j) \in \mathcal{A}$. Setting $B_j = \mathcal{R}$ for $j = 2, 3, \ldots, n$, we see that X_1 is a random variable. Similarly, $X_2, \ldots, X_n$ are also random variables.

Conversely, suppose that each X_j is a random variable. Then for every $B \in \mathcal{R}^n$, which is of the form $B = \Pi_{j=1}^n B_j, B_j \in \mathcal{B}, j = 1, \ldots, n, \mathbf{X}^{-1}(B) \in \mathcal{A}$. Let $\mathcal{E}$ be the class of all subsets of $\mathcal{R}^n$ for which $\mathbf{X}^{-1}(B) \in \mathcal{A}$. It can be verified that $\mathcal{E}$ is a σ-field, so that $\mathcal{B}_n \subset \mathcal{E}$. Hence, $\mathbf{X}$ is a random vector. $\square$

Note 10.2.1: Let $\mathbf{X} : (\mathcal{S}, \mathcal{A}) \to (\mathcal{R}^n, \mathcal{B}_n)$ be a random vector and let $\mathbf{g} : \mathcal{R}^n \to \mathcal{R}^k$ be a measurable mapping. Then $\mathbf{g}(\mathbf{X}) \to \mathcal{R}^k$ is a random vector.

DEFINITION 10.2.2 (*Probability Distribution of* $\mathbf{X}$): The random vector $\mathbf{X}$ defined on $(\mathcal{S}, \mathcal{A}, P)$ induces a probability space $(\mathcal{R}^n, \mathcal{B}_n, P_\mathbf{X})$. The probability measure $P_\mathbf{X}$ is given by

$$P_\mathbf{X}(B) = P[\mathbf{X}^{-1}(B)] \text{ for every } B \in \mathcal{B}_n. \qquad (10.2.5)$$

It can be verified that $P_\mathbf{X}$ satisfies all the three axioms of a probability measure. $P_\mathbf{X}$ is known as the probability distribution of $\mathbf{X}$.

DEFINITION 10.2.3 (*Joint Distribution Function of* $\mathbf{X}$): Let $\mathbf{X} = (X_1, \ldots, X_n)$ be a random vector defined on a probability space $(\mathcal{S}, \mathcal{A}, P)$. Define a function $F_\mathbf{X}$ on $\mathcal{R}^n$ by the relation

$$F_\mathbf{X}(x_1, \ldots, x_n) = P\{\omega : X_1(\omega) \le x_1, X_2(\omega) \le x_2, \ldots, X_n(\omega) \le x_n\},$$

$$x_j \in \mathcal{R}, \ 1 \le j \le n. \qquad (10.2.6)$$

The function $F_\mathbf{X}$ is known as the joint distribution function of the random vector $\mathbf{X}$.

Note 10.2.2: Let $F_\mathbf{X}(.) = F_{(X_1,\ldots,X_n)}(.)$ be the distribution function of the random vector $\mathbf{X} = (X_1, \ldots, X_n), n \ge 2$. We set

$$\lim_{x_n \to \infty} F_{(X_1,\ldots,X_n)}(x_1, \ldots, x_n) = F_{(X_1,\ldots,X_{n-1})}(x_1, \ldots, x_{n-1}).$$

Then $F_{(X_1,\ldots,X_{n-1})}(.)$ is a distribution function and is known as the *marginal distribution function* of the random vector $(X_1, \ldots, X_{n-1})$. In general, the joint distribution function of any subset of $X_1, \ldots, X_n$, obtained by setting all the remaining variables to be ∞ is referred to as a marginal distribution function.

The following results can be proved along the lines of proof of Theorem 4.4.1.

Theorem 10.2.2: Let **X** be a random vector with distribution function $F_{\mathbf{X}}$. Then

  (i) $F_{\mathbf{X}}(x_1, \ldots, x_n) \to 0$ as $x_i \to -\infty$ for at least one i;

  $F(x_1, \ldots, x_n) \to 1$ if $x_i \to \infty$ for each i, $1 \le i \le n$.

  (ii) $F_{\mathbf{X}}$ is right continuous in each argument; that is

$$\lim_{\epsilon_1 \downarrow 0} F(x_1 + \epsilon_1, x_2, \ldots, x_n) = \ldots$$

$$= \lim_{\epsilon_n \downarrow 0} F_{\mathbf{X}}(x_1, \ldots, x_{n-1}, x_n + \epsilon_n) = F_{\mathbf{X}}(x_1, \ldots, x_n).$$

(iii)

$$F(x_1 + h_1, x_2 + h_2, \ldots, x_n + h_n) - [F(x_1, x_2 + h_2, \ldots, x_n + h_n)$$

$$\ldots + F(x_1 + h_1, x_2 + h_2, \ldots, x_{n-1} + h_{n-1}, x_n)]$$

$$+ [F(x_1, x_2, x_3 + h_3, \ldots, x_n + h_n) +$$

$$\ldots + F(x_1 + h_1, x_2 + h_2, \ldots + x_{n-2} + h_{n-2}, x_{n-1}, x_n)]$$

$$- \ldots + (-1)^{n-1} F(x_1, \ldots, x_n) \ge 0. \qquad (10.2.7)$$

Conversely, any function F defined on $\mathcal{R}^n$ and satisfying conditions (i), (ii) and (iii) above determines uniquely a probability measure P_F on $\mathcal{B}_n$, the σ-field of Borel sets on $\mathcal{R}^n$.

Note 10.2.3: Condition (10.2.7) implies that F is non-decreasing in each argument. However, a function F which is nondecreasing in each argument does not necessarily satisfy (10.2.7) (vide Exercise 10.46).

Note 10.2.4: As in the univariate case (vide (4.8.37) and (4.8.38)), we may consider three types of joint distribution functions F, discrete, absolutely continuous and singular.

DEFINITION 10.2.4: We say that a joint distribution function is discrete, if there exists a countable subset B of $\mathcal{R}^n$ such that $P_F(B) = 1$ where $P_F(.)$ is the probability measure over $\mathcal{R}^n$ determined by F, $P_F(\mathbf{a}, \mathbf{b}] = F(\mathbf{b}) - F(\mathbf{a}) = P_{\mathbf{X}}(\mathbf{a}, \mathbf{b}]$ where $(\mathbf{a}, \mathbf{b}]$ is a n-dimensional semi-closed interval.

Let $\mathbf{x}_i = (x_{1i}, \ldots, x_{ni}), i = 1, 2, \ldots$ be the values assumed by $\mathbf{X} = (X_1, \ldots, X_n)$ with probabilities $P[\mathbf{X} = \mathbf{x}_i] = p(\mathbf{x}_i)$ i.e. $P[X_1 = x_{1i}, \ldots, X_n = x_{ni}] = p(x_{1i}, x_{2i}, \ldots, x_{ni}), i = 1, 2, \ldots$. Then $p(\mathbf{x}_i) = p(x_{1i}, \ldots, x_{ni})$ is called the *pmf* of $\mathbf{X} = (X_1, \ldots, X_n)$ at the mass point $\mathbf{x}_i = (x_{1i}, \ldots, x_{ni})$ if

(i) $p(\mathbf{x}_i) \geq 0$ for all $i = 1, 2, \ldots$

(ii) $\sum_i p(\mathbf{x}_i) = 1$.

The distribution function F is given by

$$F_{\mathbf{X}}(\mathbf{x}) = P[X_1 \leq x_1, \ldots, X_n \leq x_n] = \sum_{x_{1i} \leq x_1, \ldots, x_{ni} \leq x_n} p(x_{1i}, x_{2i}, \ldots, x_{ni}).$$

(10.2.8)

Every point $(x_{1i}, \ldots, x_{ni})$ where $p(x_{1i}, \ldots, x_{ni}) > 0$ is a jump-point of $F_{\mathbf{X}}(\mathbf{x})$ and $p(\mathbf{x}_{1i}, \ldots \mathbf{x}_{ni})$ is its jump at $(x_{1i}, \ldots, x_{ni})$. In this case, $\mathbf{X}$ is a discrete random vector.

DEFINITION 10.2.5: The distribution F is said to be absolutely continuous if there exists a Borel measurable function f defined on $\mathcal{R}^n$ such that

$$F(x_1, \ldots, x_n) = \int_{-\infty}^{x_1} \cdots \int_{-\infty}^{x_n} f(t_1, \ldots, t_n) dt_1 \ldots dt_n \qquad (10.2.9)$$

for all $x_i \in \mathcal{R}(i = 1, \ldots, n)$. In this case, the random vector $\mathbf{X}$ is said to be of absolutely continuous (or, simply, of continuous) type. Also, f is known as the *joint probability density* function of $\mathbf{X}$ and is given by

$$\frac{\partial^n f}{\partial x_1 \ldots \partial x_n} = f(x_1, \ldots x_n)$$

which exists everywhere except for a set of values of $\mathbf{x}$ of probability measure zero. Also

$$\int_{-\infty}^{\infty} \cdots \int_{-\infty}^{\infty} f_{\mathbf{X}}(x_1, \ldots, x_n) dx_1 \ldots dx_n = F(\infty, \ldots, \infty) = 1. \quad (10.2.10)$$

Since $F_{\mathbf{X}}(\mathbf{x})$ is continuous everywhere, it has no point of jump. Thus, the probability that $\mathbf{X}$ takes the value $\mathbf{x}$ is $P[\mathbf{X} = \mathbf{x}] = 0$. For any real vectors $\mathbf{a}, \mathbf{b}, \mathbf{a} < \mathbf{b}$,

$$P[\mathbf{a} \leq \mathbf{X} \leq \mathbf{b}] = \int_{\mathbf{a}+0}^{\mathbf{b}} f_{\mathbf{X}}(\mathbf{x}) d\mathbf{x} \qquad (10.2.11)$$

It follows from the mean value theorem of calculus

$$P[\mathbf{x} \leq \mathbf{X} \leq \mathbf{x} + d\mathbf{x}] \approx f(\mathbf{x}) d\mathbf{x} = f(x_1, \ldots, x_n) dx_1 \ldots dx_n. \qquad (10.2.12)$$

We consider the results for $n = 2$, writing X_1 as X and X_2 as Y. Clearly,

$$F_{X,Y}(\infty, \infty) = \lim_{x \to \infty; y \to \infty} \int_{-\infty}^{x} \int_{-\infty}^{y} f(u,v) du dv = \int_{-\infty}^{\infty} \int_{-\infty}^{\infty} f(u,v) du dv = 1.$$
(10.2.13)

Again,

$$\lim_{\delta x \to 0, \delta y \to 0} \frac{P[x < X \le x + \delta x, y < Y \le y + \delta y]}{\delta x \delta y} = f(x,y)$$

or

$$P[x < X \le x + dx, y < Y \le y + dy] = f(x,y) dx dy.$$
(10.2.14)

Also,

$$P[x < X \le x + dx, y \le Y] = F(x + dx, y) - F(x,y) = \frac{\partial F(x,y)}{\partial x \partial y} dx.$$
(10.2.15)

Similarly,

$$P[X \le x, y < Y \le y + dy] = \frac{\partial F(x,y)}{\partial x \partial y} dy.$$
(10.2.16)

Again,

$$\lim_{\delta x \to 0, \delta y \to 0} \frac{F(x + \delta x, y + \delta y) - F(x, y + \delta y) - F(x + \delta x, y) + F(x,y)}{\delta x \delta y}$$

$$= \frac{\partial^2 F(x,y)}{\partial x \partial y}.$$
(10.2.17)

For any two pairs of real numbers $(a_1, b_1), (a_2, b_2), a_i < b_i (i = 1, 2)$,

$$P[a_1 < X \le b_1, a_2 < Y \le b_2] = \int_{a_2}^{b_2} \left[\int_{a_1}^{b_1} f(u,v) du \right] dv.$$
(10.2.18)

DEFINITION 10.2.6: We say that the distribution function F is continuous singular if it is continuous and there exists a Borel set B in $\mathcal{R}^N$ with Lebesgue measure zero such that $P_F(B) = 1$.

Note 10.2.5: We know that the distribution function of a random variable has at most a countable number of discontinuity points. Let F be the joint distribution function of a random vector **X**. Then F is continuous everywhere in $\mathcal{R}^n$ except over the union of a countable set of hyperplanes of the form $x_k = c(1 \le k \le n)$.

DEFINITION 10.2.7 (*Marginal Distributions*): For discrete **X**, marginal *pmf* of a subset $(X_{j_1}, \ldots, X_{j_m})$ of **X** is

$$P(x_{j_1}, x_{j_2}, \ldots, x_{j_m}) = P[X_{j_1} = x_{j_1}, \ldots, X_{j_m} = x_{j_m}, X_i < \infty, i \neq (j_1, \ldots, j_m)]$$
$$= \sum \cdots \sum_{\substack{\text{all values of } x_i \\ X_i, i \neq (j_1, \ldots, j_m)}} p[X_{j_1} = x_{j_1}, \ldots X_{j_m} = x_{j_m},$$

$$\tag{10.2.19}$$

For continuous **X**, marginal *pdf* of a subset $(X_{j_1}, \ldots, X_{j_m})$ of **X** is

$$f_{X_{j_1}, \ldots, X_{j_m}}(x_{j_1}, \ldots, x_{j_m}) = \int_{-\infty}^{\infty} \cdots \int_{-\infty}^{\infty} f_{\mathbf{X}}(x_1, \ldots, x_n) \Pi_{i \neq (j_1, \ldots, j_m)} dx_i.$$

$$\tag{10.2.20}$$

For $n = 2$, if (X, Y) is a discrete random vector with joint *pmf* $P(X = x_i, Y = y_j) = p_{ij}, i = 1, 2, \ldots; j = 1, 2, \ldots$ then $\{p_{i0}\}$, where $p_{i0} = \sum_j p_{ij}$ is the marginal *pmf* of X. Clearly, $p_{i0} = P(X = x_i) \geq 0$ and $\sum_i p_{i0} = 1$. Similarly, $\{p_{0j}\}$ where $p_{0j} = \sum_i p_{ij} = P(Y = y_j)$ is the marginal *pmf* of Y.

If (X, Y) is a continuous random vector with joint *pdf* $f(x, y)$ then

$$f_X(x) = \int_{-\infty}^{\infty} f(x, y) dy, \tag{10.2.21}$$

$$f_Y(y) = \int_{-\infty}^{\infty} f(x, y) dx \tag{10.2.22}$$

respectively, are the marginal *pdf* of X and Y.

DEFINITION 10.2.8 (*Conditional Distributions*): For discrete **X**, the conditional *pmf* of $X_{j_1}, \ldots, X_{j_k}$ given $X_{i_1} = x_{i_1}, \ldots, X_{i_{n-k}} = x_{i_{n-k}}$ is

$$P[X_{j_1} = x_{j_1}, \ldots, X_{j_k} = x_{j_k} | X_{i_1} = x_{i_1}, \ldots, X_{i_{n-k}} = x_{i_{n-k}}]$$

$$= p_{X_{j_1}, \ldots X_{j_k} | X_{i_1}, \ldots, X_{i_{n-k}}}(x_{j_1}, \ldots, x_{j_k} | x_{i_1}, \ldots, x_{i_{n-k}}] \quad \text{(say)}$$

$$= \frac{p(x_1, \ldots, x_n)}{p(x_{j_1}, \ldots, x_{j_{n-k}})}. \tag{10.2.23}$$

The quantities in (10.2.23) are non-negative and their summation over all values of $X_{j_1}, \ldots, X_{j_k}$ is unity.

For continuous **X**, the conditional *pdf* of $X_{j_1}, \ldots, X_{j_k}$, given $X_{i_1}, \ldots, X_{i_{n-k}}$ is

$$f_{X_{j_1}, \ldots, X_{j_k} | X_{i_1}, \ldots, X_{i_{n-k}}}(x_{j_1}, \ldots, x_{j_k} | x_{i_1}, \ldots, x_{i_{n-k}})$$

$$= \frac{f_{X_1, \ldots, X_n}(x_1, \ldots, x_n)}{f_{X_{i_1}, \ldots, X_{i_{n-k}}}(x_{i_1}, \ldots, x_{i_{n-k}})}. \tag{10.2.24}$$

For $n = 2$, if (X, Y) is a discrete random vector with $P(Y = y_j) = p_{0j} > 0$, then

$$P(X = x_i | Y = y_j) = \frac{P(X = x_i, Y = y_j)}{P(Y = y_j)} = \frac{p_{ij}}{p_{0j}} = p_{X|Y}(i|j) \text{ (say)}.$$

(10.2.25)

The set of values $\{\frac{p_{ij}}{p_{0j}}\}$ is the conditional *pmf* of X given $Y = y_j$. Clearly,

$$\frac{p_{ij}}{p_{0j}} \geq 0, \quad \sum_j \frac{p_{ij}}{p_{0j}} = 1.$$

Similarly, the set of values $\{\frac{p_{ij}}{p_{i0}}\}$ is the conditional *pmf* of Y given that $X = x_i$, provided $p_{i0} > 0$. We shall denote $\frac{p_{ij}}{p_{i0}}$ as $p_{Y|X}(j|i)$. Note that the joint *pmf*

$$p_{ij} = p_{X|Y}(i|j)p_{0j} = p_{i0}p_{Y|X}(j|i).$$

(10.2.26)

Consider now that X, Y are continuous random variables. At every point (x, y) where f is continuous and $f_Y(y) > 0$ and is continuous, the conditional *pdf* of X given $Y = y$ is

$$f_{X|Y}(x|y) = \frac{f_{X,Y}(x, y)}{f_Y(y)} \quad (\geq 0).$$

(10.2.27)

The conditional *pdf* (10.2.27) remains undefined at points where $f_Y(y) = 0$. Similarly, the conditional *pdf* of Y given $X = x$ is

$$f_{Y|X}(y|x) = \frac{f(x, y)}{f_X(x)} \quad (\geq 0)$$

(10.2.28)

provided $f_X(x) > 0$. Clearly,

$$\int_{-\infty}^{\infty} f_{X|Y}(x|y)dx = \int_{-\infty}^{\infty} f_{Y|X}(y|x)dy = 1.$$

Note that

$$f(x, y) = f_{X|Y}(x|y)f_Y(y) = f_{Y|X}(y|x)f_X(x).$$

(10.2.29)

In Note 10.2.2 we have defined the marginal distribution function of a subset of random variables of **X**. Now, we define conditional distribution function.

DEFINITION 10.2.9 (*Conditional Distribution Function*): Conditional distribution function of a subset $(X_{j_1}, \ldots, X_{j_k})$ of **X** given $X_{i_1} = x_{i_1}, \ldots, X_{i_{n-k}} = x_{i_{n-k}}$ is

$$F_{X_{j_1}, \ldots X_{j_k} | X_{i_1}, \ldots, X_{i_{n-k}}}(x_{j_1}, \ldots, x_{j_k} | x_{i_1}, \ldots, x_{i_{n-k}})$$

$$= \sum_{X_{j_1} \le x_{j_1}, \ldots, X_{j_k} \le x_{j_k}} P[X_{j_1}, \ldots, X_{j_k} | X_{i_1} = x_{i_1}, \ldots, X_{i_{n-k}} = x_{i_{n-k}}].$$

$$(10.2.30)$$

For $n = 2$, if X, Y are discrete random variables, with $P(Y = y_j) > 0$, the conditional distribution function of X given $Y = y_j$ is

$$F_{X|Y}(x|y_j) = P(X \le x | Y = y_j) = \frac{1}{p_{0j}} \sum_{x_i \le x} p_{ij}. \qquad (10.2.31)$$

Similarly, the conditional distribution function of Y given $X = x_i$ is

$$F_{Y|X}(y|x_i) = P(Y \le y | X = x_i) = \frac{1}{p_{i0}} \sum_{y_j \le y} p_{ij} \qquad (10.2.32)$$

provided $p_{i0} > 0$.

If (X, Y) is a continuous random vector, the conditional distribution function of X given $Y = y$ is given by

$$F_{X|Y}(x|y) = \int_{-\infty}^{x} f_{X|Y}(u|y) du \qquad (10.2.33)$$

for all y with $f_Y(y) > 0$. Similarly, the conditional distribution function of Y given $X = x$ is

$$F_{Y|X}(y|x) = \int_{-\infty}^{y} f_{Y|X}(v|x) dv \qquad (10.2.34)$$

for all x with $f_X(x) > 0$.

Note 10.2.6: If some of the variables are discrete and others are continuous, then also the joint densities, marginal densities and conditional densities are similarly defined.

EXAMPLE 10.2.2: The variables, X, Y have joint *pdf* $f(x, y) = xe^{-x(y+1)}$ ($x \ge 0, y \ge 0$). Find the marginal *pdf*s and the conditional *pdf*s.

$$f_X(x) = xe^{-x} \int_0^\infty e^{-xy} dy = e^{-x},$$

$$f_Y(y) = \int_0^\infty xe^{-x(y+1)} dx = \frac{1}{(y+1)^2} \int_0^\infty xe^{-z} dx \quad \text{(where } x(y+1) = z\text{)}$$
$$= \frac{1}{(y+1)^2}.$$

Hence,

$$f_{X|Y}(x|y) = (y+1)^2 xe^{-x(y+1)},$$

$$f_{Y|X}(y|x) = xe^{-xy}.$$

EXAMPLE 10.2.3: (X, Y) is distributed with a constant density inside a square R of side b. Find $f(x, y), F(x, y)$ and the marginal *pdf*s.

$$f(x, y) = \begin{cases} \frac{1}{b^2} & \text{for } (x, y) \in R \\ 0 & \text{(otherwise)} \end{cases}.$$

$F(x, y) = 0$ for $x \le 0, y \le 0$.

$F(x, y) = \int_0^x \int_0^y \frac{dudv}{b^2} = \frac{xy}{b^2}$ for $0 < x \le b, 0 < y \le b$,

$F(x, y) = \frac{by}{b^2} = \frac{y}{b}$ for $x > b, 0 < y \le b$,

$F(x, y) = \frac{x}{b}$ for $0 < x \le b, y > b$,

$F(x, y) = 1$ for $x > b, y > b$,

$f_X(x) = \int_0^b \frac{dy}{b^2} = \frac{1}{b}$ for $x \in (0, b)$,

$f_Y(y) = \int_0^b \frac{dx}{b^2} = \frac{1}{b}$ for $y \in (0, b)$.

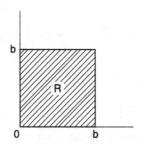

Fig. 10.2.1

EXAMPLE 10.2.4: For a given λ, X follows a Poisson distribution with parameter λ. Again, λ follows a gamma distribution

$$f_\lambda(\lambda) = \frac{\gamma^p}{\Gamma(p)} e^{-\gamma\lambda}\lambda^{p-1}, \quad \lambda > 0, \gamma > 0, p \text{ (integer) } > 0.$$

Find the overall distribution of X, its mean and variance.

For given λ, the *pmf* of X is

$$f_X(x|\lambda) = e^{-\lambda}\frac{\lambda^x}{x!}.$$

The joint *pdf* of (X, λ) is

$$f_{X,\lambda}(x, \lambda) = \frac{\gamma^p}{\Gamma(p)x!}\lambda^{x+p-1}e^{-\lambda(x+1)}.$$

Hence, marginal distribution of X is

$$f_X(x) = \int_0^\infty f_{X,\lambda}(x,\lambda)d\lambda = \frac{\gamma^p \Gamma(x+p)}{x!\Gamma(p)(\gamma+1)^{x+p}}$$

$$= \left(\frac{\gamma}{\gamma+1}\right)^p \frac{1}{(\gamma+1)} \frac{(x+p-1)!}{(p-1)!x!} = \binom{x+p-1}{x}\left(\frac{1}{(\gamma+1)}^x\right)\left(\frac{\gamma}{\gamma+1}\right)^p,$$

a negative binomial distribution with parameters p and $\left(\frac{\gamma}{(\gamma+1)}\right)$.

$$E(X) = \frac{p}{\gamma}, \ V(X) = \frac{p(\gamma+1)}{\gamma^2}. \ \square$$

We now compute the conditional probability of an event A given the value of a random variable X, both X and A being defined on the same probability space $(\mathcal{S}, \mathcal{A}, P)$.

If X is discrete,

$$P(A|X = x) = \frac{P[A \cap (X = x)]}{P(X = x)}. \tag{10.2.35}$$

Hence,

$$P(A) = \sum_x P[A|X = x]P[X = x] \ \text{(by Theorem 3.6.2)}. \tag{10.2.36}$$

Also, the joint probability

$$P[A; X \in B] = \sum_{x \in B} P[A|X = x]P[X = x]. \tag{10.2.37}$$

If X is continuous, $P(X = x) = 0$. In this case,

$$P[A|X = x] = \lim_{x \downarrow 0} P[A|x - \epsilon < X < x + \epsilon]$$

provided $P[x - \epsilon < X < x + \epsilon] > 0$ for every ϵ and the limit in above exists. Thus

$$P(A) = \int_{-\infty}^{\infty} P[A|X = x]f_X(x)dx. \tag{10.2.38}$$

Again,

$$P(A; X \in B) = \int_{x \in B} P[A|X = x]f_X(x)dx. \tag{10.2.39}$$

EXAMPLE 10.2.5: What is the probability that three points selected at random on the circumference of a circle will lie on a semi-circle?

Without loss of generality assume that the first point P_1 falls on the positive X-axis. The position of the second point P_2 is θ radian measured from OP_1 in the anti-clockwise direction (Fig. 10.2.2). The *pdf* of θ is

$$f(\theta) = \frac{1}{2\pi}, \ 0 < \theta \leq 2\pi,$$

since points are located at random. Let A be the event P_1, P_2, P_3 are on a semi-circle.

Now

$$P(A|\theta) = \begin{cases} \frac{2\pi - \theta}{2\pi}, \ (0 < \theta < \pi) \text{ since A occurs if and only if } P_3 \\ \quad \text{falls between } P_1' \text{ and } P_2' \text{ i.e., } P_3 \text{ falls on an} \\ \quad \text{arc of length } \pi - (\theta - \pi) = 2\pi - \theta. \\ \frac{\theta}{2\pi} \quad (\pi < \theta < 2\pi). \end{cases}$$

Hence,

$$\begin{aligned} P(A) &= \int_0^{2\pi} P(A|\theta) f(\theta) d\theta \\ &= \frac{1}{2\pi} \left[\int_0^\pi \frac{2\pi - \theta}{2\pi} d\theta + \int_0^{2\pi} \frac{\theta}{2\pi} d\theta \right] \\ &= \frac{3}{4}. \end{aligned}$$

Stochastic Independence: In Section 4.6 we have defined independence of

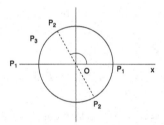

Fig. 10.2.2

random variables in terms of independence of events on which the random variables are defined. In Theorems 6.2.3, 6.2.4 and its corollary we have evaluated the expectation of product of independent random variables in terms of the expectation of the random variables. Here we shall define independence among a set of random variables in terms of the distribution function of random variables.

Theorem 10.2.3: Let $X_1, \ldots, X_n$ be random variables defined on $(\mathcal{S}, \mathcal{A}, P)$. Let F_i be the distribution function of $X_i, i = 1, \ldots, n$ and F the distribution function of $\mathbf{X} = (X_1, \ldots, X_n)$. Then, $X_1, \ldots, X_n$ are stochastically independent *iff*

$$F(x_1, \ldots, x_n) = F_1(x_1) \ldots F_n(x_n) \text{ for all } x_1, \ldots, x_n. \qquad (10.2.40)$$

Proof. If $X_1, \ldots, X_n$ are independent,

$$F(x_1, \ldots, x_n) = P[X_1 \le x_1, \ldots, X_n \le x_n] = \Pi_{i=1}^n P\{X_i \le x_i\} = \Pi_{i=1}^n F_i(x_i).$$

Conversely, assume $F(x_1, \ldots, x_n) = \Pi_{i=1}^n F_i(x_i) \; \forall \; x_1, \ldots, x_n$. Then

$$P_{\mathbf{X}}(\mathbf{a}, \mathbf{b}] = F(\mathbf{a}, \mathbf{b}] = \Pi_{i=1}^n [F_i(b_i) - F_i(a_i)]$$
$$= \Pi_{i=1}^n P_{X_i}(a_i, b_i].$$

Thus,

$$P[X_1 \in B_1, \ldots, X_n \in B_n] = P[X_1 \in B_1] \ldots P[X_n \in B_n]$$

where the B_i's are right semi-closed intervals of $\mathcal{R}$. Hence the proof. $\square$

We may also characterize the independence in terms of density functions. In this direction we state a theorem without proof.

Theorem 10.2.4: If $\mathbf{X} = (X_1, \ldots, X_n)$ has a density function f, and each X_i has a density function f_i, then $X_1, \ldots, X_n$ are independent *iff*

$$f(x_1, \ldots, x_n) = f_1(x_1) \ldots f_n(x_n) \; \forall \; (x_1, \ldots, x_n) \qquad (10.2.41)$$

except possibly for a Borel subset of $\mathcal{R}^n$ of Lebesgue measure zero.

We note that if $X_1, \ldots, X_n$ each have a density function, it does not follow that $(X_1, \ldots, X_n)$ has a density function.

If $\mathbf{X}$ is a discrete random vector, $X_1, \ldots, X_n$ are stochastically independent *iff*

$$P[X_1 = x_1, \ldots, X_n = x_n] = P[X_1 = x_1] \ldots P[X_n = x_n] \; \forall \; (x_1, \ldots, x_n) \qquad (10.2.42)$$

except possibly for a set of probability measure zero.

In case of two random variables, (10.2.41) implies X, Y are independent *iff*

$$f_{X|Y}(x|y) = f_X(x), \; f_{Y|X}(y|x) = f_Y(y) \; \forall \; (x, y). \qquad (10.2.43)$$

For discrete random variables, (10.2,42) implies

$$\begin{aligned} P(X = x | Y = y) &= P(X = x) \\ P(Y = y | X = x) &= P(Y = y) \end{aligned} \qquad (10.2.44)$$

for all (x, y).

EXAMPLE 10.2.6: (X, Y) have joint distribution function

$$F_{X,Y}(x, y) = \begin{cases} (1 - e^{-ax})y^2 & (0 \le x < \infty, 0 \le y < 1) \\ 1 - e^{-ax} & (0 \le x < \infty, 1 < y < \infty) \\ 0 & \text{(otherwise)}, \end{cases}$$

where $a(>0)$ is a constant. Here

$$F_X(x) = \begin{cases} 1 - e^{-ax} & (0 \le x < \infty) \\ 0 & \text{(otherwise)} . \end{cases}$$

$$F_Y(y) = \begin{cases} y^2 & (0 \le y \le 1) \\ 1 & (1 < y < \infty) \\ 0 & \text{(otherwise)} . \end{cases}$$

$$f(x,y) = \begin{cases} 2aye^{-ax} & (0 \le x < \infty, 0 \le y \le 1) \\ 0 & \text{(otherwise)} . \end{cases}$$

$$f_X(x) = \begin{cases} ae^{-ax} & (0 \le x < \infty) \\ 0 & \text{(otherwise)} . \end{cases}$$

$$f_Y(y) = \begin{cases} 2y & (0 \le y \le 1) \\ 0 & \text{(otherwise)} . \end{cases}$$

Note that

$$\begin{cases} F_{X,Y}((x,y) = F_X(x)F_Y(y) \\ f_{X,Y}(x,y) = f_X(x)f_Y(y) \end{cases}$$

for all (x,y). Hence, X, Y are independent.

EXAMPLE 10.2.7: Let X_1, X_2 be independently and identically distributed random variables with $P(X_i = +_-1) = \frac{1}{2}, i = 1, 2$. Let $X_3 = X_1 X_2$. Show that X_1, X_2, X_3 are pairwise independent, but not independent.

$$P(X_3 = 1) = P(X_1 = 1, X_2 = 1) + P(X_1 = -1, X_2 = -1)$$
$$= P(X_1 = 1)P(X_2=1)+P(X_1=-1)P(X_2=-1)=\tfrac{1}{4} + \tfrac{1}{4} = \tfrac{1}{2}.$$

Similarly, $P(X_3 = -1) = \frac{1}{2}$.

$$P(X_1 = 1, X_3 = 1) = P(X_1 = 1)P(X_3 = 1|X_1 = 1)$$

$$= P(X_1 = 1)\frac{P(X_1=1,X_3=1)}{P(X_1=1)}$$

$$= P(X_1 = 1)\frac{P(X_1=1,X_2=1)}{P(X_1=1)}$$

$$= \tfrac{1}{4} = P(X_1 = 1)P(X_3 = 1).$$

Thus, it can be proved that X_1, X_3 and X_2, X_3 are independent. However,

$$P(X_1 = 1, X_2 = 1, X_3 = -1) = P(X_1 = 1, X_2 = 1)$$

$$P(X_3 = -1|X_1 = 1, X_2 = 1) = 0.$$

Hence, X_1, X_2, X_3 are not independent.

10.3 Expectation and Moments of a Random Vector

Let $(\mathcal{S}, \mathcal{A}, P)$ be a probability space and let $\mathbf{X} = (X_1, \ldots, X_n)$ be a random vector defined on it. Let $g : \mathcal{R}^n \to \mathcal{R}^k$ be a Borel measurable function. Then $g(\mathbf{X})$ is a random variable.

DEFINITION 10.3.1: We say that the mathematical expectation of $g(\mathbf{X})$ exists if $g(\mathbf{X})$ is integrable over $\mathcal{S}$ with respect to P. In this case, we write

$$E\{g(\mathbf{X})\} = \int_{\mathcal{S}} g(\mathbf{X}) dP$$

$$= \int_{\mathcal{R}^n} g(x_1, \ldots, x_n) dP_{\mathbf{X}}(x_1, \ldots, x_n) \qquad (10.3.1)$$

$$= \int_{-\infty}^{\infty} \cdots \int_{-\infty}^{\infty} g(x_1, \ldots, x_n) dF_{\mathbf{X}}(x_1, \ldots, x_n).$$

Integral on the first line is the Lebesgue-Stieltjes integral of $g(\mathbf{X})$ with respect to $(\mathcal{S}, \mathcal{A}, P)$ over $\mathcal{S}$; the integral on the second line is the Lebesgue-Stieltjes integral of $g(X)$ with respect to the induced probability space $(\mathcal{R}^n, \mathcal{B}_n, P_{\mathbf{X}})$ over $\mathcal{R}^n$; the integral on the third line is the Riemann-Stieltjes integral of $g(\mathbf{X})$ with respect to the distribution function $F_{\mathbf{X}}$.

In case, $\mathbf{X}$ is a discrete random vector,

$$E\{g(\mathbf{X})\} = \sum_{x_1} \cdots \sum_{x_n} g(x_1, \ldots, x_n) P[X_1 = x_1, \ldots, X_n = x_n] \qquad (10.3.2)$$

provided the series (10.3.2) is absolutely convergent.

In case X is continuous,

$$E\{g(\mathbf{X})\} = \int_{-\infty}^{\infty} \cdots \int_{-\infty}^{\infty} g(x_1, \ldots, x_n) f(x_1, \ldots, x_n) dx_1 \ldots dx_n \qquad (10.3.3)$$

provided

$$\int_{-\infty}^{\infty} \cdots \int_{-\infty}^{\infty} |g(x_1, \ldots, x_n)| f(x_1, \ldots, x_n) dx_1 \ldots dx_n < \infty.$$

When $g(\mathbf{X}) = X_1^{r_1} \ldots X_n^{r_n}$, $E\{g(\mathbf{X})\}$ defines the non-central moment

$$\mu'_{r_1, \ldots, r_n} \qquad (10.3.4)$$

of the distribution $F_{\mathbf{X}}(x_1, \ldots, x_n)$.

In the case of two variables X, Y, we have the bivariate non-central moments

$$\mu'_{r,s} = E(X^r Y^s) \qquad (10.3.5)$$

when

$$E(X) = \mu'_{1,0} = \mu_X \text{ (say) }, \quad E(Y) = \mu'_{0,1} = \mu_Y \text{ (say)}$$

which are the mean of X and mean of Y respectively.

The bivariate central moments are

$$\mu_{r,s} = E\{(X - \mu_X)^r(Y - \mu_Y)^s\}. \tag{10.3.6}$$

We have

$$\mu_{2,0} = E(X - \mu_X)^2 = \sigma_X^2, \quad \mu_{0,2} = E(Y - \mu_Y)^2 = \sigma_Y^2 \tag{10.3.7}$$

which are variances of X, Y respectively.

The product moment

$$\mu_{11} = \mu_{X,Y} = E(X - \mu_X)(Y - \mu_Y) = E(XY) - \mu_X\mu_Y \tag{10.3.8}$$

is called the covariance between X and Y, often denoted as Cov (X, Y) or $\sigma_{X,Y}$.

It has been proved in (6.4.3) that

$$\mu_{X,Y}^2 \leq \sigma_X^2\sigma_Y^2.$$

The quantity

$$\rho(X, Y) = \frac{\sigma_{X,Y}}{\sigma_X\sigma_Y} \tag{10.3.9}$$

is defined as the correlation coefficient between X and Y, provided $\mu_{X,Y}, \sigma_X, \sigma_Y$ exist and $\sigma_X > 0, \sigma_Y > 0$.

Hence,

$$\rho^2 \leq 1 \quad \text{or} \quad -1 \leq \rho \leq 1, \tag{10.3.10}$$

$\rho^2 = 1$ iff X or Y is an exact linear function of the other a.s. (vide Section 10.7).

EXAMPLE 10.3.1:

$$f_{X,Y}(x, y) = \frac{3x + y}{4}e^{-x-y}, \quad x > 0, y > 0.$$

Find the marginal densities, conditional densities and $\rho(X, Y)$.

$$f_X(x) \quad = e^{-x} \int_0^\infty \frac{3x+y}{4} e^{-y} dy = \frac{3x+1}{4} e^{-x},$$

$$f_Y(y) \quad = e^{-y} \int_0^\infty \frac{3x+y}{4} e^{-x} dx = \frac{y+3}{4} e^{-y},$$

$$f_{X|Y}(x|y) = \frac{3x+y}{y+3} e^{-x}, \quad x > 0$$

$$f_{Y|X}(y|x) = \frac{3x+y}{3x+1} e^{-y}, \quad y > 0,$$

$$\mu'_{1,0} \quad = \mu_X = \frac{1}{4} \int_0^\infty x(3x+1)e^{-x} dx = \frac{7}{4},$$

$$\mu'_{2,0} \quad = 5, \quad \sigma_X^2 = \frac{31}{36}, \quad \mu'_{0,1} = \frac{5}{4},$$

$$\mu'_{0,2} \quad = 3, \quad \sigma_Y^2 = \frac{23}{16},$$

$$\mu'_{1,1} \quad = \frac{1}{4} \int_0^\infty \int_0^\infty xy(3x+y)e^{-x-y} dx dy = 2,$$

$$\mu_{1,1} \quad = -\frac{3}{16},$$

$$\rho(X, Y) \quad = -\frac{3}{\sqrt{31 \times 23}} = -0.112.$$

Theorem 10.3.1: If $X_1, \ldots, X_n$ are any given set of random variables with finite mean and variances, and

$$Y = c_0 + c_1 X_1 + \ldots + c_n X_n$$

where $c_0, c_1, \ldots, c_n$ are any given set of constants, then

$$E(Y) = c_0 + c_1 E(X_1) + \ldots + c_n E(X_n). \tag{10.3.11}$$

Proof.

$$E(Y) = \int_{-\infty}^\infty \cdots \int_{-\infty}^\infty \left(c_0 + \sum_{i=1}^n c_i X_i \right) dF_{\mathbf{X}}(x_1, \ldots, x_n)$$

$$= c_0 + \sum_{i=1}^n c_i \int_{-\infty}^\infty x_i dF_{\mathbf{X}}(x_1, \ldots, x_n) = c_0 + \sum_{i=1}^n c_i E(X_i)$$

by (10.3.1) and by the fact

$$\int_{-\infty}^\infty \cdots \int_{-\infty}^\infty dF_{\mathbf{X}}(x_1, \ldots, x_n) = 1.$$

Theorem 10.3.2: If $X_1, \ldots, X_n$ are independent random variables with finite means

$$E(X_1 \ldots X_n) = E(X_1) \ldots E(X_n).$$

An elaborate proof has been given in Theorem 6.2.3 and Corollary 6.2.4.1. A short proof based on the properties of integrals is given below.

$$E(X_1 \ldots X_n) = \int_{-\infty}^{\infty} x_1 \ldots x_n dF_{\mathbf{X}}(x_1, \ldots, x_n)$$

$$= \int_{-\infty}^{\infty} \cdots \int_{-\infty}^{\infty} x_1 \ldots x_n dF_{X_1}(x_1) \ldots dF_{X_n}(x_n) \quad \text{(by 10.2.40)}$$

$$= [\int_{-\infty}^{\infty} x_1 dF_{X_1}(x_1] \ldots [\int_{-\infty}^{\infty} x_n dF_{X_n}(x_n)] = E(X_1) \ldots E(X_n).$$

Corollary 10.3.2.1: If $X_1, \ldots X_n$ are independent random variables and if $g_i(X_i)$ is a Borel measurable function of X_i such that $g_i(X_i)(i = 1, \ldots, n)$ is a random variable, then

$$E[g_1(X_1) \ldots g_n(X_n)] = E(g_1(X_1)) \ldots E(g_n(X_n)). \tag{10.3.12}$$

Proof follows from the fact that $g_1(X_1), \ldots, g_n(X_n)$ are independent random variables. An elaborate proof of the result has been given in Theorem 6.2.4. □

Two random variables X, Y are said to be uncorrelated if

$$E(XY) = E(X)E(Y) \tag{10.3.13}$$

i.e., if $\text{Cov}(X, Y) = 0$.

It follows from Theorem 10.3.2 that if X, Y are independent with finite means and if $\mu'_{X,Y}$ exists then X, Y are uncorrelated. The converse is not necessarily true, i.e. X, Y may be uncorrelated, yet they may not be independent (vide Example 5.4.5 and Exercises 10.11 and 10.12).

If X, Y are uncorrelated, then in general,

$$E\{g(X)h(Y)\} \neq E\{g(X)\}E\{h(Y)\}.$$

However, if they are independent

$$E\{g(X)h(Y)\} = E\{g(X)\}E(h(Y)\}.$$

DEFINITION 10.3.2: Two random variables X, Y are orthogonal if

$$E(XY) = 0. \tag{10.3.14}$$

If X, Y are uncorrelated,

$$E(X - E(X))(Y - E(Y)) = 0$$

i.e., $(X - E(X))$ and $(Y - E(Y))$ are orthogonal.

If X, Y are orthogonal,

$$E(X + Y)^2 = E(X^2) + E(Y^2).$$

If X, Y are uncorrelated and mean of at least one of them is zero, then they are also orthogonal.

Theorem 10.3.3: Let X, Y be random variables with finite variances σ_X^2, σ_Y^2. Then

$$V(X + Y) = V(X) + V(Y) + 2 \operatorname{Cov}(X, Y). \qquad (10.3.15)$$

Proof.

$$
\begin{aligned}
V(X + Y) &= \int_{-\infty}^{\infty} \int_{-\infty}^{\infty} \{x + y - E(X) - E(Y)\}^2 dF_{X,Y}(x, y) \\
&= \int_{-\infty}^{\infty} \int_{-\infty}^{\infty} \{(x - E(X)) + (y - E(Y))\}^2 dF_{X,Y}(x, y) \\
&= \int_{-\infty}^{\infty} \int_{-\infty}^{\infty} (x - E(X))^2 dF(x, y) + \int_{-\infty}^{\infty} \int_{-\infty}^{\infty} (y - E(Y))^2 dF(x, y) \\
&\quad + 2 \int_{-\infty}^{\infty} \int_{-\infty}^{\infty} (x - E(X))(y - E(Y)) dF(x, y) \\
&= V(X) + V(Y) + 2 \operatorname{Cov}(X, Y),
\end{aligned}
$$

since means and covariance exist as variances exist.

Theorem 10.3.4: Let $X_1, \ldots, X_n$ be random variables with $E(X_i) = \mu_i, V(X_i) = \sigma_i^2$, $\operatorname{Cov}(X_i, X_j) = \sigma_{ij}(i \neq j)$, all finite. Let

$$Y = c_0 + \sum_{i=1}^{n} c_i X_i$$

where $c_j (j = 0, 1, \ldots, n)$ are arbitrary constants. Then

$$V(Y) = \sum_{i=1}^{n} c_i^2 \sigma_i^2 + \sum_{i \neq j=1}^{n} \sum c_i c_j \sigma_{ij}. \qquad (10.3.16)$$

Proof. Follows as in the proof of Theorem 10.3.3.

Theorem 10.3.5: Let

$$U = \sum_{i=1}^{n} c_i X_i, \quad V = \sum_{i=1}^{n} d_i X_i$$

be two linear functions of $X_1, \ldots, X_n, c_i, d_i (i = 1, \ldots, n)$ being arbitrary constants. Then

$$\operatorname{Cov}(U, V) = \sum_{i=1}^{n} \sum_{j=1}^{n} c_i d_j \sigma_{ij} \qquad (10.3.17)$$

where $\sigma_{ii} = V(X_i), \sigma_{ij} = \operatorname{Cov}(X_i, X_j), i \neq j$, provided the variances and the covariances exist. The proof is straightforward.

Corollary 10.3.5.1: If $U = a + cX, V = b + dY$,

(i) $\operatorname{Cov}(U, V) = cd \operatorname{Cov}(X, Y)$.

(ii) $\rho(U, V) = \{\operatorname{sign}(cd)\}\rho(X, Y)$.

EXAMPLE 10.3.2: Let $X_1, \ldots, X_n$ be independent random variables with $EX_i^2 < \infty$ and $V(X_i) = \sigma_i^2, i = 1, \ldots, n$. Let $S = \sum_{i=1}^n a_i X_i, a_1, \ldots, a_n$ are constants, $\sum_{i=1}^n a_i = 1$. Find a_i's so that $V(S)$ is minimum.

$V(S) = \Sigma$ (say) $= \sum_{i=1}^n a_i^2 \sigma_i^2 = \sum_{i=1}^{n-1} a_i^2 \sigma_i^2 + (1 - a_1 - \ldots - a_{n-1})^2 \sigma_n^2$.

For minimizing Σ with respect to a_i's we have the following equations.

$$\frac{\partial \Sigma}{\partial a_1} = 2a_1 \sigma_1^2 - 2(1 - a_1 - \ldots - a_{n-1})\sigma_n^2 = 0$$

$$\cdots\cdots\cdots\cdots$$

$$\frac{\partial \Sigma}{\partial a_{n-1}} = 2a_{n-1}\sigma_{n-1}^2 - 2(1 - a_1 - \ldots - a_{n-1}) = 0.$$

It follows that

$$a_j \sigma_j^2 = a_n \sigma_n^2, \quad j = 1, \ldots, n-1$$

i.e., $a_j = \frac{\lambda}{\sigma_j^2}, \lambda$ being the constant of proportionality. Now $\sum_{j=1}^n a_j = 1$ gives

$$\lambda = \left(\sum_{j=1}^n \frac{1}{\sigma_j^2}\right)^{-1}.$$

The minimum value of Σ is $\lambda^2 \sum_{j=1}^n \frac{1}{\sigma_j^2} = \left(\sum_{j=1}^n \frac{1}{\sigma_j^2}\right)^{-1}.$ $\square$

As in the proof of Theorem 6.8.6, the following theorem can be proved.

Theorem 10.3.6: For a function $g(X, Y)$ whose mean and variance exist,

$$E[g(X, Y)] \approx g(\mu_X, \mu_Y) + \frac{1}{2}\left[\mu_{20}\frac{\partial^2 g}{\partial x^2} + 2\mu_{11}\frac{\partial^2 g}{\partial x \partial y} + \mu_{02}\frac{\partial^2 g}{\partial y^2}\right]_{\mu_X, \mu_Y},$$
$$(10.3.18)$$

$$V[g(X, Y)] = \left[\left(\frac{\partial g}{\partial x}\right)^2 \mu_{20} + \left(\frac{\partial g}{\partial y}\right)^2 \mu_{02} + 2\left(\frac{\partial g}{\partial x}\right)\left(\frac{\partial g}{\partial y}\right)\mu_{11}\right]_{\mu_X, \mu_Y}, \quad (10.3.19)$$

where $\mu_X = E(X), \mu_Y = E(Y)$.

10.4   Multivariate Probability Generating Function

Let $X_1, \ldots, X_n$ be n discrete random variables taking values in $\mathcal{N}$, the set of non-negative integers and let

$$P(X_1 = i_1, \ldots, X_n = i_n) = p_{i_1 \ldots i_n} \qquad (10.4.1)$$

be the distribution of the discrete random vector $\mathbf{X} = (X_1, \ldots, X_n)$. The probability generating function (*pgf*) of $\mathbf{X}$ is defined as

$$
\begin{aligned}
P_{\mathbf{X}}(\mathbf{t}) = P_{(X_1,\ldots,X_n)}(t_1,\ldots,t_n) &= E(t_1^{X_1} \ldots t_n^{X_n}) \\
&= \sum_{i_1=1}^{\infty} \cdots \sum_{i_n=1}^{\infty} t_1^{i_1} \ldots t_n^{i_n} P(X_1 = i_1, \ldots, X_n = i_n) \quad (10.4.2) \\
&= \sum_{i_1=1}^{\infty} \cdots \sum_{i_n=1}^{\infty} t_1^{i_1} \ldots t_n^{i_n} p_{i_1 \ldots i_n}.
\end{aligned}
$$

Since, when $|t_i| \leq 1$ for all $i (1 \leq i \leq n)$,

$$
\sum_{i_1=1}^{\infty} \cdots \sum_{i_n=1}^{\infty} |t_1^{i_1} \ldots t_n^{i_n}| P(X_1 = i_1, \ldots, X_n = i_n)
$$

$$
\leq \sum_{i_1=1}^{\infty} \cdots \sum_{i_n=1}^{\infty} P(X_1 = i_1, \ldots, X_n = i_n) = 1,
$$

the domain of definition of P contains the subset of n-tuples $(t_1, \ldots, t_n)$ such that

$$
|t_1| \leq 1, \ldots, |t_n| \leq 1. \quad (10.4.3)
$$

If $t_2 = \ldots t_n = 1$,

$$
E[t_1^{X_1} t_2^{X_2} \ldots t_n^{X_n}] = E[t_1^{X_1}] = P_1(t_1)
$$

where P_1 is the *pgf* of X_1. Therefore,

$$
P(t_1, 1, \ldots, 1) = P_1(t_1). \quad (10.4.4)
$$

Similar relations hold for $X_2, \ldots, X_n$.

EXAMPLE 10.4.1: Consider

$$
P(t_1, t_2) = e^{\mu_1(t_1-1) + \mu_2(t_2-1) + \lambda(t_1 t_2 - 1)}, \quad \mu_1, \mu_2 > 0, |t_1| \leq 1, |t_2| \leq 1.
$$

This has a power series expansion in $t_1^i t_2^j$ with coefficients adding to unity. $P(t_1, 1)$ (or $P(1, t_2)$) is the *pgf* of a Poisson distribution with parameter $\mu_1 + \lambda$ (or $\mu_2 + \lambda$). $\square$

If $X_1, \ldots, X_n$ are n independent discrete random variables with values in $\mathcal{N}$, then

$$
\begin{aligned}
P_{\mathbf{X}}(\mathbf{t}) = E(t_1^{X_1} \ldots t_n^{X_n}) = E(t_1^{X_1}) \ldots E(t_n^{X_n}) \\
= P_1(t_1) \ldots P_n(t_n)
\end{aligned} \quad (10.4.5)
$$

where P_i is the *pgf* of $X_i (i = 1, \ldots, n)$. Setting $t_1 = \ldots = t_n = t$, we have

$$
P(t, \ldots, t) = E(t^{X_1 + \ldots + X_n}) \quad (10.4.6)
$$

which is the *pgf* of $\sum_{i=1}^{n} X_i$.

EXAMPLE 10.4.2: Using *pgf*, show that if X_1, X_2 are independent Poisson random variables with parameters λ_1, λ_2 respectively, then $X_1 + X_2$ follows Poisson $(\lambda_1 + \lambda_2)$.

We have $P_i(t) = e^{\lambda_i(t-1)}, i = 1, 2$. Therefore, the *pgf* of $X_1 + X_2$ is $P(t, t) = P_1(t)P_2(t) = e^{(\lambda_1+\lambda_2)(t-1)}$. Since the generating function characterizes the distribution, $X_1 + X_2 \cap$ Poisson $(\lambda_1 + \lambda_2)$.

EXAMPLE 10.4.3: $\{X_n\}$ is a sequence of independently and identically distributed (*iid*) random variables with *pmf* $P[X_k = j] = p_j, j = 1, 2, \ldots$ and the *pgf* $P_X(t) = \sum_j p_j t^j, |t| < 1$. Let

$$S_N = X_1 + \ldots + X_N$$

where N is a random variable with *pmf* $P[N = n] = g_n, n = 0, 1, 2, \ldots$ and *pgf*, $G_N(t) = \sum_0^\infty g_n t^n, |t| < 1$. Find the *pgf* of S_N.

In particular, find the *pgf* of S_N when the X_i's are Bernoulli variables with $P(X_i = 1) = p, P(X_i = 0) = q$ and N is a Poisson variable with mean λ.

We have

$$P(S_N = j) = \sum_{n=0}^{\infty} P(N = n)P(S_n = j) = h_j \quad \text{(say)}.$$

Hence *pgf* of S_N is

$$H_{S_N}(t) = \sum_{j=0}^{\infty} h_j t^j = \sum_{j=0}^{\infty} \{\sum_{n=0}^{\infty} g_n P(S_n = j)\} t^j$$

$$= \sum_{n=0}^{\infty} g_n \{\sum_{j=0}^{\infty} P(S_n = j) t^j\} = \sum_{n=0}^{\infty} g_n P_{S_n}(t)$$

$$\text{(where } P_{S_n}(t) = pgf \text{ of } S_n) \tag{10.4.7}$$

$$= \sum_{n=0}^{\infty} g_n [P(t)]^n \quad \text{(since } X_1, \ldots, X_n \text{ are } iid)$$

$$= G_N(P(t)).$$

When X is a Bernoulli variable, its *pgf* is $P_X(t) = q + pt, |t| < 1. N$ being a Poisson variable, its *pgf* $G_N(t) = e^{-\lambda(1-t)}$. Hence, *pgf* of S_N is

$$G_N(P(t)) = e^{-\lambda(1-q-pt)} = e^{-\lambda p(1-t)}.$$

Therefore S_N is a Poisson variable with mean λp. □

EXERCISE 10.4.4: For the above example, prove Wald's equality

$$E(S_N) = E(X)E(N). \tag{10.4.8}$$

By formula (7.2.11), we have

$$E(S_N) = \frac{\partial H_{S_N}(t)}{\partial t}\bigg]_{t=1} = \sum_{n=0}^{\infty} n g_n [P(1)]^{n-1} P'(1)$$

$$= P'(1) \sum_{n=1}^{\infty} n g_n = E(X)E(N).$$

10.5 Multivariate Moment Generating Function

DEFINITION 10.5.1 *Joint Moment Generating Function* The joint *mgf* of the random vector $\mathbf{X} = (X_1, \ldots, X_n)$ is defined as

$$M_{\mathbf{X}}(\mathbf{t}) = M_{\mathbf{X}}(t_1, \ldots, t_n) = E[e^{\sum_{i=1}^{n} t_i X_i}] \quad \text{(where } \mathbf{t} = (t_1, \ldots, t_n)\text{)}$$
(10.5.1)

if the expectation exists for all values of $t_1, \ldots, t_n$ such that $|t_j| < T$ for some $T(> 0), j = 1, \ldots, n$.

The raw moments are obtained as

$$\mu'_{r_1, \ldots, r_n} = \frac{\partial^{r_1 + \ldots + r_n}}{\partial t_1^{r_1} \ldots \partial t_n^{r_n}} M_{\mathbf{X}}(\mathbf{t})\bigg]_{\mathbf{t}=0}$$
(10.5.2)

where $\mathbf{0} = (0, \ldots, 0)$, provided the required partial derivatives exist and provided the order of differentiation and integration are interchangeable. In particular,

$$\mu'_r = \frac{\partial^r}{\partial t_1^r} M_{\mathbf{X}}(\mathbf{t})\bigg]_{\mathbf{t}=0}$$

$$\mu'_{r,s} = \frac{\partial^{r+s}}{\partial t_1^r \partial t_2^s} M_{\mathbf{X}}(\mathbf{t})\bigg]_{\mathbf{t}=0}.$$
(10.5.3)

Thus, expanding $M_{\mathbf{X}}(\mathbf{t})$ in Maclaurin's series, one obtains

$$M_{\mathbf{X}}(\mathbf{t}) = \sum_{r_1, \ldots, r_n} \frac{t_1^{r_1} \ldots t_n^{r_n}}{r_1! \ldots r_n!} \mu'_{r_1, \ldots, r_n}.$$
(10.5.4)

The marginal *mgf*s are obtained as

$$M_{X_1}(t_1) = M_{\mathbf{X}}(t_1, 0, \ldots, 0) = \lim_{t_i \to 0; i=2, \ldots, n} M_{\mathbf{X}}(t_1, t_2, \ldots, t_n)$$
$$M_{X_2}(t_2) = M_{\mathbf{X}}(0, t_2, 0, \ldots, 0) = \lim_{t_i \to 0; i(\neq 2)=1, \ldots, n} M_{\mathbf{X}}(t_1, \ldots, t_n),$$
(10.5.5)

etc. If the variables $X_1, \ldots, X_n$ are mutually independent,

$$M_{\mathbf{X}}(\mathbf{t}) = \Pi_{i=1}^n E(e^{t_i X_i}) = \Pi_{i=1}^n M_{X_i}(t_i)$$
(10.5.6)

i.e., the joint *mgf* is the product of separate *mgf*s.

We now consider two theorems on bivariate *mgfs*. The theorems can be extended to $n(\geq 2)$ random variables.

Theorem 10.5.1: The *mgf* $M(\mathbf{t})$, when it exists, determines the joint distribution of $\mathbf{X}$ uniquely and conversely, if the *mgf* exists, it is unique.

Theorem 10.5.2: The jointly distributed random variables $X_1, \ldots, X_n$ are independent *off*

$$M_{\mathbf{X}} = \Pi_{i=1}^n M_{X_i}(t_i) \qquad (10.5.7)$$

for all $t_1, \ldots, t_n$ such that $|t_i| < T(> 0), i = 1, \ldots, n$.

The proof of the first theorem is outside the scope of this book. The second theorem is proved below for $n = 2$. The proof can be extended for general n.

If X, Y are independent, (10.5.7) follows from (10.5.6). If (10.5.6) holds,

$$\int \int e^{t_1 X + t_2 Y} dF(x, y) = \int e^{t_1 x} dF(x) \int e^{t_2 y} dF(y)$$

$$= \int \int e^{t_1 x + t_2 y} dF(x) dF(y).$$

Hence,

$$dF(x, y) = dF(x) dF(y)$$

by the uniqueness property of the *mgf* (Theorem 10.5.1).

EXAMPLE 10.5.1: Let (X, Y) have a bivariate uniform distribution over the rectangle (R) defined by

$$a \leq x \leq a + k, \ b \leq y \leq b + l, \ k > l > 0.$$

The *pdf* is

$$f_{X,Y}(x, y) = \begin{cases} \frac{1}{kl} & \text{for } (x, y) \in R \\ 0 & \text{otherwise.} \end{cases}$$

The *mgf* is

$$M_{X,Y}(t_1, t_2) = \int_a^{a+k} \int_b^{b+l} e^{t_1 x + t_2 y} \frac{1}{kl} dx dy$$

$$= \frac{\exp[at_1 + bt_2]}{klt_1 t_2} (e^{kt_1} - 1)(e^{lt_2} - 1).$$

EXAMPLE 10.5.2: X_1, X_2, X_3, X_4 are independently and identically random variables with *pdf* $e^{-x}, x > 0$. Find the *mgf* of the joint distribution of U, V, W where $U = X_1 + X_2, V = X_2 + X_3, W = X_3 + X_4$.

$$\text{Mgf} = \int_0^\infty \int_9^\infty \int_0^\infty e^{ut_1 + vt_2 + wt_3} e^{-x_1 - x_2 - x_3 - x_4} \Pi_{i=1}^4 dx_i$$

$$= \int_0^\infty \int_0^\infty \int_0^\infty \int_0^\infty e^{x_1(t_1 - 1) + x_2(t_1 + t_2 + t_3 - 1) + x_3(t - 2 - 1) + x_4(t_3 - 1)} \Pi_{i=1}^4 dx_i$$

$$= \int_0^\infty e^{-x_1(1 - t_1)} dx_1 \int_0^\infty e^{-x_2(1 - t_1 - t_2 - t_3)} dx_2 \int_0^\infty e^{-x_3(1 - t_2)} dx_3$$
$$\int_0^\infty e^{-x_4(1 - t_3)} dx_4$$

$$= [(1 - t_1)(1 - t_2)(1 - t_3)(1 - t_1 - t_2 - t_3)]^{-1}.$$

10.6 Multivariate Characteristic Function

Let $\mathbf{X} = (X_1, \dots, X_n)$ be a random vector. The characteristic function of $\mathbf{X}$ is defined as

$$\psi_{\mathbf{X}}(\mathbf{t}) = E(e^{i\mathbf{t}'\mathbf{X}}) = E(e^{\sum_{i=1}^n t_i X_i}) \tag{10.6.1}$$

where $\mathbf{t} = (t_1, \dots, t_n)$ is any real vector and $i = \sqrt{-1}$.

Again, suppose that each component X_j of $\mathbf{X}$ has finite moments up to order m. Then all product moments of $\mathbf{X}$ of total order m exist. Moreover,

$$\frac{\partial^m \psi(\mathbf{t})}{\partial t_1^{m_1} \dots \partial t_n^{m_n}} \bigg]_{\mathbf{t}=0} = i^m \mu'_{m_1, \dots, m_n}, \quad \sum_{j=1}^n m_j = m. \tag{10.6.2}$$

Theorem 10.6.1 (*Inversion Theorem on $\mathcal{R}^n$*): Let F be a distribution function on $\mathcal{R}^n$ and let ψ be its characteristic function. Let $\mathbf{a} = (a_1, \dots, a_n)$ and $\mathbf{b} = (b_1, \dots, b_n)$ be two points in $\mathcal{R}^n$ such that $a_j < b_j$ and a_j, b_j are continuity points of the marginal distribution function,

$$F_j(x_j) = \lim_{x_i \to \infty; i(\neq j) = 1, \dots, n} F(x_1, \dots, x_j, \dots, x_n), \quad j = 1, \dots, n.$$

Then

$$F(\mathbf{b}) - F(\mathbf{a})$$
$$= \lim_{T_i \to \infty; i = 1, \dots, n} \frac{1}{\pi^n} \int_{-T_n}^{T_n} \dots \int_{-T_1}^{T_1} \Pi_{j=1}^n \frac{e^{-it_j b_j} - e^{-it_j a_j}}{it_j} \psi(t_1, \dots, t_n) dt_1 \dots dt_n.$$
$$\tag{10.6.3}$$

Corollary 10.6.1.1: The distribution function F is determined uniquely by its characteristic function ψ.

Corollary 10.6.1.2 (*Fourier Inversion Theorem on $\mathcal{R}^n$*): Suppose that ψ is absolutely integrable on $\mathcal{R}^n$. Then the corresponding distribution function

F is absolutely continuous on $\mathcal{R}^n$. Moreover, the probability density function $f = \frac{\partial^n F}{\partial x_1 \ldots \partial x_n}$ exists, is bounded and uniformly continuous on $\mathcal{R}^n$ and is given by

$$f(x_1, \ldots, x_n) = \frac{1}{(2\pi)^n} \int_{\mathcal{R}^n} \exp(-it'\mathbf{x})\psi(\mathbf{t})d\mathbf{t} \qquad (10.6.4)$$

where $\mathbf{t} = (t_1, \ldots, t_n)$.

We now consider the following important application of the above results.

Theorem 10.6.2: Let $\mathbf{X} = (X_1, \ldots, X_n)$ be a random vector with distribution function F and characteristic function ψ. Let F_j and ψ_j be the distribution function and the characteristic function of $X_j, j = 1, \ldots, n$. Then the following statements are equivalent.

(i) $X_1, \ldots, X_n$ are independent.

(ii) $\psi(t_1, \ldots, t_n) = \Pi_{j=1}^n \psi_j(t_j) \; \forall \; t_1, \ldots, t_n \in \mathcal{R}$.

(iii) $F(x_1, \ldots, x_n) = \Pi_{j=1}^n F_j(x_j) \; \forall \; x_1, \ldots, x_n \in \mathcal{R}$.

Proof. That $(i) \Rightarrow (ii)$ follows from the definition of independence.

To show that $(ii) \Rightarrow (iii)$ we use (10.6.3) and Fubini's Theorem (Theorem A.2.4) to conclude that

$$F(\mathbf{b}) - F(\mathbf{a}) = \Pi_{j=1}^n [F_j(b_j) - F_j(a_j)]$$

where $\mathbf{a}, \mathbf{b}$ are defined as in the statement of the Theorem 10.6.1. Taking the limits of both sides as $a_j \to -\infty, j = 1, \ldots, n$, we obtain

$$F(\mathbf{b}) = \Pi_{j=1}^n F_j(b_j)$$

which holds for all continuity points b_j of $f_j, j = 1, \ldots, n$. Since by the definition of independence, $(iii) \Rightarrow (i)$, the proof is complete.

10.7 Conditional Expectation, Variance, Regression

We shall state the results for two variables only.

DEFINITION 10.7.1: Let (X, Y) be two random variables. The conditional expected value of $g(X, Y)$ given $X = x$ is defined as

$$E[g(X, Y)|X = x] = \int_{-\infty}^{\infty} g(x, y)f_{Y|x}(y|x)dy \qquad (10.7.1)$$

if (X, Y) is a continuous random vector. For discrete (X, Y),

$$E[g(X, Y)|X = x] = \sum_y g(x, y)P[Y = y|X = x] \qquad (10.7.2)$$

where the summation is over all possible values of Y.

In particular, if $g(X, Y) = Y$, (10.7.1), (10.7.2) define respectively the conditional expectation of Y given $X = x$,

$$E(Y|x) = \mu_{Y|x} = \int_{-\infty}^{\infty} y f_{Y|x}(y|x) dy \quad \text{or} \quad \sum_y y P[Y = y|X = x] \quad (10.7.3)$$

according as (X, Y) is continuous or discrete.

Similarly, the conditional expectation of X given $Y = y$, $E(X|Y) = \mu_{X|y}$ can be defined.

In general, $\mu_{Y|x}$ is a function of x and $\mu_{X|y}$ is a function of y.

DEFINITION 10.7.2: The graph of $E(Y|x)$ as a function of x,

$$\mathcal{Y} = E(Y|x) \qquad (10.7.4)$$

is called the *regression curve* of Y on x. The graph of $E(X|y)$ as a function of y,

$$\mathcal{X} = E(X|y)$$

is called the regression curve of X on y.

Theorem 10.7.1: For two random variables (X, Y),

$$E[E\{g(X)|Y\}] = E[g(X)]. \qquad (10.7.5)$$

Proof. We prove the result when (X, Y) is jointly continuous. The left hand side

$$= \int_{-\infty}^{\infty} \left[\int_{-\infty}^{\infty} g(x) f_{X|y}(x|y) dx \right] f_Y(y) dy$$

$$= \int_{-\infty}^{\infty} \int_{-\infty}^{\infty} g(x) f(x, y) dx dy = E[g(X)].$$

As a particular case,

$$E[E(X|Y)] = E(X).$$

Similarly,

$$E[E(Y|X)] = E(Y).$$

Theorem 10.7.2: If one of the regression curves is a constant, $\sigma_{X,Y} = 0$, $\rho(X, Y) = 0$.

Proof. Suppose $E(Y|x) = k$ (a constant). Then by Theorem 10.7.1, $E(Y) = k$. Hence, assuming X, Y are both continuous variables,

$$\begin{aligned}
\sigma_{X,Y} &= E(X - \mu_X)(Y - \mu_Y) = E(XY) - k\mu_X \\
&= \int_{-\infty}^{\infty} \int_{-\infty}^{\infty} xy f(x,y)\,dx\,dy - k\mu_X \\
&= \int_{-\infty}^{\infty} x \left[\int_{-\infty}^{\infty} y f_{Y|x}(y|x) \right] f_X(x)\,dx - k\mu_X = k(\mu_X - \mu_X) = 0.
\end{aligned}$$

The case when X, Y are discrete can be proved similarly.

Theorem 10.7.3: $\rho(X, Y) = +_{-}1$ *iff* there exist constants a, b such that

$$P(Y = aX + b) = 1$$

i.e. $Y = aX + b$ a.s. Further, if $a < 0$, then $\rho = -1$ and if $a > 0, \rho = 1$.

Proof. (i) Let $Y = aX + b$ a.s. Then $V(Y) = a^2 V(X)$, Cov $(X, Y) = aV(X)$ and hence $\rho = \frac{a}{|a|} = +_{-}1$.

(ii) If $\rho = +_{-}1$, $[$ Cov $(X, Y)]^2 = V(X)V(Y)$ which holds by the conditions of equality in Cauchy-Schwarz inequality (6.4.3), i.e., *iff* $t(X - \mu_X) = Y - \mu_Y$ a.s. for some constant t, i,e. *iff* $Y = aX + b$ a.s. for some constants a and b i.e. X, Y are perfectly linearly related a.s.

DEFINITION 10.7.3 *Conditional Variance:* The conditional variance of Y given $X = x$ is

$$V(Y|X = x) = E[Y^2|X = x] - [E(Y|X = x)]^2. \tag{10.7.6}$$

Theorem 10.7.4:

$$V(Y) = E[V(Y|X)] + V[E(Y|X)]. \tag{10.7.7}$$

Proof.

$$\begin{aligned}
E[V(Y|X)] &= E[E(Y^2|X)] - E[E(Y|X)]^2 \\
&= [E(Y^2) - (E(Y))^2] - [E\{E(Y|X)\}^2 - (E(Y))^2] \\
&= V(Y) - [E\{E(Y|X)\}^2 - \{E(E(Y|X))\}^2] \\
&= V(Y) - V[E(Y|X)].
\end{aligned}$$

Theorem 10.7.5: The following theorem is easy to prove.

(i) $E[g_1(X) + g_2(X)|Y] = E[g_1(X)|Y] + E[g_2(X)|Y]$

(ii) $E[g_1(X)g_2(Y)|X = x] = g_1(X)E[g_2(Y)|X = x]. \tag{10.7.8}$

The definitions of conditional expectation, conditional variance, etc. can be extended to more than two dimensions. For example

$$E[g(X_1, \ldots, X_n)|X_1 = x_1, \ldots, X_m = x_m], \ m < n,$$

$$= \int_{-\infty}^{\infty} \cdots \int_{-\infty}^{\infty} g(x_1, \ldots, x_n) f_{x_{m+1}, \ldots, x_n | x_1, \ldots, x_m}$$

$$(x_{m+1}, \ldots, x_n | x_1, \ldots, x_m) dx_{m+1} \ldots dx_n,$$

in case $(X_1, \ldots, X_n)$ is a n-dimensional continuous random vector.

EXAMPLE 10.7.1: For the *pdf*

$$f_{X,Y}(x, y) = 2, \ 0 < x < y < 1$$

$$
\begin{aligned}
f_X(x) &= \int_0^1 f(x, y) dy = 2(1 - x), \ 0 < x < 1, \\
f_Y(y) &= \int_0^y 2 dx = 2y, \ 0 < y < 1. \\
f_{X|y}(x|y) &= \frac{1}{y}, \ 0 < x < 1, \\
f_{Y|x}(y|x) &= \frac{1}{1-x}, \ 0 < y < 1, \\
E(X|y) &= \int_0^1 \frac{x}{y} dx = \frac{x^2}{2y}\Big]_0^1 = \frac{1}{2y}, \\
E(Y|X) &= \int_0^1 y \frac{dy}{1-x} = \frac{y^2}{2(1-x)}\Big]_0^1 = \frac{1}{2(1-x)}.
\end{aligned}
$$

DEFINITION 10.7.4 *Linear Regression:* Consider (X, Y). Regression of Y on x is linear if the corresponding regression curve is a straight line, i.e., if

$$\mathcal{Y} = E(Y|x) = \eta_x \text{ (say) } = \alpha + \beta x \tag{10.7.9}$$

where α, β are suitable constants. Similarly, regression of X on y is linear if

$$\mathcal{X} = E(X|y) = \xi_y \text{ (say) } = \alpha' + \beta' y, \tag{10.7.10}$$

α', β' being suitable constants.

Theorem 10.7.6: $E(Y - \eta_x) = E(X - \xi_y) = 0.$

Proof. Follows readily from Theorem 10.7.1.

Theorem 10.7.7: If the regression of Y on x is linear and if $\sigma_X^2, \sigma_Y^2, \rho$ exist, then the constants α, β in (10.7.9) are given by

$$\alpha = \mu_Y - \rho \frac{\sigma_Y}{\sigma_X} \mu_X, \ \beta = \rho \frac{\sigma_Y}{\sigma_X}. \tag{10.7.11}$$

Proof. Assume X, Y are continuous.

$$\eta_x = \alpha + \beta x, \ E(Y|X) = \alpha + \beta X. \tag{10.7.12}$$

Taking expectation of both sides with respect to the distribution of X, by Theorem 10.7.1,

$$E(Y) = \mu_Y = \alpha + \beta E(X) = \alpha + \beta \mu_X. \tag{10.7.13}$$

Again multiplying both sides of (10.7.12) by X and taking expectation with respect to the distribution of X, assuming without loss of generality (X, Y) are continuous variables,

$$\int_{-\infty}^{\infty} x \left[\int_{-\infty}^{\infty} y f_{Y|x}(y|x) dy \right] f_X(x) dx = \alpha E(X) + \beta E(X^2)$$

i.e.,

$$E(XY) = \alpha E(X) + \beta E(X^2). \qquad (10.7.14)$$

Solving equations (10.7.13) and (10.7.14),

$$\beta = \frac{\mu_{X,Y}}{\sigma_X^2} = \rho \frac{\sigma_Y}{\sigma_X}, \quad \alpha = \mu_Y - \beta \mu_X = \mu_Y - \rho \frac{\sigma_Y}{\sigma_X} \mu_X.$$

Corollary 10.7.7.1: Similarly, α', β' in (10.7.10) are given by

$$\alpha' = \mu_X - \rho \frac{\sigma_X}{\sigma_Y} \mu_Y, \quad \beta' = \frac{\mu_{X,Y}}{\sigma_Y^2}. \qquad (10.7.15)$$

Theorem 10.7.8: If regression of Y on x is linear and if the conditional variance $V(Y|X = x)$ is independent of x (i.e., conditional distribution of Y given x is *homoscedastic*)), then

$$V(Y|x) = \sigma_Y^2 (1 - \rho^2). \qquad (10.7.16)$$

Proof. We have

$$E[V(Y|X)] = E[E\{(Y - E(Y|X))^2 | X\}]$$

$$= \int_{-\infty}^{\infty} [\int_{-\infty}^{\infty} \{y - \mu_y - \rho \frac{\sigma_y}{\sigma_x}(x - \mu_y)\}^2 f_{Y|X}(y|x) dy] f_X(x) dx$$

$$= \int_{-\infty}^{\infty} \int_{\infty}^{\infty} \{y - \mu_y - \rho \frac{\sigma_y}{\sigma_x}(x - \mu_y)\}^2 f_{X,Y}(x, y) dx dy$$

$$= \sigma_y^2 (1 - \rho^2).$$

Since $V(Y|x)$ is constant for all x, $E[V(Y|X)] = \sigma_Y^2(1 - \rho^2)$ implies $V(Y|x) = \sigma_Y^2(1 - \rho^2)$ for all x $\qquad \square$.

We conclude this chapter by considering two important probability distributions, one is a discrete Multinomial distribution (on $\mathcal{R}^n$) and the other is continuous - a bivariate Normal distribution.

10.8 The Multinomial Distribution

Suppose an experiment may result in k possible outcomes, $O_1, \ldots, O_k$ with probabilities $p_1, \ldots, p_k$ respectively, $\sum_{i=1}^{k} p_i = 1$. N such experiments are performed. Let X_i be the number of experiments that result in outcomes $O_i (i = 1, \ldots, k)$, $\sum_{i=1}^{k} X_i = N$.

$$P(X_1 = x_1, \ldots, X_k = x_k) = \frac{N!}{\Pi_{i=1}^{k} x_i!} \Pi_{i=1}^{k} p_i^{x_i}$$
$$= P(x_1, \ldots, x_k) \tag{10.8.1}$$

where $x_i = 0, 1, \ldots, N; \sum_{i=1}^{k} x_i = N$.

By virtue of independence, any specific set of N outcomes with O_i occurring x_i times $(i = 1, \ldots, k)$ has probability $\Pi_{i=1}^{k} p_i^{x_i}$ and since there are $\frac{N!}{\Pi_{i=1}^{k} x_i!}$ such orderings, (10.8.1) follows. The random vector $\mathbf{X} = (X_1, \ldots, X_k)$ is said to follow a Multinomial distribution with parameters $N, p_1, \ldots, p_{k-1}$.

The expression $P(x_1, \ldots, x_k)$ shows that it can be regarded as the coefficient of $\Pi_{j=1}^{k} t_j^{x_j}$ in the expansion of

$$(t_1 p_1 + \ldots + t_N p_N)^N.$$

The probability generating function is

$$P_{\mathbf{X}}(\mathbf{t}) = E(t_1^{X_1} \ldots t_k^{X_k}) = (p_1 t_1 + \ldots + p_k t_k)^N. \tag{10.8.2}$$

The moment generating function of the distribution is

$$M_{X_1, \ldots, X_k}(t_1, \ldots, _k) = E(e^{t_1 X_1 + \ldots + t_k X_k})$$

$$= \sum_{x_1, \ldots, x_k} \frac{N!}{\Pi_{i=1}^{k} x_i!} \Pi_{i=1}^{k} (p_i e^{t_i})^{x_i} = (p_1 e^{t_1} + \ldots + p_k e^{t_k})^N. \tag{10.8.3}$$

The marginal *mgf* of $X_i (i = 1, \ldots, k)$ is

$$M_{X_i}(t_i) = M(0, \ldots, 0, t_i, 0, \ldots, 0) = (p_i e^{t_i} + q_i)^N, \quad q_i = 1 - p_i \tag{10.8.4}$$

which is the *mgf* of a binomial distribution with parameters (N, p_i). Hence, marginal distribution of X_i is binomial (N, p_i).

The marginal *mgf* of a set $(X_1, \ldots, X_m), m < k$ is

$$M_{X_1, \ldots, X_m}(t_1, \ldots, t_M) = M(t_1, \ldots, t_m, 0, \ldots, 0)$$

$$= (p_1 e^{t_1} + \ldots + p_m e^{t_m} + 1 - p_1 - \ldots - p_m)^N \tag{10.8.5}$$

so that $(X_1, \ldots, X_m)$ follows a $(m+1)$-nomial distribution with parameters $N, p_1, \ldots, p_m$.

Calculating directly, the marginal *pmf* of $X_1, \ldots, X_m), m < k$ is

$$f_{X_1,\ldots,X_m}(x_1,\ldots,x_m) = \sum_{x_{m+1},\ldots,x_k} \frac{N!}{x_1!\ldots x_k!} p_1^{x_1} \ldots p_k^{x_k}$$

$$= \frac{N!}{x_1!\ldots x_m!(N-\sum_{j=1}^m x_j)!} p_1^{x_1} \ldots p_m^{x_m} Q^{N-\sum_{j=1}^m x_j}$$

$$\times \left[\sum_{x_{m+1}\ldots x_k} \frac{(N-\sum_{j=1}^m x_j)!}{x_{m+1}!\ldots x_k!} \left(\frac{p_{m+1}}{Q}\right)^{x_{m+1}} \ldots \left(\frac{p_m}{Q}\right)^{x_m} \right]$$

$$(10.8.6)$$

where $Q = 1 - \sum_{j=1}^m p_j$. The quantity in [ ] being the sum of the probabilities of a Multinomial distribution with parameters $(N - \sum_{j=1}^m x_j)$, $\frac{p_{m+j}}{Q} (\geq 0), j = 1, \ldots, m$, is unity. Hence, the marginal distribution of $X_1, \ldots, X_m$ is Multinomial with parameters $N, p_1, \ldots, p_m$.

We now consider the conditional distribution of $X_{m+1}, \ldots, X_k$ given $(X_1 = x_1, \ldots, X_m = x_m)$.

$$f_{X_{m+1},\ldots,x_n|x_1,\ldots,x_m}(x_{m+1},\ldots,x_k|x_1,\ldots,x_m)$$

$$= \frac{f_{X_1,\ldots,X_k}(x_1,\ldots,x_k)}{f_{X_1,\ldots,X_m}(x_1,\ldots,x_m)}$$

$$= \frac{\frac{N!}{\Pi_{i=1}^k x_i!} \Pi_{i=1}^k p_i^{x_i}}{\frac{N!}{x_1!\ldots x_m!(N-\sum_{j=1}^m x_j)!} \Pi_{j=1}^m p_j^{x_j} \left(1 - \sum_{j=1}^m p_j\right)^{N-\sum_{j=1}^m x_j}}$$

$$= \frac{(N - \sum_{j=1}^m x_j)!}{x_{m+1}!\ldots x_k!} \left(\frac{p_{m+1}}{Q}\right)^{x_{m+1}} \ldots \left(\frac{p_k}{Q}\right)^{x_k} \qquad (10.8.7)$$

This is a Multinomial distribution with parameters $(N - \sum_{j=1}^m x_j), \frac{p_{m+j}}{Q}, j = 1, \ldots, k - m$.

The modes of the distribution (10.8.1) are near the expected values, are at $x_{1m} \approx Np_1, \ldots, x_{km} \approx Np_k$.

We now consider the moments of the multinomial distribution from its *mgf* (10.8.3).

$$\frac{\partial M(t_1,\ldots,t_k)}{\partial t_i} = Np_i e^{t_i} 9 p_1 e^{t_1} + \ldots + p_k e^{t_k})^{N-1}$$

$$\frac{\partial^2}{\partial t_i^2} M(t_1,\ldots,t_k) = N(N-1)p_i^2 e^{2t_i}(p_1 e^{t_1} + \ldots + p_k e^{t_k})^{N-2}$$
$$+ Np_i e^{t_i}(p_1 e^{t_1} + \ldots + p_k e^{t_k})^{N-1}$$
$$\frac{\partial^2}{\partial t_i t_j} M(t_1,\ldots,t_k) = N(N-1)p_i p_j e^{t_i+t_j}(p_1 e^{t_1} + \ldots + p_k e^{t_k})^{N-2},$$
$$(\text{for } i \neq j).$$

Hence, evaluating the derivatives at $(0, \ldots, 0)$,

$$
\begin{aligned}
E(X_i) &= Np_i, \; E(X_i^2) = Np_i + N(N-1)p_i^2 \\
E(X_i X_j) &= N(N-1)p_i p_j \\
V(X_i) &= Np_i(1-p_i) \\
\text{Cov}\,(X_i, X_j) &= -Np_i p_j \\
\rho(X_i, X_j) &= -\sqrt{\frac{p_i p_j}{(1-p_i)(1-p_j)}}.
\end{aligned}
\tag{10.8.8}
$$

Note that since $p_i + p_j \le 1$, $p_i p_j \le (1-p_i)(1-p_j)$. The variance-covariance matrix of the distribution, $\mathbf{V}_{k\times k} = ((\,\text{Cov}\,(p_i, p_j)))$ is singular with rank $k-1$. The matrix $\mathbf{V}$ is of the form

$$
\mathbf{V} = \text{Diag.}(\mathbf{p}) - \mathbf{p}\mathbf{p}'
\tag{10.8.9}
$$

where $\mathbf{p} = (p_1, \ldots, p_k)'$, $p_i > 0$, $\sum_i p_i = 1$. For matrix of this form, if we order the components

$$
p_1 \ge p_2 \ge p_3 \ge \ldots \ge p_k > 0
$$

and denote by $\lambda_1(\mathbf{V}) \ge \lambda_2(\mathbf{V}) \ge \ldots \ge \lambda_k(\mathbf{V})$, the ordered eigenvalues of $\mathbf{V}$, the following inequalities hold:

$$
p_1 \ge \lambda_1(\mathbf{V}) \ge p_2 \ge \lambda_2(\mathbf{V}) \ge \ldots \ge p_k \ge \lambda_k(\mathbf{V}) > 0.
\tag{10.8.10}
$$

The regression of X_i on X_j is linear:

$$
E[X_i | X_j = x_j] = (N - x_j)\frac{p_i}{1 - p_j}.
\tag{10.8.11}
$$

Also, the corresponding conditional variance

$$
\text{Var}\,(X_i | X_j = x_j) = (N - x_j)\frac{p_i}{1 - p_j}\left\{1 - \frac{p_i}{1 - p_j}\right\}.
\tag{10.8.12}
$$

Hence the regression of X_i on X_j is not homoscedastic.

The joint factorial moments of this distribution is

$$
\mu_{(r_1, \ldots, r_k)} = E[(X_1)_{r_1} \cdots (X_k)_{r_k}] = N^{\sum_i r_i} p_1^{r_1} \cdots p_k^{r_k}.
\tag{10.8.13}
$$

The cumulant generating function is

$$
\log_e M(t_1, \ldots, t_k) = N \log_e \sum_{j=1}^{k} p_j e^{t_j}.
\tag{10.8.14}
$$

10.9   The Bivariate Normal Distribution

The two-dimensional random vector (X, Y) is said to follow a bivariate normal distribution with parameters $\mu_X, \mu_Y, \sigma_X, \sigma_Y$ and ρ if their joint *pdf* is given by

$$f_{X,Y}(x, y) = \frac{1}{2\pi\sigma_X\sigma_Y\sqrt{1-\rho^2}} \times$$
$$\exp\left[-\frac{1}{2(1-\rho^2)}\left\{\left(\frac{x-\mu_X}{\sigma_X}\right)^2 - 2\rho\left(\frac{x-\mu_X}{\sigma_X}\right)\left(\frac{y-\mu_Y}{\sigma_Y}\right) + \left(\frac{y-\mu_Y}{\sigma_Y}\right)^2\right\}\right],$$

$$\text{(10.9.1)}$$

$-\infty < x < \infty, -\infty < y < \infty$, when the parameters satisfy $-\infty < \mu_X < \infty, -\infty < \mu_Y < \infty, \sigma_X > 0, \sigma_Y > 0, |\rho| < 1$.

To show that

$$\int_{-\infty}^{\infty}\int_{-\infty}^{\infty} f(x, y)dxdy = 1 \tag{10.9.2}$$

we substitute

$$\frac{x-\mu_X}{\sigma_X} = u, \quad \frac{y-\mu_Y}{\sigma_Y} = v. \tag{10.9.3}$$

The left hand side of (10.9.1) becomes

$$\frac{1}{2\pi\sqrt{1-\rho^2}}\int_{-\infty}^{\infty}\int_{-\infty}^{\infty} e^{-\frac{1}{2(1-\rho^2)}[u^2-2\rho uv+v^2]}dudv \tag{10.9.4}$$

$$= \frac{1}{2\pi\sqrt{1-\rho^2}}\int_{-\infty}^{\infty}\int_{-\infty}^{\infty} e^{-\frac{1}{2(1-\rho^2)}[(u-\rho v)^2+(1-\rho^2)v^2]}dudv. \tag{10.9.5}$$

Setting $w = \frac{u-\rho v}{\sqrt{1-\rho^2}}$, (10.9.5) becomes

$$\left[\frac{1}{\sqrt{2\pi}}\int_{-\infty}^{\infty} e^{-\frac{w^2}{2}}dw\right]\left[\frac{1}{\sqrt{2\pi}}\int_{-\infty}^{\infty} e^{-\frac{v^2}{2}}dv\right] = 1.$$

Note 10.9.1: If $\rho^2 = 1$, the *pdf* (10.9.1) is not defined. In this case, it is known that there exist constants a, b such that $P(Y = aX + b) = 1$ a.s. The distribution then is said to be a *bivariate degenerate (or singular) normal distribution*.

We shall here consider non-degenerate bivariate normal distribution, where $\rho^2 < 1$.

The moment generating function is

$$M_{X,Y}(t_1, t_2) = E(e^{t_1 X+t_2 Y}) = E(e^{t_1(\mu_X+U\sigma_X)+t_2(\mu_Y+V\sigma_Y)})$$

where U and V are given in (10.9.3).

$$= \frac{e^{t_1\mu_X+t_2\mu_Y}}{2\pi\sqrt{1-\rho^2}} \int_{-\infty}^{\infty}\int_{-\infty}^{\infty} e^{t_1u\sigma_X+t_2v\sigma_Y-\frac{1}{2(1-\rho^2)}[u^2-2\rho uv+v^2]} du\, dv. \quad (10.9.6)$$

The exponent under the integral sign in (10.9.6) is

$$-\frac{1}{2(1-\rho^2)}[u^2-2\rho uv+v^2-2(1-\rho^2)\{t_1u\sigma_X+t_2v\sigma_Y\}]$$

$$= -\frac{1}{2}s^2 - \frac{1}{2}z^2 + \frac{1}{2}(t_1^2\sigma_X^2+2\rho t_1 t_2\sigma_X\sigma_Y+t_2^2\sigma_Y^2)$$

where $s = \frac{u-\rho v-(1-\rho^2)t_1\sigma_X}{\sqrt{1-\rho^2}}, z = v-\rho t_1\sigma_X-t_2\sigma_Y.$

Hence the expression (10.9.6) becomes

$$\exp[t_1\mu_X+t_2\mu_Y+\frac{1}{2}(t_1^2\sigma_X^2+2\rho t_1 t_2\sigma_X\sigma_Y+t_2^2\sigma_Y^2)]$$

$$\left[\frac{1}{\sqrt{2\pi}}\int_{-\infty}^{\infty}e^{-\frac{s^2}{2}}ds\right]\left[\frac{1}{\sqrt{2\pi}}\int_{-\infty}^{\infty}e^{-\frac{z^2}{2}}dz\right]$$

$$= \exp[t_1\mu_X+t_2\mu_Y+\frac{1}{2}(t_1^2\sigma_X^2+2\rho t_1 t_2\sigma_X\sigma_Y+t_2^2\sigma_Y^2)]. \quad (10.9.7)$$

The *mgf* of the distribution (10.9.1) about the point (μ_X,μ_Y) is

$$M_{X-\mu_X,Y-\mu_Y}(t_1,t_2) = e^{-\mu_X t_1-\mu_Y t_2}M_{X,Y}(t_1,t_2)$$
$$= \exp\left[\frac{1}{2}(t_1^2\sigma_X^2+2\rho t_1 t_2\sigma_X\sigma_Y+t_2^2\sigma_Y^2)\right]. \quad (10.9.8)$$

From (10.9.7), the following results follow:

$$
\begin{aligned}
E(X) &= \left.\frac{\partial M(t_1,t_2)}{\partial t_1}\right|_{t_1=t_2=0} = \mu_X \\
E(X^2) &= \left.\frac{\partial^2 M(t_1,t_2)}{\partial t_1^2}\right|_{t_1=t_2=0} = \mu_X^2+\sigma_X^2 \\
E(XY) &= \left.\frac{\partial^2 M(t_1,t_2)}{\partial t_1\partial t_2}\right|_{t_1=t_2=0} = \rho\sigma_X\sigma_Y+\mu_X\mu_Y.
\end{aligned}
\quad (10.9.9)
$$

Hence, $E(Y)=\mu_Y, V(X)=\sigma_X^2, V(Y)=\sigma_Y^2$, Cov $(X,Y)=\rho\sigma_X\sigma_Y$, Corr. Coeff. $(X,Y)=\rho$.

Writing $\frac{(X-\mu_X)}{\sigma_X}=U$ and $\frac{(Y-\mu_Y)}{\sigma_Y}=V$, we have the standardized bivariate normal density with parameter ρ,

$$f(u,v) = \frac{1}{2\pi\sqrt{1-\rho^2}}e^{-\frac{1}{2(1-\rho^2)}(u^2-2\rho uv+v^2)},$$

$$-\infty < u < \infty, \quad -\infty < v < \infty.$$

The *mgf* of a standardized bivariate normal density is

$$M_{U,V}(t_1, t_2) = e^{\frac{1}{2}(t_1^2 + 2\rho t_1 t_2 + t_2^2)}.$$

For this distribution

$$\mu_{12} = \mu_{21} = 0, \qquad \qquad \mu_{22} = 1 + 2\rho^2, \ \mu_{31} = \mu_{13} = 3\rho.$$
$$\mu_{32} = \mu_{23} = 0, \qquad \qquad \mu_{33} = 3\rho(3 + 2\rho^2), \ \mu_{41} = \mu_{14} = 0,$$
$$\mu_{42} = \mu_{24} = 3(1 + 4\rho^2), \ \mu_{31} = \mu_{13} = 15\rho_1.$$

All the cumulants of order higher than two vanish.

The marginal *mgfs* of the distribution (10.9.1) are

$$M_X(t_1)M_{X,Y}(t_1, 0) = e^{\mu_x t_1 - \frac{1}{2}t_1^2 \sigma_X^2}$$

$$M_Y(t_2) = M_{X,Y}(0, t_2) = e^{\mu_Y t_2 + \frac{1}{2}\sigma_Y^2}$$

so that the marginal distribution of X is univariate normal $N(\mu_X, \sigma_X^2)$ and similarly for Y.

We have already stated that if two random variables are uncorrelated, they are not necessarily independent.

Theorem 10.9.1: If (X, Y) have a bivariate normal distribution, then X, Y are independent *iff* they are uncorrelated.

Proof. If X, Y are uncorrelated, i.e. if $\rho = 0$, the bivariate density (10.9.1) can be written as the product of two univariate normal densities so that X, Y are independent. The converse is always true.

Theorem 10.9.2: If (X, Y) have a bivariate normal distribution, the marginal density of X is $N(\mu_X, \sigma_X)$ and of Y is $N(\mu_Y, \sigma_Y)$.

Proof. A proof has been given above by using the *mgf*. However, a direct proof is given below.

$$f_X(x) = \int_{-\infty}^{\infty} f(x, y) dy.$$

Substituting $v = \frac{y - \mu_Y}{\sigma_Y}$ in (10.9.1),

$$f_X(x) = \frac{1}{2\pi\sqrt{1 - \rho^2}} \int_{-\infty}^{\infty} \exp\left[-\frac{1}{2}\left(\frac{x - \mu_X}{\sigma_X} \right)^2 - \frac{1}{2(1 - \rho^2)}\left(v - \rho\frac{x - \mu_X}{\sigma_X} \right)^2 \right] dv.$$

Substituting

$$w = \left\{ v - \rho\left(\frac{x - \mu_X}{\sigma_X} \right) \right\}(1 - \rho^2)^{-1/2},$$

$$f_X(x) = \frac{1}{\sigma_X \sqrt{2\pi}} \exp\left[-\frac{1}{2}\left(\frac{x - \mu_X}{\sigma_X}\right)^2\right].$$

Similar result follows for the marginal distribution of Y.

Theorem 10.9.3: If (X, Y) have a bivariate normal distribution, then the conditional distribution of Y given $X = x$ is univariate normal $N\left(\mu_Y + \rho\frac{\sigma_Y}{\sigma_X}(x - \mu_X), \sigma_Y^2(1 - \rho^2)\right)$.

Similarly, the conditional distribution of X given $Y = y$ is univariate normal $N\left(\mu_X + \rho\frac{\sigma_X}{\sigma_Y}(y - \mu_Y), \sigma_X^2(1 - \rho^2)\right)$.

Proof.

$$f_{Y|X}(y|x) = \frac{f_{X,Y}(x,y)}{f_X(x)}$$

$$= \frac{1}{\sigma_Y\sqrt{2\pi(1-\rho^2)}} \exp\left[\frac{1}{2\sigma_Y^2(1-\rho^2)}\{y - \mu_Y - \rho\frac{\sigma_Y}{\sigma_X}(x - \mu_X)\}^2\right]$$

(10.9.10)

obtained by the substitutions as in (10.9.3) and proceeding as in (10.9.5). Hence the conditional distribution of Y is univariate normal with parameters as stated above. The result for conditional distribution of Y follows similarly.

Corollary 10.9.3.1: The regression curve of Y on X is

$$y = \mu_Y + \rho\frac{\sigma_Y}{\sigma_X}(x - \mu_X) \tag{10.9.11}$$

which is a linear function of x. More specifically, the regression curve of Y on X is a straight line with gradient $\rho\frac{\sigma_Y}{\sigma_X}$ and intercept $\left(\mu_Y - \rho\frac{\sigma_Y}{\sigma_X}\mu_X\right)$.

Similarly, the regression curve of X on Y is a straight line (Fig. 10.9.1)

$$x = \mu_X + \rho\frac{\sigma_X}{\sigma_Y}(y - \mu_Y). \tag{10.9.12}$$

Note 10.9.2: The variance of the conditional distribution of Y given X (X given Y) does not depend on $x(y)$. The conditional distributions are homoscedastic.

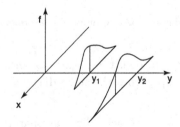

Fig. 10.9.1   Conditional distribution of y given x

The locus of the points in the (x, y)-plane such that $f(x, y)$ is a constant is an ellipse,

$$\frac{(x - \mu_X)^2}{\sigma_X^2} - 2\rho \frac{(x - \mu_X)(y - \mu_Y)}{\sigma_X \sigma_Y} + \frac{(y - \mu_Y)^2}{\sigma_Y^2} = k^2 \quad \text{(a constant)}$$

$$(10.9.13)$$

with center (μ_X, μ_Y). Thus if the probability density surface (10.9.1) is cut by the plane at a height $z = z_0$ (a sufficiently small positive quantity), parallel to the (x, y)-plane, the resulting contour curves are similar and concentric ellipses (10.9.13) with center (μ_X, μ_Y). The curves (10.9.13) are called *equi-probability contours*. The constant k is a decreasing function of z_0. The density $f(x, y)$ is maximum at the center and is equal to

$$f(\mu_X, \mu_Y) = \frac{1}{2\pi \sigma_X \sigma_Y \sqrt{1 - \rho^2}}.$$

The line of regression of y on x(x on y) passes through the two points where the tangent to the ellipse are vertical (horizontal) [Fig. 10.9.2].

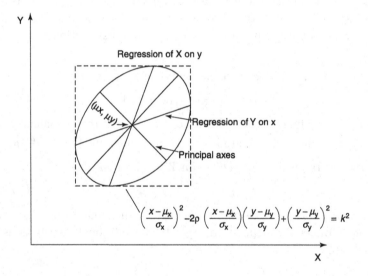

Fig. 10.9.2   An equiprobability contour of a bivariate normal density

As the value of $|\rho|$ tends to one, the ellipse becomes elongated and both the lines of regression approaches the same diagonal of the circumscribed rectangle.

If $\rho = 0$, the principal axes of each ellipse are parallel to the co-ordinate axes.

Consider a bivariate normal random vector (X, Y) with $\rho(X, Y)$ not necessarily equal to zero. For the transformed variables

$$U = X \cos\theta + Y \sin\theta, \quad V = -X \sin\theta + Y \cos\theta \qquad (10.9.14)$$

$\mathrm{Cov}(U, V) = \frac{1}{2}(\sigma_Y^2 - \sigma_X^2) \sin 2\theta + \rho \sigma_X \sigma_Y \cos 2\theta$. Hence, U, V are uncorrelated if

$$2\theta = \tan^{-1}\left(\frac{2\rho\sigma_X\sigma_Y}{\sigma_X^2 - \sigma_Y^2}\right). \qquad (10.9.15)$$

Therefore, by the transformation of co-ordinates from (X, Y) to (U, V) as in (10.9.14) and (10.9.15), the new axes of co-ordinates become parallel to the principal axes of the ellipse (10.9.13).

10.10 Exercises and Complements

10.1 (X, Y) have the joint *pdf*

$$f(x, y) = \begin{cases} \frac{1}{2} & \text{for } |x| + |y| \le 1, \\ 0 & \text{otherwise.} \end{cases}$$

Find the marginal *pdf*s and the conditional *pdf*s.

$$\left[\begin{aligned} f_X(x) &= \int_{|x|+|y|\le 1} \tfrac{1}{2} dy = \int_{|x|-1}^{1-|x|} \tfrac{1}{2} dy = 1 - |x|, \ |x| \le 1 \\ &= 0 \ \text{(otherwise)}. \end{aligned}\right.$$

$$f_{X|Y}(x|y) = \begin{cases} \frac{1}{2(1-|y|)} & \text{for } |x| \le 1 - |y| \text{ if } |y| \le 1, \\ 0 & \text{for } |x| > 1 - |y| \text{ if } |y| \le 1, \\ \text{undefined} & \text{for all } x \text{ if } |y| > 1. \end{cases}$$

10.2 The vector (X, Y) have joint *pdf* $f(x, y) = 3x^2y + 3y^2x, 0 \le x \le 1, 0 \le y \le 1; = 0$ elsewhere. Find the marginal *pdf*s. Find the conditional $P(\frac{1}{2} \le Y \le \frac{3}{4}|\frac{1}{3} \le X \le \frac{2}{3})$.

$$\left[f_X(x) = x + \frac{3}{2}x^2; \ f_Y(y) = y + \frac{3}{2}y^2; \ \text{Prob.} = \frac{515}{4096}.\right]$$

10.3 The variables (X, Y) have joint *pdf*

$$f(x, y) = \begin{cases} e^{-y}(1 - e^{-x}), & 0 < x \le y, 0 \le y < \infty, \\ e^{-x}(1 - e^{-y}), & 0 < y \le x, 0 \le x < \infty. \end{cases}$$

Find (a) marginal densities, (b) $E[Y|X = x]$ for $x > 0$, (c) $P[X \le 2, Y \le 2]$, (d) $\rho(X, Y)$.

10.4 For each of the following *pdf*s find $F_{X,Y}(x, y), F_X(x),$ $F_Y(y), f_X(x), f_Y(y), F_{Y|X}(y|x), \mu_{20}, \mu_{02}$ and ρ.

(a) $f(x,y) = 2 - x - y$ $(0 \le x \le 1, 0 \le y \le 1)$
(b) $f(x,y) = 2$ $(0 < x < y < 1)$
(c) $f(x,y) = \frac{6-x-y}{8}$ $(0 < x < 2, 2 < y < 4)$
(d) $f(x,y) = 2e^{-x-y}$ $(0 < x < y < \infty)$
(e) $f(x,y) = \frac{(n-1)(n-2)}{n(1+x+y)}$, $x > 0, y > 0, n$ (a fixed constant) > 2
(f) $f(x,y) = x + y$ $(0 < x < 1, 0 < y < 1)$
(g) $f(x,y) = x^2 + \frac{xy}{3}$ $(0 < x < 1, 0 < y < 2)$.

10.5 (X,Y) have bivariate cumulative distribution function

$$F_{X,Y}(x,y) = \begin{cases} 1 - e^{-x} - e^{-y} + e^{-(x+y)}, & x, y > 0 \\ 0 & \text{otherwise.} \end{cases}$$

Find the joint *pdf* and the marginal *pdf*s.

10.6 X has a discrete probability distribution $P(X = x) = \frac{x}{3}, x = 1, 2$; $f_{Y|X}(y|x)$ is binomial with parameters $(x, \frac{1}{2}), y = 0, 1, \ldots, x$. Find $f(x,y), E(X), V(X), E(Y), V(Y)$.

10.7 The joint *pdf* of bivariate gamma variable (X,Y) is

$$f(x,y) = \frac{\lambda^{r+p}}{\Gamma(r)\Gamma(p)} e^{-\lambda x} y^{p-1} (x - y)^{r-1}, \quad 0 < y < x < \infty; \lambda, p, r > 0.$$

Find the marginal *pdf*s and the conditional *pdf*s.

10.8 Consider (X,Y) with $Y = 2X$. Show that

$$F_{X,Y}(x,y) = \begin{cases} F_X(x) & \text{for } x \le \frac{y}{2} \\ F_X(\frac{y}{2}) = F_Y(y) & \text{for } x > \frac{y}{2}. \end{cases}$$

[(a) Suppose $2x \le y$. The events, $\{X \le x\} \Leftrightarrow \{Y \le 2x\}$. Hence,

$$P\{X \le x, Y \le y\} = P\{X \le x, Y \le 2x\} + P\{X \le x, 2x < Y \le y\}$$
$$= P\{X \le x, Y \le 2x\} = P\{X \le x\} = F_X(x).$$

(b) Suppose $y < 2x$. The events, $\{Y \le y\} \Leftrightarrow \{X \le \frac{y}{2}\}$.

$$P\{X \le x, Y \le y\} = P\{X \le \frac{y}{2}, Y \le y\} + P\{\frac{y}{2} < X \le x, Y \le y\}$$
$$= P\{X \le \frac{y}{2}, Y \le y\} = P\{X \le \frac{y}{2}\} = F_X(\frac{y}{2})].$$

10.9 Consider (X,Y) with $Y = g(X)$. Show that if

(i) $g(t)$ is monotonically increasing

$$F_{X,Y}(x,y) = \begin{cases} F_Y(y) & \text{if } y < g(x) \\ F_X(x) & \text{if } y \ge g(x). \end{cases}$$

(ii) $g(t)$ is monotonically decreasing

$$F_{X,Y}(x,y) = \begin{cases} 0 & \text{if } y < g(x) \\ F_X(x) + F_Y(y) - 1 & \text{if } y \geq g(x). \end{cases}$$

10.10 The random vector (X, Y) have joint *pdf* $f(x, y)$.

(A) Show that for the following *pdf*s

(a) X, Y are stochastically independent:

(i) $f(x, y) = 6xy^2$, $0 < x, y < 1$
(ii) $f(x, y) = 12xy(1 - y)$, $0 < x, y < 1$
(iii) $f(x, y) = 4xy$, $0 < x, y < 1$.

(b) X, Y are stochastically dependent:

(iv) $f(x, y) = \frac{1}{2}$, $|x| + |y| \leq 1$
(v) $f(x, y) = 8xy$, $0 \leq y \leq x < 1$
(vi) $f(x, y) = 2e^{-(x+y)}$, $0 < x < y < \infty$
(vii) $f(x, y) = \frac{1}{4\pi^2}[1 - \sin(x + y)]$, $-\pi \leq x, y \leq \pi$
(viii) $f(x, y) = x + y$, $0 < x, y < 1$.

(B) Verify if X, Y are stochastically independent when

(ix) $f(x, y) = p_1^x(1 - p_1)^{1-x}p_2^y(1 - p_2)^y[1 + \lambda(x - p_1)(y - p_2)], 0 < x, y < 1, 0 < p_1, p_2 < 1, |\lambda| \leq 1$
(x) $f(x, y) = \frac{\sqrt{3}}{2\pi} \exp[-\{(x - 7)^2 + (x - 7)y + y^2\}]$.

10.11 Show that the random variables X, Y with joint *pdf*

$$f(x, y) = \frac{1}{2\pi\sqrt{1 + y^2}} \exp\{-\frac{x^2}{2(1 + y^2)} - \frac{y^2}{2}\}$$

have covariance zero and yet X, Y are not independent.

10.12 Show that X, Y are dependent random variables, but are uncorrelated, when

(a) X, Y are discrete random variables with

$$P(X = x, Y = y) = \frac{1}{8}, \quad i, j = 1, -1, \quad P(X = 0, Y = 0) = \frac{1}{2}.$$

(b) X, Y are continuous random variables with $f(x, y) = \frac{1}{2}$ inside the square with corners at points $(1, 0), (0, 1), (-1, 0), (0, -1)$.

(c) $X = -\sin 2\pi R, Y = \cos 2\pi R$ when R has a uniform distribution over $(0,1)$.

10.13 A pair of fair dice are tossed. Let X denote the face value of the first dice, Y the larger of the two face values. Find the joint *pmf* of (X, Y). Check that $P(X = 2, Y = i) = \frac{1}{36}, i = 3, \ldots, 6; P(X = j, Y = j) = \frac{j}{36}, j = 1, \ldots, 6; P(X = i, Y = j) = 0, i > j = 1, \ldots, 6.$

10.14 Let (X, Y) have the joint *pdf*

$$f(x, y) = \begin{cases} \frac{1+xy}{4}, & |x| < 1, |y| < 1 \\ 0 & \text{elsewhere.} \end{cases}$$

Show that X, Y are not independent, though X^2, Y^2 are independent.

10.15 Suppose $g(X)$ is a monotonic function of the random variable X. Show that X and $g(X)$ are not independent.

10.16 Random variables X and θ are independently and uniformly distributed in the intervals $(0, a)$ and $(0, \frac{\pi}{2})$ respectively. Show that

$$P[X < b \cos \theta] = \frac{2b}{\pi a}, \quad b < a.$$

10.17 The variables X, Y have the joint *pdf* $f(x, y) = \frac{3}{2\sqrt{x}}, 0 < y < x < 1$. Find

$$f_{Y|X}(y|x = \tfrac{1}{2}), f_{X|Y}(x|y = \tfrac{1}{2}).$$

10.18 (X, Y) have joint *pdf* $f(x, y) = A\frac{x}{y}, 0 < x < 1, 1 < y < 2; = 0$ otherwise; A, a constant. Evaluate A. Find $f_{Y|X}(y|x), f_{X|Y}(x|y)$.

$$\left[\frac{2}{\log_e 2}; \ 2x, 0 < x < 1; \frac{1}{y \log_e 2}, 1 < y < 2. \right].$$

10.19 A random variable X has the *pdf* $f(x) = \frac{e^{-\frac{x}{2}}}{\sqrt{2\pi x}}, x > 0$. The conditional *pdf* of Y given $X = x$ is

$$f_{X|Y}(y|x) = \frac{\sqrt{x}}{\sqrt{2\pi}} e^{-\frac{xy^2}{2}}, \quad -\infty < y < \infty.$$

Show that the marginal density of Y is

$$f_Y(y) = \frac{1}{\pi(1 + y^2)}, \quad -\infty < y < \infty.$$

10.20 For the bivariate distribution

$$f(x,y) = 1, \ |y| < x, \ 0 < x < 1,$$

find the conditional densities, $E(Y|X), \rho(X,Y)$.

$$[f_{X|Y}(x|y) = (1+y)^{-1}, -1 < y \leq 0; (1-y)^{-1}, 0 < y < 1,$$

$$f_{Y|X}(y|x) = \frac{1}{2x}, 0 < x < 1,$$

$$E(Y|X) = 0, \ \rho = 0.]$$

10.21 Show that for the distribution

$$f(x,y) = x+y, \ 0 < x < 1, 0 < y < 1,$$

$$\mu'_{r,s} = \frac{1}{(r+2)(s+1)} + \frac{1}{(r+1)(s+2)}$$

r, s, being positive integers; hence, show that $\rho(X,Y) = -\frac{1}{11}$.

10.22 (X,Y) follows a bivariate normal distribution with parameters $(0,0,\sigma_X^2,\sigma_Y^2,\rho)$. Show that $U = \frac{X}{\sigma_X} - \rho\frac{Y}{\sigma_Y}$ and $V = \frac{(\sqrt{1-\rho^2})Y}{\sigma_Y}$ are independent variables.

10.23 Let X_1, X_2 be independent normal with parameters $(\mu_1, \sigma_1^2), (\mu_2, \sigma_2^2)$ respectively. Show that the correlation coefficient between U and V where $U = X\cos\theta + Y\sin\theta, V = -X\sin\theta + Y\cos\theta$ is given by

$$\rho^2 = 1 - \frac{4\sigma_1^2\sigma_2^2}{4\sigma_1^2\sigma_2^2 + (\sigma_1^2 - \sigma_2^2)^2\sin^2 2\theta}.$$

10.24 X, Y are standard normal variables with correlation coefficient ρ. Show that the correlation coefficient between X^2 and Y^2 is ρ^2.

[Hints: The *mgf* $M_{X,Y}(t_1,t_2) = \exp\left[\frac{1}{2}(t_1^2 + 2\rho t_1 t_2 + t_2^2)\right].E(X^2Y^2) =$ coefficient of $\frac{t_1^2}{2!}\frac{t_2^2}{2!}$ in $M_{X,Y}(t_1,t_2) = 2\rho^2 + 1$. Similarly, $E(X^4) = E(Y^4) = 3.$]

10.25 Let (X,Y) have a bivariate normal distribution $N_2(\mu_1, \mu_2, \sigma_1^2, \sigma_2^2, \rho)$. Show that $Z = aX + bY + c$ is a univariate normal variable, $N(a\mu_1 + b\mu_2 + c, a^2\sigma_1^2 + 2ab\rho\sigma_1\sigma_2 + b^2\sigma_2^2)$.

[Use *mgf* of (X,Y) to find *mgf* of Z.]

10.26 (X, Y) follow bivariate $N(\mu_X, \mu_Y, \sigma_X^2, \sigma_Y^2, \rho)$-distribution. Show that $U = \frac{X-\mu_X}{\sigma_X} + \frac{Y-\mu_Y}{\sigma_Y}$ and $V = \frac{Y-\mu_X}{\sigma_X} - \frac{Y-\mu_Y}{\sigma_Y}$ are independently normally distributed with zero means and variances $2(1+\rho)$ and $2(1-\rho)$ respectively.

10.27 *Multivariate Chebychev inequality:* Let $(X_1, \ldots, X_n)$ be jointly distributed with $E(X_i) = \mu_i, V(X_i) = \sigma_i^2, i = 1, \ldots, n$. Define $A_i = (|X_i - \mu_i| \leq \sqrt{n}t\sigma_i)$. Prove that $P(\cap_{i=1}^n A_i) \geq 1 - \frac{1}{t^2}$ for $t > 0$.

[We have by Chebychev's inequality $P(A_i) \geq 1 - \frac{1}{nt^2}$; hence $P(A^c) \leq \frac{1}{nt^2}$. Now by Bonferroni's inequality (Theorem 2.6.4)

$$P(\cap_{i=1}^n A_i) \geq 1 - \sum_{i=1}^n P(A_i^c) \geq 1 - \frac{n}{nt^2} = 1 - \frac{1}{t^2}, t > 0.]$$

10.28 X_1, X_2 are random variables with means zeros, variances unity and correlation coefficient ρ. Show that

$$E[\max(X_1^2, X_2^2)] \leq 1 + \sqrt{1 - \rho^2}.$$

Hence show that for random variables X, Y that are jointly distributed with means μ_X and μ_Y and variances σ_X^2 and σ_Y^2 respectively and correlation coefficient ρ and for any $t > 0$,

$$P[|X - \mu_X| \geq t\sigma_X \text{ or } |Y - \mu_Y| \geq t\sigma_Y] \leq \frac{1}{t^2}[1 + \sqrt{1 - \rho^2}].$$

[First part: we have $2\max(X_1^2, X_2^2) = |X_1^2 - X_2^2| + X_1^2 + X_2^2$. Hence

$$E\{\max(X_1^2, X_2^2)\} = \frac{1}{2}E|X_1^2 - X_2^2| + \frac{1}{2}E(X_1^2 + X_2^2). \qquad (i)$$

Now

$$(E|X_1^2 - X_2^2|)^2 = (E\{|X_1 - X_2||X_1 + X_2|\}^2 \leq E\{|X_1 - X_2|\}^2 E\{|X_1 + X_2|^2\}$$
$$= \{2(1-\rho)\}\{\{2(1+\rho)\}, \qquad (ii)$$

the inequality in (ii) follows from Cauchy-Schwarz inequality. Hence the result follows from (i).

Second part: Let $\frac{X-\mu_X}{\sigma_X} = Z_1, \frac{Y-\mu_Y}{\sigma_Y} = Z_2$. Hence, $E(Z_1) = E(Z_2) = 0, V(Z_1) = V(Z_2) = 1, \rho(Z_1, Z_2) = \rho$.

Let $\max(Z_1^2, Z_2^2) = Z$, which is a random variable. By Markov inequality

$$P(Z \geq t^2) \leq \frac{E(Z)}{t^2} \leq \frac{1 + \sqrt{1 - \rho^2}}{t^2}$$

i.e.

$$P[|X - \mu_X| \geq t\sigma_X \text{ or } |Y - \mu_Y| \geq t\sigma_Y] \leq \frac{1 + \sqrt{1 - \rho^2}}{t^2}.]$$

10.29 For the bivariate normal distribution $N(0, 0, 1, 1, \rho)$, show that the moments obey the recurrence relation

$$\mu_{r,s} = (r + s - 1)\rho\mu_{r-1,s-1} + (r-1)(s-1)(1 - \rho^2)\mu_{r-2,s-2}.$$

Hence or otherwise, show that $\mu_{r,s} = 0$ if $r + s$ is odd; $\mu_{3,1} = 3\rho$, $\mu_{2,2} = 1 + 2\rho^2$.

[Consider $mgf\ M(t_1, t_2) = \exp\{\frac{1}{2}(t_1^2 + 2\rho t_1 t_2 + t_2^2)\}$. Find $\frac{\partial^2 M}{\partial t_1 \partial t_2}$, $t_1 \frac{\partial M}{\partial t_1} + t_2 \frac{\partial M}{\partial t_2}$ and get relationship between these quantities. Identify the coefficients.]

10.30 (X, Y) is a discrete random variable with $pmf\ P(X = i, Y = j) = p_{ij}, i, j = 0, 1$. Find the mgf of Y, mgf of (X, Y). Show that X, Y are independent *iff*

$$p_{00}p_{11} - p_{01}p_{10} = 0.$$

10.31 Find the mgf of

(i) $f(x, y) = x + y, 0 < x, y > 1$,
(ii) $f(x, y) = e^{-(x+y)}, 0 < x, y < \infty$.

10.32 Determine the mgf of the random variable with pdf

$$f(x) = \begin{cases} k_1 & \text{for } a \leq x \leq a + c_1 \\ k_2 & \text{for } b \leq x \leq b + c_2 \\ 0 & \text{otherwise,} \end{cases}$$

where a, b, c_1, c_2, k_1, k_2 are constants such that

$$c_1 > 0, c_2 > 0, k_1 > 0, k_2 > 0, a + c_1 \leq b, k_1 c_1 + k_2 c_2 = 1.$$

10.33 If $f_{X,Y}(x, y) = \frac{1}{8}(6 - x - y), 0 \leq x \leq 2, 2 \leq y \leq 4$, calculate (i) $P(X + Y < 3)$, (ii) $P(X < \frac{3}{2}, Y < \frac{5}{2})$. $[\frac{3}{8}, \frac{9}{32}]$

10.34 A random variable X has an exponential density $f(x) = \lambda e^{-\lambda x}, x > 0, \lambda > 0$. Find (a) $E(Y), V(Y)$ where $Y = e^{-X}$; (b) $E(Z), V(Z)$ when $Z = e^X$ and examine the conditions for their existence.

[(a) $\frac{\lambda}{\lambda+1}$, $\frac{\lambda}{(\lambda+1)(\lambda+2)}$ (b) $\frac{\lambda}{\lambda-1}$ (provided $\lambda > 1$), $\frac{\lambda}{\lambda-2} - (\frac{\lambda}{\lambda-1})^2$ (provided $\lambda > 2$).]

10.35 X has a Poisson distribution with parameter λ. Conditional distribution of Y given $X = x$ is binomial with parameter (x, p). Show that the marginal distribution of Y is Poisson with parameter λp.

$[f(x, y) = h(x)g(y|x) = \{e^{-\lambda}\frac{\lambda^x}{x!}\}\{\binom{x}{y}p^y q^{x-y}\}$, $x = 0, 1, 2, \ldots$; $y = 0, 1, \ldots, x$; $\lambda > 0, p > 0$.

Let $x - y = z, z = 0, 1, 2, \ldots$

$$f(x, y) = e^{-\lambda}\frac{\lambda^{y+z}}{(y+z)!}\binom{y+z}{y}p^y q^z.$$

Marginal distribution of y is

$$g(y) = \sum_{x=0}^{\infty} f(z, y) = \frac{(\lambda p)^y}{y!}e^{-\lambda}\sum_{j=0}^{\infty}\frac{(\lambda q)^j}{j!} = e^{-\lambda p}\frac{(\lambda p)^y}{y!}, \quad y = 0, 1, \ldots]$$

10.36 $X_1, \ldots, X_k$ are k independent Poisson variables with parameters $\lambda_1, \ldots, \lambda_k$ respectively. Show that the conditional distribution of $X_1, \ldots, X_k$ given $\sum_{i=1}^{k} X_i = x$ is multinomial.

$$P(X_1 = x_1, \ldots, X_k = x_k | \sum_{i=1}^{k} X_i = x)$$

$$= \frac{P(X_1 = x_1, \ldots, X_k = x_k; \sum_{i=1}^{k} X_i = x)}{P(\sum_{i=1}^{k} X_i = x)}$$

$$= \frac{(e^{-\lambda_1}\frac{\lambda_1^{x_1}}{x_1!}) \ldots (e^{-\lambda_k}\frac{\lambda_k^{x_k}}{x_k!})}{e^{\lambda}\frac{\lambda^x}{x!}}, \quad \lambda = \sum_{i=1}^{k}\lambda_i$$

(by the reproductive property of the Poisson distribution)

$$= \frac{x!}{x_1! \ldots x_k!}(\frac{\lambda_1}{\lambda})^{x_1} \ldots (\frac{\lambda_k}{\lambda})^{x_k}$$

which is multinomial with parameters x and $\frac{\lambda_i}{\lambda}, i = 1, \ldots, k - 1$.

10.37 Suppose the conditional density (*pmf*)

$$f_{X|Y}(x|y) = e^{-y}\frac{y^x}{x!}, \quad x = 0, 1, 2, \ldots,$$

a Poisson with parameter y. The random variable Y (with a particular value y) is continuous with *pdf* $f_Y(y) = e^{-y}, y > 0$. Show that the marginal distribution of X is $f_X(x) = 2^{-(x+1)}, x = 0, 1, \ldots$

10.38 X, Y are jointly distributed discrete random variables with

$$p(x, y) = \frac{2}{n(n+1)}(0), \quad x = 1, \ldots, n; \ y = 1, \ldots, x \quad \text{(Otherwise)}.$$

Show that X, Y are not independent. Find regression of Y on X and of X on Y.

[marginal densities: $f_X(x) = \frac{2x}{n(n+1)}, \ f_Y(y) = \frac{2}{n+1}$.

$$E(X|Y = y) = \frac{n+1}{2}, \quad E(Y|X = x) = \frac{x+1}{2}.$$

Hence, regression of Y on X is $y = \frac{x+1}{2}$.]

10.39 X, Y have joint *pdf* $f(x, y) = 1, 0 < x < 1, |y| < x$. Show that on the set of positive probability density, the graph of $E(Y|X = x)$ is a straight line, while that of $E(X|Y = y)$ is not a straight line.

10.40 Show that if (X, Y) have a bivariate normal distribution, the angle θ between the regression lines (of Y on X and of X on Y) is

$$\tan^{-1}\left(\frac{1 - \rho^2}{\rho} \frac{\sigma_X \sigma_Y}{\sigma_X^2 + \sigma_Y^2}\right).$$

What can be concluded from this result about the relative location of the two lines of regression for ρ close to 0 and for ρ close to 1?

10.41 (X, Y) have a *pdf* $f(x, y) = K.e^{-(ax+by)}, 0 \le y \le x+1, a(> 0), b(> 0)$ are given constants. Find the normalizing constant K. Find the regression curve of Y on X. Sketch this regression curve for $a = 2, b = 1$.

10.42 Let $X_i = 1(0)$ if the ith trial is a success (failure) and $\mathbf{X} = (X_1, \ldots, X_n)$. Consider random Bernoulli trials where the probability of success itself is a random variable T having a Beta distribution on $(0, 1)$ with parameters (α_1, α_2). Show that the joint density of $(T, \mathbf{X})$ is

$$B(\alpha_1, \alpha_2)^{-1} t^{\alpha_1 + \sum_{i=1}^n x_i - 1}(1 - t)^{n + \alpha_2 - \sum_{i=1}^n x_i - 1}, \ t \in (0, 1); x_i \in (0, 1),$$

$i = 1, \ldots, n$.Integrating on t, obtain the joint density of $\mathbf{X}$ as

$$B(\alpha_1, \alpha_2)^{-1} B\left(\alpha_1 + \sum_{i=1}^n x_i, \alpha_2 + n - \sum_{i=1}^n x_i\right).$$

Therefore, prove that the conditional distribution of T given $\mathbf{X} = \mathbf{x} = (x_1, \ldots, x_n)$ is the Beta distribution with parameters $(\alpha_1 + \sum_{i=1}^{n} x_i, \alpha_2 + n - \sum_{i=1}^{n} x_i)$.

10.43 X, Y, Z have the distribution
$$dF = A x^{t-1} y^{m-1} z^{n-1}, x, y, z \geq 0, x + y + z \leq 1.$$
Show that
$$A = \frac{\Gamma(l + m + n + 1)}{\Gamma(l)\Gamma(m)\Gamma(n)}.$$

10.44 Let X, Y be two discrete random variables such that
$$P[(X,Y) = (1,1)] = P[(X,Y) = (1,-1)] = \frac{1}{3},$$
$$P[(X,Y) = (-1,1)] = P[(X,Y) = (-1,-1)] = \frac{1}{6}.$$
Find the characteristic function of (X,Y).

10.45 Find the *pgf* of the multinomial distribution $(X_1, \ldots, X_n)$. Show that X_i follows a Binomial distribution with parameters (n, p_i). Also, show that the variance of $X_1 + \ldots + X_n$ is not equal to the sum of the variances of the X_i's, as would be the case if the X_i's were independent.

10.46 Consider the function
$$F(x,y) = \begin{cases} 0 & \text{if } x < 0 \text{ or } x + y < 1 \text{ or } y < 0, \\ 1 & \text{otherwise.} \end{cases}$$
Hence, observe that F is non-decreasing in each argument, but
$$F(1,1) - F(1, \frac{1}{3}) - F(\frac{1}{3}, 1) + F(\frac{1}{3}, \frac{1}{3}) = -1$$
and hence (10.2.7) does not hold.

10.47 (X,Y) have joint *pdf* $f(x,y) = \frac{1}{2}xy, 0 < y < x, 0 < x < 2$. Show that X, Y are not stochastically independent.

10.48 Consider the multinomial distribution (10.8.1). Let $X_{j1}, \ldots, X_{js}$ be $s(< t)$ random variables. Show that
$$E[X_i | \cap_{t=1}^{s} (X_{jt} = x_{jt})] = (N - \sum_{t=1}^{s} x_{jt}) \frac{p_i}{1 - \sum_{t=1}^{s} p_{jt}}$$
and
$$\text{Var}\,[X_i | \cap_{t=1}^{s} (X_{jt} = x_{jt})] = (N - \sum_{t=1}^{s} x_{jt}) \frac{p_i}{1 - \sum_{t=1}^{s} p_{jt}} \{1 - \frac{p_i}{1 - \sum_{t=1}^{s} p_{jt}}\}.$$
Hence, show that the multiple regression of X_i on $X_{j1}, \ldots, X_{jt}$ is linear but is not homoscedastic.

Chapter 11

Probability Distribution of Functions of Random Variables

11.1 Introduction

In this chapter we consider the problem of finding the joint probability distribution of several functions $Y_1 = g_1(X_1, \ldots, X_n), \ldots, Y_n = g_n(X_1, \ldots, X_n)$ of random variables $(X_1, \ldots, X_n)$, starting from the joint distribution of $(X_1, \ldots, X_n)$. Section 11.2 considers the function $Y = g(X)$ of a single random variable X. The probability integral transformation is the subject matter of the next section. Section 11.4 considers functions $Y_1 = g_1(X_1, X_2), Y_2 = g_2(X_1, X_2)$ of two random variables (X_1, X_2) - first dealing with some simple univariate functions, $X + Y, X - Y, XY, X/Y$ (Subsection 11.4.1) and then considering the technique of transformation of variables (Subsection 11.4.2). Subsequently, functions of n random variables are dealt with. Section 11.6 considers the distribution of maximum and minimum of several random variables. Since the moment generating function, when it exists, uniquely determines the distribution of random variables, such functions can be used to find the joint distribution of $Y_1, \ldots, Y_n$, starting from the distribution of $(X_1, \ldots, X_n)$. The technique has been discussed in Section 11.7.

11.2 Functions of One Random Variable

Let X be a random variable and $g(X)$ a Borel measurable function of X. Then $Y = g(X)$ is a random variable. We want to find the distribution function and the *pdf* of Y. Now

$$F_Y(y) = P[Y \leq y] = P[X \in I_y] = \int_{I_y} dF_X(x)$$

where $I_y = \{x : g(x) \le y\}$ and $X \in C$ denotes the event $\{\omega : X(\omega) \in C\}$. Note that $x \in I_y$ *iff* $g(x) \le y$.

If X is a discrete random variable

$$P[Y = y] = P[g(X) = y] = \sum_x P[X(\omega) = x : g(x) = y]$$

provided g has a finite number of inverses for each y.

EXAMPLE 11.2.1: $Y = aX + b$.

(i) If $a > 0$, $ax + b \le y$ if and only if $x \le \frac{y-b}{a}$;

(ii) If $a < 0$, if and only if $x \ge \frac{y-b}{a}$.

Hence, in case (i),

$$F_Y(y) = F_X(\frac{y-b}{a});$$

in case (ii),

$$F_Y(y) = 1 - F_X(\frac{y-b}{a} - 0).$$

If $a = 0$,

$$F_Y(y) = \begin{cases} 0, y < b \\ 1, y \ge b. \end{cases}$$

EXAMPLE 11.2.2: $Y = X^2$.

$F_Y(y) = 0$ for $y < 0$;

$F_Y(y) = P(X^2 \le y) = P(-\sqrt{y} \le X \le \sqrt{y}) = F_X(\sqrt{y}) - F_X(-\sqrt{y} - 0)$, $y \ge 0$.

If X is discrete, *pmf* of Y is

$$f_Y(y) = f_X(-\sqrt{y}) + f_X(\sqrt{y}).$$

In particular, if X is a standard normal variable,

$$F_Y(y) = \Phi(\sqrt{y}) - \Phi(-\sqrt{y}).$$

EXAMPLE 11.2.3: Let X be a $N(\mu, \sigma^2)$-variable and $Y = g(X) = |X - \mu|$.

$$F_Y(y) = P(|X - \mu| \le y) = P(\mu - y \le X \le \mu + y)$$
$$= \frac{1}{\sigma\sqrt{2\pi}} \int_{\mu-y}^{\mu+y} e^{\frac{-(x-\mu)^2}{2\sigma^2}} dx$$
$$= \frac{2}{\sigma\sqrt{2\pi}} \int_0^y e^{-\frac{t^2}{2\sigma^2}} dt, \ y > 0.$$

Hence,

$$f_Y(y) = \begin{cases} 0, & y \le 0 \\ \frac{\sqrt{2}}{\sigma\sqrt{\pi}}e^{-\frac{y^2}{2\sigma^2}}, & y > 0. \end{cases}$$

EXAMPLE 11.2.4: Let $Y = g(X) = aX^2 + bX + c$, $a \ne 0$.

(i) Let $y > 0$, $g(x) < y$ iff $ax^2 + bx + (c - y) \le 0$, i.e. *iff* $\alpha \le x \le \beta$ where

$$a = \frac{-b - \sqrt{b^2 - 4a(c - y)}}{2a}, \quad \beta = \frac{-b + \sqrt{b^2 - 4a(c - y)}}{2a}.$$

Hence, $F_Y(y) = P(\alpha \le X \le \beta) = F_X(\alpha) - F_X(\beta - 0)$.

(ii) Similarly, if $\alpha < 0$,

$$F_Y(y) = 1 - F_X(\alpha - 0) + F_X(\beta).$$

Theorem 11.2.1: Let X be a random variable of continuous type with *pdf* $f_X(x)$. Let $g(x)$ be a continuous function of x and the equation $y = g(x)$ have a finite number of solutions $x_1, \ldots, x_k$ for every y. Assume that $F_X(x)$ is differentiable at all these points. Then

$$f_Y(y) = \sum_{i=1}^{k} \frac{f_X(x_i)}{|g'(x_i)|} \tag{11.2.1}$$

where $g'(x_i) = \frac{dg(x)}{dx}\Big]_{x_i}$.

If, in particular, there is one solution of the equation $y = g(x)$, $f_Y(x) = \frac{f_X(x)}{|g'(x)|}$.

Proof. Since $y = g(x)$ whenever $x = x_1, \ldots, x_k$,

$$\begin{aligned} f_Y(y)dy &= P[y < Y < y + dy] \\ &= P\{x_1 < X < x_1 + dx_1\} + \ldots + P\{x_n < X < x_n + dx_n\} \\ &= f(x_1)|dx_1| + \ldots + f(x_n)|dx_n|, \end{aligned}$$

$$\tag{11.2.2}$$

where in some of these terms in (11.2.2), where y is a decreasing function of x around x_i, dx_i will be negative (and the inequality will be $x_i + dx_i < X < x_i$).

Now

$$g'(x_i)dx_i = dy, \quad i = 1, \ldots, k.$$

Hence

$$f_Y(y) = \sum_{i=1}^{k} \frac{f_X(x_i)}{|g'(x_i)|}.$$

Note: The theorem is valid if the equation $y = g(x)$ has only a countable number of solutions $x_1, x_2, \ldots$

EXAMPLE 11.2.5 : If $Y = aX + b, a \neq 0, f_Y(y) = \frac{1}{|a|} f_X(\frac{y-b}{a})$.

In particular, if X is $N(\mu, \sigma^2)$-variable and $Y = a + bX, b \neq 0$,

$$f_Y(y) = \frac{1}{|b|\sigma\sqrt{2\pi}} \exp[-\frac{1}{2\sigma^2}\{\frac{1}{b}(y-a) - \mu\}^2]$$
$$= \frac{1}{|b|\sigma\sqrt{2\pi}} \exp[\frac{1}{2b^2\sigma^2}\{y - (a + b\mu)\}^2], \quad -\infty < y < \infty.$$

Hence, Y is normally distributed with mean $a + b\mu$ and variance $b^2\sigma^2$.

EXAMPLE 11.2.6 : Let $g(x) = |x|$. Here the equation $y = g(x)$ has no solution if $y < 0$ and has two solutions

$$x_1 = x, \ x_2 = -x \ \text{ for } y > 0.$$

Hence,

$$f_Y(y) = f_X(x) + f_X(-x), y > 0$$
$$f_Y \quad = f_X(0).$$

Again,

$$f_Y(y) = \begin{cases} P\{|X| \le y\} = F_X(y) - F_X(-y - 0), & y \ge 0 \\ 0, & y < 0. \end{cases}$$

EXAMPLE 11.2.7: $g(x) = x(0)$ if $x > 0$ (otherwise).

If $y > 0$,

$$P[Y \le y] = P[g(X) \le y] = P[g(X) = 0] + P[0 < g(X) \le y]$$
$$= P[X \le 0] + P[0 < X \le y] = F_X(y).$$

If $y < 0$, $F_Y(y) = 0$.

If $y = 0$, $F_Y(0) = P(Y = 0) = P(X \le 0) = F_X(0)$. (Fig. 11.2.1). Thus, $F_Y(y)$ is discontinuous with discontinuity at $y = 0$,

$$f_Y(y) = \begin{cases} f_X(y) & \text{for } y > 0 \\ f_X(0) & \text{for } y = 0 \\ 0 & \text{for } y < 0. \end{cases}$$

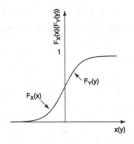

Fig. 11.2.1

EXAMPLE 11.2.8: $y = a\cos(x + \theta), a > 0$.

If $|y| < a$, the equation $y = a\cos(x + \theta)$ has infinitely many solutions $x_n, n = \ldots, -1, 0, 1, \ldots,$

$$|g'(x_n)| = a\sin(x_n + \theta) = \sqrt{a^2 - y^2}.$$

Hence

$$f_Y(y) = \begin{cases} \frac{1}{\sqrt{a^2 - y^2}} \sum_{n=-\infty}^{\infty} f_X(x_n) & |y| < a \\ 0, & |y| > a. \end{cases} \qquad (11.2.3)$$

Note that $f_Y(+_-a) = \infty$.

In particular, if X is uniformly distributed over $(-\pi, \pi)$, $f_X(x) = \frac{1}{2\pi}(0)$ for $|x| \leq \pi$ (otherwise). The equation $y = a\cos(x + \theta)$ has just two solutions over $(-\pi, \pi)$, $f_X(x) = \frac{1}{2\pi}$ at both these solutions. Hence, from (11.2.3) we obtain,

$$f_Y(y) = \frac{2}{\sqrt{a^2 - y^2}} \cdot \frac{1}{2\pi} \text{ for } |y| < a. \qquad (11.2.4)$$

The distribution function obtained by integrating (11.2.4) is

$$F_Y(y) = \begin{cases} 0, & y < -a \\ \frac{1}{\pi}\arccos\frac{y}{a} - 1, & |y| < a \\ 1, & y > a. \end{cases}$$

Note that $f_Y(+_-a) = \infty$; however $P(Y = +_-a) = 0$.

EXAMPLE 11.2.9: $Y = a\tan X$, $a > 0$.

The equation $y = a\tan x$ has infinitely many solutions

$$\begin{aligned} x_n &= \arctan\frac{y}{a}, \; n = \ldots, -1, 0, 1, \ldots \\ g'(x) &= \frac{a}{\cos^2 x} = \frac{a^2 + y^2}{a} \\ f_Y(y) &= \frac{a}{a^2 + y^2} \sum_{n=-\infty}^{\infty} f_X(x_n). \end{aligned}$$

If in particular, X is uniformly distributed over $(-\infty, \infty)$ and $f_Y(y) = \frac{2a}{a^2 + y^2} \cdot \frac{1}{2\pi} = \frac{a}{\pi(a^2 + y^2)}$, a Cauchy distribution.

EXAMPLE 11.2.10:

$$\begin{aligned} g(x) &= k.e^{-\beta x}, x \geq 0, \beta > 0, k > 0 \\ &= 0, \; x < 0 \quad \text{(Fig. 11.2.2)} \end{aligned}$$

The equation $y = g(x)$ has no solution for $y < 0$ and $y > k$. For $0 < y < k$,

$$\begin{aligned} x &= -\frac{1}{\beta}\ln\frac{y}{ak} \\ g'(x) &= -\beta y \\ f_Y(y) &= \begin{cases} \frac{1}{\beta y} f_X(-\frac{1}{\beta}\ln\frac{y}{k}), & 0 < y < k \\ 0, & \text{otherwise.} \end{cases} \end{aligned}$$

For $0 < y < k, g(x) \leq y$ for $x \leq 0$ and $x > -\frac{1}{\beta} \log(\frac{y}{k})$.

$F_Y(y) = P[X \leq 0] + P[X > -\frac{1}{\beta} \log \frac{y}{k}] = F_X(0) + 1 - F_X(-\frac{1}{\beta} \log \frac{y}{k})$, $0 < y < k$.

In particular, if X has a rectangular distribution on (a, b), $f_X(x) = (b - a)^{-1}$, $a < x < b$,

$$f_Y(y) = \frac{1}{\beta y}, \quad ke^{-b\beta} < y < ke^{-a\beta}.$$

and is zero elsewhere.

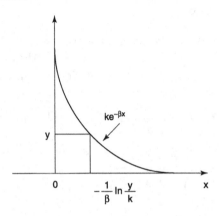

Fig. 11.2.2

EXAMPLE 11.2.11: X is a $N(\mu, \sigma^2)$ variable and $Y = bX^2$, $b > 0$.

$$f_Y(y) = \frac{1}{2a\sqrt{\frac{y}{b}}} \left[f_X(\sqrt{\tfrac{a}{b}}) + f_X(-\sqrt{\tfrac{y}{b}}) \right]$$

$$= \frac{1}{2\sigma\sqrt{2\pi by}} \left[\exp\{-\tfrac{1}{2\sigma^2}(\sqrt{\tfrac{y}{b}} - \mu)^2\} + \exp\{-\tfrac{1}{2\sigma^2}(-\sqrt{\tfrac{y}{b}} - \mu)^2\} \right]$$

$$= \frac{1}{2\sigma\sqrt{2\pi by}} e^{-\frac{1}{2\sigma^2}(\mu^2 + \frac{y}{b})} \left[e^{\frac{\mu}{\sigma^2}\sqrt{\frac{y}{b}}} + e^{-\frac{\mu}{\sigma^2}\sqrt{\frac{y}{b}}} \right]$$

$$= \frac{e^{-\frac{1}{2\sigma^2}(\mu^2 + \frac{y}{b})}}{\sigma\sqrt{2\pi b}} y^{-\frac{1}{2}} \cosh[\tfrac{\mu}{\sigma^2}\sqrt{\tfrac{y}{b}}], \quad 0 < y < \infty.$$

In particular, if $\mu = 0$,

$$f_Y(y) = \frac{y^{-\frac{1}{2}} e^{-\frac{y}{2b\sigma^2}}}{\Gamma(\frac{1}{2})\sqrt{2b\sigma^2}}, \quad y > 0$$

so that Y has a gamma distribution with parameters $(\frac{1}{2b\sigma^2}, \frac{1}{2})$. If $b = 1, \sigma^2 = 1, Y$ follows a χ^2-distribution with one degree of freedom. If $\mu \neq 0, b = 1, \sigma^2 = 1$, the resulting distribution (11.2.5) is called a non-central χ^2-distribution with one d.f.

11.3 Probability Integral Transformation

Consider a random variable X with distribution function $F_X(x), 0 \leq F_X(x) \leq 1$. We define an inverse function F_X^{-1} as $0 \leq y \leq 1$ by setting $F_X^{-1}(y)$ equal to the smallest value of x for which $F_X(x) \geq y$.

Theorem 11.3.1: If X is a random variable with continuous distribution function $F_X(x)$, then the random variable $Y = F_X(y)$ has a continuous rectangular distribution over $[0, 1]$.

Proof. The distribution function of Y is $F_Y(y) = P[Y \leq y] = P[F_X(x) \leq y] = P[X \leq F_X^{-1}(y)] = F_X(F_X^{-1}(y)) = y, 0 \leq y \leq 1$. Hence, Y follows a uniform distribution over $[0, 1]$.

Conversely, suppose that Y has a uniform distribution over $[0, 1]$. The distribution function of $X, P[X \leq x] = P[F_X^{-1}(Y) \leq x] = P[Y \leq F_X(x)] = F_X(x)$, so that X has distribution function $F_X(x)$.

The transformation $Y = F_X(x)$ is called the *probability integral transformation*.

The Theorem 11.3.1 is of fundamental importance and states that every continuous probability distribution can be transformed into a normalized uniform distribution. Theoretically, therefore, every continuous distribution can be transformed into every other distribution via the above-mentioned transformation. The transformation $Y = F_X(x)$ is very useful in drawing a random sample from the distribution of X. Since $0 \leq Y \leq 1$, if one draws a random number y between $[0, 1]$, and calculates $x = F_X^{-1}(y)$, then x is a random sample from the distribution of X.

EXAMPLE 11.3.1: $X_1, \ldots, X_n$ are independent continuous uniform variables,

$$dF = dx, \ 0 \leq x \leq 1.$$

Show that

$$Z = -2 \sum_{i=1}^{n} \log_e X_i \text{ has a } \chi^2_{(2n)} \text{ distribution.}$$

Consider the transformed variable

$$Y = -2 \log_e X; \ 0 < Y < \infty.$$

Hence

$$x = e^{-\frac{y}{2}}, \ dx = -\frac{1}{2} e^{-\frac{y}{2}} dy.$$

The *pdf* of Y is

$$f_Y(y) = \frac{1}{2}e^{-\frac{y}{2}}, \ 0 < y < \infty$$

so that Y follows a $\chi^2_{(2)}$ distribution. Hence $Z = \sum_{i=1}^{n} Y_i$ follows a $\chi^2_{(2n)}$-distribution by the reproductive property of this distribution.

EXAMPLE 11.3.2: A variable X has a uniform distribution over $[0, 1]$. Find a function of X, say, Y, with probability element

$$f_Y(y)dy = e^{-y}dy, \ 0 \le y \le \infty.$$

It is known that the distribution function of Y follows a uniform distribution over $[0, 1]$. Now,

$$F_Y(y) = \int_0^y e^{-t}dt = 1 - e^{-y}.$$

Hence, $1 - e^{-Y}$ is uniform over $[0, 1]$.

Therefore, $X = 1 - e^{-Y}$ with probability one, when $Y = \log_e(\frac{1}{1-X})$.

EXERCISE 11.3.3: $X_1, \ldots, X_n$ are random samples from a uniform distribution over $(0, 1)$. Find the *pdf* of $u = (X_1 \ldots X_n)^{1/n}$.

It is known that $Z = -2 \sum_{i=1}^{n} \log_e X_i$ has a χ^2 distribution with $2n$ degrees of freedom,

$$f_Z(z) = \frac{1}{2^n \Gamma(n)} e^{-\frac{1}{2}z} z^{n-1} dz, \ 0 \le z < \infty.$$

Make the transformations

$$u = e^{-\frac{z}{2n}}, \ z = 2n \log_e \left(\frac{1}{u}\right)$$

$$\frac{du}{dz} = -\frac{1}{2n} e^{-\frac{z}{2n}} = -\frac{u}{2n}, \ 0 < u < 1.$$

$$f_U(u)du = \frac{1}{2^n \Gamma(n)} e^{-\frac{1}{2}[2n \log_e \frac{1}{u}]} (2n \log_e \frac{1}{u})^{n-1} \frac{2n}{u} du$$

$$= \frac{n^n u^{n-1}}{\Gamma(n)} (\log_e \frac{1}{u})^{n-1} du, \ 0 < u < 1.$$

11.4 Functions of Two Random Variables

11.4.1 *Using distribution functions*

Let (X, Y) be two jointly continuous random variables with *pdf* $f_{X,Y}(x, y)$. We first consider the problem of finding probability distributions of some simple functions, $U = X + Y, V = X - Y, W = XY, Z = X/Y$ by using the procedure of finding the marginal distributions.

Theorem 11.4.1: Let (X, Y) be jointly distributed continuous random variables with *pdf* $f_{X,Y}(x, y)$. The *pdf* of $U = X + Y, V = X - Y$ are respectively given by

$$(i)\, f_U(u) = \int_{-\infty}^{\infty} f_{X,Y}(x, u - x)dx = \int_{-\infty}^{\infty} f_{X,Y}(u - y, y)dy, \qquad (11.4.1)$$

$$(ii)\, f_V(v) = \int_{-\infty}^{\infty} f_{X,Y}(x, x - v)dx = \int_{-\infty}^{\infty} f_{X,Y}(v + y, y)dy. \qquad (11.4.2)$$

Proof. The distribution function of $U = X + Y$ is

$$F_U(u) = P(X + Y \le u)$$
$$= \int\int_{x+y \le u} f_{X,Y}(x, y)dxdy = \int_{-\infty}^{\infty} \left[\int_{-\infty}^{u-x} f_{X,Y}(x, y)dy\right] dx$$
$$= \int_{-\infty}^{\infty} \left[\int_{-\infty}^{u} f_{X,Y}(x, z - x)dz\right] dx \quad \text{(Putting } y = z - x)$$
$$= \int_{-\infty}^{u} \left[\int_{-\infty}^{\infty} f_{X,Y}(x, z - x)dx\right] dz.$$
$$(11.4.3)$$

Differentiating both sides we get the *pdf* of U. The change from a double integral to a repeated integral, as well as the change of order of integration are justified since $f(x, y)$ is absolutely integrable over the xy-plane. The second expression in (11.4.1) can be derived similarly. (11.4.2) can be proved in an analogous manner.

Corollary 11.4.1.1 If X, Y are independent continuous random variables, the *pdf* of $U = X + Y$ is

$$f_U(u) = \int_{-\infty}^{\infty} f_X(x)f_Y(u - x)dx = \int_{-\infty}^{\infty} f_X(u - y)f_Y(y)dy. \qquad (11.4.4)$$

The density $f_U(.)$ in (11.4.4) is called the *convolution* of the densities $f_X(.), f_Y(.)$.

EXAMPLE 11.4.1: X, Y are independent uniform random variables over $[0, 1]$. Find the *pdf* of (i) $U = X + Y$, (ii) $V = X - Y$, (iii) $W = |X - Y|$.

$$(i) f_U(u) = \int_{-\infty}^{\infty} f_X(x) f_Y(u - x) dx, \ 0 \le u \le 2.$$

When $0 \le u \le 1$, $0 \le x \le u$,

$$f_U(u) = \int_0^u dx = u.$$

When $1 \le u \le 2$, $u - 1 \le x \le 1$,

$$f_U(u) = \int_{u-1}^1 dx = 2 - u.$$

Hence,

$$f_U(u) = \begin{cases} u, & 0 \le u \le 1 \\ 2 - u, & 1 \le u \le 2. \end{cases}$$

This is known as the *triangular distribution*.

$$(ii) f_V(v) = \int_{-\infty}^{\infty} f_X(x) f_Y(x - y) dx, \ -1 \le v \le 1.$$

When $-1 \le v \le 0$, $0 \le x \le 1 + v$,

$$f_V(v) = \int_v^{1+v} dx = 1 + v.$$

When $0 \le v \le 1, v \le x \le 1$,

$$f_V(v) = \int_0^1 dx = 1 - v.$$

Hence,

$$f_V(v) = \begin{cases} 1 + v, & -1 \le v \le 0 \\ 1 - v, & 0 \le v \le 1, \end{cases}$$

as in Fig. 11.4.1.

(iii) Let $w = |x - y|; 0 \le w \le 1$. Solving the equations $w = |v|, w = +v$ or $-v, |\frac{dw}{du}| = 1$ at each solution.

Hence, from (i),

$$f_W(w) = 2(1 - w), \ 0 \le w \le 1,$$

as in Fig. 11.4.2.

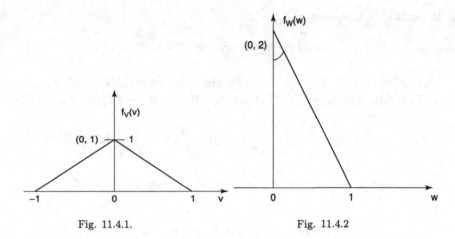

Fig. 11.4.1.   Fig. 11.4.2

Theorem 11.4.2: Let X, Y be jointly distributed continuous random variables with *pdf* $f_{X,Y}(x, y)$. The *pdf* of $W = XY$ and $Z = \frac{X}{Y}$ are, respectively,

$$(i)\, f_W(w) = \int_{-\infty}^{\infty} \frac{1}{|x|} f_{X,Y}\left(x, \frac{w}{x}\right) dx = \int_{-\infty}^{\infty} \frac{1}{|y|} f_{X,Y}\left(\frac{w}{y}, y\right) dy, \qquad (11.4.5)$$

$$(ii)\, f_Z(z) = \int_{-\infty}^{\infty} |y| f_{X,Y}(yz, y) dy, \quad \text{provided } P(Y = 0) = 0. \qquad (11.4.6)$$

In particular, if $f_{X,Y}(x, y) = f_{X,Y}(-x, -y)$, (11.4.6) reduces to

$$f_Z(z) = 2 \int_0^{\infty} y f_{X,Y}(yz, y) dy. \qquad (11.4.7)$$

EXAMPLE 11.4.2: (X, Y) follows bivariate normal distribution with parameters $(0, 0, \sigma_1^2, \sigma_2^2, \rho)$. Find the *pdf* of $Z = \frac{X}{Y}$.

By (11.4.7),

$$f_Z(z) = \frac{1}{\pi \sigma_1 \sigma_2 \sqrt{1-\rho^2}} \int_0^{\infty} y \exp\left[-\frac{y^2}{2(1-\rho^2)}\left\{\frac{z^2}{\sigma_1^2} - \frac{2\rho z}{\sigma_1 \sigma_2} + \frac{1}{\sigma_2^2}\right\}\right] dy$$

$$= \frac{\sqrt{1-\rho^2}}{\pi \sigma_1 \sigma_2 [\frac{z^2}{\sigma_1^2} - \frac{2\rho z}{\sigma_1 \sigma_2} + \frac{1}{\sigma_2^2}]}$$

$$= \frac{\sigma_1 \sigma_2 \sqrt{1-\rho^2}}{\pi [\sigma_2^2 (z - \rho \frac{\sigma_1}{\sigma_2})^2 + \sigma_1^2 (1-\rho^2)]}$$

so that Z is a Cauchy variable with parameters $\lambda = \frac{\sigma_1}{\sigma_2}\sqrt{1 - \rho^2}$ and $\theta = \rho \frac{\sigma_1}{\sigma_2}$.

In particular, if $\sigma_1 = \sigma_2 = 1, \rho = 0$,

$$f_Z(z) = \frac{1}{\pi(z^2+1)}, \quad -\infty < z < \infty.$$

EXAMPLE 11.4.3: X, Y are independent random variables, each having uniform distribution over $(0, 1)$. Find the *pdf* of (i) $W = XY$, (ii) $Z = X/Y$.

$$(i)\, f_W(w) = \int_{-\infty}^{\infty} \frac{1}{|x|} f_X(x) f_Y\left(\frac{w}{x}\right) dx = \int_w^1 \frac{1}{x} dx$$

(since $x = \dfrac{w}{y}$ and hence $w < x < 1$) $= -\log_e w, \ 0 < w < 1.$

$$(ii)\, f_Z(z) = \int_{-\infty}^{\infty} |y| f_X(yz) f_Y(y) dy.$$

When $0 < z < 1, 0 < y < 1$,

$$f_Z(z) = \int_0^1 y\, dy = \frac{1}{z}.$$

When $1 \le z < \infty, 0 < y < \frac{1}{z}$,

$$f_Z(z) = \int_0^{\frac{1}{z}} y\, dy = \frac{1}{2z^2}.$$

Hence,

$$f_Z(z) = \begin{cases} \frac{1}{z}, & 0 < z < 1 \\ \frac{1}{2z^2}, & 1 \le z < \infty. \end{cases} \quad \square$$

We will now consider the problem of finding the distribution of functions of two random variables by using the method of transformation of variables.

11.4.2 *Transformation of variables*

Consider random variables (X_1, X_2) having joint densities $f_{X_1,X_2}(x_1, x_2)$. Let $Y_1 = g_1(X_1, X_2), Y_2 = g_2(X_1, X_2)$ be two functions of (X_1, X_2). We want to find the joint distribution of (Y_1, Y_2).

The joint distribution function of (Y_1, Y_2) is

$$F_{Y_1,Y_2}(y_1, y_2) = P(Y_1 \le y_1,\ Y_2 \le y_2) = \int\int_{\mathcal{R}_{y_1,y_2}} f_{X_1,X_2}(x_1, x_2) dx_1 dx_2$$

where $\mathcal{R}_{y_1,y_2} = \{(x_1, x_2) : Y_1 \le y_1,\ Y_2 \le y_2\}.$

Define

$$J = \begin{vmatrix} \frac{\partial x_1}{\partial y_1} & \frac{\partial x_2}{\partial y_1} \\ \frac{\partial x_1}{\partial y_2} & \frac{\partial x_2}{\partial y_2} \end{vmatrix}$$

as the Jacobian of transformation.

To find the joint *pdf* of (Y_1, Y_2), we solve the simultaneous equations

$$g_1(x_1, x_2) = y_1, \quad g_2(x_1, x_2) = y_2 \tag{11.4.8}$$

giving (x_1, x_2) in terms of (y_1, y_2). Let

$$x_{1i} = g_{1i}^{-1}(y_1, y_2), \quad x_{2i} = g_{2i}^{-1}(y_1, y_2), \quad i = 1, \ldots, m$$

be the m sets of solutions, all real. Consider the Jacobians

$$J_i = \begin{vmatrix} \frac{\partial g_{1i}^{-1}}{\partial y_1} & \frac{\partial g_{2i}^{-1}}{\partial y_1} \\ \frac{\partial g_{1i}^{-1}}{\partial y_2} & \frac{\partial g_{2i}^{-1}}{\partial y_2} \end{vmatrix}.$$

The joint *pdf* of (Y_1, Y_2) is

$$f_{Y_1, Y_2}(y_1, y_2) = \sum_{i=1}^{m} f_{X_1, X_2}(x_{1i}, x_{2i}) |J_i|. \tag{11.4.9}$$

We have assumed that the first partial derivatives of g_{1i}^{-1}, g_{2i}^{-1} are continuous on the domain of (Y_1, Y_2) and $|J_i| \neq 0$ on the domains of $(Y_1, Y_2), i = 1, \ldots, m$.

If for certain (y_1, y_2), the equations (11.4.8) have no real solution $f(y_1, y_2) = 0$. The proof of this result is omitted.

If $(x_{1i}, x_{2i}) = (x_1, x_2) \; \forall \; i = 1, \ldots, m$,

$$f_{Y_1, Y_2}(y_1, y_2) = f(x_1, x_2) m |J|.$$

If only distribution of Y_1 is desired, a variable $Y_2 = g_2(X_1, X_2)$ is suitably chosen, joint distribution of (Y_1, Y_2) is obtained and distribution of Y_1 is found out by integrating out Y_2.

EXAMPLE 11.4.4: X_1, X_2 are independent gamma variables with parameters $(\alpha, p_1), (\alpha, p_2)$ respectively. Show that $U = X_1 + X_2, V = X_1/(X_1 + X_2)$ are independently distributed, U following a gamma distribution with parameters $(\alpha, p_1 + p_2), V$ a beta distribution with parameters (p_1, p_2). Also find the distribution of $W = \frac{X_1}{X_2}$.

The probability element of (X_1, X_2) is

$$dF = \frac{\alpha^{p_1 + p_2}}{\Gamma(p_1)\Gamma(p_2)} e^{-\alpha(x_1 + x_2)} x_1^{p_1 - 1} x_2^{p_2 - 1} dx_1 dx_2, \; 0 < x_1, x_2 < \infty.$$

Use transformations

$$u = x_1 + x_2, \quad v = \frac{x_1}{x_1 + x_2},$$

i.e.

$$x_1 = uv, \quad 0 \le u < \infty,$$

$$x_2 = u(1 - v) \quad 0 \le v < 1.$$

Here $|J| = u$.

Joint probability element of (U, V) is

$$dF = \frac{\alpha^{p_1+p_2}}{\Gamma(p_1)\Gamma(p_2)} e^{-\alpha u} (uv)^{p_1-1} [u(1-v)]^{p_2-1} u \, du \, dv$$

$$= \{\frac{\alpha^{p_1+p_2}}{\Gamma(p_1+p_2)} u^{p_1+p_2-1} e^{-uv} du\} \{\frac{1}{B(p_1,p_2)} v^{p_1-1}(1-v)^{p_2-1} dv\}$$

$$= f_U(u) f_V(v).$$

Again, $w = \frac{v}{1-v}$; hence, $v = \frac{w}{1+w}$; $dv = \frac{dw}{(1+w)^2}$.

The *pdf* of W is obtained from the *pdf* of V as

$$f_W(w) = \frac{1}{B(p_1,p_2)} [\frac{w}{1+w}]^{p_1-1} \frac{1}{(1+w)^{p_2+1}}$$

$$= \frac{1}{B(p_1,p_2)} \frac{w^{p_1-1}}{(1+w)^{p_1+p_2}}, \quad 0 \le w < \infty$$

so that W follows a beta distribution of the second kind with parameters p_1, p_2.

[Alternatively, put $X_1 = r\cos^2\theta, X_2 = r\sin^2\theta, |J| = 2r\cos\theta\sin\theta, X_1 + X_2 = r, \frac{X_1}{(X_1+X_2)} = \cos^2\theta$; show that r and $\cos^2\theta$ are independently distributed.]

EXAMPLE 11.4.5: Let X, Y be independent beta variables with parameters $(a, b), (c, d)$ respectively such that $a = c + d$, Show that XY is beta variable with parameter $(c, b + d)$.

$$f_{X,Y}(x, y) = k.x^{a-1}(1-x)^{b-1}y^{k-1}(1-y)^{d-1} \text{ where } k = [B(a,b)B(c,d)]^{-1}.$$

Let $u = x, z = xy$; hence $x = u, y = \frac{x}{u}, 0 < u, z < 1, |J| = \frac{1}{|u|}$.

The joint density of U and Z is

$$f_{U,Z}(u, z) = k.u^{a-c-d}(1-u)^{b-1}z^{c-1}(u-z)^{d-1}$$

$$= k(1-u)^{b-1}z^{c-1}(u-z)^{d-1} \text{ (since } a = c+d).$$

The marginal density of Z is

$$f_Z(z) = kz^{c-1} \int_0^1 (1-u)^{b-1}(u-z)^{d-1} du$$

$$= kz^{c-1}(1-z)^{b+d-1} \int_0^1 t^{b-1}(1-t)^{d-1} dt \quad \text{where } t = \frac{1-u}{1-z}$$

$$= \frac{B(b,d)}{B(a,b)B(c,d)} z^{a-1}(1-z)^{c+d-1} = \frac{1}{B(c,b+d)} z^{c-1}(1-z)^{b+d-1}.$$

In general product of Beta variables with parameters $(a_1, b_1), \ldots, (a_k, b_k)$ such that $a_i = a_{i+1} + b_{i+1}, i = 1, \ldots, k-1, \Pi_{i=1}^k X_i$ is distributed as a beta variable with parameters $(a_k, \sum_{i=1}^k b_i)$.

EXAMPLE 11.4.6: Let X, Y be independent, each being a $N(0, \sigma^2)$-variable. Find the joint *pdf* of $Z = \sqrt{X^2 + Y^2}, W = \frac{X}{Y}$.

Solving the equations

$$z = \sqrt{x^2 + y^2}, \quad w = \frac{x}{y},$$

we have two solutions

$$\left(x_1 = -\frac{zw}{\sqrt{1+w^2}}, \quad y_1 = \frac{z}{\sqrt{1+w^2}} \right), \quad (-x_1, -y_1)$$

for any z, w. Here

$$\begin{vmatrix} \frac{\partial z}{\partial x} & \frac{\partial z}{\partial y} \\ \frac{\partial w}{\partial x} & \frac{\partial w}{\partial y} \end{vmatrix} = \begin{vmatrix} \frac{x}{\sqrt{x^2+y^2}} & \frac{y}{\sqrt{x^2+y^2}} \\ \frac{1}{y} & -\frac{x}{y^2} \end{vmatrix} = \frac{1+w^2}{-z}; \quad \text{hence } |J_i| = \frac{|z|}{1+w^2}$$

for each point $(x_i, y_i), i = 1, 2.$

$$f_{X,Y}(x, y) = \frac{1}{2\pi\sigma^2} e^{-\frac{1}{2\sigma^2}(x^2+y^2)}.$$

Since $f(x_1, y_1) = f(x_2, y_2)$

$$f_{Z,W}(z, w) = \frac{2}{2\pi\sigma^2} e^{-\frac{z^2}{2\sigma^2}} \left(\frac{z}{1+w^2} \right) = \left\{ \frac{z}{\sigma^2} e^{\frac{-z^2}{2\sigma^2}} \right\} \left\{ \frac{1}{\pi(1+w^2)} \right\}$$

which shows that Z, W are independently distributed, Z having a Raleigh distribution and W a Cauchy distribution.

Alternative Solution: Consider a polar transformation from (x, y) to (z, θ) : $x = z\cos\theta, y = z\sin\theta, z > 0, 0 < \theta < 2\pi, w = \cot\theta, J = z.$

The joint probability element of (Z, θ) is

$$dF(z, \theta) = \frac{z}{2\pi\sigma^2} e^{-\frac{z^2}{2\sigma^2}} dz d\theta,$$

Hence, *psf* of Z is

$$f_Z(z) = \frac{z}{\sigma^2} e^{-\frac{z^2}{2\sigma^2}};$$

pdf of θ is

$$f(\theta) = \frac{1}{2\pi}, \quad 0 < \theta < 2\pi. \tag{i}$$

For each value of w there are two values of θ in $(0, 2\pi)$ satisfying $\cot \theta = w$.

$$\frac{dw}{d\theta} = \csc^2 \theta = 1 + w^2.$$

Hence, from (i), *pdf* of W is

$$f_W(w) = \frac{1}{\pi(1 + w^2)}, \quad -\infty < w < \infty.$$

11.5   Functions of n Random Variables

Let $X_1, \ldots, X_n$ be jointly continuous random variables with joint *pdf* $f_{X_1, \ldots, X_n}(x_1, \ldots, x_n)$ and let $\mathcal{X} = \{(x_1, \ldots, x_n) : f(x_1, \ldots, x_n) > 0\}$. Let $Y_1 = g_1(X_1, \ldots, X_n), \ldots, Y_n = g_n(X_1, \ldots, X_n)$. Let $\mathcal{Y} = $ domain of $(Y_1, \ldots, Y_n)$.

Suppose $\mathcal{X}$ can be decomposed into m disjoint subsets $\mathcal{X}_1, \ldots, \mathcal{X}_m$ such that $(x_1 \ldots, x_n) \to (y_1, \ldots, y_n)$ is a one-to-one transformation of $\mathcal{X}_i$ onto $\mathcal{Y}$. Let $x_1 = g_{1i}^{-1}(y_1, \ldots, y_n), \ldots, x_n = g_{ni}^{-1}(y_1, \ldots, y_n)$ denote the inverse transformation of $\mathcal{Y}$ onto $\mathcal{X}_i, i = 1, \ldots, m$. Define

$$J_i = \begin{vmatrix} \frac{\partial g_{1i}}{\partial y_1} & \cdots & \frac{\partial g_{ni}^{-1}}{\partial y_1} \\ \cdot & \cdots & \cdot \\ \frac{\partial g_{1i}}{\partial y_{1i}} & \cdots & \frac{\partial g_{ni}^{-1}}{\partial y_n} \end{vmatrix}.$$

We assume that all the partial derivatives exist and are continuous over $\mathcal{Y}$ and $J_i \neq 0, i = 1, \ldots, m$. Then

$$f_{Y_1, \ldots, Y_n}(y_1, \ldots, y_n) = \sum_{i=1}^{m} f_{X_1, \ldots, X_n}(x_{1i}, \ldots, x_{ni}) |J_i| \tag{11.5.1}$$

for $(y_1, \ldots, y_n) \in \mathcal{Y}$.

If the joint distribution of only $k(< n)$ variables $Y_1, \ldots, Y_k$ is desired, additional $(n - k)$ variables $Y_{k+1} = g_{k+1}(X_1, \ldots, X_n), \ldots, Y_n = g_n(X_1, \ldots, X_n)$ are suitably chosen, joint *pdf* of $(Y_1, \ldots, Y_n)$ is obtained and then the distribution of $(Y_1, \ldots, Y_k)$ is found by integrating out the variables $Y_{k+1}, \ldots, Y_n$.

EXAMPLE 11.5.1: Let X_1, X_2, X_3 be three independently and identically distributed random variables, each with *pdf* $f(x) = e^{-x}(x > 0)$. Show that the variables

$$Y_1 = X_1 + X_2 + X_3, \quad Y_2 = \frac{X_1 + X_2}{X_1 + X_2 + X_3}, \quad Y_3 = \frac{X_1}{X_1 + X_2}$$

are independently distributed.

Using the transformation

$$y_1 = x_1 + x_2 + x_3, \quad y_2 = \frac{x_1 + x_2}{x_1 + x_2 + x_3}, \quad y_3 = \frac{x_1}{x_1 + x_3},$$

$$x_1 = y_1 y_2 y_3, \quad x_2 = y_1 y_2 (1 - y_3), \quad x_3 = y_1 (1 - y_2), \quad |J| = y_1^2 y_2,$$

$$f_{X_1, X_2, X_3}(x_1, x_2, x_3) = e^y.$$

Hence,

$$f_{Y_1, Y_2, Y_3}(y_1, y_2, y_3) = e^{-y} y_1^2 y_2$$

which shows that Y_1, Y_2, Y_3 are independently distributed.

EXAMPLE 11.5.2: Let (X, Y, Z) have joint *pdf*

$$f(x, y, z) = 6(1 + x + y + z)^{-4}, \quad x, y, z > 0.$$

Find the *pdf* of $X + Y + Z$.

Use the transformation: $u = x + y + z$, $uv = y + z$, $uvw = z$. Then $J = u^2 v, 0 < u < \infty, 0 < v < 1, 0 < w < 1$.

$$f(u, v, w) = \frac{6u^2 v}{(1 + u)^4}.$$

Hence, integrating over v and w,

$$f_U(u) = \frac{C.u^2}{(1 + u)^4}, \quad 0 < u < \infty$$

where C is a suitable constant.

EXAMPLE 11.5.3: Let $X_i, i = 1, \ldots, n$ be n independently distributed normal random variables with parameters (μ_i, σ_i^2). Find the *pdf* of $Y = a + \sum_{i=1}^n b_i X_i$ where $a, b_1, \ldots, b_n$ are arbitrary constants.

The probability element of the joint distribution of $(X_1, \ldots, X_n)$ is

$$dF = \frac{1}{(2\pi)^{\frac{n}{2}} \sigma_1 \ldots \sigma_n} \exp \left\{ -\frac{1}{2} \sum_{i=1}^n \left(\frac{x_i - \mu_i}{\sigma_i} \right)^2 \right\}. \qquad (11.5.2)$$

We make the following transformation from $(x_1, \ldots, x_n)$ to $(y_1, \ldots, y_n)$

$$y_1 = \frac{b_1\sigma_1(\frac{x_1-\mu_1}{\sigma_1})+\ldots+b_n\sigma_n(\frac{x_n-\mu_n}{\sigma_n})}{\sqrt{\sum_{i=1}^n b_i^2\sigma_i^2}}$$

$$y_2 = l_{21}(\tfrac{x_1-\mu_1}{\sigma_1}) + \ldots + l_{2n}(\tfrac{x_n-\mu_n}{\sigma_n}) \qquad (11.5.3)$$

$$\cdots \quad \cdots$$

$$y_n = l_{n1}(\tfrac{x_1-\mu_1}{\sigma_1}) + \ldots + l_{nn}(\tfrac{x_n-\mu_n}{\sigma_n})$$

where $(l_{i1}, \ldots, l_{in}), i = 2, \ldots, n$ are mutually orthogonal vectors of unit length and each is orthogonal to the vector

$$\left(\frac{b_1\sigma_1}{\sqrt{\sum_{i=1}^n b_i^2\sigma_i^2}}, \ldots, \frac{b_n\sigma_n}{\sqrt{\sum_{i=1}^n b_i^2\sigma_i^2}} \right).$$

The transformation (11.5.3) can be written as

$$\begin{bmatrix} y_1 \\ \cdot \\ \cdot \\ \cdot \\ y_n \end{bmatrix} = \mathbf{L} \begin{bmatrix} \frac{(x_1-\mu_1)}{\sigma_1} \\ \cdot \\ \cdot \\ \frac{(x_n-\mu_n)}{\sigma_n} \end{bmatrix}, \text{ i.e., } \mathbf{Y} = \mathbf{LZ} \text{ (say)}$$

where $\mathbf{L}$ is an orthogonal matrix, $\mathbf{LL'} = \mathbf{L'L} = \mathbf{I} = $ Diag. $(1, \ldots, 1)$. Hence,

$$\sum_{i=1}^n y_i^2 = \mathbf{Y'Y} = \mathbf{Z'L'LZ} = \mathbf{Z'Z} = \sum_{i=1}^n \frac{(x_i - \mu_i)^2}{\sigma_i^2},$$

$$J(\tfrac{x_1 \ldots x_n}{y_1 \ldots y_n}) = \sigma_1 \ldots \sigma_n |\mathbf{L}| = +_-\sigma_1 \ldots \sigma_n.$$

Hence, *pdf* of $y_1, \ldots, y_n$ is

$$\frac{1}{(2\pi)^n} \exp\{-\frac{1}{2}\sum_{i=1}^n y_i^2\}, \quad -\infty < y_i < \infty, \ i = 1, \ldots, n,$$

which shows that $Y_1, \ldots, Y_n$ are independent, each $N(0, 1)$-random variable. In particular Y_1 has a $N(0, 1)$-distribution. Now

$$Y = a + \sum_{i=1}^n b_i X_i = a + \sum_{i=1}^n b_i\mu_i + Y_1\sqrt{\sum_{i=1}^n b_i^2\sigma_i^2},$$

$$E(Y) = a + \sum_{i=1}^{n} b_i \mu_i, \ V(Y) = \sum_{i=1}^{n} b_i^2 \sigma_i^2.$$

Hence Y is a $N(a + \sum_{i=1}^{n} b_i \mu_i, \sum_{i=1}^{n} b_i^2 \sigma_i^2)$-variable, since if Z is $N(\mu, \sigma^2)$, any linear function $\alpha Z + \beta (\alpha \neq 0)$ itself follows a $N(a\mu + \beta, \alpha^2 \sigma^2)$-distribution (vide Example 7.6.7)).

[The results in this example were proved in Theorem 9.7.1 by using the characteristic function.]

11.6 Distributions of Maxima and Minima

Let $X_1, \ldots, X_n$ be random variables, each defined on $(\mathcal{S}, \mathcal{A}, P)$. Let

$$Y_1 = \max(X_1, \ldots, X_n), \ Y_2 = \min(X_1, \ldots, X_n).$$

Clearly, Y_1, Y_2 are random variables, because

$$\{\omega : Y_2(\omega) \leq z\} = \{\omega : X_1(\omega) \geq z, \ldots, X_n(\omega) \geq z\}$$
$$= \cap_{i=1}^{n} \{\omega : X_i(\omega) \geq z\} \in \mathcal{A}.$$

Thus Y_2 is a $\mathcal{A}$-measurable function and hence is a random variable. Similarly, Y_1 is a random variable.

Theorem 11.6.1: If $X_1, \ldots, X_n$ are independent random variables, X_i having distribution function $F_{X_i}(x)$, the distribution function of Y_1, Y_2, respectively are

(i) $F_{Y_1}(x) = \Pi_{i=1}^{n} F_{X_i}(x)$

(ii) $F_{Y_2}(x) = 1 - \Pi_{i=1}^{n}[1 - F_{X_i}(x)]$.

Proof.

$$\begin{aligned} F_{Y_1}(x) = P(Y_1 \leq x) &= P\{X_1 \leq x, \ldots, X_n \leq x\} \\ &= P(X_1 \leq x) \ldots P(X_n \leq x) \ \text{(by independence)} \\ &= \Pi_{i=1}^{n} F_{X_i}(x). \end{aligned}$$

$$\begin{aligned} F_{Y_2}(x) = P(Y_2 \leq x) = 1 - P(Y_2 > x) &= 1 - P\{X_1 > x, \ldots, X_n > x\} \\ &= 1 - P(X_1 > x) \ldots P(X_n > x) \ \text{(by independence)} \\ &= 1 - \Pi_{i=1}^{n}[1 - F_{X_i}(x)]. \end{aligned}$$

Corollary 11.6.1.1: If $X_1, \ldots, X_n$ are *iid* random variables, each having distribution function $F_X(x)$,

$$F_{Y_1}(x) = (F_X(x))^n, \ F_{Y_2}(x) = 1 - (1 - F_X(x))^n.$$

Corollary 11.6.1.2: If further, $X_1, \ldots, X_n$ are continuous random variables with common *pdf* f, the *pdf* of Y_1, Y_2, are given respectively by

(i) $f_{Y_1}(x) = nF^{n-1}(x)f(x)$

(ii) $f_{Y_1}(x) = n[1 - F(x)]^{n-1}f(x)$.

EXAMPLE 11.6.1 : Let X, Y be independent random variables with negative exponential distributions, having parameters λ_1, λ_2 respectively. Find $E[\max(X, Y)]$.

Let $Z = \max(X, Y)$. The distribution function of Z is

$$F_Z(z) = F_X(z)F_Y(z) = (1 - e^{-\lambda_1 z})(1 - e^{-\lambda_2 z})$$
$$f_Z(z) = \frac{d}{dz}(F_Z(z)) = \frac{d}{dz}[1 - e^{-\lambda_1 z} - e^{-\lambda_2 z} + e^{-(\lambda_1 + \lambda_2)z}]$$
$$= \frac{e^{-\lambda_1 z}}{\lambda_1} + \frac{e^{-\lambda_2 z}}{\lambda_2} - \frac{e^{(-\lambda_1 + \lambda_2)z}}{(\lambda_1 + \lambda_2)}$$
$$E(Z) = \int_0^\infty z f_Z(z)dz.$$

Now,

$$\int_0^\infty z \frac{e^{-\lambda_1 z}}{\lambda_1}dz = \int_0^\infty e^{-\lambda_1 z} z(\lambda_1)^{-1}dz = \frac{\Gamma 2}{\lambda_1^2} = \frac{1}{\lambda_1^2}.$$

Thus,

$$E(Z) = \frac{1}{\lambda_1^3} + \frac{1}{\lambda_2^3} - \frac{1}{(\lambda_1 + \lambda_2)^3}.$$

EXAMPLE 11.6.2: Show that a necessary and sufficient condition for independent random variables $X_1, \ldots, X_n$ to be distributed with parmeter p is that $Y = \min(X_1, \ldots, X_n)$ has a geometric distribution with parameter $(1 - q^n), q = 1 - p$.

Suppose X has a geometric distribution with parameter p. The distribution function is $F_X(x) = P(X \le x) = 1 - q^{n+1}$. Hence, distribution function of Y is

$$F_Y(x) = 1 - [1 - F_X(x)]^n = 1 - q^{n(x+1)} \tag{11.6.1}$$

so that Y has a geometric distribution with parameter $(1 - q^n)$.

Conversely, suppose that the distribution function of Y is (11.6.1). Now

$$F_Y(x) = 1 - [1 - F_X(x)]^n.$$

Hence,

$$1 - F_X(x) = q^{x+1} \quad \text{when } F_X(x) = 1 - q^{x+1}$$

so that each X is a geometric variable with parameter p.

Theorem 11.6.2: If $X_1, \ldots, X_n$ are each independent random variables, the joint distribution function of Y_1, Y_2 is

$$F_{Y_1,Y_2}(y_1, y_2) = \begin{cases} \Pi_{i=1}^n F_{X_i}(y_1) - \Pi_{i=1}^n [F_{X_1}(y_1) - F_{X_1}(y_2)], & y_1 > y_2 \\ \Pi_{i=1}^n F_{X_i}(y_i), & y_1 \le y_2. \end{cases}$$

Proof.

$$F_{Y_1,Y_2}(y_1, y_2) = P[Y_1 \le y_1, Y_2 \le y_2] = P[Y_1 \le y_1] - P[Y_1 \le y_1, Y_2 \ge y_2]$$

$$= P[X_1 \le y_1, \ldots, X_n \le y_1] - P[y_2 \le X_1, \ldots, X_n \le y_1]$$

$$= \Pi_{i=1}^n F_{X_i}(y_1) - \Pi_{i=1}^n [F_{X_1}(y_1) - F_{X_1}(y_2)]$$
(by independence) .

$$(11.6.2)$$

If $y_1 \le y_2$, the last term in the right side of (11.6.2) vanishes. Hence the result. $\square$

Corollary 11.6.2.1: If $X_1, \ldots, X_n$ are *iid* with common distribution function F, then

$$F_{Y_1,Y_2}(y_1, y_2) = \begin{cases} F^n(y_1) - [F(y_1) - F(y_2)]^n, & y_1 \le y_2 \\ F^n(y_1), & y_1 \le y_2. \end{cases}$$

Corollary 11.6.2.2: If $X_1, \ldots, X_n$ are continuous random variables with common *pdf* f, the joint *pdf* of Y_1, Y_2 is

$$F_{Y_1,Y_2}(y_1, y_2) = \begin{cases} n(n-1)[F(y_1) - F(y_2)]^{n-1} f(y_1) f(y_2), & y_1 > y_2 \\ 0, & y_1 \le y_2. \end{cases}$$

EXAMPLE 11.6.3: $X_1, \ldots, X_n$ are independent exponential random variables with the same parameter λ. Find the joint distribution of Y_1, Y_2.

$$F_{Y_1,Y_2}(y_1, y_2) = \begin{cases} (1 - e^{-\lambda y_1})^n - [e^{-\lambda y_2} - e^{-\lambda y_1}]^n, & y_1 > y_2 \\ (1 - e^{-\lambda y_1})^n, & y_1 \le y_2. \end{cases}$$

The joint *pdf* is

$$f_{Y_1,Y_2}(y_1, y_2) = \begin{cases} n(n-1)\lambda^2 [e^{-\lambda y_2} - e^{-\lambda y_1}]^{n-2} e^{-\lambda(y_1+y_2)}, & y_1 > y_2 \\ 0, & y_1 \le y_2. \end{cases}$$

11.7 Use of Moment Generating Function

Since moment generating function, when it exists, uniquely determines the distribution function, one may use the *mgf* to determine the distribution. Thus to find the joint distribution of $Y_1 = g_1(X_1, \ldots, X_n), \ldots, Y_n = g_n(X_1, \ldots, X_n)$, one may find the joint *mgf* of $(Y_1, \ldots, Y_n)$ with the help of the joint *pdf* of $(X_1, \ldots, X_n)$. If the *mgf* so obtained can be identified to be the *mgf* of a known distribution ξ, then ξ would be the joint distribution of $(Y_1, \ldots, Y_n)$. However, since only the *mgf*s of some one-dimensional random variables and some two-dimensional random variables are easily recognizable, the use of *mgf* to obtain the distribution will be limited in use to univariate functions or bivariate functions of random variables only.

Mostly by using the property that the *mgf* of a sum of n independent random variables is the product of their *mgf*s, we have been able to prove several properties (including the reproductive property) of binomial, poisson, geometric, gamma, Cauchy and normal variables and hence have been able to identify the distribution of the relevant functions of these random variables. Here we consider the *mgf* of arbitrary functions of random variables, restricting our discussion to the case of jointly continuous two-dimensional variables only.

Let (X_1, X_2) be two continuous random variables having joint *pdf* $f_{X_1, X_2}(x_1, x_2)$. The joint moment generating function of $Y_1 = g_1(X_1, X_2), Y_2 = g_2(X_1, X_2)$ is

$$M_{Y_1, Y_2}(t_1, t_2) = E(e^{t_1 Y_1 + t_2 Y_2}) = \int_{-\infty}^{\infty} \int_{-\infty}^{\infty} e^{t_1 g_1(x_1, x_2) + t_2 g_2(x_1, x_2)} f(x, y) dx dy.$$

$$(11.7.1)$$

The case of discrete random variables can be treated analogously.

EXAMPLE 11.7.1: Let X, Y be independent standard normal variables. Find the *mgf* of XY.

$$M_{X,Y}(t_1, t_2) = E(e^{txy}) = \tfrac{1}{2\pi} \int_{-\infty}^{\infty} \int_{-\infty}^{\infty} e^{txy - \frac{x^2 + y^2}{2}} dx dy$$

$$= \tfrac{1}{2\pi} \int_{-\infty}^{\infty} e^{-\frac{y^2}{2}(1-t^2)} \left[\int_{-\infty}^{\infty} e^{-\frac{1}{2}(x-ty)^2} dx \right] dy = \tfrac{1}{\sqrt{1-t^2}}.$$

EXAMPLE 11.7.2: (X, Y) follows a bivariate normal distribution with parameters $(\mu_1, \mu_2, \sigma_1, \sigma_2, \rho)$. Show that $Z = aX + bY + c$ where a, b, c are constants, is $N(a\mu_1 + b\mu_2 + c, \sqrt{a^2 \sigma_1^2 + b^2 \sigma_2^2 + 2ab\rho\sigma_1\sigma_2})$.

Proof. Moment generating function of Z is

$$M_Z(t) = \int_{-\infty}^{\infty} \int_{-\infty}^{\infty} e^{t(ax+by+z)} f_{X,Y}(x,y) dx dy. \qquad (11.7.2)$$

Now, it is known that

$$M_{X,Y}(t_1,t_2) = E(e^{t_1 X + t_2 Y}) = e^{\mu_1 t_1 + \mu_2 t_2 + \frac{1}{2}(\sigma_1^2 t_1^2 + \sigma_2^2 t_2^2 + 2\rho\sigma_1\sigma_2 t_1 t_2)}. \qquad (11.7.3)$$

Putting $t_1 = at, t_2 = bt$ in (11.7.3), we obtain

$$M_Z(t) = \exp[c + at\mu_1 + bt\mu_2 + \frac{t^2}{2}(a^2\sigma_1^2 + b^2\sigma_2^2 + 2\rho\sigma_1\sigma_2 ab)]$$

which is the *mgf* of a $N(a\mu_1 + b\mu_2 + c, a^2\sigma_1^2 + b^2\sigma_2^2 + 2\rho ab\sigma_1\sigma_2)$. Hence the result follows by the uniqueness property of the *mgf*.

11.8 Exercises and Complements

11.1 X is a random variable with distribution function $F(x)$. Let

$$g(X) = \begin{cases} x, & |x| \geq c, \\ 0, & |x| < c. \end{cases}$$

Show that the distribution function of $Y = g(X)$ is $F_Y(y) =$ (i) $F_X(y)$ for $y \geq c$, (ii) $F_X(c) = 0$ for $0 \leq y < c$, (iii) $F_X(-c)$ for $-c \leq y < 0$, (iv) $F_X(y)$ for $y < -c$.

11.2 X is uniformly distributed over $(-a, a)$. Find the distribution of $Y =$ (i) $\frac{1}{X^2}$, (ii) $\sqrt{|X|}$, (iii) $\cos X$, (iv)e^X, (v) $bX^2 + cX + d$.

11.3 X has a uniform distribution over $(0, 1)$. Find the probability distribution of $Y =$ (i) X^2, (ii)$\frac{1}{X}$, (iii) $\sqrt{X}$, (iv) $\sin t(\pi X)$, (v) $-2\log_e X$, (iv)$\tan X$ (vii) $aX^2 + bX + c$, (viii)e^X (ix) $\frac{X}{(1+X)}$.

11.4 The *pdf* of X is

$$f(x) = \begin{cases} \lambda e^{-\lambda x}, & x > 0 \\ 0, & x \leq 0, \lambda > 0. \end{cases}$$

Find the *pdf* of Y where Y is (i) $\sin X$, (ii) $\cos X$, (iii) $\tan X$.

11.5 X, Y are independent, each $N(0, \sigma^2)$-variable. Find the *pdf* of $w = \tan^{-1}(\frac{x}{y})$.

$$\left[f(w) = \frac{1}{\pi}, |w| < \frac{\pi}{2}.\right]$$

11.6 The *pdf* of X is $f_X(x) = C.e^{-|x|}$. Find the constant C and the *pdf* of $Y = X^2$.

11.7 X has the *pdf*

$$f_X(x) = \begin{cases} 0, & x \le -k, \\ \frac{1}{k^2}(k+x), & -k < x \le 0, \\ \frac{1}{k^2}(k-x), & 0 < x \le k, \\ 0, & x > k. \end{cases}$$

Find the *pdf* of $Y = X^2$.

11.8 Let X be a random variable with *pdf* $f(x) = \frac{2x}{\pi^2}(0), 0 < x < \pi$ (otherwise). Find the distribution function of $Y = \sin X$.

11.9 Let X have a binomial distribution with parameters (n, p). Find the *pdf* of (i) $Y = (X - np)^2$, (ii) $Y = X^2$.

$$\left[(ii) \frac{n!}{(\sqrt{y})!(n-\sqrt{y})!} p^{\sqrt{y}} q^{n-\sqrt{y}}, p + q = 1. \right]$$

11.10 X, Y are independent standard normal variables. Find the *pdf* of $T = |X - Y|$.

11.11 Let X_1, X_2 be independent $\chi^2_{(2)}$ variables. Show that the *pdf* of $Y = \frac{1}{2}(X_1 - X_2)$ is

$$g_Y(y) = \frac{1}{2} e^{-|x|}, \quad -\infty < y < \infty.$$

11.12 X_1, X_2 are random samples from the rectangular population

$$dF = dx, \ 0 \le x \le 1.$$

Find the *pdf* of standard deviation of $S = \frac{1}{2}|X_1 - X_2|$.

11.13 Let X, Y, Z be independent random variables with the same geometric distribution (p). Find

(a) $P\{X = Y\}$, (b)$P\{X \ge 2\}$, (c) $P\{X + Y \le Z\}$.

$$\left[(a) \frac{p}{1+q}, (b) \frac{1}{1+q+q^2}, (c) \frac{1}{(1+q)^2} . \right]$$

11.14 The random variables X, Y are uniformly distributed over the circle C of radius 1 so that

$$f_{X,Y}(x, y) = \frac{1}{\pi}(0) \text{ if } (x, y) \in C \ (\text{otherwise}).$$

Find the *pdf* of $R = +_-\sqrt{X^2 + Y^2}$.

[Hints: Use polar transformation.]

11.15 The variables X, Y have the joint density

$$f_{X,Y}(x,y) = \frac{1}{4}, \quad -1 \le x, y \le 1.$$

Find (i) $P(X + Y \le t)$ where $0 < t \le 1$, (ii) $P(X^2 + Y^2 \le r^2)$, (iii) the distribution function and *pdf* of $R^2 (= X^2 + Y^2)$.

$$\left[(i) \ \tfrac{t^2}{4}, \ (ii) \ r^2, \ (iii) \ df = \tfrac{\pi}{4} d(r^2). \right].$$

11.16 X has a Raleigh distribution $f(x) = \frac{x}{\sigma^2} e^{-\frac{x^2}{2\sigma^2}}, x > 0$. Find the *pdf* of $Y = e^{-x^2}$.

$$\left[\tfrac{1}{2y\sigma^2} y^{-\frac{1}{2\sigma^2}}, \ 0 < y < 1. \right]$$

11.17 X has a Cauchy distribution $f(x) = \frac{1}{\pi(1+x^2)}, -\infty < x < \infty$. Find the *pdf* of $Y = \frac{1}{X}$.

$$\left[\tfrac{1}{\pi(1+y^2)}, \ -\infty < y < \infty. \right]$$

11.18 The *pdf* of X is $f_1(x) = \sqrt{\frac{2}{\pi}} e^{-\frac{x^2}{2}}, 0 \le x < \infty$, that of Y is $f_2(y) = 2y^3 e^{-y}, 0 < y < \infty$. Find the distribution of $U = \frac{X}{Y}$.

$$\left[\tfrac{6}{(u^2+2)^{3/2}}, \ 0 < u < \infty. \right]$$

11.19 X, Y are each independent binomial variables with parameters (n, p). Find the *pmf* of (i) $X + Y$, (ii) $X - Y$, (iii) $U = \frac{X}{X+1}$.

11.20 Two independent variables X, Y are each uniformly distributed over the range $(-a, a)$. Show that the sum $Z = X + Y$ has a triangular density

$$f(z) = \begin{cases} \frac{2a+z}{4a^2}, & -2a \le z \le 0 \\ \frac{2a-z}{4a^2}, & 0 \le z \le 2a. \end{cases} \tag{i}$$

Show that the *mgf* of Z is $(\frac{1}{at} \sinh at)^2$.

11.21 (X, Y) are continuous random variables with joint *pdf*

$$f_{X,Y}(x,y) = \begin{cases} \frac{1}{4a^2} [1 + xy(x^2 - y^2)], & |x|, |y| \le a, a > 0 \\ 0, & \text{(otherwise)} \end{cases}$$

show that the *pdf* of $Z = X + Y$ is triangular with *pdf* given in (i).

11.22 Show that the *df* of $Z = X_1 + X_2$, when X_1, X_2 are independent, each rectangular over (a, b) is

$$f_Z(z) = \begin{cases} \frac{z-2a}{(b-a)^2}, & 2a \leq z \leq a+b, \\ \frac{2b-z}{(b-a)^2}, & a+b \leq z \leq 2b. \end{cases}$$

11.23 X is a continuous random variable with *pdf* $f_X(x)$. Another random variable Y, distributed independently of X, has a continuous uniform distribution over $(-a, a)$. Show that the *pdf* of $Z = X + Y$ is

$$f_Z(z) = \frac{1}{2a}[F_X(z+a) - F_X(z-a)].$$

11.24 X, Y are independent standard normal variables. Show that

(i) $Z = \sqrt{X^2 + Y^2}$ and $W = \arctan \frac{X}{Y}$ are independently distributed, Z having a Raleigh distribution (vide Exercise 11.17) and W a uniform distribution over $(-\frac{\pi}{2}, \frac{\pi}{2})$.

(ii) Z and $V = \frac{X}{Y}$ are independently distributed, Z having a Raleigh distribution and V a cauchy distribution. (Note that if X, Y are normal, but their expected values are not zeros, Z and V are not independent.)

(c) Find the joint density of $(X^2 + Y^2, X)$.

[Let $U = X^2 + Y^2, V = X$. Then

$$f_{U,V}(u, v) = \frac{1}{\pi} \cdot \frac{1}{\sqrt{u - v^2}} e^{-\frac{u^2}{2}}, \quad -\infty < v < \infty; 0 < v^2 < u < \infty.]$$

11.25 Let X be a non-negative random variable of the continuous type with *pdf* $f_X(x)$. Show that *pdf* of $Y = X^m, m > 0$ is

$$f_Y(y) = \begin{cases} \frac{1}{m} y^{\frac{1}{m}-1} f(y^{\frac{1}{m}}), & y > 0 \\ 0, & y \leq 0. \end{cases}$$

11.26 Let X be a standard normal variable. Show that *pdf* of $Y = X^{2m}$ is

$$f_Y(y) = \begin{cases} \frac{1}{\sqrt{2\pi} m y^{1-\frac{m}{2}}} \exp\{-\frac{y^{\frac{1}{m}}}{2}\}, & y > 0 \\ 0, & y \leq 0. \end{cases}$$

11.27 Lt $X_1, \ldots, X_n$ be *iid* rectangular random variables over $(0, 1)$. Let $X_{(1)} = \min(X_1, \ldots, X_n), X_{(n)} = \max(X_1, \ldots, X_n)$.

(a) Find the distribution function of (i) $X_{(1)}$, (ii) $X_{(n)}$.
(b) Find the joint distribution function and joint *pdf* of $X_{(1)}, X_{(n)}$.
(c) Find the distribution function and *pdf* of $X_{(n)} - X_{(1)}$.
(d) Find $E[X_{(1)}], \sigma^2_{X_{(1)}}, E[X_{(n)}], \sigma^2_{X_{(n)}}, \rho_{X_{(1)}, X_{(n)}}$.
(e) Find the conditional *pdf* and conditional mean and variance of $X_{(1)}$ given $X_{(n)}$.

Extend all the results if $X_1, \ldots, X_n$ are *iid* rectangular variables over (a, b).

11.28 Let $X_1, \ldots, X_n$ be independent gamma variables, X_i being $G(\alpha_i, \beta), i = 1, \ldots, n$. Show that $\sum_{i=1}^n X_i$ is $G(\sum_{i=1}^n \alpha_i, \beta)$.

11.29 Let $X_1, \ldots, X_n$ be *iid* variables, each uniform over $(0, 1)$. Let $M_t = \max\{m(\geq 1) : \Pi_{i=1}^m X_i \geq t\}$. Show that $[M_t = m] = -\sum_{i=1}^{m-1} \ln X_i \leq -\ln t < -\sum_{i=1}^m \ln X_i$ and M_t is a Poisson variable with parameter $-\ln t$.

11.30 Suppose X has a Beta distribution of type I with parameters (α_1, α_2). Show that

(i) $Y = \frac{1}{X} - 1$ has the density $f_Y(y) = B(\alpha_1, \alpha_2)^{-1}(1 + y)^{-(\alpha_1 + \alpha_2)} y^{\alpha_2 - 1}, y > 0$;

(ii) $Z = \frac{a}{X}$ where $a(> 0)$ is a constant, has the *pdf* $f_Z(z) = B(\alpha_1, \alpha_2)^{-1} a^{\alpha_1} z^{-(\alpha_1 + \alpha_2)}(z - a)^{\alpha_2 - 1}, z > 0$;

(iii) $\frac{X}{1-X}$ has the inverse Beta distribution on $(0, \infty)$ with parameters (α_2, α_1).

11.31 Let X, Y be independent with $X \cap$ gamma (α_1, β) and $Y \cap$ gamma (α_2, β). Show that $\frac{X}{Y}$ follows inverse Beta on $(0, \infty)$ with parameters (α_1, α_2).

11.32 Let $X_1 \ldots, X_k$ be independent Beta random variables with $X_i \cap B(\alpha_i, \beta_i), i = 1, \ldots, k$. Let $\alpha_i = \alpha_{i+1} + \beta_{i+1}, i = 1, \ldots, k - 1$. Then, show that $\Pi_{i=1}^k X_i$ is a Beta variable with parameters $(\alpha_k, \beta_1 + \ldots + \beta_k)$.

11.33 The joint *pdf* of X, Y, Z is

$$g(x, y, z) = \frac{1}{\sqrt{xyz}} f(x + y + z), \ 0 < x, y, z < \infty.$$

Derive the distribution of (i) $U = X + Y + Z$, (ii) $Z = \frac{Y}{X}$, (iii) $T = \frac{Z}{X+Y}$.

[Hints: (i) transformation : $u = x + y + z, uv = y + z, uvw = z$.

(ii) and (iii): transformation: $u = x + y + z, uv = x + y, uvw = x.$]

11.34 Find the distribution of $\bar{X} = \frac{1}{n}\sum_{i=1}^{n} X_i$, where $X_1, \ldots, X_n$ are *iid* random variables with common *pdf* $f(x) = \frac{1}{\theta}e^{\frac{-x}{2}}$, $0 < x < \infty, \theta > 0$.

$$\left[f(x) = \frac{n}{\theta}\exp(-\frac{n\bar{x}}{2}).\right]$$

11.35 Let X, Y be independently distributed with a common *pdf* $f(x) = \frac{1}{\lambda}e^{-x}, x > 0, \lambda > 0$. Define $U = X + Y, V = \frac{X}{(X+Y)}, W = \frac{X}{Y}$. Check if (i) U is independent of V (ii) U is independent of W.

11.36 X_1, X_2 are independent standard normal variables. Let $Y = \min(X_1, X_2), Z = \max(X_1, X_2)$. Find $E(Y), E(Z), V(Y), V(Z)$.

11.37 X is a $N(0,1)$ random variable. Find the moment generating function of $Y = X^2$.

$$\left[M_Y(t) = E(e^{tX^2}) = \frac{1}{\sqrt{1-2t}}, \ t < \frac{1}{2}.\right]$$

11.38 The joint *pdf* of (X, Y) is

$$f(x, y) = \frac{y}{(1+x)^4}\exp(-\frac{y}{1+x}), \ x, y > 0.$$

Show that the joint density of U and V where $U = \frac{Y}{1+X}$, $V = \frac{1}{1+X}$ is $dF = ue^{-u}, 0 \le u < \infty, 0 \le v \le 1$.

11.39 X_1, X_2, X_3 are each independent standard normal variables. By using polar transformations

$$x_1 = r\cos\theta_1\cos\theta_2, x_2 = r\cos\theta_1\sin\theta_2, x_3 = r\sin\theta_1,$$

$$(0 \le r < \infty, -\frac{\pi}{2} < \theta_1 < \frac{\pi}{2}, 0 < \theta_2 < 2\pi),$$

show that the variables r, θ_1, θ_2 are independently distributed and the distribution of r is

$$dF = \sqrt{\frac{2}{\pi}}e^{-\frac{r^2}{2}}r^2 dr, \ r \ge 0.$$

11.40 Let X, Y be independent standard normal variables and let $U = aX + bY, V = cX + dY$, where a, b, c, d are constants. (i) Show that U, V are independent if $ac + bd = 0$. (ii) Find the joint density of U and V if $ad - bc \ne 0$.

11.41 X_1, X_2, X_3 are independent random samples from the distribution $f(x) = \frac{x}{\lambda^2}e^{-\frac{x}{\lambda}}, x > 0, \lambda > 0$. Find the joint *pdf* of $Y_1 = X_1 + X_2 + X_3, Y_2 = X_2, Y_3 = X_3$ and the marginal *pdf* of Y_1.

$$\left[\frac{y_1^5}{120\lambda^6}e^{-\frac{y_1}{\lambda}}.\right]$$

11.42 Let X_1, X_2 be independent random variables, each rectangular over $(0, 1)$. Show that

$$Y = \sqrt{-2\log_e X_1}\cos 2\pi X_2 \text{ and } Z = \sqrt{-2\log_e X_1}\sin 2\pi X_2$$

are independent random variables, each $N(0, 1)$.

11.43 (X, Y) have joint *pdf*

$$f_{X,Y}(x, y) = \frac{e^{-y}}{\Gamma(1, k_1)\Gamma(1, k_2)}x^{k_1-1}(y - x)^{k_2-1},\ 0 < x < y < \infty$$

which is zero otherwise. Show that marginal distribution of X and Y are gamma distributions $G(k_1)$ and $G(k_1 + k_2)$ respectively. Find the correlation coefficient between X and Y.

$$\left[\sqrt{\frac{k_1}{k_2 + k_1}}.\right]$$

11.44 X, Y are independent random variables with *pdf* $f_X(x) = \alpha e^{-\alpha x}, x \geq 0$, and $f_Y(y) = \beta e^{-\beta y}, y \geq 0$. Show that the *pdf* of $Z = X + Y$ is

$$f_Z(z) = \begin{cases} \frac{\alpha\beta}{\beta-\alpha}(e^{-\alpha z} - e^{-\beta z}), & \beta \neq \alpha \\ \alpha^2 z e^{-\alpha z}, & \beta = \alpha. \end{cases}$$

11.45 X, Y are independent random variables, each with Raleigh density

$$f_X(x) = \frac{x}{\alpha^2}e^{-\frac{x^2}{2\alpha^2}},\ x \geq 0$$
$$f_Y(y) = \frac{y}{\beta^2}e^{-\frac{y^2}{2\beta^2}},\ y \geq 0.$$

Show that the *pdf* and the distribution function of $Z = \frac{X}{Y}$ are, respectively,

$$f_Z(z) = \frac{2\alpha^2}{\beta^2}\frac{z}{(z^2 + \frac{\alpha^2}{\beta^2})^2},\ z \geq 0,$$
$$F_Z(z) = \frac{z^2}{z^2 + \frac{\alpha^2}{\beta^2}},\ z \geq 0.$$

11.46 Let X, Y, Z be mutually independent random variables with *pd*

$$f(u) = \begin{cases} \alpha e^{-\alpha u}, 0 \leq u < \infty, \alpha > 0 \\ 0, \qquad \text{otherwise .} \end{cases}$$

Find the *pdf* of (i) $X + Y + Z$, (ii) $\frac{(X+Y)}{Z}$, (iii) $\frac{(X+Y)}{X}$.

11.47 Consider the one-one transformation $(x_1, \ldots, x_n) \to (r, \theta_1, \ldots, \theta_{n-1})$ such that

$$
\begin{aligned}
x_1 &= r \cos \theta_1, \quad 0 \le r < \infty \\
x_2 &= r \sin \theta_1 \cos \theta_2 \quad -\frac{\pi}{2} < \theta_1 < \frac{\pi}{2}, i = 1, \ldots, n-2 \\
x_3 &= r \sin \theta_1 \sin \theta_2 \cos \theta_3, \; 0 < \theta_{n-1} < 2\pi \\
&\cdots \quad \cdots\cdots\cdots \\
x_{n-1} &= r \sin \theta_1 \sin \theta_2 \sin \theta_3 \ldots \sin \theta_{n-2} \cos \theta_{n-1} \\
x_n &= r \sin \theta_1 \sin \theta_2 \sin \theta_3 \ldots \sin \theta_{n-2} \sin \theta_{n-1}.
\end{aligned}
$$

Show that the Jacobian of transformation is

$$
J = r^{n-1} (\sin \theta_1)^{n-2} (\sin \theta_2)^{n-3} \ldots \sin \theta_{n-2}.
$$

11.48 Suppose the random vector $\mathbf{X}' = (X_1, \ldots, X_n)$ has the joint density $f_{\mathbf{X}}(\mathbf{x})$, Find the *pdf* of $\mathbf{S} = (S_1, S_1, \ldots, S_n)'$ where $S_i = \sum_{j=1}^{i} (X_1 + \ldots + X_j)$.

[Here

$$
\mathbf{S} = \mathbf{A}\mathbf{X}
$$

where

$$
A = \begin{bmatrix}
1 & 0 & 0 & \ldots & 0 \\
1 & 1 & 0 & \ldots & 0 \\
1 & 1 & 1 & \ldots & 0 \\
\cdot & \cdot & \cdot & \ldots & \cdot \\
1 & 1 & 1 & \ldots & 1
\end{bmatrix}.
$$

It follows that

$$
f_{\mathbf{S}}(\mathbf{s}) = f(s_1, s_2 - s_1, s_3 - s_2, \ldots, s_n - s_{n-1}).]
$$

11.49 Let $X_1, \ldots, X_n$ be *iid* random variables, each exponential with parameter $\lambda = 1$. Let $S_0 = 0, S_n = \sum_{i=1}^{n} X_i, n \ge 1$. Let $M_x = \max\{m (\ge 0) : S_m \le x\}$. Show that $[M_x = m] = [S_m \le x < S_{m+1}]$. Also, show that for $x > 0$, M_x is a Poisson variable with parameter x.

Chapter 12

Convergence of a Sequence of Random Variables

12.1 Introduction

In this chapter we consider a sequence of random variables all defined on the same probability space $(\mathcal{S}, \mathcal{A}, P)$ and study their different modes of convergence. For each point ω, we have therefore, an infinite sequence of values $\{X_n(\omega)\}$. Subsection 12.2.1 considers convergence in probability, Subsection 12.2.2, almost sure convergence or strong convergence, Subsection 12.2.3, convergence in the rth mean, Subsection 12.2.4, convergence in distribution or weak convergence, Subsection 12.2.5, complete convergence and 12.2.6, gradation among different modes of convergence. Weak Laws of Large Numbers and Strong Laws of Large Numbers are considered in Sections 12.3 and 12.4 respectively. The Section 12.5 deals with the celebrated Central Limit Theorem, which discusses the situations when a sequence of random sums converges in distribution to a normal law.

12.2 Various Modes of Stochastic Convergence

It is known that a sequence of numbers x_n converges to a limit x, if, given any $\epsilon > 0$, we can find an integer n_0 such that

$$|x_n - x| < \epsilon \text{ for every } n > n_0. \qquad (i)$$

Consider now the sequence of random variables

$$X_1, \ldots, X_n, \ldots \qquad (ii)$$

For each experimental outcome ω, we have, now, a sequence of numbers

$$X_1(\omega), \ldots, X_n(\omega), \ldots \qquad (iii)$$

359

and (ii) represents a family of such sequences. Hence, we can no longer use (i) to define convergence of the entire family.

Again, consider the case of sequence of ordinary numbers. In many cases, the limit x may remain unknown. In order to test for convergence, we must use a criterion that avoids x. Such a criterion is the existence of the limit (Cauchy)

$$|x_n - x_{n+m}| \to 0 \text{ for } n \to \infty \text{ and any } m > 0.$$

Against this background we now define various modes of convergence of a sequence of random variables.

12.2.1   *Convergence in probability*

DEFINITION 12.2.1: A sequence $\{X_n\}$ of random variables is said to converge in probability to a (possibly degenerate) random variable X, if for every (given) positive number ϵ and η, there exists a positive integer $n_0 = n_0(\epsilon, \eta)$ such that

$$P\{|X_n - X| > \epsilon\} < \eta \text{ for all } n \geq n_0 \qquad (12.2.1)$$

i.e.

$$\lim_{n \to \infty} P\{|X_n - X| > \epsilon\} = 0 \qquad (12.2.2)$$

for every $\epsilon(> 0)$. This mode of convergence is usually written as

$$X_n \to^P X \text{ or } p \lim X_n = X. \qquad (12.2.3)$$

The sequence X_n is said to converge stochastically to X. Note that X may be non-stochastic also.

EXAMPLE 12.2.1: Consider the probability space $(\mathcal{S}, \mathcal{A}, P)$ where $\mathcal{S} = [0, 1]$ and P is such that $P[a \leq \omega \leq b] = (b - a)^{-1}, 0 \leq a < b \leq 1$. Let $\{X_n\}$ be such that $X_n(\omega) = 1(0)$ if $\omega \in [0, \frac{1}{n}]$ (otherwise). Then $P(X_n = 1) = \frac{1}{n}, P(X_n = 0) = 1 - \frac{1}{n}$. Hence,

$$P(|X_n| \leq \epsilon) = \begin{cases} P(X_n = 0) = 1 - \frac{1}{n} & \text{if } \epsilon < 1, \\ 1 & \text{if } \epsilon \geq 1. \end{cases}$$

Thus, $P(|X_n| \geq \epsilon) \to 0$ as $n \to \infty$ for every $\epsilon(> 0)$. Hence, $X_n - X \to^P 0$ where X is a degenerate random variable with $P(X = 0) = 1$.   □

It can be shown that

$$X_n \to^P X \Rightarrow X_n - X \to^P 0. \qquad (12.2.4)$$

Convergence in probability has the Cauchy property of convergence. This is stated in the following theorem.

Theorem 12.2.1 If $\{X_n\}$ is a sequence of random variables such that given any $\epsilon > 0, \eta > 0$, there exists a n_0^* such that
$$P(|X_n - X_m| > \epsilon) < \eta, \qquad (12.2.5)$$
for all $m, n \geq n_0^*$, then there exists a random variable X such that
$$P(|X_n - X| > \epsilon)$$
converges to zero for every $\epsilon > 0$.

This is, because,
$$P[|X_n - X_m| > \epsilon] \leq P[|X_n - X| > \frac{\epsilon}{2}] + P[|X_m - X| > \frac{\epsilon}{2}].$$
The following results hold.

Theorem 12.2.2 Let $\{X_n\}, \{Y_n\}$ be two sequences of random variables with $X_n \to^P X, Y_n \to^P Y$. Then

(a) $P[X = Y] = 1$, because,
$$P|X - Y| > \epsilon] \leq P[|X_n - X| > \frac{\epsilon}{2}] + P[|X_n - Y| > \frac{\epsilon}{2}]$$
and therefore $P[|X - Y| > \epsilon] = 0$ for every ϵ.

(b) $X_n \to^P 1 \Rightarrow X_n^{-1} \to^P 1$; because
$$P\{|X_n^{-1} - 1| \geq \epsilon\} = P\{X_n^{-1} \geq 1 + \epsilon\} + P\{X_n^{-1} \leq 1 - \epsilon\}$$
$$= P\{X_n^{-1} \geq 1 + \epsilon\} + P\{X_n^{-1} \leq 0\} + P\{0 \leq X_n^{-1} \leq 1 - \epsilon\};$$
each of these terms $\to 0$ as $n \to \infty$.

(c) $cX_n \to cX$ where $'c'$ is a constant;

(d) $X_n + Y_n \to^P X + Y$;

(e) $[X_n \to^P c] \Rightarrow X_n^2 \to^P c^2$;

(f) $[X_n \to^P c, Y_n \to^P d] \Rightarrow [X_n Y_n \to^P cd]$, where $'c', 'd'$ are constants; this is because,
$$X_n Y_n = \frac{1}{4}[(X_n + Y_n)^2 - (X_n - Y_n)^2] \to^P \frac{1}{4}[(a+b)^2 - (a-b)^2] = ab.$$

(g) $[X_n \to^P c, Y_n \to^P d, d \neq 0] \Rightarrow [\frac{X_n}{Y_n} \to^P \frac{c}{d}]$;

(h) $[X_n \to^P X; Y$ a random variable$] \Rightarrow [X_n Y \to^P XY]$.

This can be seen as follows. Since Y ia random variable, given $\delta > 0$, there exists a $'k'$ such that $P\{|Y| > k\} < \frac{\delta}{2}$. Therefore,
$$P\{|X_n Y - XY| > \epsilon\} = P\{|X_n - X||Y| > \epsilon, |Y| > k\} +$$
$$P\{|X_n - X||Y| > \epsilon, |Y| \leq k\}$$
$$< \frac{\delta}{2} + P\{|X_n - X| > \frac{\epsilon}{k}\}.$$

(i) $X_n Y_n \to^P XY$; this follows, because $(X_n - X)(Y_n - Y) \to^P 0$ and the result follows on multiplication and using (h).

(j) $\frac{X_n}{Y_n} \to \frac{X}{Y}$ if $P[Y_n = 0] = 0$ and $P[Y = 0] = 0$.

We now give a necessary and sufficient condition for convergence in probability.

Theorem 12.2.3: Let $\{X_n\}$ be a sequence of random variables. Then $X_n \to^P 0$ *iff*

$$E\left[\frac{|X_n|^r}{1 + |X_n|^r}\right] \to 0 \text{ for some } r > 0. \tag{12.2.6}$$

Proof. Let $\epsilon > 0$. Recall the inequality (6.10.6),

$$[\frac{|X|^r}{1 + |X_n|^r}] - \frac{\epsilon^r}{1 + \epsilon^r} \le P\{|X_n| \ge \epsilon\} \le \frac{1 + \epsilon^r}{\epsilon^r} E[\frac{|X_n|^r}{1 + |X_n|^r}].$$

The proof follows immediately on letting $n \to \infty$.

12.2.2   *Almost sure convergence*

We say that a sequence $\{X_n\}$ converges to X with probability 1 or almost surely, if the set of outcomes ω such that

$$\lim_{n \to \infty} X_n(\omega) = X(\omega)$$

has probability equal to 1. We have, therefore, the following definition.

DEFINITION 12.2.2 : A sequence $\{X_n\}$ of random variables is said to converge *almost surely* (or *almost everywhere* or *strongly*) to a (possibly degenerate) random variable X, if

$$P\{\lim_{n \to \infty} X_n = X\} = 1, \tag{12.2.7}$$

i.e., if

$$P\{\omega : \lim_{n \to \infty} X_n(\omega) = X(\omega)\} = 1,$$

i.e., if

$$P\{|X_m - X| < \epsilon; m \ge n\} \to 1, \text{ as } n \to \infty \tag{12.2.8}$$

for arbitrarily small positive ϵ,

i.e., if for every (given) positive ϵ and η, there exists a positive integer $n_0 = n_0(\epsilon, \eta)$ such that

$$P\{|X_m - X| < \epsilon \ \forall \ m \in [n_0, N]\} > 1 - \eta, \ \forall \ N(> n_0). \tag{12.2.9}$$

In symbol, we write this as

$$X_n \to^{a.s.} X \text{ or } \lim X_n = X. \tag{12.2.10}$$

To verify if $X_n \to^P X$, one has to calculate the probability, $P\{|X_n - X| < \epsilon\} = p_n(\epsilon)$ and show that $p_n(\epsilon) \to 1$ as $n \to \infty$. To verify if $X_n \to^{a.s.} X$, one has to show that the set of points ω for which $\lim_{n\infty} X_n(\omega) = X(\omega)$ has probability one.

The following results elucidate the definition.

Theorem 12.2.4 : $X_n \to^{a.s.} X$ *iff*

$$\lim_{n\to\infty} P\{ \sup_{m\geq n} (|X_m - X| \geq \epsilon)\} = 0 \text{ for all } \epsilon > 0, \tag{12.2.11}$$

i.e.

$$\lim_{n\to\infty} P\{\cup_{m=n}^\infty (|X_m - X| \geq \epsilon)\} = 0 \text{ for every } \epsilon > 0. \tag{12.2.12}$$

Theorem 12.2.5 (Cauchy Criterion): $X_n \to^{a.s.} X$ *iff*

$$\lim_{n\to\infty} P\{\sup_m (|X_{m+n} - X_n| \leq \epsilon)\} = 1 \text{ for all } \epsilon > 0, \tag{12.2.13}$$

i.e.

$$\lim_{n\to\infty} P\{\cup_{m=1}^\infty (X_{m+n} - X_n| \geq \epsilon)\} = 0 \text{ for every } \epsilon > 0. \tag{12.2.14}$$

Corollary 12.2.4.1: The sequence $\{X_n\}$ converges *a.s.* to the random variable X *iff* $\sum_{n=1}^\infty P\{|X_n - X| \geq \epsilon\} < \infty$ for all $\epsilon > 0$.

Proof. Since

$$P\{\cup_{m=n}^\infty (|X_m - X| \geq \epsilon)\} \leq \sum_{m=n}^\infty P\{|X_m - X| \geq \epsilon\},$$

the result follows by (12.2.12). (Also, see subsection 12.2.5)

The following theorems hold.

Theorem 12.2.6 : If $X_n \to^{a.s.} X$ and $Y_n \to^{a.s.} Y$, then $X_n + Y_n \to^{a.s.} X + Y$ and $X_n Y_n \to^{a.s.} XY$.

Theorem 12.2.7: (*The Slutsky-Fre'chet Theorem*): If $f(x)$ is continuous everywhere and $\{X_n\}$ converges to X in probability or almost surely, $f(X_n)$ also converges in probability or almost surely to the random variable $f(X)$.

Proof. Given any $\epsilon(> 0)$, we choose an a such that

$$P(|X| > a) < \epsilon.$$

Again, since the function $f(x)$ is uniformly continuous in any finite range, given any $\eta(> 0)$, there exists a ξ such that

$$|f(x) - f(y)| < \eta$$

for all y whenever $|x - y| < \xi$ and $x \in [-a, a]$. Therefore,

$$P[|f(X_n) - f(X)| < \eta, |X| \le a]] \ge P[|X_n - X| \le \xi, |X| < a] \quad (12.2.15)$$

which can be made $\ge (1 - 2\epsilon)$ by choosing n sufficiently large, provided X_n converges in probability to X.

The proof for almost sure convergence is similar. $\qquad\qquad\square$

Theorem 12.2.8: If a sequence $\{X_n\}$ converges almost surely to a random variable X, it also converges in probability to X.

Proof. $X_n \to^{a.s.} X$ means $P[X_n \to X] = 1$, i.e., $P[|X_n - X| < \epsilon$ for $n = n_0+1, n_0+2, \ldots, n_0+r] > 1-\eta$ for all r. Putting $r = 1$, we get $P\{|X_n - X| < \epsilon\} > 1 - \eta$ for $n \ge n_0$, which is convergence in probability.

However, the converse of the theorem is not true, i.e., it is possible to have a sequence $\{X_n\}$ which converges in probability to a random variable X, but does not converge almost surely to the random variable X.

Thus *a.s.* convergence is a stronger mode of convergence than convergence in probability.

EXAMPLE 12.2.2 : We first note that any integer n can be written as $n = 2^k + m$, $m = 0, 1, \ldots, 2^k$.

On the sample space $S = [0, 1]$, define a random variable X_n such that

$$X_n(\omega) = \begin{cases} 2^k, & \text{for } \omega \in [\frac{m}{2^k}, \frac{m+1}{2^k}] \\ 0, & \text{otherwise .} \end{cases}$$

Also, let the *pmf* of X_n be such that $P[X_n = j] = $ length of the interval I for which $X_n(\omega) = j \; \forall \; \omega \in I$. Thus,

$$P[X_n = 2^k] = \frac{1}{2^k}, \quad P[X_n = 0] = 1 - \frac{1}{2^k}.$$

Hence, $\lim_{n\infty} X_n(\omega)$ does not exist. Therefore X_n does not converge almost surely.

However,

$$P\{|X_n| > \epsilon\} = \begin{cases} 0, & \text{if } \epsilon \ge 2^k \\ \frac{1}{2^k}, & \text{if } 0 < \epsilon < 2^k. \end{cases}$$

Therefore, $P(|X_n| > \epsilon) \to 0$ as $n \to \infty$, i.e., $X_n \to^P 0$. $\qquad\square$

Although the converse of Theorem 12.2.8 is not true in general, there is a partial converse, which is stated in the following theorem.

Theorem 12.2.9 : If a sequence $\{X_n\}$ converges in probability to a random variable X, there exist a subsequence $\{X_{n_k}\}(k = 1, 2, \ldots)$ which converges almost surely to X. Moreover, for this subsequence

$$\sum_{k=1}^{\infty} P(|X_{n_k} - X| > 2^{-k}) < \infty. \qquad (12.2.16)$$

For a proof, see Example 12.4.2.

12.2.3 Convergence in the rth mean

DEFINITION 12.2.3: Let $\{X_n\}$ and X be random variables having moments of rth order. Then $|X_n - X|$ has also moments of rth order. The sequence $\{X_n\}$ is said to converge in the rth mean to a (possibly degenerate) random variable X if

$$E|X_n - X|^r \to 0 \text{ as } n \to \infty, \; r > 0, \qquad (12.2.17)$$

that is, if for any given positive number ϵ, there exists a positive integer $n_0(\epsilon)$ such that

$$E|X_n - X|^r < \epsilon \text{ for } n \geq n_0. \qquad (12.2.18)$$

This is also called L_r-convergence, $X_n \to^{L_r} X$. For $r = 2$, this is called the mean-square convergence. $X_n \to^{m.s.} X$ if $E|X_n - X|^2 \to 0$.

Theorem 12.2.10: Convergence in rth mean satisfies the Cauchy criterion, that is, $X_n \to^{L_r} X$ if

$$E(|X_n - X_m|^r) \to 0.$$

Theorem 12.2.11: If $X_n \to^{L_r} X$, then

$$E|X_n|^r \to E|X|^r.$$

Proof. Suppose $r \leq 1$. Then by X_r-inequality (6.4.21),

$$E|X_n|^r \leq E(|X_n - X|^r + E|X|^r),$$

$$|E|X_n|^r - E|X|^r| \leq E|X_n - X|^r.$$

For $r \geq 1$, the result follows from Minkowski's inequality (6.4.20).

EXAMPLE 12.2.3: Consider $\{X_n\}$ where

$$P(X_n = n) = \frac{1}{n^r}, \ P(X_n = 0) = 1 - \frac{1}{n^r}, \ r > 0.$$

Then

$$P\{|X_n| > \epsilon\} = \begin{cases} 0, & \epsilon \geq n, \\ P(X_n = n) = \frac{1}{n^r}, & \epsilon < n. \end{cases}$$

Therefore, $P\{|X_n| > \epsilon\} \to 0$ as $n \to \infty$.

However, $E|X_n|^r = 1$, so that $X_n \not\to^{L_r} 0$.

Theorem 12.2.12: Let $\{X_n\}$ be a sequence of random variables and X another random variable. Then

$$[X_n \to^{L_r} X] \Rightarrow [X_n \to^{L_s} X; 0 \leq s \leq r] \Rightarrow [X_n \to^P X]. \qquad (12.2.19)$$

Proof. (a) L_r-convergence implies L_s-convergence ($r \geq s$): We have

$$[X_n \to^{L_r} X] \Rightarrow E|X_n - X|^r \to 0.$$

Similarly,

$$[X_n \to^{L_s} X] \Rightarrow E|X_n - X|^s \to 0.$$

Now,

$$(E|X_n - X|^r)^{\frac{1}{r}} \geq (E|X_n - X|^s)^{\frac{1}{s}}, \ r > s > 0 \ \text{(by Theorem 6.4.3)}. \qquad (12.2.20)$$

Hence, if $X_n \to^{L_r} X$, the left hand side of (12.2.20)$\to 0$, which in turn implies $X_n \to^{L_s} X$.

(ii) L_r-convergence implies P-convergence.

By Markov's inequality (6.10.2),

$$P\{|X_n - X| > \epsilon\} \leq \frac{E|X_n - X|^r}{\epsilon^r}, \ r > 0,$$

because $|X_n - X|^r$ is a positive increasing function of $|X_n - X|$; but $E|X_n - X|^r \to 0$; hence $P\{|X_n - X| > \epsilon\} \to 0$ for every ϵ.

Note 12.2.1: Convergence in probability does not imply convergence in rth mean.

We give two examples in this context.

EXAMPLE 12.2.4: Let $X_n \cap$ Bin. $(n, \pi), 0 < \pi < 1$. Then, $E(T_n) = \pi, V(T_n) = \frac{\pi(1-\pi)}{n} < \frac{1}{4n}$ where $T_n = \frac{X_n}{n}$. Thus, $T_n \to^P \pi$.

Suppose we are interested in the parameter $\theta = \frac{1}{\pi}$. Since $Z_n = \frac{1}{T_n}$ is a continuous function of T_n (except at $T_n = 0$) and θ is a continuous function of π (except at $\pi = 0$), it follows by Theorem 12.2.6, that $Z_n \to^P \theta$. However, $P(X_n = 0) = (1 - \pi)^n > 0$. Therefore, for $r > 0, E(Z_n^r)$ is not finite. Hence, Z_n does not converge in rth mean to θ for any $r > 0$.

EXAMPLE 12.2.5: Let $\{X_n\}$ be a sequence of random variables with

$$P(X_n = 0) = 1 - \frac{1}{n}, P(X_n = n^{\frac{2}{r}}) = \frac{1}{n}.$$

Then

$$P\{|X_n| \le \epsilon\} = P(X_n = 0) = 1 - \frac{1}{n} > 1 - \delta \text{ for } n > \frac{1}{\delta}.$$

Hence, $X_n \to^P 0$. Now $E|X_n|^r = n \to \infty$. Therefore, $X_n \not\to^{L-r} 0$.

Theorem 12.2.13: Although convergence in rth mean is stronger than P-convergence, it is neither implied by nor implies convergence *a.s.*

Proof. To show this we give an example. Let X_n be a sequence of independent random variables with $P(X_n = 0) = 1 - \frac{1}{n}, P(X_n = n^{\frac{1}{2r}}) = \frac{1}{n}$. Let X be a degenerate random variable with $P(X = 0) = 1$.

Therefore, $E|X_n|^r = n^{-\frac{1}{2}} \to 0$; but

$$P(X_n = 0 \,\forall\, m \le n \le N) = \Pi_{n=m}^N (1 - \frac{1}{n}) \qquad (12.2.21)$$

which diverges to zero as N increases for all values of m. Hence, X_n does not converge *a.s.* to X.

Suppose now, X_n is another random variable with $P(X_n = 0) = 1 - \frac{1}{n^2}, P(X_n = e^n) = \frac{1}{n^2}$. Then, for any $r > 0$,

$$E|X_n|^r = \frac{e^{nr}}{n^2} \to \infty.$$

Therefore, X_n does not converge to zero in the rth mean. However,

$$P(X_n = 0 \,\forall\, m \le n \le N) = \Pi_{n=m}^N (1 - \frac{1}{n^2}). \qquad (12.2.22)$$

As both m and N increase, subject to the condition $m < N$, this infinite product converges to unity. Thus X_n converges to zero a.s. □

Although, convergence in probability does not in general imply convergence in the rth mean, there is a special type of random variables, for which convergence in probability implies rth mean convergence, $r > 0$. This is stated in the following theorem.

Theorem 12.2.14: Convergence in probability for *almost surely bounded* random variables implies convergence in the rth mean, for every $r > 0$.

Proof. A random variable X is said to be bounded, if there exists a positive constant $K(< \infty)$ such that $P\{|X| < K\} = 1$.

Suppose that $X_n - X \to^P 0$ so that $P\{|X_n - X| < K\} = 1$ for some $K \in (0, \infty)$. Therefore, writing $P\{|X_n - X| \le t\} = F(t)$, we have,

$$E|X_n - X|^r = \int_{|t| \le K} |t|^r dF(t) = \int_{|t| \le \epsilon} |t|^r dF(t) + \int_{K \ge |t| > \epsilon} |t|^r dF(t)$$

$$< \epsilon^r dP\{|X_n - X| \le \epsilon\} + K^r P\{|X_n - X| > \epsilon\}.$$
$$(12.2.23)$$

Since $X_n - X \to^P 0$ and ϵ is arbitrary, the right hand side of (12.2.23) can be made arbitrarily small by choosing ϵ sufficiently small and n sufficiently large. This ensures $|X_n - X|$ converges to zero in the rth mean. Since $r(> 0)$ is arbitrary, the proof is complete. $\square$

Theorem 12.2.15: Let $w = g(t), t \in (0, \infty)$ be a monotone non-decreasing continuous and bounded function of t, $g(0) = 0$. Let $W_n = g(|X_n - X|), n \ge 1$. Then $X_n - X \to^P 0$ *iff* $W_n \to^{L_r} 0$ for some $r > 0$.

Proof. Suppose $X_n - X \to^P 0$. Then $W_n = g(|X_n - X|) \to^P 0$ (Theorem 12.2.6).

By construction W_n is *a.s.* bounded. Therefore, by Theorem 12.2.14, $(W_n \to^P 0) \Rightarrow (W_n \to^{L_r} 0)$ for any $r(> 0)$.

Alternatively, if $W_n \to^{L_r} 0$ for some r, then by Theorem 12.2.12, $W_n \to^P 0$, and this in turn implies $|X_n - X| \to^P 0$. Hence the proof. $\square$

In practice, we may choose $W_n = g(|X_n - X|)$ by taking

$$W_n^r = \frac{|X_n - X|^r}{1 + |X_n - X|^r}$$

for an appropriate choice of $r(> 0)$.

EXAMPLE 12.2.6: Consider the example of Binomial variables under note 12.2.1. We are, here, interested in checking convergence of $Z_n = \frac{1}{T_n} = \frac{n}{X_n}$ to $\theta = \frac{1}{\pi}$.

We consider

$$W_n^2 = \frac{(Z_n - \theta)^2}{1 + (Z_n - \theta)^2} = \frac{(n\pi - X_n)^2}{\pi^2 X_n^2 + (n\pi - X_n)^2}$$

$$= \frac{(T_n - \pi)^2}{(T_n - \pi)^2 + \pi^2 T_n^2}.$$

Now, $E(T_n - \pi)^2 \to 0$, i.e., $T_n - \pi \to^{L_2} 0$. This implies $W_n \to^{L_2} 0$ and hence by Theorem 12.2.15, $Z_n - \theta \to^P 0$.

12.2.4 Convergence in distribution

Let $\{X_n\}$ be a sequence of random variables, all defined on the probability space $(\mathcal{S}, \mathcal{A}, P)$ with $F_n(x)$ as the distribution function of $X_n, n = 1, 2, \ldots$

DEFINITION 12.2.4: Let $\{F_n\}$ be a sequence of distribution functions. If there exists a distribution function F such that

$$F_n(x) \to F(x)$$

at every point x where $F(x)$ is continuous, then F_n is said to *converge weakly* to F and we write

$$F_n \to^w F. \tag{12.2.24}$$

If $\{X_n\}$ be a sequence of random variables with $\{F_n\}$ as the corresponding sequence of distribution functions and if there exists a random variable X with distribution function F such that $F_n \to^w F$, then we say that X_n converges in law or distribution to X and we write

$$X_n \to^L X \text{ or } \mathcal{L}(X_n) \to \mathcal{L}(X) \text{ or } X_n \to^D X. \tag{12.2.25}$$

To establish if in a certain situation (12.2.25) holds, one, therefore, has to verify, if for any given $\epsilon(> 0)$, there exists an integer $n_0 = n_0(\epsilon)$, such that at every continuity point x of F,

$$|F_n(x) - F(x)| < \epsilon, \ n \geq n_0(\epsilon). \tag{12.2.26}$$

When $X_n \to^L X$ where X has a degenerate distribution with $P[X = c] = 1$, $'c'$ being a constant, we will write $X_n \to^L c$.

Note 12.2.2: The variables X_n and X need not be defined on the same sample space.

Note 12.2.3: If the sequence $\{F_n\}$ converges to F, it is not necessary that F will be a distribution function always.

Note 12.2.4: As the definition 12.2.4 implies, it is not necessary that $\{F_n(x)\}$ should converge to $F(x)$ at every discontinuity point of $F(x)$, even if $F_n \to^w F$.

EXAMPLE 12.2.7: Suppose X is a degenerate random variable (at the point t) so that its distribution function is

$$F(x) = \begin{cases} 0, & x < t, \\ 1, & x \geq t. \end{cases}$$

Therefore, if $F_n \to^w F$, then (12.2.25) entails that

$$|F_n(x) - 0| < \epsilon, \quad \text{for } x < t, \text{ i.e. } F_n(x) \to 0 \text{ for } x < t,$$

$$|F_n(x) - 1| < \epsilon \text{ for } x \geq t, \text{ i.e. } F_n(x) \to 1 \text{ for } x \geq t;$$

that is,

$$X_n \to^P t \text{ as } n \to \infty.$$

We have, therefore, the following lemma.

Lemma 12.2.1: For a sequence $\{X_n\}$ of random variables, convergence in law to a random variable X, degenerate at a point t, ensures that $X_n \to t$ in probability as $n \to \infty$.

EXAMPLE 12.2.8: Consider $\{X_n\}$ where $P(X_n = n) = 1, n = 1, 2, \ldots$ Here

$$F_n(x) = \begin{cases} 0, x < n \\ 1, x \geq n. \end{cases}$$

Therefore, $\lim_{n \to \infty} F_n(x) = 0 \ (-\infty < x < \infty)$. Thus, $\{F_n\}$ does not converge to a distribution function.

EXAMPLE 12.2.9: Let $\{X_n\}$ be a sequence of independent random variables, each having a uniform distribution over $(0, \theta)$. Let $X_{(n)} = \max_{1 \leq i \leq n} X_i$. The distribution function of X_n is

$$F_{(n)}(x) = \begin{cases} 0, & x \leq 0 \\ (\frac{x}{\theta})^n, & 0 < x < \theta, \\ 1, & x \geq \theta. \end{cases}$$

Consider $Y_n = \frac{X_{(n)}}{n}$. The distribution function of Y_n is

$$F_n(x) = P(Y_n \leq x) = P(X_{(n)} \leq nx) = \begin{cases} 0, & x \leq 0, \\ (\frac{x}{\theta})^n, & 0 < x < \frac{\theta}{n}, \\ 1, & x \geq \frac{\theta}{n}. \end{cases}$$

Hence,

$$\lim_{n \to \infty} F_n(x) = F(x) = \begin{cases} 0, x < 0, \\ 1, x \geq 0. \end{cases}$$

$F(x)$ is the distribution function of a degenerate random variable X with $P(X = 0) = 1$. Hence, $F_n \to^w F$ and $X_n \to^L X$. Note that $x = 0$ is a point of discontinuity of $F(x)$; $F_n(0) = 0$ for all n and $F(0) = 1$, so that F_n does not converge to F at the point of discontinuity of F, namely, $x = 0$. $\quad\square$

The following theorems hold.

Theorem 12.2.16: Let $\{X_n\}$ be a sequence of random variables such that $X_n \to^L X$. Then, for any constant $'a'$,

$$a + X_n \to^L a + X, \; aX_n \to^L aX \; (a \neq 0). \tag{12.2.27}$$

Theorem 12.2.17: Let $\{X_n\}, \{Y_n\}$ be two sequences of random variables with $X_n \to^L X, Y_n \to^L c$ where X is a random variable and $'c'$ is a constant. Then

(i) $X_n + Y_n \to^L X + c$,

(ii) $X_n Y_n \to^L cX$,

(iii) $\frac{X_n}{Y_n} \to^L \frac{X}{c} \; (c \neq 0)$.

By virtue of Lemma 12.2.1, the Theorem 12.2.17 holds if we replace $'Y_n \to^L c'$ by $Y_n \to^P c$.

Theorem 12.2.18: P-convergence implies convergence in law.

To prove this theorem, first we consider the following lemma.

Lemma 12.2.2: Let X, Y be two random variables with distribution functions $F(x), G(y)$ respectively. Let ϵ, δ be two positive constants such that

$$P(|X - Y| \leq \epsilon) \geq 1 - \delta. \tag{12.2.28}$$

Then

$$|F(x) - G(x)| \leq |F(x + \epsilon) - F(x - \epsilon)| + \delta. \tag{12.2.29}$$

Proof of Lemma 12.2.2: Let A be the event $[|X - Y| \leq \epsilon], B$ be the event $[Y \leq x]$. Then

$$P[|X - Y| \leq \epsilon; \; Y \leq x] \geq P[Y \leq x] - \delta, \; [\text{ since } P(AB) \geq P(A) + P(B) - 1].$$

Hence,

$$P[X \leq x + \epsilon] \geq P[Y \leq x] - \delta. \tag{12.2.30}$$

Similarly, considering the events A and B^c,

$$P[X > x - \epsilon] \geq P[Y > x] - \delta. \tag{12.2.31}$$

From (12.2.30) and (12.2.31) it follows that

$$F(x - \epsilon) - \delta \leq G(x) \leq F(x + \epsilon) + \delta.$$

Since $F(x - \epsilon) \leq F(x) \leq F(x + \epsilon)$,

$$F(x - \epsilon) - F(x + \epsilon) - \delta \leq F(x) - G(x) \leq F(x + \epsilon) - F(x - \epsilon) + \delta.$$

The proof is complete.

Proof of Theorem 12.2.18: Suppose $X_n \to^P X$ so that

$$P(|X_n - X| \leq \epsilon) \geq 1 - \delta$$

for every $\epsilon(> 0), \delta(> 0)$ and n sufficiently large.

By Lemma 12.2.2,

$$|F(x) - F_n(x)| \leq F(x + \epsilon) - F(x - \epsilon) + \delta. \tag{12.2.32}$$

If x is a continuity point of F, the right hand side of (12.2.32) can be made arbitrarily small by choosing ϵ and δ sufficiently small. Hence, $F_n \to^w F$ and $X_n \to^L X$, where F is the distribution function of X. □

However, the converse of Theorem 12.2.18 is not true, except in a special case, which is depicted below.

Theorem 12.2.19: If $X_n \to^P a$, where $'a'$ is a constant, then $[X_n \to^P a] \Leftrightarrow [X_n \to^L a]$.

Proof. (a) $[X_n \to^P a] \Rightarrow [X_n \to^L a]$ follows by Theorem 12.2.18. This can also be seen as follows. We have

$$[X_n \to^P a] \Rightarrow P\{|X_n - a| \leq \epsilon\} \to 1 \text{ for any } \epsilon(> 0)$$

i.e., $P\{X_n \leq a - \epsilon; \ X_n \geq a + \epsilon\} \to 0$.

i.e., $P\{X_n \leq a - \epsilon\}.P\{X_n \geq a + \epsilon\} \to 0$ (the events being independent),

i.e. $F_n(a - \epsilon) \to 0$ and $1 - F_n(a + \epsilon - 0) \to 0$, \hfill (i)

where F_n is the distribution function of X_n. Since ϵ is arbitrary, taking limit of (i) as $n \to \infty$

$$\lim_{n \to \infty} F_n(x) = F(x) \text{ (say)}) = \begin{cases} 0, & x < a \\ 1, & x > a, \end{cases} \tag{ii}$$

i.e., $X_n \to^L X$ where $P(X = a) = 1$.

(ii) For the converse part, a proof has been given in Lemma 12.2.1. An alternative proof is given below.

Let $X_n \to^L a$, i.e., limiting distribution of X_n is $F(x)$ in (ii). Now,

$$P\{|X_n - a| \geq \epsilon\} = P\{X_n \leq a - \epsilon\}.P\{X_n \geq a + \epsilon\} = F_n(a - \epsilon)\{1 - F_n(a + \epsilon - 0)\}. \tag{iii}$$

Taking limit of both sides of (iii),

$$\lim P\{|X_n - a| \geq \epsilon\} = F(a - \epsilon)\{1 - F(a + \epsilon - 0)\} = 0(1 - 1) = 0. \quad \square$$

The following results hold.

Let the sequence $F_n \to^w F$, the distribution function of X. Then

$$\lim_{n \to \infty} P(a < X_n \leq b) = F(b) - F(a), \qquad (12.2.33)$$

for any two points of continuity a, b of $F(x), a < b$.

Note 12.2.5: Whenever we say that a standardized variable

$$\frac{T_n - E(T_n)}{\sqrt{V(T_n)}} = Z_n \text{ (say) } \to \tau \text{ as } n \to \infty,$$

where τ is a standard $N(0, 1)$ variable, we actually mean $Z_n \to^L \tau$, i.e., $F_{T_n}(z) \to^w \Phi(z)$ for all $z \in (-\infty, \infty)$ (since $\Phi(z)$ has no point of discontinuity).

Theorem 12.2.20: (a) If X_n, X are continuous random variables, with density functions $f_n(x)(n = 1, 2, \ldots), f(x)$ respectively, then $X_n \to^K X$ *iff*

$$f_n(x) \to f(x) \text{ almost everywhere as } n \to \infty.$$

(b) If X_n, X are discrete random variables with *pmf*s

$$P(X_n = j) = p_n(j), \ P(X = j) = p_j, \ j = 0, 1, 2, \ldots; \ n = 1, 2, \ldots,$$

then $X_n \to^L X$ *iff*

$$p_n(j) \to p(j) \text{ for all } j = 0, 1, 2, \ldots$$

It will be often tedious to determine if $F_n \to^w F$ by calculating the limit of $F_n(x)$ at every continuity point of $F(x)$. The following theorems stated without proof show that this difficulty can be overcome by part by the use of characteristic function. The proof can be seen in the treatise by Lukcas (1960), Moran (1968), among others.

Theorem 12.2.21: Let $\{F_n(x)\}$ be a sequence of distribution functions with characteristic functions $\psi_n(t)$. Then a necessary and sufficient condition that the sequence $F_n(x)$ converges weakly to a proper distribution function $F(x)$ is that the sequence $\psi_n(t)$ converges for each real t to a function $\psi(t)$ which is continuous at $t = 0$. $\psi(t)$ is then the characteristic function of $F(x)$.

A closely related theorem imposes uniform convergence of $\psi_n(t)$ in some interval instead of continuity at the origin and may be stated as follows.

Theorem 12.2.22 (*Levy-Cramer*): A necessary and sufficient condition that the sequence of distribution functions $F_n(x)$ converges weakly to a proper distribution function $F(x)$, is that the sequence $\psi_n(t)$ converges to a function $\psi(t)$ for all real t, and the convergence is uniform in some finite interval containing $t = 0$ in its interior. The limiting function $\psi(t)$ is the characteristic function of the limiting distribution function $F(x)$.

The above theorems relate the convergence of the sequence $\{F_n(x)\}$ to the convergence of $\psi_n(t)$ on the real axis. Since we know $\psi_n(t)$ always exists, this is a very useful theorem. It is, however, useful to have a corresponding theorem for moment generating functions and this is stated in the following theorem.

Theorem 12.2.23: Let $\{X_n\}$ be a sequence of random variables and let $\{F_n(x)\}$ and $M_n(t)$ be respectively the distribution functions and moment generating function of $F_n(x)$, where $M_n(t)$ exists for $|t| < T$ for every n. If there exists a distribution function F with corresponding *mgf* $M(t)$ which exists for $|t| \le T_1 \le T$ such that $M_n(t) \to M(t)$ as $n \to \infty$ for every $|t| \le T_1$, then $F_n(x)$ converges to the distribution function $F(x)$.

EXAMPLE 12.2.10: Consider $\{X_n\}$ of Example 12.2.7, where $F_n(x) = 0(1)$ for $x < n (\ge n)$. Here $\psi_n(t) = e^{itn} = \cos nt + i \sin nt$.

For $t = \frac{\pi}{2}, \psi_n(t)$ takes the values $1, i, -i, 1$ only and hence $\psi_n(t)$ does not converge for all real values of t. Hence, $F_n(x)$ does not converge.

EXAMPLE 12.2.11: Let $\{X_n\}$ be a sequence of independent Cauchy random variables. Consider the sequence $Y_n = \frac{S_n}{n^2}$ where $S_n = \sum_{i=1}^{n} X_i$. The characteristic function of Y_n is $\psi_n(t) = e^{-\frac{|t|}{n}}$ (vide Example 7.6.5). Hence

$$\lim_{n \to \infty} \psi_n(t) = \psi(t) \text{ (say) } = \begin{cases} 1, t = 0, \\ 0, t \ne 0. \end{cases}$$

The function $\psi(t)$ is not continuous at $t = 0$ and cannot be a characteristic function. Therefore, $\frac{S_n}{n^2}$ does not converge in law.

EXAMPLE 12.2.12: Let $\{X_n\}$ be a sequence of random variables such that

$$P(X_n = \frac{k}{n}) = \frac{1}{n}, \; k = 1, \ldots, n.$$

The characteristic function of X_n is

$$\psi_n(t) = E(e^{itX_n}) = \frac{1}{n} \sum_{k=1}^{n} e^{\frac{itk}{n}} = \frac{e^{\frac{it}{n}}}{n} \left(\frac{1 - e^{it}}{1 - e^{\frac{it}{n}}} \right).$$

Now,

$$\lim_{n \to \infty} \frac{n(1 - e^{\frac{it}{n}})}{e^{\frac{it}{n}}} = -it.$$

Hence,

$$\lim_{n \to \infty} \psi_n(t) = \frac{e^{it} - 1}{it} = \psi(t),$$

which is the characteristic function of the uniform distribution over $[0, 1]$. Hence X_n converges in law to a uniform distribution. □

We now apply Theorems 12.2.21 and 12.2.22 to show that for many distributions, the limiting distribution is a standard normal distribution.

EXAMPLE 12.2.13: Let $\{X_n\}$ be a sequence of binomial variables

$$P(X_n = r) = \binom{n}{r} p^r q^{n-r}, \ 0 < p < 1, \ q = 1 - p; \ r = 0, 1, \dots, n.$$

Here $E(X_n) = np, V(X_n) = npq$. Consider the sequence of standardized variables

$$Y_n = \frac{(X_n - np)}{\sqrt{npq}}.$$

The characteristic function of Y_n is

$$\psi_n(t) = E(e^{it(X - np)/\sqrt{npq}})$$
$$= e^{-it\sqrt{npq}}(pe^{it/\sqrt{npq}} + q)^n.$$

$$\log_e \psi_n(t) = -it\sqrt{np/q} + n \log_e(pe^{it/\sqrt{npq}} + q)$$
$$= -it\sqrt{np/q} + n \log_e \left[1 + p\frac{it}{\sqrt{npq}} - \frac{pt^2}{2npq} + o(\frac{t^2}{n})\right]$$
$$\left(\text{since } \log_e(1 + x) = x - \frac{x^2}{2} + \frac{x^3}{3} - \dots, \ |x| < 1 \right)$$
$$= -it\sqrt{np/q} + n\left[\frac{ipt}{\sqrt{npq}} - \frac{pt^2}{2npq} + \frac{p^2t^2}{2npq} + 0(\frac{t^3}{n^{3/2}})\right]$$
$$= -\frac{t^2}{2} + 0(\frac{t^3}{\sqrt{n}}).$$

Thus, for any finite $t, \log_e \psi_n(t)$ converges to $-t^2/2$ uniformly and hence $\psi_n(t) \to e^{-t^2/2}$ uniformly in t. The latter being the characteristic function of the standard normal distribution, it follows that the distribution function of Y_n converges to that of a standard normal distribution.

We have thus proved De-Moivre Laplace Theorem (Theorem 9.7.2). The same theorem is stated below in a slightly different form.

Theorem 12.2.24: Let $\{X_n\}$ be a sequence of binomial variables with parameters (n, p),

$$Y_n = \frac{X_n - np}{\sqrt{npq}}$$

and let $F_n(Y)$ be the distribution function of Y_n. If $0 < p < 1$, the relation

$$\lim_{n \to \infty} F_n(Y) = \frac{1}{\sqrt{2\pi}} \int_{-\infty}^{y} e^{-\frac{z^2}{2}} dz \qquad (12.2.34)$$

holds for every value of y (since the distribution function of a normal distribution has no point of discontinuity, convergence considered in (12.2.34) holds for every value of y).

Corollary 12.2.24.1 The sequence $\{\frac{X_n}{n}\}$ is asymptotically $N(p, \sqrt{\frac{pq}{n}})$ (vide Definition 12.5.1).

For a yet-another proof, see Example 12.5.2.

EXAMPLE 12.2.14: Let $\{X_n\}$ be a sequence of binomial variables with parameters (n, p). Let $n \to \infty$ in such a way that np remains a constant at λ.

The characteristic function of X_n is

$$\psi_n(t) = (q + pe^{it})^n = \left(1 - \frac{\lambda}{n} + \frac{\lambda}{n} e^{it}\right)^n$$
$$= \left(1 + \frac{\lambda}{n}(e^{it} - 1)\right)^n \to \exp\{\lambda(e^{it} - 1)\} = \psi(t),$$

as $n \to \infty$ $\psi(t)$ is the characteristic function of a Poisson distribution with parameter λ. Hence, the binomial distribution approaches a Poisson distribution. (An alternative proof was given in Lemma 8.6.2.)

EXAMPLE 12.2.15: Let $\{X_n\}$ be a sequence of Poisson variables with parameters λ. Consider the standardized variables $Y_n = \frac{X_n - \lambda}{\sqrt{\lambda}}$. The characteristic function of Y_n is

$$\psi_n(t) = e^{-it\sqrt{\lambda}} \exp\{\lambda(e^{it\sqrt{\lambda}} - 1)\}].$$

$$\log_e \psi_n(t) = -it\sqrt{\lambda} + \lambda(e^{it/\sqrt{\lambda}} - 1) = -it\sqrt{\lambda} + \lambda\left(\frac{it}{\sqrt{\lambda}} - \frac{t^2}{2\lambda} - \frac{it^3}{6\lambda^{\frac{3}{2}}} + \ldots\right)$$
$$= -\frac{t^2}{2} - \frac{it^3}{6\lambda^{\frac{1}{2}}} + \ldots$$

Hence,

$$\psi_n(t) \to \exp(-t^2/2) \text{ as } n \to \infty.$$

Thus Y_n tends in distribution to a $N(0, 1)$-variable.

EXAMPLE 12.2.16: Let $\{X_n\}$ be a sequence of independently and identically distributed $\chi^2_{(1)}$ random variables. Let $Y_n = \frac{S_n - n}{\sqrt{2n}}$, where $S_n = \sum_{i=1}^{n} X_i$.

The *m.g.f.* of Y_n is

$$M_n(t) = E(e^{tY_n}) = \left(1 - \sqrt{\frac{2}{n}}t\right)^{-\frac{n}{2}} \exp[(-t\sqrt{\frac{n}{2}})], \ |t| < \sqrt{\frac{n}{2}}.$$

$$\log_e M_n(t) = -t\sqrt{\frac{n}{2}} - \frac{n}{2}\log_e(1 - \sqrt{\frac{2}{n}}t)$$
$$= -t\sqrt{\frac{n}{2}} + \frac{n}{2}\left[t\sqrt{\frac{2}{n}} + \frac{1}{2}(t\sqrt{\frac{2}{n}})^2 + \ldots\right]$$
$$= \frac{t^2}{2} + 0(\frac{1}{n}).$$

Hence, $M_n(t) \to e^{\frac{t^2}{2}}$ and $Y_n \to^L \tau$ where τ is a $N(0,1)$-variable. $\square$

An important question regarding the weak convergence of F_n to F is whether the point-wise convergence holds uniformly. The following result is quite useful.

Theorem 12.2.25: (*Polya*): If $F_n \to^w F$ and F is continuous, then

$$\lim_{n\to\infty} \sup_t |F_n(t) - F(t)| = 0.$$

We consider without proof two relevant results in the theory of Stieltjes integrals known as the first and the second theorems of Helly.

Theorem 12.2.26 (*Helly's First Theorem*): Every sequence, $\{F_n(x)\}$, of uniformly bounded non-decreasing functions contains a subsequence $F_n(x)$ which converges weakly to a non-decreasing bounded function $F(x)$.

Theorem 12.2.27 (*Helly's Second Theorem*): Suppose that $g(x)$ is a continuous function, and that $\{F_n(x)\}$ is a sequence of distribution functions converging weakly to a function $F(x)$, for all values of x in a finite closed interval $[a, b]$ for which $F(x)$ is continuous at $x = a$ and $x = b$. Then

$$\int_a^b g(x)dF_n(x) \to \int_a^b g(x)dF(x)$$

in the sense of ordinary convergence.

12.2.5 *Complete convergence*

There is another mode of convergence, which is stronger than almost sure convergence.

DEFINITION 12.2.5: A sequence $\{X_n\}$ of random variables is said to converge completely to a (possibly degenerate) random variable X, if for every $\epsilon(> 0)$,

$$\sum_{n=1}^{\infty} P\{|X_n - X| > \epsilon\} < \infty. \tag{12.2.35}$$

This is denoted as $X_n \to^c X$.

Theorem 12.2.28: Complete convergence implies almost sure convergence.

Proof. We note that the convergence of the series in (12.2.35) implies that $\sum_{n \geq n_0} P\{|X_n - X| > \epsilon\} \to 0$ as $n_0 \to \infty$.

Denoting the event $[|X_k - X| > \epsilon]$ as A_k, we note that $P(\cup_{k=n_0}^{\infty} A_k) \leq \sum_{k=n_0}^{\infty} P(A_k)$ (Boole's inequality).

Now, from (12.2.8), a.s. convergence entails $P(\cup_{n_0}^N A_n) \to 0$ as $n_0 \to \infty$. Hence, if $X_n \to^c X, X_n \to^{a.s.} X$. Thus $[X_n \to^c X] \Rightarrow [X_n \to^{a.s.} X]$. □

However, the converse may not be true.

EXAMPLE 12.2.17: Consider the Binomial variable of Example 12.2.3 with $T_n = \frac{X_n}{n}$ and $T = \pi$. We note that for $\pi \in [0, 1]$,

$$E(T_n - \pi)^4 = \frac{\pi(1-\pi)}{n^3}[1 + 3\pi(1 - \pi)(n - 2)]$$
$$\leq \frac{3}{16n^2}. \tag{12.2.36}$$

By Chebychev's inequality,

$$P\{|T_n - \pi| > \epsilon\} = P\{|T_n - \pi|^4 > \epsilon^4\}$$
$$\leq \frac{E(T_n - \pi)^4}{\epsilon^4} \leq \frac{3}{n^2 16\epsilon^4}.$$

Therefore, the series in (12.2.35) converges and hence $T_n \to^c \pi$.

12.2.6 *Gradation of different modes of convergence*

In Theorems 12.2.7, 12, 18, 25 we have proved the implications of different modes of convergence on the others. The results are summarized in the following theorem.

Theorem 12.2.29: Let $\{X_n\}$ be a sequence of random variables and X another random variable. Then

(a) $[X_n \to^c X] \Rightarrow [X_n \to^{a.s.} X] \Rightarrow [X_n \to^P X] \Rightarrow [X_n \to^L X]$

(b) $[X_n \to^{L_r} X] \Rightarrow [X_n \to^{L_s} X; \ 0 \leq s \leq r] \Rightarrow [X_n \to^P x]$.

The converses are not true in general.

12.3 Weak Law of Large Numbers

We shall now consider the convergence of sequence of partial sums, $S_n = X_1 + \ldots + X_n, n \geq 1$, where $X_1, X_2, \ldots$ is the a sequence of random variables, in the various senses of convergence, depicted in the previous section. The convergence of S_n in the sense of probability is related to the Weak Law of Large Numbers. A precise definition is given below.

DEFINITION 12.3.1: Let $\{X_n\}$ be a sequence of random variables and $S_n = \sum_{k=1}^{n} X_k, n = 1, 2, \ldots$. We say that $\{X_n\}$ obeys the weak law of large numbers (WLLN) with respect to a set of non-negative constants $\{B_n\}$ with $B_n \to \infty$ as $n \to \infty$, if there exists a sequence of real constants $\{A_n\}$ such that

$$\frac{X_n - A_n}{B_n} \to^P 0 \text{ as } n \to \infty.$$

The constants A_n are called 'centering constants' and B_n the 'normalizing constants'.

In this context we have the following theorem.

Theorem 12.3.1: Let $\{X_i\}$ be a sequence of pairwise uncorrelated random variables with $E(X_i) = \mu_i, V(X_i) = \sigma_i^2, i = 1, 2, \ldots$. If $\sum_{i=1}^{n} \sigma_i^2 \to \infty$ as $n \to \infty$, choosing $A_n = \sum_{i=1}^{n} \mu_i, B_n = \sum_{i=1}^{n} \sigma_i^2$, we have

$$\frac{S_n - \sum_{i=1}^{n} \mu_i}{\sum_{i=1}^{n} \sigma_i^2} \to^P 0 \text{ as } n \to \infty. \tag{12.3.1}$$

Proof. First we recall Chebychev's inequality: If X is a random variable with $E(X) = \mu, V(X) = \sigma^2$,

$$P\{|X - \mu| \leq \epsilon) \geq 1 - \frac{\sigma^2}{\epsilon^2} \tag{12.3.2}$$

for any arbitrary real $\epsilon(> 0)$. By Chebychev's inequality,

$$P\{|S_n - \sum_{i=1}^{n} \mu_i| > \epsilon \sum_{i=1}^{n} \sigma_i^2\} \leq \frac{E\{\sum_{i=1}^{n}(X_i - \mu_i)\}^2}{\epsilon^2 (\sum_{i=1}^{n} \sigma_i^2)^2}$$

$$= \frac{1}{\epsilon^2 \sum_{i=1}^{n} \sigma_i^2} \to 0 \text{ as } n \to \infty. \quad \square$$

From now on, we shall consider $B_n = n$ and say that $\{X_n\}$ satisfies the WLLN if $\bar{X}_n \to c$ where $'c'$ is a constant and $\bar{X}_n = S_n/n..$

We shall now consider different situations where WLLN is satisfied.

(a) **Bernoulli's WLLN**: Consider n independent Bernoulli trials each with probability of success p. Let $X_i = 1(0)$ if the ith trial is a success (failure) and $\bar{X}_n = \frac{S_n}{n}$. Then $E(\bar{X}_n) = p, V(\bar{X}_n) = \frac{pq}{n}, q = 1 - p$. Applying (12.3.2) to the sequence $\{\bar{X}_n\}$,

$$P\{|\bar{X}_n - p| \le \epsilon\} \ge 1 - \frac{pq}{n\epsilon^2} \quad \ge 1 - \frac{1}{4n\epsilon^2}$$

$$\text{i.e. } \lim_{n \to \infty} P\{|\bar{X}_n - p| \le \epsilon\} = 1$$

for every $\epsilon (> 0)$. Hence, $\bar{X}_n \to^P p$ (i.e., $\bar{X}_n \to^L X$ where X is a degenerate random variable at $x = p$). This is known as Bernoulli's WLLN. Here $A_n = 0, B_n = n$.

Bernoulli's WLLN is the simplest expression of the WLLN. It shows that in a large number of independent trials of the same experiment, the probability that the relative frequency with which a particular result occurs will differ from the probability of this result by more than a given small amount is small and can be made as small as we please by taking n large. The relative frequency is said to converge in probability to $'p'$. The result therefore relates the relative frequency of an event with its probability; but it cannot be used to provide a definition of probability in terms of the relative frequency, since the Weak Law of Large Numbers is itself a probability statement.

(b) **Chebychev's WLLN**: Let $\{X_n\}$ be a sequence of uncorrelated random variables, for which $E(X_i) = \mu_i, V(X_i) = \sigma_i^2$. Then

$$E(\bar{X}_n) = \frac{1}{n} \sum_{i=1}^{n} \mu_i = \bar{\mu}_n, \ V(\bar{X}_n) = \frac{1}{n^2} \sum_{i=1}^{n} \sigma_i^2.$$

If the condition

$$\frac{1}{n^2} \sum_{i=1}^{n} \sigma_i^2 \to 0 \text{ as } n \to \infty \tag{12.3.3}$$

is satisfied, the $\bar{X}_n \to^P \bar{\mu}_n$.

Proof. By (12.3.2)

$$P\{|\bar{X}_n - \bar{\mu}_n| \le \epsilon\} \ge 1 - \frac{\sum_{i=1}^{n} \sigma_i^2}{n^2 \epsilon^2}$$

for every $\epsilon (> 0)$. Hence the proof.

The assumption (12.3.3) is immediately satisfied in $V(X_i) = \sigma^2 (< \infty)$ for all i.

We note that if the condition (12.3.3) is satisfied, $E|\bar{X}_n - \bar{\mu}_n|^2 \to 0$ and hence $\bar{X}_n \to^{m.s.} \bar{\mu}_n$.

The condition (12.3.3) is a sufficient (not necessary) conditions for $\{X_n\}$ to satisfy the WLLN.

(c) **Poisson's WLLN**: Consider Poisson's scheme of experiments: n independent Bernoulli trials are performed, the ith trial with probability of success p_i. Let $X_i = 1(0)$ if the ith trial results in success (failure). Here

$$E(\bar{X}_n) = \frac{\sum_{i=1}^n p_i}{n} = \bar{p}; \ V(\bar{X}_n) = \frac{1}{n^2} \sum_{i=1}^n p_i q_i \leq \frac{1}{4n} \to 0 \text{ as } n \to \infty.$$

Hence, by inequality (12.3.1), $\bar{X}_n \to^P \bar{p}$.

(d) **Khintchein's WLLN**: Assume $\{X_n\}$ to be a sequence of independently and identically distributed random variables with $E(X_i) = \mu \ \forall \ i$. Then the sequence $\bar{X}_n \to^P \mu$.

Proof. Let $\Psi(t)$ be the characteristic function of $X_i(i = 1, \ldots, n)$. Then the characteristic function of $\bar{X}_n$ is

$$\psi_n(t) = [\Psi(\frac{t}{n})]^n.$$

Now, expanding $\Psi(t)$ in Maclaurin series about $t = 0$,

$$\Psi(t) = 1 + i\mu t + o(t)$$

where $o(t)$ denotes terms with power of t greater then one. Hence,

$$\Psi_n(t) = [1 + \frac{\mu it}{n} + o(\frac{t}{n})]^n, \ \log_e \Psi_n(t) = n[\frac{\mu it}{n} + o(\frac{t}{n})] = \mu it + no(\frac{t}{n}).$$

Hence, $\log_e \Psi_n(t) \to \mu it$ and $\Psi_n(t) \to \exp(\mu it)$ as $n \to \infty$. Now, $\exp(\mu it)$ is the characteristic function of a degenerate random variable Y with $P(Y = \mu) = 1$. Hence, by Lemma 12.2.1, $\bar{X}_n \to^P \mu$. Note that we require only $E|X| < \infty$ and nothing is required about the variance.

(e) **Markov's WLLN**: Let $\{X_n\}$ be a sequence of variables not necessarily uncorrelated. Then

$$\bar{X}_n \to^P \bar{\mu}_n \text{ if } V(\bar{X}_n) \to 0 \text{ as } n \to \infty.$$

EXAMPLE 12.3.1: Let $X_1, X_2, \ldots$ be *iid* random variables with common *pdf*

$$f(x) = \begin{cases} \frac{1+t}{x^{2+t}}, & x \geq 1 \\ 0, & x < 1, \text{ where } t > 0. \end{cases}$$

Then

$$E|X| = (1+t) \int_1^\infty \frac{1}{x^{1+t}} dx = \frac{1+t}{t} < \infty.$$

Therefore, By Khintchein's theorem WLLN holds and

$$\bar{X}_n \to^P \frac{1+t}{t} \quad \text{as } n \to \infty.$$

EXAMPLE 12.3.2: Let $X_1, X_2, \ldots$ be *iid* random variables, each having a Cauchy distribution $f(x) = \frac{1}{\pi(1+x^2)}$. Since the Cauchy distribution has no finite mean, Khintchein's result does not apply. Also, $\bar{X}_n$ has the same distribution as X_1. Therefore, $\bar{X}_n \not\to^P 0$. Hence WLLN does not hold.

Theorem 12.3.2 (Rohatgi, 1990): Let $\{X_n\}$ be a sequence of random variables and $Y_n = \bar{X}_n = \frac{S_n}{n}$ where $S_n = \sum_{i=1}^n X_i$. A necessary and sufficient condition for $\{X_n\}$ to satisfy the WLLN is that

$$E\{\frac{\bar{X}_n^2}{1 + \bar{X}_n^2}\} \to 0 \text{ as } n \to \infty. \tag{12.3.4}$$

Proof. Sufficiency: We know that for two positive integers a, b with $a \geq b > 0$,

$$(\frac{a}{1+a})(\frac{b+1}{b}) \geq 1. \tag{12.3.5}$$

Let $A \in \mathcal{A}$ be such that $A = \{\omega : |\bar{X}_n| > \epsilon\}$. Then $\omega \in A \Rightarrow \bar{X}_n^2 > \epsilon^2 > 0$. By virtue of (12.3.5),

$$\omega \in A \Rightarrow \frac{\bar{X}_n^2}{1 + \bar{X}_n^2} \cdot \frac{1 + \epsilon^2}{\epsilon^2} \geq 1.$$

Hence,

$$P(A) \leq P\{\frac{\bar{X}_n^2}{1+\bar{X}_n^2} \cdot \frac{1+\epsilon^2}{\epsilon^2} \geq 1\}$$

$$= P\{\frac{\bar{X}_n^2}{1+\bar{X}_n^2} \geq \frac{\epsilon^2}{1+\epsilon^2}\}$$

$$\leq \frac{1+\epsilon^2}{\epsilon^2} E[\frac{\bar{X}_n^2}{1+\bar{X}_n^2}] \quad \text{(by Markov's inequality)}$$

$$\to 0 \text{ as } n \to \infty.$$

Hence, $\bar{X}_n \to^P 0$.

Necessity: Suppose that the *pdf* of $Y_n = \bar{X}_n$ is $f_n(y)$. Then,

$$E\{\tfrac{Y_n^2}{1+Y_n^2}\} = \int_{-\infty}^{\infty} (\tfrac{y^2}{1+y^2}) f(y) dy = [\int_{|y|>\epsilon} + \int_{|y|\le\epsilon}] \tfrac{y^2}{1+y^2} f_n(y) dy$$

$$\le P\{|Y_n| > \epsilon\} + \int_{-\epsilon}^{\epsilon} (1 - \tfrac{1}{1+y^2}) f_n(y) dy$$

$$\le P\{|Y_n| > \epsilon\} + \tfrac{\epsilon^2}{1+\epsilon^2} \le P\{|Y_n| \ge \epsilon\} + \epsilon^2$$

$$\to 0 \text{ since } Y_n \to^P 0.$$

Hence the proof is complete. $\qquad\square$

EXAMPLE 12.3.3: Let $(X_1, \ldots, X_n)$ have a multivariate distribution with $E(X_i) = 0, V(X_i) = 1, i = 1, \ldots, n$ and $\text{Cov}(X_i, X_j) = \rho[0]$ if $|i - j| = 1 [> 1]$. Then show that $\bar{X}_n \to^P 0$.

Here $S_n = \sum_{i=1}^n X_i$ is univariate normal $N(0, \sigma^2)$ where $\sigma^2 = \text{Var}(S_n) = n + 2(n-1)\rho$. Therefore,

$$E\{\tfrac{\bar{X}_n^2}{1+\bar{X}_n^2}\} = E\{\tfrac{S_n^2}{n^2+S_n^2}\}$$

$$= \tfrac{2}{\sigma\sqrt{2\pi}} \int_0^\infty \tfrac{x^2}{n^2+x^2} e^{-\frac{x^2}{2\sigma^2}} dx = \sqrt{\tfrac{2}{\pi}} \int_0^\infty \tfrac{y^2\sigma^2}{n^2+y^2\sigma^2} e^{-\frac{y^2}{2}} dy$$
$$\text{where } y = \tfrac{x}{\sigma}$$

$$= \{n + 2(n-1)\rho\} \sqrt{\tfrac{2}{\pi}} \int_0^\infty \tfrac{y^2}{n^2+y^2\{n+2(n-1)\rho\}} e^{-\frac{y^2}{2}} dy$$

$$\le \tfrac{n+2(n-1)\rho}{n^2} \int_0^\infty \sqrt{\tfrac{2}{\pi}} y^2 e^{-\frac{y^2}{2}} dy$$
$$\to 0 \text{ as } n \to \infty.$$

Hence, by Theorem 12.3.2, $\bar{X}_n \to^P 0$.

12.3.1 *Weierstrass approximation*

In this section we shall show how the WLLN can be used to approximate a continuous function $f(p)$ defined over $p \in [0, 1]$. For this first we consider a definition.

DEFINITION 12.3.2 *Bernstein Polynomial:* Let $f(x)$ be a bounded function on $[0, 1]$ and let $X_1, \ldots, X_n$ be independent Bernoulli variables, each with probability of success p. Let $S_n = \sum_{i=1}^n X_i$. Clearly, $0 \le \tfrac{S_n}{n} \le 1$. Then using the binomial distribution

$$E[f(\frac{S_n}{p})] = \sum_{j=0}^{n} f(\frac{j}{n})\binom{n}{j}p^j(1-p)^{n-j} = B_n(p) \quad \text{(say)}. \qquad (12.3.6)$$

Clearly $B_n(p)$ is a polynomial in p with degree at most n. $B_n(p)$ is called the Bernstein polynomial.

It is known that when n is large most of the weights of the binomial distribution is concentrated in the region of values of j such that $\frac{j}{n}$ is close to p. Using this idea, $B_n(p)$ is used to approximate $f(p)$.

Theorem 12.3.3 (*Weierstrass Approximation Theorem*): If f is continuous on the closed interval $[0, 1]$, then

$$\lim_{n \to \infty} \max_{0 \le p \le 1} |B_n(p) - f(p)| = 0. \qquad (12.3.7)$$

Proof. Since f is continuous in a compact space, it is bounded in absolute value, by A say, and is uniformly continuous. Thus, given any $\epsilon(> 0)$ we can choose a positive δ such that for all x and y in $[0, 1]$,

$$|f(x) - f(y)| < \epsilon \quad \text{whenever } |x - y| < \delta. \qquad (12.3.8)$$

Now,

$$|B_n(p) - f(p)| = \left| \sum_{j=0}^{n} [f(\frac{j}{n}) - f(p)]\binom{n}{j}p^j(1-p)^{n-j} \right|$$

$$\le \sum_{j=0}^{n} |f(\frac{j}{n}) - f(p)| \binom{n}{j}p^j(1-p)^{n-j}. \qquad (12.3.9)$$

We divide the last sum into two parts: the first part is the sum of the terms containing j for which $|\frac{j}{n} - p| < \delta$ and the second part is the sum of the remaining terms. Because of (12.3.8) and the fact that the sum of the terms of the binomial distribution is one, the first term must be less than ϵ.

Now, the second term

$$\sum_{j: |\frac{j}{n} - p| \ge \delta} |f(\frac{j}{n}) - f(p)| \binom{n}{j}p^j(1-p)^{n-j}$$

$$\le 2A \sum_{j: |\frac{j}{n} - p| \ge \delta} \binom{n}{j}p^j(1-p)^{n-j}$$

$$= 2AP(|\frac{S_n}{n} - p| \ge \delta) \le \frac{2A}{4n\delta^2} \qquad (12.3.10)$$

by the weak law of large numbers. Therefore, for large n, the sum in (12.3.10) can be made less than ϵ and the entire sum in (12.3.9) less than 2ϵ. Hence the proof. □

Corollary 12.3.3.1 If f is continuously differentiable on $[0,1]$, then

$$\max_{0 \leq p \leq 1} |B_n(p) - f(p)| = 0(\frac{1}{\sqrt{n}}). \qquad (12.3.11)$$

Proof. We have,

$$B_n(p) - f(p) = E\left[f(\frac{S_n}{n}) - f(p)\right]. \qquad (12.3.12)$$

If f is continuously differentiable on $[0,1]$, then by the mean value theorem,

$$|f(x) - f(p)| \leq \max_{0 \leq \xi \leq 1} |f'(\xi)||x - p| \qquad (12.3.13)$$

for any $x \in [0,1]$. Therefore, from (12.3.13), for any given p,

$$|B_n(p) - f(p)| \leq \max_{0 \leq \xi \leq 1} |f'(\xi)|E(|\frac{S_n}{n} - p|). \qquad (12.3.14)$$

Again, by Cauchy-Schwarz inequality inequality,

$$E(|\frac{S_n}{n} - p|) \leq \{E[(\frac{S_n}{n} - p)^2]\}^{\frac{1}{2}} = \sqrt{\text{Var}\,(\frac{S_n}{n})}. \qquad (12.3.15)$$

Also, $\text{Var}(\frac{S_n}{n}) = \frac{pq}{n} \leq \frac{1}{4n}$. Therefore,

$$\max_{0 \leq p \leq 1} |B_n(p) - f(p)| \leq \frac{\max_{0 \leq \xi \leq 1} |f'(\xi)|}{2\sqrt{n}}.$$

Hence the proof. □

12.4 Strong Law of Large Numbers

We will now consider a stronger form of the law of large numbers. Let $X_1, X_2, \ldots$ be a sequence of random variables defined on the same probability space.

DEFINITION 12.4.1: A sequence $\{X_n\}$ of random variables is said to obey the Strong Law of Large Numbers (SLLN) with respect to the sequence of constants $\{B_n\}$, with $B_n > 0$ and $B_n \to \infty$ as $n \to \infty$, if there exists a sequence of constants $\{A_n\}$ such that the sequence

$$\frac{S_n - A_n}{B_n} \to^{a.s} 0 \text{ as } n \to \infty \qquad (12.4.1)$$

where $S_n = \sum_{i=1}^{n} X_i$.

In what follows we shall take $B_n = n \;\forall\; n$. We shall henceforth say that a sequence $\{X_n\}$ obeys the SLLN if $\bar{X}_n \to^{a.s.} c$ where $'c'$ is a constant and $\bar{X}_n = \frac{S_n}{n}$.

We first consider a Lemma on the sequence of events, the so-called Borel-Cantelli Lemma. For this we first note a definition and a few relations.

DEFINITION 12.4.2: Let A_n be a sequence of subsets of $S, A_n \in \mathcal{A}$. We define (vide (3.2.10), (3.2.11))

$$\overline{\lim} A_n = \limsup_n A_n = \cap_{n=1}^{\infty} \cup_{k=n}^{\infty} A_k,$$
$$\underline{\lim} A_n = \liminf_n A_n = \cup_{n=1}^{\infty} \cap_{k=n}^{\infty} A_k.$$

Note that

$$\liminf_n A_n = (\limsup_n A_n^c)^c. \tag{12.4.2}$$

Lemma 12.4.1: (i) A point ω belongs to $\limsup_n A_n$ *iff* it belongs to infinitely many terms of the sequence $\{A_n, n \geq 1\}$.

(ii) A point ω belongs to $\liminf_n A_n$ *iff* it belongs to all terms of the sequence from a certain term on (i.e., all but finitely many terms).

Proof. (i) If ω belongs to infinitely many A_n's, then ω belongs to

$$C_n = \cup_{k=n}^{\infty} A_k \quad \text{for every } k;$$

hence, it belongs to

$$\cap_{n=1}^{\infty} C_n = \limsup_n A_n.$$

Conversely, if ω belongs to $\cap_{n=1}^{\infty} C_n$, then ω belongs to C_n for every n. If ω were to belong to only a finite number of A_n's, there would be an n such that $\omega \notin A_k$ for $k \geq n$, so that

$$\omega \notin \cup_{k=n}^{\infty} A_k = C_n.$$

This contradiction proves that ω must belong to infinite number of A_n's.

(ii) If ω belongs to all but a finite number of A_n's, then ω belongs to

$$B_n = \cap_{k=n}^{\infty} A_k$$

for at least one n and hence ω belongs to $\cup_{n=1}^{\infty} B_n$.

Conversely, if ω belongs to $\cup_{n=1}^{\infty} B_n$, then ω belongs to B_n for at least one n. Let n_0 be the smallest of all such integers n. Then ω belongs to all A_n's except for $A_1, \ldots, A_{n_0-1}$.

Proof of the lemma is now complete. □

Note 12.4.1: We say that the event $(\limsup_n A_n)$ occurs *iff* the events ω (and the events containing all such ω's) occur infinitely often (i.o). Thus

$$P(\limsup_n A_n) = P(A_n \text{ i.o.}).$$

Lemma 12.4.2: If each $A_n \in \mathcal{A}$, then

$$P(\limsup_n A_n) = \lim_{n\to\infty} P(\cup_{k=1}^n A_k)$$
$$P(\liminf_n A_n) = \lim_{n\to\infty} P(\cap_{k=n}^\infty A_k).$$

Proof. Note that C_n is a decreasing sequence and B_n is an increasing sequence. Hence, by the monotonicity property of probability measures (vide Continuity theorem, Theorem 3.5.1),

$$P(\limsup_n A_n) = P(\cap_{n=1}^\infty C_n) = \lim_{n\to\infty} P(C_n);$$
$$P(\liminf_n A_n) = P(\cup_{n=1}^\infty B_n) = \lim_{n\to\infty} P(B_n).$$

We now consider the celebrated Borel-Cantelli Lemmas (Borel, 1909; Cantelli, 1917).

Theorem 12.4.1 (i) *First Borel-Cantelli Lemma:* Let $\{A_n\}$ be a sequence of events in $(\mathcal{S}, \mathcal{A}, P)$ such that $\sum_{n=1}^\infty P(A_n) < \infty$. Then

$$P(A_n \text{ i.o}) = 0. \tag{12.4.3}$$

(ii) *Second Borel-Cantelli Lemma:* (a) Let $\{A_n\}$ be a sequence of pair-wise independent events in $(\mathcal{S}, \mathcal{A}, P)$ such that $\sum_{n=1}^\infty P(A_n) = \infty$. Then

$$P(A_n \text{ i.o.}) = 1. \tag{12.4.4}$$

(b) If $\{A_n\}$ is a sequence of mutually independent events, then $P(A_n \text{ i.o.}) = 0$ or 1 according as $\sum_{n=1}^\infty P(A_n) < \infty$ or $= \infty$.

Proof. (i) By Lemma 12.4.2,

$$\begin{aligned} P(\limsup_n A_n) &= \lim_{n\to\infty} P(\cup_{k=n}^\infty A_k) \\ &\le \lim_{n\to\infty} \sum_{k=n}^\infty P(A_k) \quad \text{(by Boole's inequality)} \\ &= 0, \end{aligned}$$

since $\sum_{n=1}^\infty P(A_n) < \infty$.

(bi) Let I_n be the indicator function of the event A_n; $I_n(\omega) = 1(0)$ if $\omega \in A_n$ (otherwise). Then $E(I_n) = P(A_n); E(I_n I_{n'}) = E(I_n)E(I_{n'})$, $n \ne n'$.

Now, if ω occurs an infinite number of times,

$$\sum_{n=1}^{\infty} I_n(\omega) = \infty.$$

Therefore,

$$P(A_n \text{ i.o.}) = P(\sum_{n=1}^{\infty} I_n(\omega) = \infty). \qquad (12.4.5)$$

Consider the partial sum

$$S_n = \sum_{k=1}^{n} I_k.$$

By Chebychev's inequality (12.3.1), we have, with $N > 0$,

$$P\{|S_n - E(S_n)| \le N\sigma(S_n)\} \ge 1 - \frac{1}{N^2},$$

i.e.,

$$P\{E(S_n) - N\sigma(S_n) \le S_n \le E(S_n) + N\sigma(S_n)\} \ge 1 - \frac{1}{N^2} \qquad (12.4.6)$$

where $\sigma^2(S_n)$ denotes the variance of S_n.

Let $p_n = P(A_n)$. Then $E(S_n) = \sum_{k=1}^{n} p_k$,

$$\sigma^2(S_n) = \sum_{k=1}^{n} p_k(1 - p_k) \le \sum_{k=1}^{n} p_k.$$

Therefore,

$$\frac{\sigma(S_n)}{E(S_n)} \le \frac{1}{\sqrt{E(S_n)}} \to 0 \text{ as } n \to \infty \qquad (12.4.7)$$

since $\sum_{k=1}^{n} p_k \to \infty$ as $n \to \infty$. It follows, therefore, that for given $N(> 1)$ and $0 < \alpha < 1$, there exists a $n_0 = n_0(N, \alpha)$ such that for $n \ge n_0$,

$$\frac{\sigma(S_n)}{E(S_n)} \le \alpha_1 = \frac{\alpha}{N}(< 1).$$

Therefore,

$$N\sigma(S_n) \le \alpha E(S_n) \text{ for } n \ge n_0.$$

From (12.4.6),

$$P\{E(S_n) - N\sigma(S_n) \le S_n\} \ge 1 - \frac{1}{N^2} \forall \, n \ge n_0$$

which implies that

$$P\{\beta E(S_n) \leq S_n\} \geq 1 - \frac{1}{N^2} \forall \, n \geq n_0 \qquad (12.4.8)$$

where $\beta = 1 - \alpha(> 0)$. Therefore, by the monotonicity of the events under P in (12.4.8),

$$P[\beta \lim_{n \to \infty} E(S_n) \leq \lim_{n \to \infty} S_n] \geq 1 - \frac{1}{N^2}.$$

Letting $n \to \infty$ and then $N \to \infty$ so that $\beta \to 1$, we get

$$P[\lim_{n \to \infty} S_n = \infty] = 1.$$

This establishes the result because of (12.4.4).

(bii) By Lemma 12.4.2,

$$P(\liminf A_n^c) = \lim_{n \to \infty} P(\cap_{k=n}^{\infty} A_k^c).$$

The events A_k^c's are independent, since A_k's are so. Hence, for any $n' > n$,

$$P(\cap_{k=n}^{n'} A_k^c) = \Pi_{k=n}^{n'} P(A_k^c) = \Pi_{k=n}^{n'} (1 - P(A_k)). \qquad (12.4.9)$$

Again, for any $t > 0, 1 - t < e^{-t}$. Hence,

$$(12.4.9) \leq \Pi_{k=n}^{n'} e^{-P(A_n)} = e^{-\sum_{k=n}^{n'} P(A_n)} \to 0 \text{ as } n' \to \infty,$$

since $\sum_{n=1}^{\infty} P(A_n) = \infty$. Therefore,

$$P(\cap_{k=n}^{\infty} A_k^c) = \lim_{n' \to \infty} P(\cap_{k=n}^{n'} A_k^c) = 0.$$

Hence,

$$P(\liminf_n A_n^c) = 0$$

and consequently,

$$P(A_n \text{ i.o.}) = 1$$

by (12.4.2).

The $P(A_n \text{ i.o.}) = 0$ part follows by (i). $\qquad \square$

EXAMPLE 12.4.1: Consider the sequence of experiments of tossing a fair coin. Let A_n denote the event that a head turns up at both the nth and $(n+1)$th toss of the coin. Let $A = \limsup_n A_n$, i.e. A is the event that two successive heads will appear i.o. in repeated tossing of a fair coin. Clearly $P(A_n) = \frac{1}{4}$. Hence $\sum_{n=1}^{\infty} P(A_n) = \infty$. However, A_n's are not independent events. Hence, no conclusion can be drawn.

However, the events $A_2, A_4, A_6, \ldots$ are independent events. Thus $\{A_{2n}, n \geq 1\}$ is a sequence of independent events. Also $\sum_{n=1}^{\infty} P(A_{2n}) = \sum_{n=1}^{\infty} \frac{1}{4} = \infty$. Hence, by Borel-Cantelli Lemma, the event $\lim \sup_n A_{2n} = A$ occurs with probability one, i.e., $P(A_{2n} \text{i.o.}) = 1$.

EXAMPLE 12.4.2: Let $\{X_n\}$ be a sequence of random variables with $X_n \to^P 0$ as $n \to \infty$. Let $\epsilon > 0$. Then $P\{|X_n| \geq \epsilon\} \to 0$. We can now choose a sequence of positive integers n_k such that $P\{|X_{n_k}| \geq 2^{-k}\} \leq 2^{-k}$ for each $k \geq 1$. Then $\sum_{k=1}^{\infty} P\{|X_{n_k}| \geq 2^{-k}\} < \infty$. Hence, the event $P\{E_k = [|X_{n_k}| \geq 2^{-k}]\text{i.o.}\} = 0$, that is $P\{E_k^c = [|X_{n_k}| \leq 2^{-k}]\text{i.o.}\} = 1$, that is $\{X_{n_k}\} \to^{a.s.} 0$. This proves the Theorem 12.2.8.

As another application of Borel-Cantelli Lemma we consider the following SLLN.

Theorem 12.4.2: If $X_1, X_2, \ldots$ are *iid* random variables with common mean μ and finite fourth order moment, then

$$P\{\lim_{n \to \infty} \frac{S_n}{n} = \mu\} = 1, \qquad (12.4.10)$$

where $S_n = X_1 + \ldots + X_n$.

Proof. We have

$$E\{\sum_{i=1}^{n} (X_i - \mu)\}^4 = nE(X_1 - \mu)^4 + 6\binom{n}{2}\sigma^4 \leq Kn^2$$

where K is a constant. By Markov's inequality

$$P\{|\sum_{i=1}^{n} (X_i - \mu)| > n\epsilon\} \leq (n\epsilon)^{-4} E\{\sum_{i=1}^{n} (X_i - \mu)\}^4 \leq \frac{Kn^2}{(n\epsilon)^4} = \frac{K'}{n^2}.$$

Therefore,

$$\sum_{n=1}^{\infty} P\{|S_n - n\mu| > n\epsilon\} \leq K' \sum_{n=1}^{\infty} \frac{1}{n^2} < \infty.$$

Hence, it follows by the Borel-Cantelli Lemma, that with probability 1 only finitely many of the events $\{\omega : |\frac{S_n}{n} - \mu| > \epsilon\}$ occur, i.e., $P(A_\epsilon) = 0$ where

$$A_\epsilon = \lim_{n \to \infty} \sup\{|\frac{S_n}{n} - \mu| > \epsilon\}.$$

As $\epsilon \to 0$, the set A_ϵ approaches the set of ω's on which $\frac{S_n}{n} \to \mu$. Letting $\epsilon \to 0$ through a countable set of values, we have,

$$P\{\frac{S_n}{n} - \mu \nrightarrow 0\} = P\{\cup_K A_{1/K}\} = 0.$$

Hence, the proof.

We now consider several inequalities.

12.4.1 *Kolmogorov's inequality and its ramifications*

Theorem 12.4.3 (*Kolmogorov's Inequality*): Suppose that

$$E(X_j) = E(X_j|X_1, X_2, \ldots, X_{j-1}) = 0 \ (j = 1, 2, \ldots). \qquad (12.4.11)$$

Then

$$P\{|S_j| \le u; j = 1, 2, \ldots, n\} \ge 1 - \frac{E(S_n^2)}{u^2} = 1 - \frac{\sum_{j=1}^n E(X_j^2)}{u^2} \qquad (12.4.12)$$

where $S_n = \sum_{j=1}^n X_j$.

Proof. Let A denote the event$[|S_j| < u; j = 1, \ldots, n]$ and B_i the event

$$[|S_1| < u, \ldots, |S_{i-1}| < u, |S_i| \ge u].$$

Then

$$\cup_{i=1}^n B_i = A^c.$$

Hence,

$$P(A) = 1 - \sum_{i=1}^n P(B_i) \qquad (12.4.13)$$

since B_i's are mutually disjoint. Now,

$$E(S_n^2) = \sum_{i=1}^n P(B_i)E(S_n^2|B_i) + P(A)E(S_n^2|A) \text{ (by Theorem 2.8.8).}$$

$$(12.4.14)$$

Again,

$$\begin{aligned} E(S_n^2|B_i) &= E[(S_n - S_i + S_i)^2|B_i] \\ &= E[(S_n - S_i)^2|B_i] + 2E[(S_n - S_i)S_i|B_i] + E(S_i^2|B_i) \ge u^2, \end{aligned}$$
$$(12.4.15)$$

(by definition of B_i), since the first term on the right-hand side of (12.4.15) is positive and the second term is zero, as shown below.

$$\begin{aligned} E[(S_n - S_i)S_i|B_i] &= E[E\{(S_n - S_i)S_i|X_1, \ldots, X_i, B_i\}|B_i] \\ &= E[S_iE\{(S_n - S_i)|X_1, \ldots, X_i, B_i\}|B_i]. \end{aligned}$$

Now,

$$E(S_n - S_i|X_1, \ldots, X_i, B_i) = E(S_n - S_i|X_1, \ldots, X_i) = 0$$

by (12.4.11). Hence, by (12.4.14) and (12.4.15),

$$\begin{aligned} E(S_n^2) &\ge u^2 \sum_{i=1}^n P(B_i) + P(A)E(S_n^2|A) \\ &\ge u^2 \sum_{i=1}^n P(B_i) \end{aligned} \qquad (12.4.16)$$

and (12.4.12) follows by (12.4.13).

Note 12.4.1: If (12.4.12) is a bound simply for $P(|S_n| \leq u)$, then one gets the Chebychev's inequality. The inequality (12.4.12) gives a lower bound for the probability that $S_1, S_2, \ldots, S_n$ are simultaneously (and not simply S_n) less than u in modulus.

Note 12.4.2: The inequality is derived on the assumption (12.4.11) that the conditional mean of X_j should have the same value as the unconditional mean $E(X_j)$, which is zero. However, if $E(X_j) = \mu_j (\neq 0)$, one can consider $X_j - \mu_j$ in place of X_j.

Note 12.4.3: We have assumed that the second moment of X_j exists ($j = 1, 2, \ldots$).

Theorem 12.4.4 (*The Ha'jek-Re'nyi generalization of Kolmogorov's inequality*): Under assumption (12.4.11),

$$P(|S_j| \leq u_j, j = 1, \ldots, n) \geq 1 - \sum_{j=1}^{n} \frac{E(X_j^2)}{u_j^2} \qquad (12.4.17)$$

for any positive increasing sequence $\{u_j\}$.

Proof. Consider the expression

$$T = \sum_{j=1}^{n} (\alpha_j - \alpha_{j+1}) S_j^2 \qquad (12.4.18)$$

where the α_j's constitute a non-negative decreasing sequence, with $\alpha_{n+1} = 0$. Then

$$E(T) = \sum_{j=1}^{n} \alpha_j E(X_j^2) \geq \sum_{i=1}^{n} P(B_i) E(T|B_i) \qquad (12.4.19)$$

where B_i is the event

$$[|S_j| \geq u_j \text{ first for } j = i]$$

and hence

$$A = (\cup_{i=1}^{n} B_i)^c = [|S_j| < u_j, j = 1, \ldots, n].$$

Now,

$$E(S_j^2|B_i) \geq \begin{cases} 0 & (j < i) \\ u_i^2 & (j \geq i). \end{cases} \qquad (12.4.20)$$

Hence, by (12.4.19) and (12.4.20),

$$\sum_{j=1}^{n} \alpha_j E(X_j^2) \geq \sum_{i=1}^{n} P(B_i) \sum_{j=1}^{n} (\alpha_j - \alpha_{j+1}) u_j^2 = \sum_{i=1}^{n} \alpha_i u_i^2 P(B_i).$$

The inequality (12.4.17) follows by choosing $\alpha_i = u_i^{-2}$. For a generalization of the Kolmogorov-Ha'jek-Re'nyi (KHR) inequality, see exercise 12.30.

12.4.2 *Different SLLN's*

We now consider different strong laws of large numbers.

Theorem 12.4.5 (*Kolmogorov's SLLN (First form of SLLN)*): Let the independent random variables X_k have zero means and variances $\sigma_k^2 (k = 1, 2, \ldots)$ such that

$$\sum_{k=1}^{\infty} \frac{\sigma_k^2}{k^2} < \infty. \tag{12.4.21}$$

Then the sequence $\bar{X}_n \to^{a.s.} 0$. That is, given $\epsilon > 0, \eta > 0$, however small, there exists an integer n_0 depending on ϵ and η such that

$$P\{|\bar{X}_j| < \epsilon; j > n_0\} > 1 - \eta.$$

Proof. For an arbitrary positive integer n, define the random variables

$$Y_m = \frac{S_m - S_{\nu_i - 1}}{\nu_i} \quad \text{for } \nu_i = n.2^{i-1} \leq m < n.2^i.$$

Then

$$E(Y_m) = 0, \quad V(Y_m) = \frac{1}{\nu_i^2} \sum_{e=\nu_i}^{m} \sigma_e^2.$$

Let

$$Z_i = \max_m Y_m^2, \quad \nu_i \leq m < \nu_{i+1}.$$

By Kolmogorov's inequality (12.4.11),

$$P(Z_i > \frac{\epsilon^2}{4}) = q_i \text{ (say)} < \frac{4}{\epsilon^2 \nu_i^2} \sum_{e=\nu_i}^{\nu_{i+1}-1} \sigma_e^2 = \frac{4}{\epsilon^2 n^2 2^{2i-2}} \sum_{e=\nu_i}^{\nu_{i+1}-1} \sigma_e^2$$

and

$$\sum_{i=1}^{\infty} q_i < 4\epsilon^{-2} \sum_{i=1}^{\infty} \frac{1}{n^2 2^{2i-2}} \sum_{e=n.2^{i-1}}^{n.2^i-1} \sigma_e^2 < 16\epsilon^{-2} \sum_{i=1}^{\infty} \sum_{e=n.2^{i-1}}^{n.2^i-1} \frac{\sigma^2}{e^2}$$

or

$$\sum_{i=1}^{\infty} q_i < 16\epsilon^{-2} \sum_{k=n}^{\infty} \frac{\sigma_k^2}{k^2}.$$

Therefore, by Boole's inequality

$$P\left[Z_i \leq \frac{\epsilon^2}{4} \ \forall \ i = 1, 2, \ldots \right] > 1 - \frac{16}{\epsilon^2} \sum_{k=n}^{\infty} \frac{\sigma^2}{k^2}.$$

Now, the inequality $\frac{|S_k|}{k} \leq \epsilon, k = n, n+1, n+2, \ldots$ are satisfied, when simultaneously

$$Z_i \leq \frac{\epsilon^2}{4}, i = 1, 2, \ldots \text{ and } |\frac{S_{n-1}}{n}| \leq \frac{\epsilon}{2}.$$

Again,

$$P(|\frac{S_{n-1}}{n}| \leq \frac{\epsilon}{2}) \geq 1 - \frac{4 \sum_{k=1}^{n} \sigma_k^2}{n^2 \epsilon^2} \quad \text{(by Markov's inequality)}.$$

Hence,

$$P\left[\frac{S_k}{k}| \leq \epsilon, k = n, n+1, \ldots\right] > 1 - \frac{16}{\epsilon^2} \sum_{k=n}^{\infty} \frac{\sigma_k^2}{k^2} - \frac{4 \sum_{k=1}^{n} \sigma_k^2}{n^2 \epsilon^2}$$

(by Corollary 2.8.3.1) $\to 1$ as $n \to \infty$, since $\sum_{k=1}^{\infty} \sigma^2/k^2 \to \infty$.

Corollary 12.4.5.1: Let the sequence of independent random variables $\{X_k\}$ be such that

$$E(X_k) = \mu_k, \ V(X_k) = \sigma_k^2, k = 1, 2, \ldots.$$

Then

$$\sum_{k=1}^{\infty} \frac{\sigma_k^2}{k^2} < \infty \Rightarrow \bar{X}_n - \bar{\mu}_n \to^{a.s.} 0$$

where $\bar{\mu}_n = \frac{1}{n} \sum_{k=1}^{n} \mu_k$. This means that under condition (12.4.21), $\{X_n\}$ obeys the SLLN.

Corollary 12.4.5.2: Every sequence $\{X_n\}$ of independent random variables with uniformly bounded variances obeys the SLLN.

Proof. If $V(X_k) \leq A \ \forall \ k$, then

$$\sum_{k=1}^{\infty} \frac{\sigma_k^2}{k^2} \leq A \sum_{k=1}^{\infty} \frac{1}{k^2} < \infty.$$

Hence $\bar{X}_n - \bar{\mu}_n \to^{a.s.} 0$.

The SLLN holds necessarily for a sequence of *iid* random variables, X_k's having finite variance, since then the condition $\sum_{k=1}^{\infty} \sigma_k^2/k^2 < \infty$ is automatically satisfied.

Theorem 12.4.6 (*Borel's SLLN*): Consider n independent Bernoulli trials each with probability p of success. Let $X_i = 1(0)$ if the ith trial results in success (failure). Then

$$\frac{S_n}{n} \to^{a.s.} p.$$

Proof. Here

$$E(X_k) = p, V(X_k) = p(1-p) \le \frac{1}{4}, \ 0 < p < 1.$$

The result follows by Corollary 12.4.5.1.

Theorem 12.4.7 (*Main SLLN*): Let $\{X_n, n \ge 1\}$ be independent random variables on $(\mathcal{S}, \mathcal{A}, P)$ with a common distribution and $S_n = \sum_{k=1}^{n} X_k$. Then $\frac{S_n}{n} \to \alpha$, a constant, almost surely, (a.s.) *iff* $E(|X_k|) < \infty$, in which case $\alpha = E(X_1)$. On the other hand, if $E(|X_1|) = \infty$, then $\limsup_n (\frac{|S_n|}{n}) = \infty$ *a.e.*

Theorem 12.4.8: Let $X_1, X_2, \ldots$ ba sequence of independent random variables on $(\mathcal{S}, \mathcal{A}, P)$ with means $\mu_1, \mu_2, \ldots$ and variances $\sigma_1^2, \sigma_2^2, \ldots$. Let

$$S_n' = \sum_{k=1}^{n} (X_k - \mu_k)$$

and $\sigma^2 = \sum_{k=1}^{\infty} \sigma_k^2$. Suppose $\sigma^2 < \infty$ and $\sum_{k=1}^{\infty} \mu_k$ converges. Then $\sum_{k=1}^{\infty} X_k$ converges *a.s.* and in the mean of order 2 to a random variable X. Moreover, $E(X) = \sum_{k=1}^{\infty} \mu_k$, $\mathrm{Var}(X) = \sigma^2$ and for any $\epsilon > 0$,

$$P\left[\sup_{n \ge 1} |S_n'| \ge \epsilon\right] \le \frac{\sigma^2}{\epsilon^2}.$$

Theorem 12.4.9: Let $X_1, X_2, \ldots$ have means $\mu_1, \mu_2, \ldots$ and variances $\sigma_1^2, \sigma_2^2, \ldots$ and covariances $\mathrm{Cov}(X_i, X_j)$ satisfying

$$\mathrm{Cov}\,(X_i, X_j) \le \rho_{j-i}\sigma_i\sigma_j \ (i \le j)$$

where $0 \le \rho_k \le 1$ for all $k = 0, 1, 2, \ldots$. If the series $\sum_{k=1}^{\infty} \rho_k$ and $\sum_{k=1}^{\infty} \sigma_k^2 (\log k)^2/k^2$ are both convergent, then

$$\bar{X}_n - \bar{\mu}_n \to^{a.s.} 0$$

where $\bar{\mu}_n = \sum_{k=1}^{n} \mu_k/n$.

For proof of these results and further details, the reader may refer to Feller (1968), vol.I and Rao (1984).

12.5　Central Limit Theorems

Let $\{X_n\}$ be a sequence of independent random variables. In laws of large numbers, we consider the cases when a sequence of random averages $\bar{X}_n$ converges to $E(\bar{X}_n)$, either in probability or almost certainly. Here, we consider some different situations, namely, when $\bar{X}_n \to^L \tau$ where τ is a standard normal variable. If the sequence $\bar{X}_n \to^L \tau, \bar{X}_n$ is said to follow the central limit law or normal convergence. In this section, we shall consider different Central Limit Theorems (CLT).

DEFINITION 12.5.1: A sequence of random variables $\{X_n\}$ is asymptotically normal with mean a_n and variance b_n^2, if $b_n > 0 \ \forall \ n$ sufficiently large and

$$\frac{\bar{X}_n - a_n}{b_n} \to^L N(0,1).$$

We write X_n is $AN(a_n, b_n^2)$.

Here, $\{a_n\}, \{b_n\}$ are sequences of constants. It is not necessary that a_n and b_n^2 be the mean and variance of X_n, not even that X_n possesses such moments. Note that if X_n is asymptotically $N(a_n, b_n^2)$, it does not necessarily mean that $\{X_n\}$ converges in distribution to anything. The sequence a_n and b_n may or may not depend on n.

Lemma 12.5.1: If X_n is $AN(a_n, b_n^2)$, then also X_n is $AN(\tilde{a}_n, \tilde{b}_n^2)$ *iff*

$$\frac{\tilde{b}_n}{b_n} \to 1 \ \text{ and } \ \frac{\tilde{a}_n - a_n}{b_n} \to 0.$$

Lemma 12.5.2: If X_n is $AN(a_n, b_n^2)$, then also $c_n X_n + d_n$ is $AN(a_n, b_n^2)$ *iff*

$$c_n \to 1, \quad \frac{a_n(c_n - 1) + d_n}{b_n} \to 0.$$

EXAMPLE 12.5.1: If X_n is $AN(n, 2n)$ then so is $\frac{n-1}{n} X_n$ but not $\frac{\sqrt{n}-1}{\sqrt{n}} X_n$.

We now consider the simplest form of CLT.

Theorem 12.5.1 (*Lindeberg-Levy Central Limit Theorem*): Let $\{X_n\}$ be a sequence of independently and identically distributed (*iid*) random variables, with

$$E(X_i) = \mu, \ V(X_i) = \sigma^2 < \infty, \ i = 1, 2, \ldots$$

Then the average $\bar{X}_n = \frac{1}{n}\sum_{i=1}^n X_i$ is asymptotically normal $(\mu, \frac{\sigma^2}{n})$.

Proof. Consider the standardized variable

$$Z_n = \frac{1}{\sigma\sqrt{n}} \sum_{i=1}^n (X_i - \mu).$$

The random variable $Y_i = (X_i - \mu)$ are *iid* with $E(Y_i) = 0, V(Y_i) = \sigma^2 (i = 1, \ldots, n)$.

Let $\Psi_Y(t)$ be the characteristic function (c.f.) of Y_i. The c.f. of Z_n is

$$\Psi_n(t) = E[e^{it(S_n - n\mu)/\sigma\sqrt{n}}] = \left[\Psi_Y\left(\frac{t}{\sigma\sqrt{n}}\right)\right]^n$$

Developing $\Psi_Y(t)$ in Mclaurin's series about $t = 0$,

$$\Psi_Y(t) = \Psi_Y(0) + t\Psi_Y'(0) + \frac{t^2}{2!}\Psi_Y''(0) + o(t^2) = 1 - \frac{t^2}{2!}\sigma^2 + o(t^2).$$

Hence

$$\Psi_n(t) = \left[1 - \frac{t^2}{2n} + o\left(\frac{t^2}{n}\right)\right]^n$$

$$\log_e \Psi_n(t) = n \log_e\left[1 + \left\{-\frac{t^2}{2n} + o\left(\frac{t^2}{n}\right)\right\}\right]$$

$$= n\left[-\frac{t^2}{2n} + o\left(\frac{t^2}{n}\right)\right] = -\frac{t^2}{2} + no\left(\frac{t^2}{n}\right).$$

Hence, $\log_e \Psi_n(t) = -\frac{t^2}{2}$ (since $\lim_{n\to\infty} no\left(\frac{t^2}{n}\right) = 0$ for every fixed t) and $\Psi_n(t) \to e^{-\frac{t^2}{2}}$, the c.f. of τ, a standard normal deviate. Therefore, by the continuity theorem, $Z_n \xrightarrow{L} \tau$ and hence $\bar{X}_n$ is asymptotically $N(\mu, \frac{\sigma^2}{n})$.

Corollary 12.5.1.1: Under the conditions of Theorem 12.5.1, the standardized sum

$$\frac{(S_n - n\mu)}{\sigma\sqrt{n}} = \frac{\sqrt{n}(\bar{X} - \mu)}{\sigma}$$

is asymptotically $N(0, 1)$.

EXAMPLE 12.5.2: Consider binomial random variables X_n with parameters (n, p). Now $X_n = Y_1 + \ldots + Y_n$ where Y_i denotes the result of the ith Bernoulli trial, $Y_i = 1(0)$ if a success (failure) occurs, $Y_1, \ldots, Y_n$ being independent with $E(Y_i) = p, V(Y_i) = p(1-p)$. Hence, by applying Theorem 12.5.1, $\bar{X}_n$ is asymptotically $N(p, \frac{pq}{n})$.

EXAMPLE 12.5.3: Let $\{X_n\}$ be a sequence of *iid* standard Cauchy variables. Since the Cauchy variable has no finite mean, Theorem 12.5.1 does not apply.

The c.f. of each X is $\Psi_X(t) = e^{-|t|}$. The c.f. of $\frac{S_n}{n} = \bar{X}_n$ is $\Psi_n(t) = e^{-|t|}$. Hence, $\bar{X}_n$ has a Cauchy distribution. $\square$

In case the independent variables $X_1, \ldots, X_n$ are not identically distributed, their sum need not converge to a normal distribution, even if all the variables have finite standard deviations.

The following theorem due to Liapounov gives a sufficient condition that a sum of independent random variables should have an asymptotically normal distribution.

Theorem 12.5.2: (*Liapounov*): Let $\{X_n\}$ be a sequence of mutually independent random variables such that for some $\delta(> 0), E\{|X_k - \mu_k|^{2+\delta}\}$ exists for every $k = 1, 2, \ldots$. Then, if the condition

$$\lim_{n \to \infty} \frac{1}{(\sum_{k=1}^n \sigma_k^2)^{1+\frac{\delta}{2}}} \sum_{k=1}^n E\{|X_k - \mu_k|^{2+\delta}\} = 0 \qquad (12.5.1)$$

is satisfied, then $\frac{(S_n - \mu_{(n)})}{\sigma_{(n)}}$ is asymptotically normal with mean 0 and variance 1, where

$$\mu_{(n)} = \sum_{k=1}^n \mu_k, \quad \sigma_{(n)}^2 = \sum_{k=1}^n \sigma_k^2.$$

The proof is omitted.

The Liapounov theorem gives only a sufficient condition for $\{S_n\}$ to follow central limit law. The Lindeberg-Feller theorem stated below (without proof) gives a necessary and sufficient condition for the sum S_n to have an asymptotically normal distribution. This is the Central Limit Theorem in the most general form.

Theorem 12.5.3 (Lindeberg-Feller): Let $\{X_n\}$ be a sequence of independent random variables with the cumulative distribution function of X_k as $F_k(x), E(X_k) = \mu_k, V(X_k) = \sigma_k^2 < \infty$. Then

(1) S_n is asymptotically $N(\mu_{(n)}, \sigma_{(n)}^2)$;

(2) $\lim_{n \to \infty} \max_{1 \le k \le n} \frac{\sigma_k}{\sigma_{(n)}} = 0$

if and only if

$$A_n(\epsilon) = \frac{1}{\sigma_{(n)}^2} \sum_{k=1}^n \int_{|x_k - \mu_k| \ge \epsilon \sigma_{(n)}} (x - \mu_k)^2 dF_k(x) \to 0 \qquad (12.5.2)$$

as $n \to \infty$, holds for every $\epsilon(> 0)$.

Corollary 12.5.3.1: If X_k's are *iid* random variables, each with finite variance σ^2, condition (12.5.2) is always satisfied; because,

$$A_n(\epsilon) = \frac{1}{n\sigma^2} \sum_{k=1}^n \int_{|x_k - \mu_k| \ge \epsilon \sigma \sqrt{n}} (x - \mu)^2 dF(x) = \frac{1}{\sigma^2} \int_{y^2 \ge n\epsilon^2 \sigma^2} y^2 dF(y) \to 0$$

as $n\epsilon^2\sigma^2 \to \infty$ as $n \to \infty$.

Corollary 12.5.3.2: If Liapounov's condition (12.5.1) is satisfied, then Lindeberg-Feller condition (12.5.2) is also satisfied. Because,

$$A_n(\epsilon) = \frac{1}{\sigma_{(n)}^2} \sum_{k=1}^n \int_{|x-\mu_k| \geq \epsilon\sigma_{(n)}} (x - \mu_k)^2 dF_k(x)$$

$$\leq \frac{1}{\epsilon^\delta \sigma_{(n)}^{2+\delta}} \sum_{k=1}^n \int_{|x-\mu_k| \geq \epsilon\sigma_{(n)}} |x - \mu_k|^{2+\delta} dF_k(x)$$

$$\leq \frac{1}{\epsilon^\delta \sigma_{(n)}^{2+\delta}} \sum_{k=1}^n E|X_k - \mu_k|^{2+\delta} \to 0 \text{ as } n \to \infty$$

by (12.5.1) and hence (12.5.2) holds.

Note 12.5.1: The condition (12.5.2) can also be stated as follows. Define truncated random variables,

$$U_k = \begin{cases} X_k - \mu_k, & \text{if } |X_k - \mu_k| \geq \epsilon\sigma_{(n)} \\ 0, & \text{if } |X_k - \mu_k| < \epsilon\sigma_{(n)}. \end{cases} \tag{12.5.3}$$

Then (12.5.2) reduces to the condition

$$B_n(\epsilon) = \frac{1}{\sigma_{(n)}^2} \sum_{k=1}^n E(U_k^2) \to 0, \text{ as } n \to \infty. \tag{12.5.4}$$

The CLT has also been studied for the case when X_k's are not all mutually independent random variables.

It is of interest to assess the error of approximation in CLT. Denote by

$$G_n(t) = P(S_n^* \leq t)$$

where

$$S_n^* = \frac{\sum_{i=1}^n X_i - E(\sum_{i=1}^n X_i)}{\sqrt{V(\sum_{i=1}^n X_i)}}.$$

For the *iid* case, an exact bound on the error of approximation is provided by the following theorem due to Berry (1941) and Esse'en (1945).

Theorem 12.5.4: Let $\{X_i\}$ be *iid* with mean μ and variance $\sigma^2 > 0$. Then

$$\sup_t |G_n(t) - \Phi(t)| \leq \frac{33}{4} \frac{E|X_i - \mu|^3}{\sigma^3 \sqrt{n}}.$$

EXAMPLE 12.5.4: Examine if the law of large numbers and CLT hold for the sequence of mutually independent random variables $\{X_k\}$, where

$$P[X_k = k] = P[X_k = -k] = \frac{1}{2\sqrt{k}}, P[X_k = 0] = 1 - \frac{1}{\sqrt{k}}.$$

Taking $\delta = 1$ in (12.5.1), CLT holds if $\lim_{n\to\infty} \frac{\rho}{\delta} = 0$ where

$$\rho^3 = \sum_{k=1}^{n} E|X_k - \mu|^3 \quad \text{and} \quad \sigma = \sqrt{\sum_{k=1}^{n} \sigma_k^2}.$$

Here, $E(X_k) = 0, E|X_k^3| = k^{\frac{5}{2}}$,

$$\lim_{n\to\infty} \frac{\rho}{\sigma} = \lim_{n\to\infty} \frac{\left(\sum_{k=1}^{n} k^{\frac{5}{2}}\right)^{\frac{1}{3}}}{\left(\sum_{k=1}^{n} k^{\frac{3}{2}}\right)^{\frac{1}{2}}} = \lim_{n\to\infty} \frac{\left(\int_0^n x^{\frac{5}{2}}\,dx\right)^{\frac{1}{3}}}{\left(\int_0^n x^{\frac{3}{2}}\,dx\right)^{\frac{1}{2}}} = \lim_{n\to\infty} \frac{A}{\sqrt{n}} = 0$$

(where $A = (\sqrt{5})/(7^{\frac{1}{3}})(2^{\frac{1}{6}})$).

Hence, CLT holds. However,

$$\lim_{n\to\infty} \frac{V(S_n)}{n^2} = \frac{2}{5}\sqrt{n}$$

does not tend to zero as $n \to \infty$. Hence, condition (12.3.3) is not satisfied.

EXAMPLE 12.5.5: For the sequence of random variables $\{X_n\}$ with

$$P(X_k = k^\lambda) = P(X_k = -k^\lambda) = \frac{1}{2}, \quad \lambda > 0,$$

determine if the WLLN holds, central limit law holds.

Here

$$E(X_k) = 0, \sigma_{(n)}^2 = \sum_{k=1}^{n} \sigma_k^2 = \sum_{k=1}^{n} k^{2\lambda} \sim \int_0^n x^{2\lambda}\,dx = \frac{n^{2\lambda+1}}{2\lambda+1}.$$

Hence, $\sigma_{(n)}^2/n^2 = n^{2\lambda-1}/(2\lambda+1) \to 0$ as $n \to \infty$ if $\lambda < \frac{1}{2}$. Hence, WLLN holds for $\lambda < \frac{1}{2}$.

To verify if the CLT holds, consider the truncated variables U_k defined in (12.5.3),

$$|X_k| = k^\lambda \leq n^\lambda.$$

Now, $n^\lambda \leq \epsilon\sigma_{(n)} \Rightarrow n \geq \frac{2\lambda+1}{\epsilon^2} = n_0(\epsilon)$ (say). Hence,

$$U_k = \begin{cases} X_k, & n \leq n_0(\epsilon), \\ 0, & n > n_0(\epsilon). \end{cases}$$

Also, $B_n(\epsilon) = \sum_{k=1}^{n_0(\epsilon)} k^{2\lambda} / \sum_{k=1}^{n} k^{2\lambda} \to 0$ as $n \to \infty$ for every $\epsilon(> 0)$ and every $\lambda(> 0)$. Hence, CLT holds for every $\lambda > 0$.

S_n is likely to be of the order of $n^{\lambda+\frac{1}{2}}$, since S_n lies within $E(S_n) +_- 3\sigma_{(n)}$ with high probability. Hence, for $\lambda \geq \frac{1}{2}$, S_n does not converge in probability to zero.

EXAMPLE 12.5.6: Show that if the variables X_k are uniformly bounded, then (i) $S_n = \sum_{k=1}^{n} X_k$ satisfies WLLN, (ii) S_n obeys the central limit property, provided $\sigma_{(n)} \to \infty$.

(1) $V(S_n) = \sum_{k=1}^{n} \sigma_k^2 \leq 2nA^2$ where $|X_k| < A \ \forall \ k$. Hence,

$$\frac{V(S_n)}{n^2} \leq \frac{2A^2}{n} \to 0 \ \text{ as } n \to \infty.$$

Therefore, WLLN holds.

(2) Suppose $\sigma_{(n)}^2 \to \infty$. Here $|X_k| < A \ \forall \ k$. Hence, the truncated variables in (12.5.3) are:

$$U_k = \begin{cases} 0 & \text{if } n \text{ is so large that } \sigma_{(n)} > \frac{2A}{\epsilon}, \text{ i.e., for } n \geq n_0(\epsilon)(\text{say}) \\ X_k - \mu_k & \text{otherwise,} \end{cases}$$

because, $|X_k - \mu_k| \leq |X_k| + |\mu_k| \leq 2A$ and hence $X_k - \mu_k \leq \epsilon\sigma_{(n)}$ holds if $2A \leq \epsilon\sigma_{(n)}$. Therefore, for

$$n \geq n_0(\epsilon), \ P\{(U_k = 0), k = 1, \ldots, n\} = 1.$$

Hence,

$$\lim_{n\to\infty} B_n(\epsilon) = 0$$

for every ϵ and the central limit law holds.

The condition $\sigma_{(n)}^2 \to \infty$ is a necessary condition for CLT to hold. If $\sigma_{(n)}^2 \not\to \infty$, there exists a constant B such that $\sigma_{(n)}^2 \uparrow B^2$. For any fixed m, we can find an ϵ such that

$$P\{|X_m - \mu_m| > \epsilon\sigma_{(n)}\} > P\{|X_m - \mu_m| > \epsilon B\} > 0.$$

For any $n \geq m$,

$$A_n(\epsilon) = \frac{1}{\sigma_{(n)}^2} \sum_{k=1}^{n} \int_{|x-\mu_k| \geq \epsilon\sigma_{(n)}} (x - \mu_k)^2 dF_k(x)$$
$$\geq \epsilon^2 \sum_{k=1}^{n} P\{|X_k - \mu_k| \geq \epsilon\sigma_{(n)}\} \geq \epsilon^2 P\{|X_m - \mu_m| \geq \epsilon B\} > 0$$

and hence condition (12.5.2) does not hold.

EXAMPLE 12.5.7: Using the central limit theorem, show that

$$\lim_{n\to\infty} e^{-n} \sum_{k=0}^{n} \frac{n^k}{k!} = \frac{1}{2}.$$

Let $\{X_n\}$ be a sequence of Poisson variables, each with parameter $\lambda = 1$. Then $\sum_{i=1}^{n} X_i$ is Poisson with parameter $\lambda = n$. Again, by CLT, $\sum_{i=1}^{n} X_i$ is asymptotically $N(n, n)$. Therefore,

$$e^{-n} \sum_{k=1}^{n} \frac{n^k}{k!} = P(\sum_{i=1}^{n} X_i - n \leq 0) = \frac{1}{\sqrt{2\pi}} \int_{-\infty}^{0} e^{-\frac{z^2}{2}} dz = \frac{1}{2}.$$

12.6 Exercises and Complements

12.1 Show that $[X_n \to^P X] \Rightarrow [X_n^2 \to^P X^2]$.

12.2 Let $\{X_n\}$ be a sequence of random variables and let X be a random variable such that $E(X_n - X)^2 < \infty$. Show that $X_n \to^{a.s.} X$.

12.3 Let $\{F_n\}$ be a sequence of distribution functions such that

$$F_n(x) = \begin{cases} 0, & x < 0 \\ 1 - \frac{1}{n}, & 0 \leq x < n, \\ 1, & n \leq x. \end{cases}$$

Show that $F_n \to^w F$, where

$$F(x) = \begin{cases} 0, & x < 0 \\ 1, & x \geq 0. \end{cases}$$

12.4 Let $\{X_n\}$ be a sequence of independent random variables, each uniformly distributed over $(0, 1)$. Let

$$Y_n = \frac{\min_{1 \leq i \leq n} X_i}{n}.$$

Show that $Y_n \to^L Z$ where Z is an exponential random variable with mean 1.

12.5 Let X be a $N(0, 1)$-variable. Define the sequence of random variables $\{X_n\}$ such that

$$X_{2m+1} = X(m = 0, 1, 2, \ldots), \quad X_{2m} = -X(m = 1, 2, \ldots).$$

Show that $X_n \to^L X$; however, X_n does not converge in probability to X.

12.6 Let $\{F_n(x)\}$ be a sequence of negative binomial distributions with parameters (r_n, q_n) with $r_n \to \infty$, $q_n \to 0$ in such a way that $r_n q_n \to \lambda$. Show that $F_n(x)$ converges weakly to a Poisson distribution.

12.7 Let $\{X_n\}$ be a sequence of independent random variables and *pmf* of X_n is $P(X_n = -(n+4)) = \frac{1}{n+4}, P(X_n = -1) = 1 - \frac{4}{n+4}, P(X_n = n+4) = \frac{3}{n+4}$. Show that $p\lim_{n\to\infty} X_n = -1$.

12.8 Let $\{X_n\}$ be a sequence of independent random variables defined by $P(X_n = n) = \frac{1}{n^r}, P(X_n = 0) = 1 - \frac{1}{n^r}, r > 2$. Then, show that $X_n \to^{a.s} 0$, but $X_n \not\to^{L_r} 0$.

[Hints: $P(X_n = 0$ for any $n \in [m, n_0]) = \Pi_{n=m}^{n_0}(1 - \frac{1}{n^r})$. As $n_0 \to \infty$, this converges to some nonzero quantity, which again converges to 1 as $m \to \infty$.]

12.9 Let $\{X_n\}$ be a sequence of random variables such that $X_n \to^{L_2} X$. Then, show that $EX_n \to EX$ and $EX_n^2 \to EX^2$ as $n \to \infty$.

12.10 Let $\{X_n\}, \{Y_n\}$ be two sequences of random variables. Then show that

$$[|X_n - Y_n| \to^P 0 \text{ and } Y_n \to^L Y] \Rightarrow [X_n \to^L Y].$$

12.11 Let $\{X_n\}$ be a sequence of *iid* random variables, each uniformly distributed over $(0, a)$. Consider the variables $Y_n, Z_n = nY_n$ where $Y_n = \min_{1 \le i \le n} X_i$. Find the limiting distribution of Y_n and Z_n.

12.12 Prove that $[X_n \to^{L_r} X] \Rightarrow [E|X_n|^s \to E|X|^s] \; \forall \; s \le r$.

12.13 Let $\{X_n\}$ be a sequence of independent random variables defined by $P(X_n = 1) = \frac{1}{n}, P(X_n = 0) = 1 - \frac{1}{n}$. Show that $X_n \to^{L_2} 0$. However, X_n does not converge *a.s* to 0.

[Hints: $E|X_n|^2 = \frac{1}{n}$; however, $P(X_n = 0$ for every $n \in [m, n_0]) = \Pi_{n=m}^{n_0}(1 - \frac{1}{n}) \approx \frac{m-1}{n_0}$ which diverges to zero as $n_0 \to \infty$.]

12.14 Let $P(X_n = -\frac{1}{n}) = P(X_n = \frac{1}{n}) = \frac{1}{2}$. Show that $X_n \to^{a.s.} 0$ and also $X_n \to^{L_r} 0$.

[Note that $\cup_{j=n}^\infty \{|X_j| > \epsilon\} = \{|X_n| > \epsilon\}$. Choosing $n > \frac{1}{\epsilon}$, we have

$$P[\cup_{j=n}^\infty \{|X_j| > \epsilon\}] = P[|X_n| > \epsilon] \le P\{|X_n| > \frac{1}{n}] = 0.$$

Taking limit of both sides,

$$\lim_{n\to\infty} P[\cup_{j=n}^\infty \{|X_j| > \epsilon\}] = 0$$

which proves that $X_n \to^{a.s.} 0$.]

12.15 Let $\{F_n\}$ be a sequence of distribution functions such that

$$F_n(x) = \begin{cases} 0, & x < \theta + \frac{1}{n} \\ 1, & x \geq \theta + \frac{1}{n}. \end{cases}$$

Show that the corresponding sequence of random variables $X_n \to^L X$ where X is a random variable, degenerate at $X = \theta$.

12.16 Prove Theorem 12.2.16.

12.17 Let $\{X_n\}$ be a sequence of geometric random variables with parameter $\frac{\lambda}{n}, n > \lambda > 0$. Let $Y_n = \frac{X_n}{n}$. Show that Y_n converges in law to a gamma variable with parameters $(1, \frac{\lambda}{n})$.

12.18 Let $X_1, \ldots, X_n$ be a sequence of random variables with common absolutely continuous distribution function F. Let $M_n = \max(X_1, \ldots, X_n)$ and $Y_n = n[1 - F(M_n)]$. Find the limiting distribution of Y_n.

12.19 Show that the sequence of characteristic functions $\psi_n(t) = \frac{\sin(nt)}{nt}$ converges to a limiting function $\psi(t)$. However, show that $F_n(x)$ does not converge to a limiting distribution. Explain why.

12.20 Let $X_1, \ldots, X_n$ be independent $N(0, 1)$-variables. Show that $W_n = \sqrt{n}(\frac{\sum_{i=1}^n X_i}{\sum_{i=1}^n X_i^2}) \to^L \tau$ where τ is a $N(0, 1)$-variable.

[Using *mgf* show that $U_n = \frac{\sum_{i=1}^n X_i}{\sqrt{n}} \to^L \tau$. Also, the *mgf* of $V_n = \frac{1}{n}\sum_{i=1}^n X_i^2$ is $(1 - \frac{2t}{n})^{-\frac{n}{2}}$, $(t < \frac{n}{2})$ which is the *mgf* of a gamma variable with parameters $p = \frac{n}{2}, \lambda = \frac{2}{n}$ (vide (9.4.12)). Then applying Chebychev's inequality, prove that $P\{|V_n - 1| > \epsilon\} \to 0$ as $n \to \infty$. Therefore, $V_n \to^P 1$. Hence $W_n = \frac{U_n}{V_n} \to^L \tau$ by Theorem 12.2.17.]

12.21 Consider the set up of Weierstrass approximation theorem. Show that if f is twice continuously differentiable,

$$\max_{0 \leq p \leq 1} |B_n(p) - f(p)| = 0(\frac{1}{n}).$$

Again, assuming that f has continuous derivatives of order 3 or more, show that

$$\lim_{n \to \infty} n[B_n(p) - f(p)] = pq\frac{f''(p)}{2}.$$

Hence, observe that if f is not linear, the rate of convergence of $B_n(p)$ to $f(p)$ is of order $0(\frac{1}{n})$.

12.22 $\{X_n\}$ is a sequence of independent random variables. Determine if they obey WLLN and /or the central limit law.

(a) $P[X_k = 2^k] = P[X_k = -2^k] = \frac{1}{2}$.

(b) $P[X_k = 2^k] = P[X_k = -2^k] = \frac{1}{2^{2k+1}}; P[X_k = 0] = 1 - \frac{1}{2^k}$.

(c) $P[X_k = 2^k] = P[X_k = -2^k] = \frac{1}{2^{k+1}}; P[X_k = 1] = P[X_k = -1] = \frac{1}{2}(1 - 2^{-k})$.

(d) $P[X_k = 2^{-k}] = P[X_k = -2^{-k}] = \frac{1}{2}$.

(e) $P[X_k = 2^{-k}] = P[X_k = -2^{-k}] = \frac{1}{2^{k+1}}; P[X_k = 1] = P[X_k = -1] = \frac{1}{2}(1 - 2^{-k})$.

12.23 Determine if the SLLN holds for the following sequence of independent random variables.

(a) $P[X_k = 2^k] = P[X_k = -2^k] = \frac{1}{2}$.

(b) $P[X_k = k] = P[X_k = -k] = \frac{1}{2\sqrt{k}}; P[X_k = 0] = 1 - \frac{1}{\sqrt{k}}$.

(c) $P[X_k = 2^{k+1}] = P[X_k = -2^{k+1}] = 2^{-(k+3)}; P[X_k = 0] = 1 - 2^{-(k+2)}$.

12.24 Let $\{X_k\}$ be a sequence of *iid* random variables with $P(X_k = 1) = P(X_k = -1) = \frac{1}{2}$. Define $Y_k = \sum_{j=1}^{k} \frac{X_j}{2^j}$. Show that $Y_k \to^L Z$ where $Z \cap$ uniform (0.1).

12.25 Let $\{X_n\}$ be a sequence of independently distributed $N(0, \frac{\sigma^2}{n})$-variables. Show that $X_n \to^L X$ where X is a degenerate random variable at $x = 0$.

12.26 Show that a sequence $\{X_n\}$ is *a.s.* convergent *iff* it is *a.s.* mutually convergent, ie., *iff* $X_m - X_n \to^{a.s.} 0$.

12.27 Prove that if $\sum P(|X_{n+1} - X_n| \geq \epsilon_n) < \infty$ where $\sum \epsilon_n$ is a convergent sum of positive terms, then $\{X_n\}$ is *a.s.* convergent.

12.28 Let $Y_1, Y_2, \ldots$ be *iid* random variables. Define $X_n = Y_n/n$. Then clearly, $X_n \to^P 0$. Show that $X_n \to^{a.s.} 0$ *iff* $E|Y_1| < \infty$.

12.29 Let $X_1, X_2, \ldots$ be a sequence of independent random variables. Then show that $X_n \to^{a.s.} 0 \Leftrightarrow \sum_{n=1}^{\infty} P[|X_n| > \epsilon] < \infty$ for all $\epsilon > 0$.

12.30 Let ϕ be a positive symmetric convex function. Then prove that under the assumption of Theorem 12.4.5,

$$P(A) = P(|S_j| < u_j, j = 1, \ldots, n)$$
$$\geq 1 - \sum_{j=1}^{n} \frac{E[\phi(S_j)] - E[\phi(S_{j-1})]}{\phi(u_j)}$$

where we set $\phi(S_0) = 0$. In particular,

$$P(A) \geq 1 - 2^{2-r} \sum_{j=1}^{n} \frac{E(|X_j|^r)}{|u_j|^r}$$

for any $r \in [0, 2]$. Here, $S_n = \sum_{j=1}^{n} X_j$.

[Hints: As in the proof of Theorem 12.4.5, consider a form

$$T = \sum (\alpha_j - \alpha_{j+1}) \phi(S_j) \quad \text{with } \alpha_j = \phi(u_j)^{-1}$$

and use the fact that by Jensen's inequality

$$E[\phi(S_j)|X_1, \ldots, X_i] \geq \phi(S_i) \geq \phi(u_i) \ (j > i).$$

if the values of $X_1, X_2, \ldots, X_i$ imply the occurrence of B_i.]

Chapter 13

Elements of Stochastic Process

13.1 Introduction

A stochastic process means a process whose results depend on some chance elements. The models of this process can, therefore, be used to describe, study and investigate many phenomena in real life such as size and composition of a population, taking into account birth, death and migration, spread of epidemics, size of a queue in a railway booking counter, growth of a bacterial population in a laboratory culture, Brownian motion of a particle in a fluid, thermal noise in an electric circuit, etc. Here we study the probability distribution of a family of vector random variables $\{X_t, Y_t, \ldots; t \in T\}$ indexed by a parameter t, like time, space, etc.

In section 13.2 we introduce definition and classification of stochastic processes. Section 13.3 considers Markov chains, Section 13.4 the simple random walk, Section 13.5, the discrete branching process. Continuous time Markov processes are dealt with in Section 13.6.

13.2 Preliminary Notions

A stochastic process may be defined as a family of random variables $\{X_t, t \in T\}$, where T is an index set. For each t, let $\mathcal{X}_t$ be the sample space of $X_t, \mathcal{X}_t = \{x_t\}$, the set of all possible values x_t of X_t, which may be finite or infinite. The random variables $X_t, X_{t+r} (r > 0)$ may be dependent or independent.

The index set T may be finite or infinite. If T is countably infinite, e.g., $T = \{0, 1, 2, \ldots\}$ or $T = \{0, -_{+}1, -_{+}2, \ldots\}$, then the stochastic process $\{X_t, t \in T\}$ is a discrete-parameter stochastic process. If T is any finite or infinite interval, e.g., $T = \{t : a < t < b\}, T = \{t : 0 \leq t < \infty\}$, then the process is said to be a continuous parameter stochastic process. Often t is considered as a time parameter. The particular value x_t of X_t is often called the *state* and the sample space $\mathcal{X}_t$, the *state space* of X_t.

For example, for a birth and death process, if we assume that the births (deaths) occur only at points $1, 2, \ldots$, then this is a discrete stochastic process. A Poisson process (e.g., number of calls entering a telephone exchange) where we assume that a change may occur at any time is a continuous time stochastic process.

Clearly, X_t may also be a vector and in this case $\{\mathbf{X}_t, t \in T\}$ is a multivariate stochastic process.

13.3   Markov Chain

Consider a stochastic process $\{X_n, n = 0, 1, 2, \ldots\}$, i.e., a sequence of random variables each defined on the sample space $\mathcal{X}$. The space $\mathcal{X}$ is called state space and the values x as different states. We shall consider here $x = 0, 1, 2, \ldots$. Hence the process is discrete both with respect to state variable x and time variable n.

Consider now the conditional probability

$$P[X_{n+1} = x_{n+1} | X_n = x_n, X_{n-1} = x_{n-1}, \ldots, X_0 = x_0]. \qquad (13.3.1)$$

If the stochastic process is such that the conditional probability (13.3.1) depends only on the value of X_n and is independent of all previous values, we say that the process has the *Markovian property* and call it a *Markov chain*. In this case, the conditional probability (13.3.1) is

$$P[X_{n+1} = x_{n+1} | X_n = x_n]. \qquad (13.3.2)$$

More generally, a stochastic process $\{X_n, n = 0, 1, \ldots\}$ is a Markov chain of order r if

$$P[X_{n+1} = x_{n+1} | X_n = x_n, \ldots, X_0 = x_0]$$

$$= P[X_{n+1} = x_{n+1} | X_n = x_n, \ldots, X_{n-r+1} = x_{n-r+1}]$$

for all $n \geq r$ for all possible values of the random variables $X_0, \ldots, X_{n-r+2}$. A stochastic process $\{X_n\}$ with property (13.3.2) may be called a Markov chain of the first order. We shall here consider Markov chains of first order only.

Consider again the special cases of conditional probabilities as follows:

$$p_{ij} = P[X_{n+1} = j | X_n = i], \quad i, j = 0, 1, 2, \ldots \quad (13.3.3)$$

where

$$p_{ij} \geq 0, \quad \sum_{j=0}^{\infty} p_{ij} = 1 \ \forall \ i = 0, 1, \ldots.$$

Note that the conditional probability p_{ij} that at time $(n+1)$ the system is at state j given that at time n it was at state i, depends only on i and j and not on n. The quantity p_{ij} is called the *transition probability* that the system passes from state i at time n to state j at time $(n+1)$ for any n. Such conditional probabilities where p_{ij}'s do not depend on n, are also called *constant* or *stationary probabilities*. Markov chains with this property are called *homogeneous Markov chains*. We shall confine our attention to homogeneous Markov chains of first order.

We shall denote the initial probabilities

$$P\{X_0 = j\} = q_j, \ j = 0, 1, \ldots,$$

$$q_j \geq 0, \sum_{j=0}^{\infty} q_j = 1. \quad (13.3.4)$$

The unconditional or absolute probabilities at step n will be denoted as

$$P\{X_n = j\} = q^{(n)}(j), \ j = 0, 1, \ldots$$

$$q^{(n)}(j) \geq 0, \ \sum_{j=0}^{\infty} q^{(n)}(j) = 1. \quad (13.3.5)$$

We shall denote $\mathbf{q} = (q_0, q_1, \ldots)'$ and $\mathbf{q}^{(n)} = (q_0^{(n)}, q_1^{(n)}, \ldots)'$, $n \geq 1$ as vectors of initial probabilities and n-step absolute probabilities respectively.

Transition probability matrix

Consider the matrix of transition probabilities p_{ij},

$$\mathbf{P} = \begin{bmatrix} p_{00} & p_{01} & p_{02} & \cdots \\ p_{10} & p_{11} & p_{12} & \cdots \\ p_{20} & p_{21} & p_{22} & \cdots \\ \cdots & \cdots & \cdots & \cdots \end{bmatrix}. \quad (13.3.6)$$

$\mathbf{P}$ is a square matrix of infinite order with non-negative elements with row-sum equal to unity, $\sum_{j=0}^{\infty} p_{ij} = 1 \ \forall \ i$. A matrix satisfying this condition is called a *stochastic matrix* or a *Markov matrix*. A Markov chain is completely specified by a matrix of transition probabilities $\mathbf{P}$ and a vector $\mathbf{q}$.

Higher order transition probabilities can be defined in a similar way. Let us denote by $p_{ij}^{(n)}$ the probability that the system passes from state i to state j in n steps. Clearly,

$$
\begin{aligned}
p_{ij}^{(1)} &= p_{ij}, \\
p_{ij}^{(n+1)} &= \sum_{\nu=0}^{\infty} p_{i\nu}^{(n)} p_{\nu j}, \quad n = 1, 2, \dots.
\end{aligned}
\tag{13.3.7}
$$

It follows by the assumption of stationarity of p_{ij} above,

$$
p_{ij}^{(k)} = P[X_{m+k} = j | X_m = i] \; \forall \; k = 1, 2, \dots,
\tag{13.3.8}
$$

whatever m. This can be seen as follows. For $k = 1$, (13.3.8) is true by definition. Assume that it is true for $k = n$. Now,

$$
\begin{aligned}
P\{X_{m+n+1} = j | X_m = j\} &= \sum_{\nu=0}^{\infty} P\{X_{m+n} = \nu | X_m = i\} \\
&\qquad P\{X_{m+n+1} = j | X_{m+n} = \nu\} \\
&= \sum_{\nu=0}^{\infty} p_{i\nu}^{(n)} p_{\nu j} = p_{ij}^{(n+1)}.
\end{aligned}
$$

Hence, (13.3.8) follows by induction. In general, we have the following relation:

$$
p_{ij}^{(m+n)} = \sum_{\nu=0}^{\infty} p_{i\nu}^{(m)} p_{\nu j}^{(n)}.
\tag{13.3.9}
$$

That (13.3.9) is true for all m can also be shown by induction. Denoting by $\mathbf{P}^{(n)} = ((p_{ij}^{(n)}))$ the matrix of n-step transition probabilities, the equation (13.3.9) can be written as

$$
\mathbf{P}^{(m+n)} = \mathbf{P}^{(m)} \mathbf{P}^{(n)}.
\tag{13.3.10}
$$

Equations (13.3.10) are called the *Chapman-Kolmogorov functional equations* and these characterize Markov chains. Clearly,

$$
\mathbf{P}^{(n)} = \mathbf{P}^{(n-1)} \mathbf{P};
\tag{13.3.11}
$$

$$
\mathbf{P}^{(n)} = \mathbf{P} \mathbf{P}^{(n-1)}.
\tag{13.3.12}
$$

Clearly, $\mathbf{P}^{(n)} = \mathbf{P}^n$, the nth power of the matrix $\mathbf{P}$.

The equations (13.3.11) and (13.3.12) respectively are called *forward* and *backward equations*. The unconditional probabilities $q^{(n)}(j)$ are given by

$$
q^{(n)}(j) = \sum_{i=0}^{\infty} q(i) p_{ij}^{(n)}.
\tag{13.3.13}
$$

Asymptotic values of $p_{ij}^{(n)}$ and $q^{(n)}(j)$ as $n \to \infty$ will be considered in subsection 13.3.2.

EXAMPLE 13.3.1: The simplest non-trivial Markov chain is one of two states, the transition matrix being necessarily of the form

$$\mathbf{P} = \begin{bmatrix} 1 - \alpha & \alpha \\ \beta & 1 - \beta \end{bmatrix}, \ 0 < \alpha, \beta < 1. \tag{13.3.14}$$

The two states may correspond to two energy levels, say 1 and 2 of a molecule.

EXAMPLE 13.3.2: (*Random walk and gambler's ruin problem*): A gambler A with initial capital z plays against an adversary with initial capital $(a-z)$. In every game, the winner will get one unit of money from the loser (if he has any money left to lose). The probability of a win is p and losing is $q = 1 - p$ at each game. The play stops as soon as one of the players has 0 or a units of money.

Let X_n be the fortune of the gambler A (the money left with A) after n games. Thus

$$P[X_{n+1} = j | X_n = i] = p_{ij} = \begin{cases} p & \text{if } j = i+1 \\ q & \text{if } j = i-1 \\ 0 & \text{otherwise}, \end{cases}$$

$(n = 1, 2, \ldots), (i, j = 1, 2, \ldots, a-1), p_{00} = p_{aa} = 1$. Thus $\{X_n\}$ is a Markov chain with stationary probabilities. The transition probability matrix is

$$\mathbf{P} = \begin{bmatrix} 1 & 0 & 0 & 0 & \ldots & 0 \\ q & 0 & p & 0 & \ldots & 0 \\ 0 & q & 0 & p & \ldots & 0 \\ \multicolumn{6}{c}{\cdots\cdots\cdots\cdots\cdots} \\ 0 & 0 & 0 & 0 & \ldots & 1 \end{bmatrix}_{(a+1) \times (a+1)}.$$

The state space is $\{0, 1, \ldots, a\}$, the states '0' and 'a' are the absorbing states, because no transition from these states ate possible. Initial probability vector is $\mathbf{q} = (0, \ldots, 0, 1, 0, \ldots, 0)'$ with 1 only at the zth place and zero elsewhere. Questions of interest are: (a) what is the probability of ultimate ruin of the gambler? (b) what is the probability that the game will end in exactly n steps? Answers to such queries will be considered in Section 13.5.

EXAMPLE 13.2.3: Let $\{X_i\}$ be a sequence of independent random variables and let $S_n = \sum_{j=1}^{N} X_j$. If X_i's can take only integer values,

$$P[S_n = j | S_1 = s_1, \ldots, S_{n-1} = i] = P[S_n = j | S_{n-1} = i]$$
$$= P[X_n = j - i].$$

Thus $\{S_n\}$ is a Markov chain. It will have stationary transition probabilities if X_n's are *iid*. The state space is the space of all positive and negative integers including zero.

EXAMPLE 13.3.4: In genetics Mandel's law states that the inherited characters depend on genes, which always occur in pairs. In the simplest case, every gene may be of two forms A or a. The possible genotypes are (AA, Aa, aa). An offspring receives one gene from each parent under the conditions of Bernoulii scheme. Suppose that the population consists of N members in each generation. Thus we have $2N$ genes in each generation. If in some generation, there are i genes of the form $A(0 \le i \le 2N)$, the generation is said to be in state i. Thus, we have here a homogeneous Markov chain with $(2N + 1)$ possible states: $0, 1, \ldots, 2N$. The probability of passing from state i to state j in the next generation is

$$p_{ij} = \binom{2N}{j} (\frac{i}{2N})^j (1 - \frac{i}{2N})^{2N-j}.$$

Note that the states 0 and $2N$ are here absorbing states. If in some generation, the population is in one of these states, it will remain there for ever.

13.3.1   *Classification of states*

Classification of states of a Markov chain is of great importance in the study of the asymptotic properties of $p_{ij}^{(n)}$. First we consider a few definitions.

DEFINITION 13.3.1: Two states i and j of a Markov chain are said to communicate $(i \to j)$ if $p_{ij}^{(n)} > 0$ and $p_{ji}^{(m)} > 0$ for some $n(\ge 1)$ and $m(\ge 1)$. The following relations hold for communicative states.

(a) $i \to j \Leftrightarrow j \to i$;
(b) $i \to j$ and $j \to k \Leftrightarrow i \to k$;
(c) if $i \to j$ for some j then $i \to i$.

DEFINITION 13.3.2: A set of states $S \subset \mathcal{X}$ is called *closed* if $p_{ij} = 0$ for all $i \in S$ and $j \in \bar{S} = \mathcal{X} - S$. If the closed set S contains only one state i, then this state is called an *absorbing state*. Clearly, a necessary and sufficient condition for state i to be an absorbing state is $p_{ii} = 1$. If the state space of a Markov chain contains two or more closed sets, the chain is called *decomposable* or *reducible*.

The transition probability matrix associated with a decomposable chain can be written in the form of a partitioned matrix. For example, suppose

$$\mathbf{P} = \begin{bmatrix} \mathbf{P}_1 & \mathbf{0} \\ \mathbf{0} & \mathbf{P}_2 \end{bmatrix}.$$

Here $\mathbf{P}_1, \mathbf{P}_2$ represent Markov matrices which describe the transition within the two closed sets of states.

A chain or matrix which is not decomposable is called *indecomposable* or *irreducible*. A chain is irreducible *iff* every state can be reached from every other state.

For an arbitrary set $\mathcal{G}$ of states the smallest closed set containing $\mathcal{G}$ is called *closure* of $\mathcal{G}$. For a state i, the closure of i will be denoted as $C(i)$. It is the set of all states that are accessible from i.

DEFINITION 13.3.3: If $i \to i$, the greatest common divisor of the set of integers n such that $p_{ii}^{(n)} > 0$ is called the *period of state i*. We denote by $w(i)$ the period of the state i and say that the state i is periodic with period $w(i)$. If $i \not\to i$, we say $w(i) = 0$ and the state i is aperiodic.

EXAMPLE 13.3.5: The simplest example of a chain with period 3 is one in which only the transitions $1 \to 2 \to 3 \to 1$ are possible. Here

$$\mathbf{P} = \begin{bmatrix} 0 & 1 & 0 \\ 0 & 0 & 1 \\ 1 & 0 & 0 \end{bmatrix}.$$

EXAMPLE 13.3.6: Consider the following transition matrix where a * denotes a positive element and states are $1, 2, \ldots, 9$.

$$\mathbf{P} = \begin{bmatrix} 0 & 0 & 0 & * & 0 & 0 & 0 & 0 & * \\ 0 & * & * & 0 & * & 0 & 0 & 0 & * \\ 0 & 0 & 0 & 0 & 0 & 0 & 0 & * & 0 \\ * & 0 & 0 & 0 & 0 & 0 & 0 & 0 & 0 \\ 0 & 0 & 0 & 0 & * & 0 & 0 & 0 & 0 \\ 0 & * & 0 & 0 & 0 & 0 & 0 & 0 & 0 \\ 0 & * & 0 & 0 & 0 & * & * & 0 & 0 \\ 0 & 0 & * & 0 & 0 & 0 & 0 & 0 & 0 \\ 0 & 0 & 0 & * & 0 & 0 & 0 & 0 & * \end{bmatrix}.$$

It can be easily checked that 5 is an absorbing state ($p_{55} = 1$). The sets $(3, 8), (1, 4, 9)$ are two closed sets. The closure of state 2 is $(2, 3, 5, 8)$.□

If a Markov chain is irreducible and has k irreducible sets, then by a suitable relabelling of the states, the transition probability matrix $\mathbf{P}$ can be written in the form

$$\mathbf{P} = \begin{bmatrix} \mathbf{P}_1 & \mathbf{0} & \mathbf{0} & \ldots & \mathbf{0} & \ldots & \mathbf{0} \\ \mathbf{0} & \mathbf{P}_2 & \mathbf{0} & \ldots & \mathbf{0} & \ldots & \mathbf{0} \\ \ldots & \ldots & \ldots & \ldots & \ldots & \ldots & \ldots \\ \mathbf{0} & \mathbf{0} & \mathbf{0} & \ldots & \mathbf{P}_k & \ldots & \mathbf{0} \\ \mathbf{A} & \mathbf{B} & \mathbf{C} & \ldots & \mathbf{G} & \ldots & \mathbf{H} \end{bmatrix}$$

where $\mathbf{P}_1, \ldots, \mathbf{P}_k$ are transition matrices for these k sets. The matrix $\mathbf{P}^{(n)}$ is of the same type with $\mathbf{P}_1, \mathbf{P}_2, \ldots, \mathbf{P}_k$ and $\mathbf{H}$ replaced by $\mathbf{P}_1^{(n)}, \mathbf{P}_2^{(n)}, \ldots, \mathbf{P}_k^{(n)}, \mathbf{H}^n$ respectively, whereas $\mathbf{A}, \mathbf{B}, \ldots, \mathbf{G}$ have more complicated forms.

We recall without proof the following theorem.

Theorem 13.3.1: If in the matrix $\mathbf{P}^{(n)}$ of a reducible chain all rows and all columns corresponding to the states outside the closed set C are omitted, the elements of the resulting stochastic matrix also satisfy the fundamental relations (13.3.11) and (13.3.12).

This means that we have a Markov chain defined on C and this subchain can be studied independently of the remaining states.

We now define several other probabilities. Let

$$f_{ij}^{(n)} = P\{X_n = j | X_0 = i, X_h \neq j, 0 < h < n\}. \tag{13.3.15}$$

Thus $f_{ij}^{(n)}$ is the conditional probability that the system reaches the state j for the first time at time n given that it started from state i at time zero. Let

$$f_{ij} = P\{X_h = j \text{ for at least one } h | X_0 = i\}. \tag{13.3.16}$$

f_{ij} is the probability that the system reaches the state j given that it was at state i at time zero. Clearly,

$$f_{ij} = \sum_{n=1}^{\infty} f_{ij}^{(n)}. \tag{13.3.17}$$

Note that

$$p_{ij}^{(n)} = \sum_{\nu=1}^{n} f_{ij}^{(\nu)} p_{jj}^{(n-\nu)}. \tag{13.3.18}$$

We have $f_{ij} \leq 1$. When $f_{ij} = 1$, $\{f_{ij}^{(n)}\}$ is a proper probability distribution and can be referred to as the *first passage distribution* for state j given that the system stated from state i at time zero.

Let also

$$
\begin{aligned}
g_{ij} &= P\{X_h = j \text{ infinitely often} \,|X_0 = i\} \\
&= P\{X_h = j \text{ for an infinite number of values of } h|X_0 = i\}.
\end{aligned}
$$
(13.3.19)

DEFINITION 13.3.4: A state i is called *recurrent (persistent)* if $f_{ii} = 1$ or $g_{ii} = 1$ (this means a return to state i is certain); a state i is called *non-recurrent (transient)* if $f_{ii} < 1$ or $g_{ii} = 0$ (that is, a return to state i is uncertain).

The equivalence of condition on f_{ii} and g_{ii} can be seen as follows. Let

$$
F_n = P\{X_h = i \text{ for at least } n \text{ values of } h|X_0 = i\}.
$$

Then

$$
F_1 = f_{ii}, F_2 = f_{ii}.F_1, \ldots, F_n = f_{ii}.F_{n-1} = (f_{ii})^n.
$$

Hence,

$$
g_{ii} = \lim_{n \to \infty} F_n = \lim_{n \to \infty} (f_{ii})^n = \begin{cases} 0, & \text{if } f_{ii} < 1 \\ 1, & \text{if } f_{ii} = 1. \end{cases}
$$

Suppose a state i is recurrent. Given that $X_n = i$, let T_i be the waiting time (recurrence time) till the next occurrence of state i. Thus $T_i = m$ if $X_{n+k} \neq i$ for $1 \leq k < m$ and $X_{n+m} = i$. Then $P[T_i = m] = f_{ii}^{(m)}$.

The mean recurrence time for state i is

$$
E(T_i) = \mu_i = \sum_{m=1}^{\infty} m f_{ii}^{(m)} \leq \infty.
$$
(13.3.20)

DEFINITION 13.3.5: A recurrent state i is *null* if $\mu_i = \infty$. A recurrent state which is neither null nor periodic is called *ergodic*. Thus, an aperiodic persistent state i with $\mu_i < \infty$ is called ergodic.

If $f_{ij} = \sum_{n=1}^{\infty} f_{ij}^{(n)} = 1$, we can define the mean first passage time from state i to state j as

$$
\mu_{ij} = \sum_{n=1}^{\infty} n f_{ij}^{(n)}.
$$
(13.3.21)

Theorem 13.3.2: If the series

$$
\sum_{n=1}^{\infty} p_{ij}^{(n)}
$$
(13.3.22)

converges, then $g_{ij} = 0$ (this always holds if the state j is nonrecurrent). If (12.3.22) diverges, $g_{ij} > 0$ and $g_{jj} = 1$. If $i \to j, g_{ij} = g_{ji} = 1$.

Proof. Let E_n denote the event $(X_{m+n} = j | X_m = i)$. Let $P(X_m = i) > 0$. Then

$$P(E_n) = P(X_m = i)p_{ij}^{(n)}. \tag{13.3.23}$$

Therefore, if (13.3.22) converges, by Borel-Cantelli lemma (Theorem 12.4.1),

$$P\{E_n \text{ occurs infinitely often}\} = 0.$$

Hence $g_{ij} = 0$. Again, from (13.3.18), summing both sides over $n = N$,

$$
\begin{aligned}
\sum_{n=1}^{N} p_{ij}^{(n)} &= \sum_{n=1}^{N} \sum_{\nu=1}^{n} f_{ij}^{(\nu)} p_{jj}^{(n-\nu)} \\
&= \sum_{\nu=1}^{N} f_{ij}^{(\nu)} \sum_{n=\nu}^{N} p_{jj}^{(n-\nu)} \le \sum_{\nu=1}^{N} f_{ij}^{(\nu)}(1 + \sum_{n=1}^{N} p_{jj}^{(n)}).
\end{aligned}
\tag{13.3.24}
$$

Allowing $N \to \infty$ and using (13.3.17),

$$\sum_{n=1}^{\infty} p_{ij}^{(n)} \le f_{ij}(1 + \sum_{n=1}^{\infty} p_{jj}^{(n)}). \tag{13.3.25}$$

Setting $j = i$ in (13.3.24) and dividing both sides by $1 + \sum_{n=1}^{N} p_{ii}^{(n)}$,

$$\frac{\sum_{n=1}^{N} p_{ii}^{(n)}}{1 + \sum_{n=1}^{N} p_{ii}^{(n)}} \le \sum_{\nu=1}^{N} f_{ii}^{(\nu)}.$$

Hence, if (13.3.22) diverges, it follows from (13.3.25) that $\sum_{n=1}^{\infty} p_{jj}^{(n)}$ diverges and hence

$$\sum_{\nu=1}^{\infty} f_{ii}^{(\nu)} = f_{ii} \ge 1.$$

Therefore, $f_{ii} = 1$ and $g_{ii} = 1$.

In general, if (13.3.22) diverges, $\sum_{n=1}^{\infty} p_{jj}^{(n)}$ diverges. Hence, $g_{jj} = 1$. This implies the existence of infinitely many $p_{ij}^{(n)} > 0$ and hence $f_{ij} > 0$ and therefore, $g_{ij} = f_{ij}g_{jj} > 0$.

If $i \to j, f_{ij} > 0, f_{ji} > 0$. It follows that if $g_{ii} = 1$ and $f_{ij} > 0$, then $g_{ij} = 1$. Hence, $g_{jj} = 1$ implies $g_{ji} = 1$. If the state j is nonrecurrent, $g_{ji} = 0$. Hence, (13.3.22) must converge.

13.3.2   Limits of higher order transition probabilities

It is of interest in many practical applications to investigate the behavior of $p_{ij}^{(n)}$ as $n \to \infty$. This is considered in this section.

We shall assume throughout that the process was at state i at time zero. We shall henceforth denote $f_{ii}^{(n)}$ as $f_i^{(n)}$, f_{ii} as f_i and $p_{ii}^{(n)}$ as $u_i^{(n)}$. From (13.3.18),

$$u_i^{(n)} = u_i^{(0)} f_i^{(n)} + u_i^{(1)} f_i^{(n-1)} + u_i^{(2)} f_i^{(n-2)} + \ldots + u_i^{(n-1)} f_i^{(1)}, \; n \geq 1, \quad (13.3.26)$$

where $u_i^{(0)} = 1$, $f_i^{(0)} = 0$. Consider the probability generating functions

$$U(s) = \sum_{n=0}^{\infty} u_i^{(n)} s^n, \quad\quad\quad\quad\quad (13.3.27)$$

$$F(s) = \sum_{n=0}^{\infty} f_i^{(n)} s^n. \quad\quad\quad\quad\quad (13.3.28)$$

Hence,

$$U(s) = 1 + \sum_{n=1}^{\infty} u_i^{(n)} s^n = 1 + \sum_{n=1}^{\infty} \{ \sum_{j=0}^{n} u_i^{(n)} f_i^{(n-j)} \} s^n = 1 + U(s)F(s),$$

i.e.,

$$U(s) = \frac{1}{1 - F(s)}. \quad\quad\quad\quad\quad (13.3.29)$$

Theorem 13.3.3: A state is transient *iff* $\sum_{n=1}^{\infty} u_i^{(n)}$ converges. In this case,

$$f_i = \frac{u_i - 1}{u_i} \quad\quad\quad\quad\quad (13.3.30)$$

where

$$u_i = \sum_{n=0}^{\infty} u_i^{(n)}. \quad\quad\quad\quad\quad (13.3.31)$$

Proof. Suppose i is transient. Then $f_i < 1$. The series $F(s)$ converges and $F(1) = f_i < 1$. The series $U(s)$ has only non-negative coefficients, and since $U(s) \to \frac{1}{1-f_i}$ as $s \to 1$ it follows from Abel's theorem that the series converges for $s = 1$ and also $U(1) = u_i = \frac{1}{1-f_i}$.

If i is persistent, $f_i = 1$. As $s \to 1$, $F(s) \to 1$ and $U(s) \to \infty$ so that in this case $\sum_{n=0}^{\infty} u_i^{(n)}$ diverges.

Remark 13.3.1: The quantity $\sum_{k=1}^{n} u_i^{(k)} = \sum_{k=1}^{n} p_{ii}^{(k)}$ may be interpreted as the expected number of returns to step i during the course of n steps. Hence $\sum_{n=0}^{\infty} u_i^{(n)} - 1 = u_i - 1$ may be interpreted as the expected number of returns to step i (excluding the initial position) in an infinitely many steps.

Proof. Omitted.

Theorem 13.3.4: (a) If i is a persistent null state, then $u_i^{(n)} \to 0$ as $n \to \infty$; (b) if i is ergodic, then $u_i^{(n)} \to 1/\mu_i$.

Theorem 13.3.5: If the state i is persistent and has period ω, then as $n \to \infty$,

$$p_{ii}^{(n\omega)} = u_i^{(n\omega)} \to \frac{\omega}{\mu_i}$$

where $u_i^{(k)} = 0$ for every k which is not divisible by ω.

Proof. Since i has period ω, $F(s)$ contains only powers of s^ω and therefore $F(s^{1/\omega}) = F_1(s)$ (say) is also a power series. Now, $F_1(1) = 1$ so that

$$F_1(s) = \sum_{n=1}^{\infty} f_{i(1)}^{(n)} s^n$$

($f_{i(1)}^{(n)}$ being suitably defined) is regarded as a probability generating function corresponding to a persistent state. Hence, applying Theorem 13.3.4, coefficients of

$$U_1(s) = \frac{1}{1 - F_1(s)} \quad \text{(by (13.3.29))}$$

tends to $1/\mu_{(1)}$ where

$$\mu_{(1)} = \frac{d}{ds} F_1(s)]_{s=1} = \frac{1}{\omega} \frac{d}{ds} F(s)]_{s=1} = \frac{\mu_i}{\omega}.$$

But the coefficient of s^n in $U_1(s)$ is the coefficient of $s^{n\omega}$ in $U(s)$, i.e., is $u_i^{(n\omega)}$. Hence the first part. The second part follows from the definition of a persistent state with period ω.

Theorems 13.3.3 - 5 indicate that except when i is periodic, $u_i^{(n)}$ has a unique limit which is null or transient and is otherwise $1/\mu_i$. In the periodic case, a limit exists for the subsequence $n = \omega, 2\omega, 3\omega, \ldots$

Theorem 13.3.6: In an ergodic chain,

$$\lim_{n\to\infty} p_{ij}^{(n)} = \pi_j \tag{13.3.32}$$

exists and is independent of the initial state i. Moreover,

$$\pi_j = \frac{1}{\mu_j} \tag{13.3.33}$$

where μ_j is the mean recurrence time of j. Again, $\{\pi_j\}$ is a probability distribution and satisfies

$$\pi_j = \sum_{i=0}^{\infty} \pi_i p_{ij}. \tag{13.3.34}$$

For a proof the reader may refer to Feller (1968), vol.1, pp. 393 - 394 or Bharucha-Reid (1960), pp. 28 - 30.

13.3.3 *Irreducible chains*

We consider in this section some theorems on the closure of sets and irreducible sets of states.

Theorem 13.3.7: If C is an irreducible closed set of states and if $i \in C$, then $C = C(i)$.

Suppose $j \in C(i)$. Then $j \in C$, for otherwise, the root from j to i passes through one state outside C. This is impossible since C is irreducible and closed. Hence, $C(i) \subset C$.

Suppose now $j \in C$. If $j \notin C(i)$, this means $C(i)$ is a closed proper subset of C. This is not possible since C is irreducible. Hence, $C \subset C(i)$.

Therefore, $C = C(i)$.

Theorem 13.3.8: If i is a persistent state then the closure of $i, C(i)$ is an irreducible set of states.

Proof. Suppose that the result is not true and there is a closed subset C of $C(i)$.

(1) Suppose $i \in C$ and let $j \in (C(i) - C)$. Then, since $j \in C(i)$, there is a finite sequence of transitions, each with positive probability, from $i \to j$. Thus, $p_{ij} > 0$, which is a contradiction.

(2) Suppose $i \in (C(i) - C)$ and $j \in C$.

Then there is a positive probability that the system will move from state i to state j. Since $j \in C$, transition from j to i is not possible. Then i cannot be a persistent state, because $f_{ii} < 1$. This is a contradiction.

Theorem 13.3.9: If C is a closed irreducible set of states, then all states in C have the same period.

Proof. Suppose i and j are both in C. Then there exist positive integers M and N such that

$$p_{ij}^{(M)} > 0, p_{ji}^{(N)} > 0.$$

Obviously,

$$p_{ii}^{(n+M+N)} \geq p_{ij}^{(M)} p_{jj}^{(n)} p_{ji}^{(N)} > 0 \tag{13.3.35}$$

and

$$p_{jj}^{(n+M+N)} \geq p_{ji}^{(N)} p_{ii}^{(n)} p_{ij}^{(M)} > 0. \tag{13.3.36}$$

Here, i, j, M, N are fixed while n is arbitrary. Let Ω_i be the set of integers n for which $p_{ii}^{(n)} > 0$. Similarly, let Ω_j be the set of integers n for which $p_{jj}^{(n)} > 0$.

Let $\lambda_i = HCF(\Omega_i), \lambda_j = HCF(\Omega_j)$ be the periodicity of state i and j respectively. (HCF means, highest common factor).

From (13.3.35), $(M + N) \in \Omega_i$; from (13.3.36), $(M + N) \in \Omega_j$. Hence, $(M + N) \in \Omega_i \cap \Omega_j$. Again, if $n \in \Omega_i$, then $(n + M + N) \in \Omega_i$; hence, n is divisible by λ_i. Thus, λ_i divides every member in Ω_j and hence divides λ_j. By the same logic, λ_j divides λ_i. Hence, $\lambda_i = \lambda_j$.□

We shall define that two states are of the same type if they agree in all the characteristics defined in subsection 13.3.1. Thus, two states of the same type have the same period or are aperiodic; both are transient or persistent. If they are persistent, both have the mean recurrence time finite or infinite.

Theorem 13.3.10: In an irreducible Markov chain all states are of the same type.

Proof. Let i and j be two arbitrary states. Since every state can be reached from every other state, there exist integers M and N such that $p_{ij}^{(M)} = \alpha > 0$ and $p_{ji}^{(N)} = \beta > 0$. Obviously,

$$p_{ii}^{(n+M+N)} \geq p_{ij}^{(M)} p_{jj}^{(n)} p_{ji}^{(N)} = \alpha\beta p_{jj}^{(n)}. \tag{13.3.37}$$

If j is transient, $\sum_{\nu=1}^{\infty} p_{jj}^{(\nu)} < \infty$ and hence $\sum_{r=1}^{\infty} p_{ii}^{(r)} < \infty$. Further, if $p_{ii}^{(n)} \to 0$, then also $p_{jj}^{(n)} \to 0$. The same results hold if i and j are interchanged and hence either both are transient or persistent. If one is a null state, the other is also so.

Suppose i has period ω. For $n = 0$, the right hand side of (13.3.37) is positive and hence $(M + N)$ is a multiple of ω. Again, left hand side vanishes unless n is a multiple of ω, i.e., j has a period ω. By interchanging i and j we see that both the states have the same period.

Note 13.3.2: For all practical purposes, it is, therefore, always possible to restrict attention to states of one particular type only.

Theorem 13.3.11: In any Markov chain the persistent states can be divided in a unique manner into non-overlapping irreducible sets $C_1, C_2, \ldots$, such that any state in a given set can be reached from any other state in that set and such that all states in a given set are of the same type. Apart from the persistent states, there will, in general, remain transient states from which the persistent states can be reached but not vice versa.

Proof. Let i be the state with the lowest suffix which is persistent. Consider closure of i, $C(i) = C_i$ (say). Now $C(i)$ is closed and since i is persistent, $C(i)$ is an irreducible set (Theorem 13.3.8). Again, all states of this set are of the same type (Theorem 13.3.10).

Next consider the state $i'(\notin C_1)$ with the lowest suffix which is persistent. Consider closure of i', $C(i') = C_2$ (say). C_2 is an irreducible set of states of the same type. Proceeding in this way, one can divide the persistent states into sets $C_1, C_2, \ldots$ having the above-mentioned properties. Again, such a division is unique. $\quad\square$

We now consider the problem of finding the nth order transition probability matrix $\mathbf{P}^{(n)} = \mathbf{P}^n$. This is done by considering the spectral decomposition of a matrix.

Let $\mathbf{A}$ be a $N \times N$ matrix with latent roots $\lambda_1, \ldots, \lambda_N$, assumed distinct. Let $\mathbf{X}_i$ be a column vector such that $(\mathbf{A} - \lambda_i \mathbf{I})\mathbf{X}_i = \mathbf{0}$ so that $\mathbf{X}_i$ is the column latent vector corresponding to $\lambda_i (i = 1, \ldots, N)$. Similarly, let $\mathbf{Y}_i$ be a column vector such that $\mathbf{Y}_i'(\mathbf{A} - \lambda_i \mathbf{I}) = \mathbf{0}$ so that $\mathbf{Y}_i'$ is the row latent vector corresponding to $\lambda_i (i = 1, \ldots, N)$. Since these vectors are determined uniquely only up to a multiplicative constant, we may choose them in such a way that $\mathbf{X}_i'\mathbf{Y}_i = 1 (i = 1, \ldots, N)$.

The matrix $\mathbf{A}_i = \mathbf{X}_i\mathbf{Y}_i'$ is called the *latent matrix* or *spectral matrix* associated with $\lambda_i (i = 1, \ldots, N)$. Their properties are: (i) $\mathbf{A}_i$'s are idempotent, $\mathbf{A}_i = \mathbf{A}_i^2$; (ii) $\mathbf{A}_i\mathbf{A}_j = \mathbf{0} (i \neq j)$; (iii) $\mathbf{A} = \sum_{i=1}^{N} \lambda_i \mathbf{A}_i$. Property (iii) is called the spectral decomposition of $\mathbf{A}$.

It follows from (i) - (iii) that $\mathbf{A}^r = \sum_{i=1}^{r} \lambda_i^r \mathbf{A}_i$ for any integer r. Hence, we get any power of $\mathbf{A}$ by knowing λ_i^r and $\mathbf{A}_i (i = 1, \ldots, N)$.

EXAMPLE 13.3.7: Consider Example 13.3.1. Here, the latent roots of $\mathbf{P}$ are $\lambda_1 = 1, \lambda_2 = 1 - \alpha - \beta$. The latent matrices associated with λ_1 and λ_2 are

$$\mathbf{P}_1 = \frac{1}{\alpha + \beta} \begin{bmatrix} \beta & \alpha \\ \beta & \alpha \end{bmatrix} \text{ and } \mathbf{P}_2 = \frac{1}{\alpha + \beta} \begin{bmatrix} \alpha & -\alpha \\ -\beta & \beta \end{bmatrix},$$

respectively. Hence

$$\begin{aligned} \mathbf{P}^n &= \mathbf{P}_1 + (1 - \alpha - \beta)^n \mathbf{P}_2 \\ &= \frac{1}{\alpha + \beta} \begin{bmatrix} \beta & \alpha \\ \beta & \alpha \end{bmatrix} + \frac{(1 - \alpha - \beta)^n}{\alpha + \beta} \begin{bmatrix} \alpha & -\alpha \\ -\beta & \beta \end{bmatrix}, \ 0 < \alpha, \beta < 1. \end{aligned} \tag{13.3.38}$$

If $|1 - \alpha - \beta| < 1$, then

$$\mathbf{P}^n \to \frac{1}{\alpha + \beta} \begin{bmatrix} \beta & \alpha \\ \beta & \alpha \end{bmatrix} \text{ as } n \to \infty \tag{13.3.39}$$

so that the process is ergodic. This means whatever the initial distribution, the process tends to an equilibrium distribution between states with

$$\pi_1 = \frac{\beta}{\alpha + \beta}, \pi_2 = \frac{\alpha}{\alpha + \beta}, \ 0 < \alpha, \beta < 1.$$

The distribution satisfies the equilibrium equation (13.3.34).

In case $\alpha = \beta = 0$ or $\alpha = \beta = 1$, the process is not ergodic. In the first case, all transitions are impossible and the molecule freezes in its initial position, whatever that is. The chain here is reducible. In the second case, transition is certain and the molecule alternates regularly between the two positions. In this case, this chain is periodic.

EXAMPLE 13.3.8: Consider the case of pure inbreeding, in which an individual is mated with itself, as in plants. An offspring gets one allele from each parent at random. Thus an individual with AA mated with itself ($AA \times AA$) produces only AA; similarly, an aa produces only aa; an Aa mated with itself produces AA, Aa, aa in the proportion $\frac{1}{4}, \frac{1}{2}, \frac{1}{4}$. If we consider only a single individual in each generation, then this constitutes a Markov chain with three states with transition matrix

$$\mathbf{P} = \begin{bmatrix} 1 & 0 & 0 \\ \frac{1}{4} & \frac{1}{2} & \frac{1}{4} \\ 0 & 0 & 1 \end{bmatrix}.$$

It can be shown that

$$\mathbf{P}^n = \begin{bmatrix} 1 & 0 & 0 \\ \frac{1 - 2^{-n}}{2} & 2^{-n} & \frac{1 - 2^{-n}}{2} \\ 0 & 0 & 1 \end{bmatrix} \to \begin{bmatrix} 1 & 0 & 0 \\ \frac{1}{2} & 0 & \frac{1}{2} \\ 0 & 0 & 1 \end{bmatrix}.$$

Thus ultimately an AA or an aa line remains unchanged while an Aa line becomes either AA or aa each with probability $\frac{1}{2}$. The effect of inbreeding is, therefore, to produce a pure (homozygous) line.

Consider now inbreeding such that all aa individuals are discarded and one continues breeding in a given generation until a non-aa is produced. Here

$$\mathbf{P} = \begin{bmatrix} & AA & Aa \\ & 1 & 0 \\ & \frac{1}{3} & \frac{2}{3} \end{bmatrix}$$

and the process is till Markovian.

$$\mathbf{P}^n = \begin{bmatrix} 1 & 0 \\ 1 - (\frac{2}{3})^n & (\frac{2}{3})^n \end{bmatrix} \rightarrow \begin{bmatrix} 1 & 0 \\ 1 & 0 \end{bmatrix}.$$

Thus, one ends with a pure AA line.

13.3.4   *Martingales*

DEFINITION 12.3.6 *Martingales*: A Markov chain is said to be a Martingale if for every j, the expectation of the conditional probability distribution $\{p_{jk}\}$ equals j, i.e., if

$$E\{X_{n+1}|X_n = j\} = \sum_k k p_{jk} = j. \tag{13.3.40}$$

For simplicity, consider a martingale with a finite number of states $0, 1, \ldots, a$. Putting $j = 0$ and $j = a$ in (13.3.40), we have

$$\sum_{k=0}^{a} k p_{0k} = 0 \Rightarrow p_{0k} = 0 \ \forall \ k \Rightarrow p_{00} = 1$$
$$\sum_{k=0}^{a} k p_{ak} = a \Rightarrow p_{aa} = 1.$$

Hence, a and 0 must be absorbing states.

Assume that the chain contains no further closed set. It follows that the interior states are transient states. Hence, the process will terminate either at 0 or at a.

From (13.3.40), it follows by induction that for all n,

$$\sum_{k=0}^{a} p_{jk}^{(n)} k = j. \tag{13.3.41}$$

But $p_{jk}^{(n)} \rightarrow 0$ for every transient state k. Hence, (13.3.41) implies that for all $j > 0$,

$$p_{ja}^{(n)} \rightarrow \frac{j}{a},$$

This means that if the process starts with j, the probabilities of ultimate absorption at 0 and a are $1 - \frac{j}{a}, \frac{j}{a}$ respectively.

13.4   Discrete Branching Process

The Markov process $\{X_n, n = 0, 1, 2, \ldots\}$ is said to represent a simple discrete branching process if the following conditions are satisfied.

(a) $X_0 = x_0 = 1$ with probability one;

(b) $p(x) = P\{X_1 = x\}, p(x) \geq 0, \sum_{x=0}^{\infty} p(x) = 1, x = 0, 1, 2, \ldots;$

(c) the conditional distribution of X_{n+1} given $X_n = j$ is the sum of j independent random variables, each having the same *pmf* as X_1.

Let us denote

$$p_m(r) = P\{X_m = r\},$$

By assumption (c),

$$p_{n+1}(k) = \sum_{j=0}^{\infty} p_n(j) P\{X_{n+1} = k | X_n = j\}$$

$$= \sum_{j=0}^{\infty} p_n(j) P[Y_1 + \ldots + Y_j = k]$$

where $Y_1, \ldots, Y_j$ are *iid* random variables, each having *pmf* $p(x), x = 0, 1, 2, \ldots$

$$= \sum_{j=0}^{\infty} p_n(j) \sum_{k_i \geq 0; \sum k_i = k} \{P(Y_1 = k_1) \ldots P(Y_j = k_j)\}$$

$$\tag{13.4.1}$$

$$= \sum_{j=0}^{\infty} p_n(j) [p(k)]^{j*}$$

where $[p(k)]^{j*}$ denotes the j-fold convolution (defined below) of $p(k)$ with itself. The expression (13.4.1) gives a recursion relation containing *pmf* of X_{n+1} and X_n.

Here X_n denotes the number of persons at time (generation) n. For human population we consider sons (fathers) only.

DEFINITION 13.4.1 *Convolution:* Let $\{a_k\}, \{b_k\}$ be two numerical sequences. Define the sequence $\{c_k\}$ where

$$c_r = a_0 b_r + a_1 b_{r-1} + \ldots + a_{r-1} b_1 + a_r b_0.$$

Then the sequence $\{c_k\}$ is called the convolution of sequences $\{a_k\}$ and $\{b_k\}$ and is denoted as

$$\{c_k\} = \{a_k\} * \{b_k\}.$$

If we have sequences $\{a_k\}, \{b_k\}, \{c_k\}, \{d_k\}, \ldots$, we can form convolution $\{a_k\} * \{b_k\}$ and then convolution of this sequence with $\{c_k\}$, etc. If we have

n such identical sequences $\{a_k\}$, then n-fold convolution of $\{a_k\}$ with itself is

$$\{a_k\}^{n*} = \{a_k\} * \ldots * \{a_k\}. \quad \square$$

In particular, if $Z_1, \ldots, Z_n$ are *iid* with *pmf* $\{q_k\}[P(Z_1 = k) = q_k]$ then the distribution of $\xi_n = Z_1 + \ldots + Z_n$ is $\{q_k\}^{n*}$.

Let $F(s)$ denote the probability generating function of X_1,

$$F(s) = \sum_{x=n}^{\infty} p(x)s^x, \quad |s| \leq 1.$$

Theorem 13.4.1: The probability generating function (*pgf*) $F_n(s)$ of X_n, given that $X_0 = 1$ is

$$F_n(s) = F[F_{n-1}(s)]. \tag{13.4.2}$$

(If $X_0 = x_0$, the pgf of X_n is $[F_n(s)]^{x_0}$.)

Proof. Consider X_2 which is determined by the number of offsprings of X_1. Let S_i be the number of offsprings of the ith individual in generation (time) 1. Hence,

$$X_2 = \sum_{i=1}^{X_1} S_i.$$

By the result in Example 10.4.3, pgf of X_2 is

$$F_2(s) = F[F_1(s)].$$

Similarly,

$$F_3(s) = F[F_2(s)]$$
$$\cdots \qquad \cdots$$
$$F_n(s) = F[F_{n-1}(s)].$$

Clearly, we can write

$$F_n(s) = F_{n-1}[F(s)]. \tag{13.4.3}$$

Moments of X_n

We have

$$\mu = E(X_1) = \frac{dF_1(s)}{ds}\bigg]_{s=1} = F_1'(1) = F'(1).$$

Differentiating (13.4.2) with respect to s,

$$F_n'(s) = F'[F_{n-1}(s)]F_{n-1}'(s).$$

Hence,

$$F_n'(s) = F'(1)F_{n-1}'(1) \ [\text{ since } F_{n-1}(1) = 1]$$
$$= \mu F_{n-1}'(1) = \mu^2 F_{n-2}'(1) = \dots = \mu^{n-1} F_1'(1) = \mu^n.$$

Therefore,

$$E(X_n) = \mu^n. \tag{13.4.4}$$

Again,

$$\sigma^2 = V(X_1) = F''(1) + F'(1) - [F'(1)]^2. \tag{13.4.5}$$

Differentiating $F_{n+1}(s) = F_n[F(s)]$ with respect to s and putting $s = 1$,

$$F_{n+1}'(s) = F_n'[F(s)]F'(s).$$

Differentiating again and putting $s = 1$,

$$F_{n+1}''(1) = F_n''(1)(F'(1))^2 + F_n'(1)F''(1). \tag{13.4.6}$$

Similarly, differentiating $F_{n+1}(s) = F[F_n(s)]$ twice and putting $s = 1$,

$$F_{n+1}''(1) = F''(1)(F_n'(1))^2 + F'(1)F_n''(1). \tag{13.4.7}$$

If $\mu \neq 1$, one can equate (13.4.6) and (13.4.7) and solve for $F_n''(1)$. This gives

$$F_n''(1) = \sigma^2 \mu^n \left(\frac{\mu^n - 1}{\mu^2 - \mu} \right) + \mu^n(\mu^n - 1). \tag{13.4.8}$$

Using this in the relation $V(X_n) = F_n''(1) + F_n'(1) - [F_n'(1)]^2$ (compare with (13.4.5)),

$$V(X_n) = \sigma^2 \mu^n \left\{ \frac{\mu^n - 1}{\mu(\mu - 1)} \right\}, \ \mu \neq 1. \tag{13.4.9}$$

If $\mu = 1$, L' Hospital rule yields

$$V(X_n) = \sigma^2 n. \tag{13.4.10}$$

Let ρ be the probability that the population will ultimately be extinct, i.e.,

$$\rho = \lim_{n \to \infty} P\{X_n = 0 | X_0 = 1\}.$$

Now,

$$\mu^n = E(X_n) = \sum_{i=1}^{\infty} iP[X_n = i] \geq \sum_{i=1}^{\infty} P[X_n = i] = P[X_n \geq 1].$$

Hence, if $\mu < 1$,

$$P[X_n = 0] \geq 1 - \mu^n \to 1 \tag{13.4.11}$$

as $n \to \infty$. Thus, if $\mu < 1$, the population will ultimately die out. In case $\mu > 1$, ρ may be found as follows.

$$\begin{aligned}\rho &= P[\text{population ultimately dies out}]\\ &= \textstyle\sum_{i=0}^{\infty} P[\text{population ultimately dies out} \,|\, X_1 = i] p(i).\end{aligned} \tag{13.4.12}$$

Now, when $X_1 = i$, the population dies out if and only if each of the i families started by i individuals in the first generation dies out. Since the individuals act independently, this probability is ρ^i. Therefore, from (13.4.12),

$$\rho = \sum_{i=0}^{\infty} \rho^i p(i). \tag{13.4.13}$$

The solution ρ is the smallest positive number satisfying (13.4.13).

The extinction probability can be calculated more accurately as follows. Assume $X_0 = 1$. Let

$$\rho_n = P[X_n = 0] = p_n(0).$$

Here ρ_n is the probability of extinction by the nth generation and is not to be confused with the event 'extinction at the nth generation', which is $\rho_n - \rho_{n-1}$. Hence, $\{\rho_n\}$ is a non-decreasing sequence and since it is bounded it has a limit.

$$\rho = \lim_{n \to \infty} \rho_n.$$

Now,

$$\rho_{n+1} = F_{n+1}(0) = F(\rho_n) \quad \text{(by (13.4.2))} \tag{13.4.14}$$

and this relation together with the initial condition $\rho_0 = 0$ determines the sequence $\{\rho_n\}$.

By letting $n \to \infty$ in (13.4.14) we see that the limiting probability must be a root of the equation

$$\rho = F(\rho). \tag{13.4.15}$$

This expression directly relates the probability of extinction to the pgf of the progeny.

Theorem 13.4.2: If $F(s)$ is not identically to s (i.e., if a man does not have exactly one son with probability one), then the equation (13.4.15) has just two real positive roots (of which $\rho = 1$ is always one). The extinction probability ρ is the smaller of the two roots: also $\rho < 1$ if $\mu > 1$ and $\rho = 1$ if $\mu \leq 1$.

Proof. Omitted. The reader may refer to Feller (1968), vol. 1, pp. 206 - 7 or Bharucha-Reid (1960), pp. 25 - 26.

The problem of branching process has its origin in 1873-74 when Galton and de Candolle became interested in many instances of family names that had become extinct. The model for this situation was first proposed by H.W.Watson (a clergyman) in 1874. The analysis was completed by J.F.Steffenson in 1930.

13.5   Simple Random Walk

Consider a Markov process $\{X_n, n = \ldots, -1, 0, 1, \ldots\}$ for which the transition probabilities are

$$
\begin{aligned}
p_{jj+1} &= p_j, \\
p_{jj} &= r_j, \\
p_{jj-1} &= q_j, \\
p_{jk} &= 0(k \neq (j-1, j, j+1))
\end{aligned}
\qquad (13.5.1)
$$

where $p_j + r_j + q_j = 1$. If j represents the co-ordinate of a point in the x-axis, then the process describes that a point moves one step right with probability p_j, one step left with probability q_j and stays where it is with probability r_j.

The forward and backward equations (13.3.11) and (13.3.12) respectively, take the form

$$
p_{jk}^{(n)} = p_{jk-1}^{(n-1)} p_{k-1} + p_{jk}^{(n-1)} r_k + p_{jk+1}^{(n-1)} q_{k+1}, \qquad (13.5.2)
$$

$$
p_{jk}^{(n)} = p_j p_{j+1k}^{n-1)} + r_j p_{jk}^{(n-1)} + q_j p_{j-1k}^{(n-1)}. \qquad (13.5.3)
$$

The system (13.5.2) is more useful if we consider $p_{jk}^{(n)}$ for a fixed initial state j and a variable current state k. The system (13.5.3) is more useful in the reverse situation.

The model (13.5.1) has been used to represent *Brownian motion*, that is the motion of a large colloidal particle which is subject to random forces. However, the model originated from the theory of games of chance.

Suppose two players A, B play a series of games of chance. At each game, the winner takes unit stake from the loser. Then p_j, q_j, r_j, respectively, can be taken as the probabilities that A wins, loses and makes a draw when he holds a capital of j units. If the rule of the game is that the game must end if $j = 0$ (because A is then ruined), then $p_0 = q_0 = 0, r_0 = 1$. Similarly, if a is the total capital of the two players, then play must also stop at $j = a$ (because then B is ruined) and hence $p_a = q_a = 0, r_a = 1$.

Generally, the probabilities p_j, q_j, r_j will be independent of j so long the game continues.

The Markov chain in this case has three distinct sets of states, $\{0\}, \{1, 2, \ldots, a - 1\}, \{a\}$. The states 0 and a are persistent, where as the states $1, 2, \ldots, a-1$ are each transient. The states $0, a$ are absorbing states.

We shall calculate the probability that

$$A_j = \lim_{n \to \infty} p_{j0}^{(n)}, \tag{13.5.4}$$

i.e., the probability that the player is ultimately ruined, given that he started with an initial capital j. We shall assume $p_j = p, q_j = q, r_j = 0, j = 1, 2, \ldots, a - 1$. We have

$$A_0 = 1, A_a = 0. \tag{13.5.5}$$

We note that in view of the independence of trials,

$$A_j = pA_{j+1} + qA_{j-1}$$

or equivalently,

$$A_j - A_{j+1} = \frac{q}{p}(A_{j-1} - A_j), \quad j = 1, \ldots, a - 1. \tag{13.5.6}$$

Using the boundary conditions (13.5.5),

$$\begin{aligned} A_1 - A_2 &= \tfrac{q}{p}(A_0 - A_1) = \tfrac{q}{p}(1 - A_1), \\ A_2 - A_1 &= \tfrac{q}{p}(A_1 - A_2) = (\tfrac{q}{p})^2(1 - A_1), \\ &\quad \ldots \ldots \quad \ldots \ldots \ldots \ldots \\ A_{a-1} - A_a &= \tfrac{q}{p}(A_{a-2} - A_{a-1}) = (\tfrac{q}{p})^{a-1}(1 - A_1). \end{aligned} \tag{13.5.7}$$

Adding the first $(j - 1)$ equations,

$$\begin{aligned} A_1 - A_j &= (1 - A_1)\textstyle\sum_{r=1}^{j-1}(\tfrac{q}{p})^r \\ &= \begin{cases} \frac{q}{p}\frac{(q/p)^{j-1}-1}{(q/p)-1}(1 - A_1), & \text{if } \frac{q}{p} \neq 1 \\ (j - 1)(1 - A_1) & \text{if } \frac{q}{p} = 1. \end{cases} \end{aligned} \tag{13.5.8}$$

Adding $(1 - A_1)$ to each side of (13.5.8),

$$1 - A_j = \begin{cases} \frac{(q/p)^j - 1}{(q/p) - 1}(1 - A_1), & \text{if } p \neq \frac{1}{2}, \\ j(1 - A_1), & \text{if } p = \frac{1}{2}. \end{cases} \qquad (13.5.9)$$

Putting $j = a$ in the above equations,

$$1 - A_1 = \begin{cases} \frac{(q/p) - 1}{(q/p)^a - 1}, & \text{if } p \neq \frac{1}{2}, \\ 1/a, & \text{if } p = \frac{1}{2}. \end{cases}$$

Hence, from (13.5.8),

$$A_j = \begin{cases} \frac{(q/p)^a - (q/p)^j}{(q/p)^a - 1}, & \text{if } p \neq \frac{1}{2}, \\ 1 - \frac{j}{a}, & \text{if } p = \frac{1}{2}. \end{cases} \qquad (13.5.10)$$

A_j gives the probability of A's ruin. The probability A'_j of A's fortune eventually reaching a is equal to the probability of B's ruin (when B starts with an initial capital of $a - j$) and is obtained by replacing p, q and j by $q, p, a - j$ respectively in (13.5.10). This gives

$$A'_j = \begin{cases} \frac{(q/p)^j - 1}{(q/p)^a - 1}, & \text{if } p \neq \frac{1}{2}, \\ j/a, & \text{if } p = \frac{1}{2}. \end{cases} \qquad (13.5.11)$$

The gambler's (A's) ultimate gain is a random variable Z. Hence, if A starts with initial capital j,

$$E(Z) = (a - j)(1 - A_j) + A_j(-j) \\ = a(1 - A_j) - j.$$

Hence, $E(Z) = 0$ iff $p = \frac{1}{2}$ so that under this system of gambling, a fair game (p = q) remains fair $(E(Z) = 0)$ and no unfair game $(p \neq q)$ can be reduced to a fair game.

One special case of interest is that in which $a \to \infty$ so that A is playing against an infinitely rich opponent. We find from (13.5.10) that in this case,

$$\lim_{a \to \infty} A_j = A_j^* = \begin{cases} (q/p)j, & (p > q) \\ 1, & (p \leq q). \end{cases} \qquad (13.5.12)$$

Thus A loses with probability one even in case $p = q$, when the game is a fair one. If the game is advantageous to him $(p > q)$, his chance of ruin decreases exponentially with the size of his original capital.

13.6 Continuous Time Markov Process

Consider now t as a continuous parameter; X_t would, however, be assumed to be discrete (i.e., $\mathcal{X}_t$ is countable) taking values x_t. We shall assume that the process is tine-homogeneous (stationary). Thus,

$$P(X_{s+t} = x_k | X_s = x_j) = p_{jk}(t), \ t \ge 0. \tag{13.6.1}$$

This gives the Chapman-Kolmogorov equation

$$p_{jk}(s+t) = \sum_m p_{jm}(s) p_{mk}(t) \ (s, t \ge 0). \tag{13.6.2}$$

We can rewrite this equation in the matrix form as

$$\mathbf{P}(s+t) = \mathbf{P}(s)\mathbf{P}(t) \tag{13.6.3}$$

where $\mathbf{P}(t)$ is the matrix of t-step transition probabilities.

Previously we considered one-step transition probabilities as given and used (13.3.11) to obtain n step transition probabilities. Here, we shall allow both s and t tend to zero and obtain continuous time analogue of forward and backward equations (13.3.11) and (13.3.12).

Assume, therefore,

$$\mathbf{P}(t) = ((p_{jk}(t))) \to \mathbf{I} \text{ as } t \to 0, \tag{13.6.4}$$

where $\mathbf{I}$ is an identity matrix, because one may reasonably expect that the process that has left the initial state j would remain at the same state j after time t as $t \to 0$. We shall further assume that $\mathbf{P}(t)$ possesses a matrix of right derivatives $\mathbf{Q} = ((q_{jk}))$ at time $t = 0$. Thus,

$$p_{jk}(t) = \delta_{jk} + q_{jk}t + 0(t), \ t \ge 0. \tag{13.6.5}$$

Since for $j \ne k, p_{jk}(0) = 0, p_{jk}(t) \ge 0$, we must have $q_{jk} \ge 0$. Hence for $j \ne k, q_{jk}$ may be interpreted as the probability intensity of transition from j to k in the sense that $q_{jk}t + 0(t)$ is the probability that the process changes from j to k in tine $(0, t)$, given that it was at state j at $t = 0$.

The probability of a transaction from state j within time period $(0, t)$ is

$$\sum_{k(\ne j)} p_{jk}(t) = t \sum_{k(\ne j)} q_{jk} + 0(t).$$

Hence probability of no transaction from state j within $(0, t)$ is

$$\begin{aligned} p_{jj}(t) &= 1 - \sum_{k(\ne j)} p_{jk}(t) \\ &= 1 - t \sum_{k(\ne j)} q_{jk} + 0(t) \\ &= 1 + q_{jj}t + 0(t) \ \text{(by (13.6.5))} . \end{aligned}$$

Hence we must have

$$q_{jj} = -\sum_{k(\neq j)} q_{jk} = -q_j \quad \text{(say)}$$

where q_j is the total intensity of transaction out of state j i.e., the rate at which the state j is left for some other state.

By letting t and s tend to zero in (13.6.3) one obtains the forward and backward equations

$$\dot{\mathbf{P}}(t) = \mathbf{P}(t)\mathbf{Q}, \tag{13.6.6}$$

$$\dot{\mathbf{P}}(t) = \mathbf{Q}\mathbf{P}(t), \tag{13.6.7}$$

where $\dot{\mathbf{P}}(t) = \frac{d\mathbf{P}(t)}{dt}$. Written in full, this leads to

$$\dot{p}_{jk}(t) = \sum_{m(\neq k)} p_{jm}(t)q_{mk}(t) - q_k p_{jk}(t), \tag{13.6.8}$$

$$\dot{p}_{jk}(t) = \sum_{m(\neq j)} q_{jm}p_{mk}(t) - q_j p_{jk}(t). \tag{13.6.9}$$

Note that the condition (13.6.5) is not necessary for a continuous time Markov process. However, almost all processes of physical interest have this property.

We shall subsequently consider X_t as the number of occurrences of an event of a given type up to time $t(\geq 0)$. Thus, the process is at state j, if the number of events up to time t, X_t is j. Clearly, j is integer-valued and $X_t \leq X_{t+s}, s \geq 0$. Such a process is called a *counting process*. We shall consider here continuous time stochastic counting process. Such a process will be said to have independent increments if the number of events recurring in disjoint time intervals are mutually independent.

13.6.1 *Poisson process*

The counting process $\{X_t, t \geq 0\}$ is said to be a Poisson process with rate (intensity) λ if

   (i) $X(0) = 0$;
   (ii) the process is stationary and has independent increments;
   (iii) $P(X_t = 1) = \lambda t + 0(t)$;
   (iv) $P(X_t \geq 2) = o(t)$.

Here

$$q_{jk}(t) = \begin{cases} \lambda & (k = j+1) \\ -\lambda & (k = j) \\ 0 & \text{otherwise.} \end{cases} \tag{13.6.10}$$

For example, X_t may be the number of registrations on a Geiger counter, the number of calls entering a telephone exchange, etc. Assumptions tacitly made in the model are : (i) the occurrence of events after time t is not affected by the number of events occurring up to time t (e.g., assumption that one telephone call can block the others); (ii) the events occur with constant intensity λ; (iii) the multiple events (two telephone calls at the same time) occur with zero probability. The forward equation (13.6.8) becomes

$$\begin{align} \frac{dP_0(t)}{dt} &= -\lambda P_0 \\ \frac{dP_k(t)}{dt} &= \lambda P_{k-1} - \lambda P_k \end{align} \tag{13.6.11}$$

where $P_k(t) = p_{0k}(t) = P(X_t = k)$. In this notation, we shall mean $P_x(t)$ as the probability that $X_t = x$ when it started with $X_0 = 0$. The initial condition $P_0(0) = 1, P_k(0) = 0, k \neq 0$. The equations (13.6.11) have already been solved in 8.6. The solution gives $P_x(t) = e^{-\lambda t}(\lambda x)^x/x!$.

13.6.2 *Birth process*

Here the only possible transitions are of the form $j \to j+1$. The probability intensities are

$$q_{jk} = \begin{cases} \lambda_j & (k = j+1) \\ -\lambda_j & (k = j) \\ 0 & \text{(otherwise).} \end{cases} \tag{13.6.12}$$

The forward equations (13.6.8) are

$$\dot{p}_{jk}(t) = \lambda_{k-1} p_{jk-1}(t) - \lambda_k p_{jk}(t). \tag{13.6.13}$$

Note that the rate λ_j depends on j, not on t. The Poisson process is a particular case ($\lambda_j = \lambda$) of this process. Another example is that of radioactive disintegration where a radioactive atom can pass through a number of states by successive disintegration. Here, however, (unlike the pure birth process) there is an absorbing state, after which further disintegration is not possible.

Suppose that the atom is at state 0 for $t = 0$, i.e., $X_0 = 0$ and m is the final state, i.e., $k = 0, 1, \ldots, m$. Let as before, $P_k(t) = P(X_t = k)$. Then

$$P_k(t + \Delta t) = (1 - \lambda_k \Delta t)P_k(t) + \lambda_{k-1} P_{k-1}(t)\Delta t + 0(\Delta t). \tag{13.6.14}$$

By transferring $P_x(t)$ on the other side, dividing by Δt and taking limit we obtain the system of differential equations

$$
\begin{aligned}
\dot{P}_0(t) &= -\lambda_0 P_0(t) \\
\dot{P}_k(t) &= \lambda_{k-1} P_{k-1}(t) - \lambda_k P_k(t), t = 1, 2, \ldots, m-1 \\
\dot{P}_m(t) &= \lambda_{m-1} P_{m-1}(t) \\
P_k(0) &= 0 \ \text{ for } k \neq 0.
\end{aligned}
\tag{13.6.15}
$$

The system can be solved recursively. For example,

$$
\begin{aligned}
P_0(t) &= e^{-\lambda_0 t} \\
P_1(t) &= \tfrac{\lambda_0}{\lambda_1 - \lambda_0}(e^{-\lambda_0 t} - e^{-\lambda_1 t}).
\end{aligned}
\tag{13.6.16}
$$

An important special case is the simple (linear) birth process, where λ_j is proportional to j, say $\lambda_j = j\lambda$. Consider a culture of j bacteria each of which reproduces independently at the rate λ. Then the probability of one division at time t is

$$
\binom{j}{1}[\lambda t + 0(t)](1 - \lambda(t) + 0(t)]^{j-1} = \lambda j t + 0(t)
$$

so that the transition $j \to j+1$ does not actually have intensity λj in this case. The probability of more than one birth is at least of order $(\lambda t)^2$ so that the transition $j \to j+r$ has zero probability, $r \geq 2$.

However, in the simple birth process we assume $\lambda_j = j\lambda (j > 1, \lambda > 0)$. Assume that $X_0 = 1$. Let $P_x(t)$ denote the probability that $X_t = x$. Then by solving the equations (13.6.14) one gets

$$
P_x(t) = e^{-\lambda x t}[\lambda(x-1)\int_0^t P_{x-1}(\tau)e^{x\tau}d\tau + C_x], \ x \geq 1.
\tag{13.6.17}
$$

From the initial condition $P_1(0) = 1, C_1 = 1$.

By induction one gets the *Yule-Furry distribution*

$$
P_x(t) = \begin{cases} e^{-\lambda t}(1 - e^{-\lambda t})^{x-1}, \ x = 1, 2, \ldots \\ 0, \hspace{2.5cm} \text{otherwise}. \end{cases}
\tag{13.6.18}
$$

It is easily seen that $P_x(t)$ is a geometric distribution with parameter $e^{-\lambda t}$. Hence

$$
\begin{aligned}
E[X_t] &= e^{\lambda t} \\
V[X_t] &= e^{\lambda t}(e^{\lambda t} - 1).
\end{aligned}
\tag{13.6.19}
$$

Consider now $X_0 = i$. In this case, Kolmogorov equations (13.6.13) has solution

$$
\begin{aligned}
P_{ij}(t) &= \binom{j-1}{i-1}e^{-i\lambda t}(1 - e^{-\lambda t})^{j-1}, j \geq 1 \\
&= 0 \text{ otherwise}.
\end{aligned}
\tag{13.6.20}
$$

In this case,

$$E[X_t] = ie^{\lambda t}$$
$$V(X_t) = ie^{\lambda t}(e^{\lambda t} - 1).$$

We note the following condition for a pure birth process to be a proper probability distribution, that is, for $\sum_{x=0}^{\infty} P_x(t) = 1$. This is given by the following theorem.

Theorem 13.6.1 (*Feller-Lindeberg*): For a pure birth process with parameter λ_x, $\sum_{x=0}^{\infty} P_x(t) = 1 \ \forall \ t$ iff

$$\sum_{x=0}^{\infty} \frac{1}{\lambda_x} = \infty. \tag{13.6.21}$$

For a proof, the reader may refer to Bharucha-Reid (1960), pp. 81 - 82.

13.6.3 *Birth and death process*

Here only possible transactions are: $j \to j +_- 1$ with

$$q_{jk} = \begin{cases} \lambda_j & (k = j + 1) \\ \mu_j & (k = j - 1) \\ -\lambda_j - \mu_j & (k = j) \\ 0 & (\text{otherwise}) . \end{cases}$$

The state $x = 0$ is an absorbing state.

First we obtain the solution for $P_x(t)$, the probability for $X_t = x$. We have

$$P_x(t+\Delta t) = \lambda_{x-1} P_{x-1}(t)\Delta(t) + [1-(\lambda_x+\mu_x)\Delta t]P_x(t) + \mu_{x+1}P_{x+1}(t)\Delta t + 0(\Delta t). \tag{13.6.22}$$

This gives

$$P_x'(t) = \lambda_{x-1} P_{x-1}(t) - (\lambda_x + \mu_x)P_x(t) + \mu_{x+1}P_{x+1}(t), \ x = 1, 2, \ldots, \tag{13.6.23}$$

$$P_0'(t) = \mu_1 P_1(t) \tag{13.6.24}$$

since $\lambda_{-1} = \lambda_0 = \mu_0 = 0$. If $X_0 = i$,

$$P_i(0) = 1; \ P_x(0) = 0, \ x \neq i. \tag{13.6.25}$$

When $\lambda_x = \lambda x$ and $\mu_x = \mu x, (\lambda, \mu > 0)$, it can be shown that

$$P_x(t) = [1 - \alpha(t)][1 - \beta(t)][\beta(t)]^{x-1}, \ x = 1, 2, \ldots$$
$$P_0(t) = \alpha(t) \tag{13.6.26}$$

where

$$\begin{aligned}
\alpha(t) &= [\mu(e^{(\lambda-\mu)t} - 1)][\lambda e^{(\lambda-\mu)t} - \mu]^{-1} \\
\beta(t) &= [\lambda(e^{(\lambda-\mu)t} - 1)][\lambda e^{(\lambda-\mu)t} - \mu]^{-1} \\
E(X_t) &= \psi_t = e^{(\lambda-\mu)t} \\
V(X_t) &= \tfrac{\lambda+\mu}{\lambda-\mu} e^{(\lambda-\mu)t}[e^{(\lambda-\mu)t} - 1].
\end{aligned} \tag{13.6.27}$$

It can be shown that

$$\lim_{t\to\infty} \psi(t) = \begin{cases} 0, & \lambda < \mu \\ 1, & \lambda = \mu \\ \infty, & \lambda > \mu. \end{cases} \tag{13.6.28}$$

Thus for $\lambda = \mu$, the mean population size ψ_t is stationary.

The probability that the process will die out by time t is $P_0(t)$.

$$\lim_{t\to\infty} P_0(t) = \begin{cases} 1, & \lambda < \mu \\ \mu/\lambda, & \lambda > \mu. \end{cases} \tag{13.6.29}$$

It can be shown that the limits

$$\lim_{t\to\infty} P_x(t) = \pi_x \tag{13.6.30}$$

exist and are independent of the initial conditions (13.6.25). They satisfy the system of equations (13.6.23), (13.6.24) on replacing the derivatives on the left by zero.

For the pure process the forward equations are

$$\dot{p}_{jk} = \lambda_{k-1}p_{jk-1} + \mu_{k+1}p_{jk+1} - (\lambda_k + \mu_k)p_{jk}. \tag{13.6.31}$$

The equations for the equilibrium probabilities π_k are

$$\lambda_{k-1}\pi_{k-1} + \mu_{k+1}\pi_{k+1} = (\lambda_k + \mu_k)\pi_k. \tag{13.6.32}$$

Assume that $\mu_0 = 0$ (so that transition to $k < 0$ is impossible). Considering the equation (13.6.33) for $k = 0, 1, 2, \ldots$ one gets the simpler equations

$$\lambda_{k-1}\pi_{k-1} = \mu_k\pi_k \, (k = 1, 2, \ldots). \tag{13.6.33}$$

Solving

$$\pi_k = \frac{\lambda_0\lambda_1 \ldots \lambda_{k-1}}{\mu_1\mu_2 \ldots \mu_k}\pi_0 \, (k = 1, 2, \ldots). \tag{13.6.34}$$

π_0 is determined from the condition $\sum \pi_k = 1$.

EXAMPLE 13.6.1 (*A simple trunking problem*): Calls are coming at a telephone exchange as a Poisson process with rate λ. The probability of a conversation ending between t and $t + h$ is $\mu h + 0(h)$. The durations of conversations are assumed to be independent. The system is at state n if n lines are busy. Assume that infinitely many trunks or channels are available.

If n lines are busy, the probability that one of them will be freed within time h is $n\mu h + 0(h)$. The probability that within this time two or more lines will be freed is $0(h^2)$ and can be neglected. The probability of a new call arriving is $\lambda h + 0(h)$. The probability of a combination of several calls or a conversation ending is $0(h)$. Hence,

$$\lambda_n = \lambda, \quad q_{n,n+1} = \lambda, \quad q_{n,n-1} = n\mu, \quad q_{n,n} = -\lambda - n\mu.$$

The limiting probability

$$\pi_n = e^{-\lambda/\mu} \frac{(\lambda/\mu)^n}{n!}. \tag{13.6.35}$$

EXAMPLE 13.6.2 (*A waiting problem*): Consider the same model as in Example 13.6.1 with modification that the number of channels is $'a'(< \infty)$. If all $'a'$ channels are busy each new call joins a waiting line and waits until a channel is freed.

The system is in state $n(> a)$ if there are $'a'$ persons being served and $(n-a)$ persons in the waiting line. If at least one line is free, the situation is the same as in previous example. If the system is at state $n, \mu_n = a\mu(n \geq a)$. The basic system of differential equations is given by (13.6.23) and (13.6.24) for $n < a$ (defining the system for $n < a$) and is

$$P'_n(t) = -(\lambda + a\mu)P_n(t) + \lambda P_{n-1}(t) + a\mu P_{n+1}(t) \tag{13.6.36}$$

for $n \geq a$. The limiting probabilities π_n satisfy (13.6.35) for $n < a$. Thus

$$\begin{aligned} \pi_n &= \pi_0 \frac{(\lambda/\mu)^n}{n!}, \quad n < a \\ \pi_n &= \frac{(\lambda/\mu)^n}{a! a^{n-a}} \pi_0 \quad \text{(by using (13.6.35))}, \quad n \geq a. \end{aligned} \tag{13.6.37}$$

The series $\sum \frac{\pi_n}{\pi_0} < \infty$ only if $\frac{\lambda}{\mu} < a$. Hence, a limiting distribution $\{\pi_n\}$ cannot exist if $\lambda \geq a\mu$. In this case, the distribution becomes concentrated on the large values of n, which means that the waiting line increases to infinity. If $\lambda < a\mu$, one can determine π_0 so that $\sum_{n=0}^{\infty} \pi_n = 1$.

EXAMPLE 13.6.3 (*A single server problem*): Consider the servicing of automatic machines. At an epoch t a machine in the working state calls for service with probability $\lambda h + 0(h)$ during the time $(t, t + h)$. Again, when the machine is being serviced, the probability that the servicing time terminates during $(t, t+h)$ is $\mu h + 0(h)$. For an efficient machine the servicing factor λ/μ should be small. Assume that m machines are working independently and are serviced by a single repairman (when required). The system is said to be in state n if n machines are not working. When $1 \leq n \leq m$, this means one machine is being repaired and $(n-1)$ machines are in the waiting line for servicing. Thus, transition $n \to n+1$ occurs by a breakdown of one of the working $(m-n)$ machines; $n \to n-1$ if the machine being serviced reverts to working state.

Thus,

$$\lambda_n = (m-n)\lambda, \ \mu_0 = 0, \mu_1 = \ldots = \mu_m = \mu.$$

EXAMPLE 13.6.4 (*A multiple server problem*): Consider the same situation as in Example 13.6.3, but now there are r repairmen $(r < m)$. It can be checked that

$$\lambda_0 = m\lambda, \ \mu_0 = 0$$
$$\lambda_n = (m-n)\lambda, \ \mu_n = n\mu(1 \leq n \leq r)$$
$$\lambda_n = (m-n)\lambda, \ \mu_n = r\mu(r \leq n \leq m).$$

In the discontinuous Markov processes with continuous time parameter considered above we have assumed that for a small time interval Δt the probability of no change far exceeds the probability of a change of state. However, if a change takes place it may be rather striking. In the Markov process which are both continuous in time and space (also called diffusion processes) it is assumed that some changes will take place in any interval Δt and if Δt is small the change of state will also be small with probability one. We shall consider only one such process, called the normal process.

13.6.4 *Normal process*

A stochastic process $\{X_t, t \in T\}$ is said to be a normal process if for any integer n and any subset $\{t_1, t_2, \ldots, t_n\}$ of T the n random variables $X_{t_1}, X_{t_2}, \ldots, X_{t_n}$ are jointly normally distributed. As a result the probability law of the process is completely specified by specifying the mean value function $E(X_t)$ and the covariance kernel $\text{Cov}(X_s, X_t)$.

An example of a normal process is *Weiner* process which provides a model for Brownian motion and thermal noise in electric circuits, among other applications. A tiny particle (of say, diameter 10^{-4} cm.) when immersed in a liquid or fluid undergoes ceaseless irregular motion due to the continuous bombardment of the molecules of the surrounding medium. This is known as Brownian motion.

A process $\{X_t, t \geq 0\}$ is said to be a Weiner process if

(i) it has stationary independent increments;
(ii) for $s < t, X_t - X_s$ is normally distributed;
(iii) $E(X_t) = 0 \ \forall \ t > 0$;
(iv) $X_0 = 0$.

The first condition means that for every choice of indices $t_0 < t_1 < \ldots < t_n$ and for every n, the n random variables $X_{t_1} - X_0, X_{t_2} - X_{t_1}, \ldots, X_{t_n} - X_{t_{n-1}}$ are independently distributed and also for every choice of $t_1 < t_2, X_{t_2+h} - X_{t_1+h}$ has the same distribution as $X_{t_2} - X_{t_1}$.

Given n points $t_1 < t_2 < \ldots < t_n$, let $Z_1 = X_{t_1} = X_{t_1} - X_{t_0}, Z_2 = X_{t_2} - X_{t_1}, \ldots, Z_n = X_{t_n} - X_{t_{n-1}}$. By hypothesis, $Z_1, Z_2, \ldots, Z_n$ are independently normally distributed. Now, $X_{t_1}, X_{t_2}, \ldots, X_{t_n}$ are linear combinations of $Z_1, \ldots, Z_n$. Hence, $X_{t_1}, \ldots, X_{t_n}$ are jointly normally distributed. Therefore, $\{X_t, t \geq 0\}$ is a normal process. It follows that $E[X_t - X_s] = 0$. The fact that X_t has independent increments shows that there is some positive constant σ^2 such that for $t \geq s \geq 0$,

$$\text{var} \ [X_t - X_s] = \sigma^2[t - s].$$

This follows because $\text{var}[X_t]$ satisfies the functional form $f(t_1+t_2) = f(t_1) + f(t_2)$. Again,

$$\begin{aligned} cov[X_s, X_t] &= cov[X_s, X_t - X_s + X_s] \\ &= cov[X_s, X_t - X_s] + var(X_s) \\ &= var(X_s). \end{aligned}$$

Therefore, the covariance kernel is

$$K(s,t) = \sigma^2 \min(s,t) \ \forall \ s,t \geq 0.$$

We now show that the Wienner process can be used as a model for Brownian motion of a particle on a line arising from the numerous random impacts with other particles. Assume that times between two successive impacts are independent random variables, exponentially distributed with mean $1/\mu$. It then follows that the number $N(t)$ of impacts during $(0, t)$ is a Poisson process with intensity μ. Assume that the effect of an impact on the particle is to change its position by either $+a$ or $-a$, each with probability $\frac{1}{2}$. The position X_t of the particle at time t may be represented as

$$X_t = \sum_{n=1}^{N(t)} Y_n,$$

where Y_n is the particle's change of position as a result of the nth impact. The random variables Y_n are assumed to be independently distributed with

$$P[Y_n = +a] = P[Y_n = -a] = \frac{1}{2}.$$

It follows that $\{X_t, t \geq 0\}$ is a stochastic process with stationary independent increments and characteristic function $\psi_X(u)$ given by

$$\ln \psi_X(u) = \mu t E[e^{iuY} - 1].$$

Now,

$$E[e^{iuY} - 1] = iuE[Y] - \frac{u^2}{2}E[Y^2] + \theta|u|^3 E[|Y|^3]$$
$$= -\frac{1}{2}u^2 a^2 + \theta|u|^3 a^3, \quad |\theta| \leq 1.$$

Defining $\sigma^2 = a^2 \mu$,

$$\ln \psi_{X_t}(u) = -\frac{1}{2}u^2 \sigma^2 t + a\theta|u|^3 \sigma^3 t.$$

If we now let $\mu \to \infty$ and $a \to 0$ such that the product μa^2 (total mean square displacement of the particle per unit time) is constant, then

$$\ln \psi_{X_t}(u) \to -\frac{1}{2}u^2 \sigma^2 t.$$

Hence $\{X_t, t \geq 0\}$ is approximately a Weinner process, since it is approximately normally distributed and has stationary independent increments. The above results are due to Parzen (1960).

13.7   Exercises and Complements

13.1 For the following stochastic matrices classify all the states.

$$(i) \ \mathbf{P} = \begin{bmatrix} \frac{1}{4} & \frac{1}{2} & 0 & \frac{1}{4} \\ \frac{1}{5} & 0 & \frac{1}{3} & \frac{7}{15} \\ 0 & \frac{2}{3} & \frac{1}{3} & 0 \\ \frac{1}{4} & \frac{1}{4} & \frac{1}{4} & \frac{1}{4} \end{bmatrix}; \quad (ii) \ \mathbf{P} = \begin{bmatrix} \frac{1}{3} & \frac{2}{3} & 0 & 0 \\ \frac{3}{4} & \frac{1}{8} & \frac{1}{8} & 0 \\ 0 & 0 & \frac{1}{2} & \frac{1}{2} \\ 0 & 0 & \frac{1}{3} & \frac{2}{3} \end{bmatrix};$$

$$(iii) \ \mathbf{P} = \begin{bmatrix} \frac{2}{5} & 0 & \frac{3}{5} & 0 & 0 \\ \frac{1}{5} & \frac{3}{5} & \frac{1}{5} & 0 & 0 \\ \frac{1}{5} & 0 & \frac{4}{5} & 0 & 0 \\ 0 & 0 & 0 & \frac{2}{5} & \frac{3}{5} \\ 0 & 0 & 0 & \frac{2}{5} & \frac{3}{5} \end{bmatrix}.$$

13.2 Consider a homogeneous Markov chain with a countable number of states with transition matrix

$$\mathbf{P} = \begin{bmatrix} \frac{1}{2} & \frac{1}{2} & 0 & 0 & 0 & \cdots & 0 \\ \frac{1}{2} & 0 & \frac{1}{2} & 0 & 0 & \cdots & 0 \\ \frac{1}{3} & 0 & 0 & \frac{2}{3} & 0 & \cdots & 0 \\ \frac{1}{4} & 0 & 0 & 0 & \frac{3}{4} & \cdots & 0 \\ \cdots & \cdots & \cdots & \cdots & \cdots & \cdots & \cdots \end{bmatrix}.$$

Show that all states are recurrent null states and hence $\lim_{n \to \infty} p_{ij}^{(n)} = 0 \, (i, j = 1, 2, \ldots)$.

[Fisz, 1963]

13.3 A matrix is said to be doubly stochastic if in addition to the property $\sum_j p_{ij} = 1$, we have $\sum_i p_{ij} = 1$. Show that a finite Markov chain with a doubly stochastic matrix has no transient state. Show also that if the chain is irreducible and aperiodic then $p_{ij}^{(n)} \to N^{-1}$ where N is the number of states. On the other hand, if an irreducible infinite chain has a doubly stochastic matrix, then its states are either null or transient.

(Prabhu, 1965)

13.4 Consider Example 13.3.8. Suppose that the initial individual is chosen randomly from a population in which three genotypes AA, Aa, aa occur in the proportion p_0, q_0, r_0 respectively. Show that the probability that the line ultimately tends to the AA-strain is $p_0 + \frac{1}{2}q_0$, which is also the A-gene frequency in the original population.

13.5 Suppose that X_n is a numerical-valued Markov process with transition probability matrix (13.3.14). Show that for the ergodic case, the correlation coefficient between X_n and X_{n+m} is $(1 - \alpha - \beta)^m, (m > 0)$ when n is so large that the equilibrium is attained.

13.6 (*A non-homogeneous random walk*) : Let $p_{00} = q_0, p_{01} = p_0, p_{i,i-1} = q_i, p_{i,i+1} = p_i (i = 1, 2, \ldots)$, where $p_i = 1 - q_i (i \geq 1)$.

Further, let $L_i = (q_1 q_2 \ldots q_i)(p_1 p_2 \ldots p_i)^{-1} (i \geq 1)$. Show that the states are (i) transient if the series $\sum L_i < \infty$; (ii) persistent null if $\sum L_i = \infty$ and $\sum (P_i L_i)^{-1} = \infty$; (iii) persistent non-null if $\sum L_i = \infty, \sum (L_i P_i)^{-1} < \infty$. Find the stationary distribution in the last case.

13.7 Let Z be a random variable whose possible values are the integers $0, +_-1, +_-2, \ldots$. Assume that $E[|Z|] < \infty$.

Let $Z_1, Z_2, \ldots$ be a sequence of independent random variables distributed independently as Z. Define $X_n = Z_1 + Z_2 + \ldots + Z_n$.

Describe the transition probability matrix of the Markov chain $\{X_n, n = 1, 2, \ldots\}$. Show that the chain is recurrent iff $E(Z) = 0$.

13.8 For a certain Markov chain,

$$\mathbf{P} = \begin{bmatrix} p_0 & p_1 & p_2 & \cdots & p_{m-1} \\ p_{m-1} & p_0 & p_1 & \cdots & p_{m-2} \\ \cdots & \cdots & \cdots & \cdots & \cdots \\ p_1 & p_2 & p_3 & \cdots & p_0 \end{bmatrix}.$$

Determine $\mathbf{P}^n$ and prove that $\lim_{n \to \infty} P[X_n = j] = \frac{1}{m}$ $(j = 1, \ldots, m)$.

13.9 Show that for the discrete branching process considered in Section 13.4, the relation (13.4.2) can be generalized to

$$F_{m+n}(s) = F_m[F_n(s)], (m, n = 0, 1, 2, \ldots).$$

13.10 Suppose that

$$F(s) = \frac{ps + q}{1 + c - cs}.$$

Show that if $p, q, c \geq 0, p + q = 1$, then this is the probability generating function of a non-negative random variable. Show also that the nth iterate of $F(s)$ is

$$F^{(n)}(s) = F[F[[F \ldots [F(s)] \ldots] \ldots] = \frac{A_n s + B_n}{C_n s + D_n}$$

where

$$\begin{bmatrix} A_n & B_n \\ C_n & D_n \end{bmatrix} = \begin{bmatrix} p & -c \\ q & 1 + c \end{bmatrix}^n.$$

13.11 Show that

$$F(s) = 1 - c(1 - s)^\delta$$

is the probability generating function of a non-negative random variable if $0 \leq c, \delta \leq 1$ and evaluate its nth derivative $F^{(n)}(s)$ (defined in Exercise 13.10).

13.12 Suppose that there are two types of individuals A and B whose numbers in the nth generation have joint pgf $F_n(s, t)$. If individuals of type A and B have progeny pgf's $G_A(s, t), G_B(s, t)$ respectively, show that

$$F_{n+1}(s, t) = F_n[G_A(s, t), G_B(s, t)].$$

Make the same type of assumptions as in Section 13.4.

13.13 Using equation (13.5.3) show that if M_j is the expected duration of game given that A started with an initial capital of j units, show that

$$M_j = 1 + pM_{j+1} + rM_j + qM_{j-1} \ (j = 1, 2, \ldots, n - 1)$$

with $M_0 = M_n = 0$. Solve for M_j.

13.14 Consider the model (13.5.1) to represent the Brownian motion of a particle in a one-dimensional container with walls at $j = 0$ and $j = a$ so that $q_0 = p_a = 0$. Show that if π_j is the equilibrium probability that the particle is at position j, then

$$\pi_j = \frac{p_0 p_1 \cdots p_{a-1}}{q_1 q_2 \cdots q_j} \pi_0 \ (j = 1, 2, \ldots, a).$$

13.15 Suppose that the Poisson process is modified to allow multiple events, so that

$$q_{jk} = \begin{cases} \lambda_{k-j} & (k > j) \\ 0 & (k \le j). \end{cases}$$

Show that the probability generating function of the number of events in $(0, t)$ is

$$P(s, t) = \exp\left[t \sum_r \lambda_r(s^r - 1)\right], \ |s| \le 1.$$

13.16 For the simple death process

$$q_{jk} = \begin{cases} \mu j & (k = j - 1) \\ -\mu j & (k = j) \\ 0 & (\text{otherwise}), \end{cases}$$

show that the probability generating functions is

$$\pi_j(s, t) = E(s^{X_j} | X_0 = j) = \sum_k p_{jk}(t) s^k = [1 + e^{\mu t}(s - 1)]^j.$$

13.17 Show that if $q_{jj+1} = \lambda(N - j), q_{jj-1} = j\mu(0 \leq j \leq N)$ and $X_0 = 0$, then the distribution of X_0 is binomial

$$\binom{N}{k} p^k q^{N-k} \ (0 \leq k \leq N)$$

when $p = \lambda[1 - e^{-(\lambda+\mu)t}]/(\lambda + \mu)$ and $q = 1 - p$.

13.18 Consider a pure birth process with $\lambda_j = \lambda + j\alpha(j \geq 0)$. Show that the transition probabilities of X_t are given by

$$p_{jk}(t) = e^{-(\lambda+jk)t} \binom{-j - (\lambda/\alpha)}{k - j} (e^{-\alpha t} - 1)^{k-j} \ (k \geq j).$$

13.19 Show that in a pure death process with $\mu_j = j\mu$, the transition probabilities are given by

$$p_{jk}(t) = \begin{cases} \binom{j}{k} e^{-k\mu t} (1 - e^{-\mu t})^{j-k}, & (k \leq j) \\ 0, & (k > j). \end{cases}$$

Show that the time for extinction has the distribution

$$e^{\mu t} (1 - e^{-\mu t})^{j-1} j\mu dt \ (0 < t < \infty).$$

13.20 Show that the time spent in states $j, j+1, \ldots,$ of a birth process are independent exponential variables with parameters $\lambda_j^{-1}, \lambda_{j+1}^{-1}, \ldots.$

13.21 Let $k(> 0)$ denote the number of customers in a queue and suppose that the customers leave the queue with intensity μ and join with intensity λ (for $k < a$) or zero (for $k \geq a$). Obtain the differential equations for the system. Show that the limiting probability

$$\pi_k = \frac{(1 - \rho)\rho^k}{1 - \rho^{k+1}}, \ k = 0, 1, \ldots, a.$$

where $\rho = \lambda/\mu$, the traffic intensity. Show that when $a \to \infty$ and $\rho < 1$, the limiting distribution is a geometric distribution. However, if $a \to \infty$ and $\rho \geq 1$, the distribution becomes concentrated over indefinitely large values of k.

13.22 For the gambler's ruin problem, let μ_i denote the expected number of games to be played by A until the gambler either goes broke or reaches a fortune a, given that he started with an initial capital i. Show that μ_i satisfies

$$\mu_0 = \mu_a = 0,$$
$$\mu_i = 1 + p\mu_{i+1} + q\mu_{i-1} \ (i = 1, 2, \ldots, a - 1).$$

Hence, show that

$$\mu_i = \begin{cases} i(a-i), & \text{if } p = \frac{1}{2} \\ \frac{i}{q-p} - \frac{a}{q-p} \times \frac{1-(\frac{q}{p})^i}{1-(\frac{q}{p})^a}, & \text{if } p \neq \frac{1}{2}. \end{cases}$$

13.23 Let the smaller root of equation (13.4.15) be ξ. Then $F'(\xi)$ exists and $0 < F(\xi) \leq 1$. Show that $\xi \geq P(X_1 = 0) = p_0$. Show also that in the range $0 \leq s \leq \xi$,

$$\xi + F'(\xi)(s - \xi) \leq F(s) \leq \xi + \frac{\xi - p_0}{\xi}(s - \xi). \qquad (i)$$

Using this inequality in (13.4.14), show that

$$(\frac{\xi - p_0}{\xi})(\xi - \rho_n) \leq (\xi - \rho_{n+1}) \leq F'(s)(s - \rho_n). \qquad (ii)$$

The first inequality in (ii) indicates that ξ is the only possible limit for the sequence $\{\rho_n\}$ and $\rho = \xi$ (ρ is defined in (13.4.12)).

Appendix

A.1 Random Variables as Limits

Let $(\mathcal{S}, \mathcal{A})$ be the measure space on which the random variable $X(.)$ is defined. Note that $X(.)$ can be expressed as

$$X(.) = X_+(.) - X_-(.) \qquad (A.1.1)$$

where

$$X_+(.) = \begin{cases} X(.) & \text{for } X(.) \geq 0 \\ 0 & \text{for } X(.) < 0 \end{cases} \qquad (A.1.2)$$

and

$$X_-(.) = \begin{cases} -X(.) & \text{for } X(.) \leq 0 \\ 0 & \text{for } X(.) > 0. \end{cases} \qquad (A.1.3)$$

Clearly, both X_+ and X_- are non-negative random variables.

Theorem A.1.1: A bounded non-negative random variable $X(.)$ can be expressed as the limit of a nondecreasing sequence of simple functions. The convergence of the sequence is uniform in ω over the whole basic space $\mathcal{S}$.

Proof. Suppose $X(.)$ is a non-negative bounded random variable with $X(.) \leq N$ for all ω. Range of values of $X(.)$ is, therefore, $\{t : 0 \leq t \leq N\} = R(N)$ (say). We now make a sequence of partitioning of $R(N)$ as follows.

At the nth partition, we divide $R(N)$ into $N2^n$ disjoint half-open intervals at points $t_{0n} = 0, t_{1n} = \frac{1}{2^n}, \ldots, t_{(N2^n)n} = N$, i.e. at points $t_{in} = \frac{i}{2^n}, i = 0, 1, \ldots, N2^n$.

At the $(n+1)$th partition we divide $R(N)$ into $N2^{n+1}$ disjoint half-open intervals at points $t_{in+1} = \frac{i}{2^{n+1}}, i = 0, 1, \ldots, N2^{n+1}$.

A typical interval M in the nth partition is $[a, a + \delta)$ where $\delta = \frac{1}{2^n}$ and a is the point of the subdivision at the left end of the interval.

At the $(n + 1)$th partition, the interval M is divided into two non-overlapping, half-open intervals M' and M'':

$$M = M' \cup M'' = [a, a + \frac{\delta}{2}) \cup [a + \frac{\delta}{2}, a + \delta).$$

We let

$$E = X^{-1}(M), \ E' = X^{-1}(M'), \ E'' = X^{-1}(M'').$$

Then, $E = E' \cup E''$ and E', E'' are disjoint sets (Lemma 3.3.2).

On the basis of the nth partitioning of $R(N)$, we form the random variable $X_n(.)$ as follows: Map each point ω of E into the point $t = a$ of the interval $[a, a + \delta)$; we do this for each interval of the partition.

For the $(n + 1)$th partition we form $X_{n+1}(.)$ as follows: Map each point ω of E' into the point $t = a$ of the interval $[a, a + \frac{\delta}{2})$; map each point ω of E'' into the point $a + \frac{\delta}{2}$ of the interval $[a + \frac{\delta}{2}, a + \delta)$; we do this for each interval of the $(n + 1)$th partition.

Therefore,

$$X_n(.) = \sum_{i=0}^{N2^n} t_{in} I_{E_{in}} \tag{A.1.4}$$

where M_{in} is the ith interval of the nth partition of $R(N)$ and $E_{in} = X_n^{-1} M_{in}$. Obviously, $X_n(.)$ is a simple function.

From the construction, it is obvious that for each ω and any n,

$$X_n(\omega) \leq X_{n+1}(\omega) \leq X(\omega) \tag{A.1.5}$$

and

$$X(\omega) - X_n(\omega) < \delta = \frac{1}{2^n}. \tag{A.1.6}$$

The sequence $\{X_n(\omega), n = 1, 2, \ldots\}$ is, therefore, a monotonically non-decreasing sequence of simple functions whose limit for any ω is $X(\omega)$. Also, because of the inequality (A.1.6), the convergence is uniform in ω, i.e., the approximation does not depend on ω. Hence the proof. $\square$

Remark A.1.1: The mode of partitioning is not unique. Any sequence of partitioning of $R(N)$ into non-overlapping intervals can be used, provided each partition divides the intervals of the previous partition in such a way that the maximum length of the interval goes to zero as $n \to \infty$.

Remark A.1.2: The first part of the theorem holds even if the hypothesis of boundedness is removed. However, for unbounded random variables, the uniformity of convergence does not hold.

Remark A.1.3: The results of the theorem can be extended to the general case, because of the representation (A.1.1).

A.1.1 Approximating simple functions for independent random variables

If $X(.)$ and $Y(.)$ are two independent random variables and if $X_i(.), Y_j(.)$, respectively, are approximating simple random variables, we must have independence of X_i and Y_j; because, inverse image $X_i^{-1}(t_i)$ of any point t_i in the range of X_i is the inverse image $X^{-1}(M)$ of an appropriate interval in the real line; similarly, the inverse image $Y_j^{-1}(u_j)$ of a point u_j in the range of values of Y_j is the inverse image $Y^{-1}(N)$ of an appropriate interval N in the real line. These two events must be independent.

We have, therefore, the following theorem.

Theorem A.1.2: Suppose $X(.)$ and $Y(.)$ are independent random variables, each of which is non-negative. Then there exist nondecreasing sequences of simple random variables $\{X_n(.) : 1 \leq n < \infty\}$ and $\{Y_m(.) : 1 \leq m < \infty\}$ such that

$$\lim_{n \to \infty} X_n(\omega) = X(\omega), \quad \lim_{m \to \infty} Y_m(\omega) = Y(\omega) \; \forall \; \omega$$

and $\{X_n(\omega), Y_m(\omega)\}$ is an independent pair for any choice of n, m.

A.2 Lebesgue-Integration of a Borel Measurable Function (a Random Variable)

Since a random Variable X is a function on $(\mathcal{S}, \mathcal{A})$ and probability P is a measure, we should be able to define integration of X with respect to P. We state without proof some important results in this regard.

Consider the measure space $(\Omega, \mathcal{F}, \mu)$ where μ is a measure on $\mathcal{F}$. If a function $f : \Omega \to \mathcal{R}$ (or $\bar{\mathcal{R}}$) is Borel measurable, we will define the Lebesgue integral of f with respect to μ, written as $\int_\Omega f d\mu, \int_\Omega f(\omega)\mu(d\omega)$ or $\int_\Omega f(\omega)d\mu(\omega)$.

DEFINITION A.2.1: *Lebesgue Integral*: First, let us consider f to be a simple function $f = \sum_{i=1}^{r} t_i I_{A_i}$, where the A_i are disjoint sets in $\mathcal{F}$ and $\cup_i A_i = \Omega$. In this case,

$$\int_\Omega f d\mu = \int_\Omega \sum_{i=1}^{r} t_i I_{A_i} d\mu = \sum_{i=1}^{r} t_i \mu(A_i) \qquad (A.2.1)$$

provided ∞ and $-\infty$ do not both appear in the sum. If they do, we say that the integral does not exist. Again, the value of the sum in (A.2.1) should be the same for any other simple function representation of f, i.e., even if $f = \sum_{j=1}^{s} q_j I_{B_j}$, the same value as in the right side of (A.2.1) should be obtained.

We have noted in Section 3.4 that a non-negative Borel measurable function is the limit of an increasing sequence of non-negative finite-valued simple functions. If f is a non-negative Borel measurable function, we define

$$\int_\Omega f d\mu = \sup\{\int_\Omega g d\mu : g \text{ is a simple function }, 0 \le g(\omega) \le f(\omega) \ \forall \ \omega \in \Omega\}.$$
$$(A.2.2)$$

Note that according to this definition, the integral of a non-negative Borel measurable function always exists. It may be $+\infty$.

For an arbitrary Borel measurable function f we have

$$f = f_+ + f_- \qquad (A.2.3)$$

where

$$f_+(\omega) = \begin{cases} f(\omega) & \text{if } f(\omega) \ge 0 \\ 0 & \text{if } f(\omega) < 0 \end{cases}$$

and

$$f_-(\omega) = \begin{cases} -f(\omega) & \text{if } f(\omega) \le 0 \\ 0 & \text{if } f(\omega) > 0. \end{cases}$$

We define

$$\int_\Omega f d\mu = \int_\Omega f_+ d\mu - \int_\Omega f_- d\mu, \qquad (A.2.4)$$

if this is not of the form $\infty - \infty$. In the latter case we say that the integral does not exist.

The function f is said to be μ-integrable *iff* $\int f d\mu < \infty$, i.e., *iff*

$$\int_\Omega f_+ d\mu < \infty \text{ and } \int_\Omega f_- d\mu < \infty. \qquad (A.2.5)$$

If $A \in \mathcal{F}$, we define

$$\int_A f d\mu = \int_\Omega f I_A d\mu. \qquad (A.2.6)$$

We now note down some properties of integral of Borel measurable functions f.

Theorem A.2.1:

(a) If $\int_\Omega f d\mu$ exists and $c \in \mathcal{R}$, then $\int_\Omega c f d\mu$ exists and equals $c \int_\Omega f d\mu$.

(b) If $\int_\Omega f d\mu$ exists, then $|\int_\Omega f d\mu| < \int_\Omega |f| d\mu$.

(c) If f is Borel measurable, f is integrable *iff* $|f|$ is integrable.

(d) (a) If $f \geq 0$ and $A \in \mathcal{F}$, then

$$\int_A f d\mu = \sup\{\int_A g d\mu : g \text{ is a simple function, } 0 \leq g \leq f\}.$$

(e) If $\int_\Omega f d\mu$ exists, so does $\int_A f d\mu$ for each $A \in \mathcal{F}$.

(f) If $\int_\Omega f d\mu$ is finite, so is $\int_A f d\mu$ for each $A \in \mathcal{F}$.

(g) *Additivity Theorem:* Let f and g be two Borel measurable functions and assume that $f + g$ is well-defined. If $\int_\Omega f d\mu$ and $\int_\Omega g d\mu$ exist and $\int_\Omega f d\mu + \int_\Omega g d\mu$ is well-defined (i.e. not of the form $+\infty - \infty$ or $-\infty + \infty$), then

$$\int_\Omega (f + g) d\mu = \int_\Omega f d\mu + \int_\Omega g d\mu. \qquad (A.2.7)$$

(h) If $f_1, f_2, \ldots$ are non-negative Borel measurable,

$$\int_\Omega (\sum_{n=1}^\infty f_n) d\mu = \sum_{n=1}^\infty \int_\Omega f_n d\mu. \qquad (A.2.8)$$

Thus any series of non-negative Borel measurable functions may be integrated term by term.

(i) If f and g are Borel measurable, with $|g| \leq f, f$ integrable, then g is integrable.

(j) If $g(\omega) \geq f(\omega) \; \forall \; \omega$, then $\int_\Omega g d\mu \geq \int_\Omega f d\mu$.

(k) If f is Borel measurable and $f = 0$ a.e. $[\mu]$, then $\int_\Omega f d\mu = 0$.

(l) if f and g are Borel measurable and $f = g$ a.e.$[\mu]$, and $\int_\Omega g d\mu$ exists, so does $\int_\Omega f d\mu$ and $\int_\Omega g d\mu = \int_\Omega f d\mu$.

(m) If f is Borel measurable and is integrable, then f is finite a.e.$[\mu]$.

(n) If $f \geq 0$ and $\int_\Omega f d\mu = 0$, then $f = 0$ a.e.$[\mu]$.

Convergence Theorems

It is frequently important to know whether the operations of integration and passage to a limit can be interchanged. The Lebesgue integral has desirable properties in this respect. Proof of these theorems can be found in Kolmogorov and Fomin (1975), Loeve (1977), Billingsley (1995), among others.

Theorem A.2.2: *Monotone Convergence Theorem*: Let $f_1, f_2, \ldots$ be an increasing sequence of nonnegative Borel measurable functions and let $f(\omega) = \lim_{n \to \infty} f_n(\omega) \; \forall \; \omega \in \Omega$. Then

$$\lim_{n \to \infty} \int_\Omega f_n d\mu = \int_\Omega f d\mu. \qquad (A.2.9)$$

Lemma A.2.1 *Fatau's Lemma*: Let $f_1, f_2, \ldots, f$ be Borel measurable.

(a) If $f_n \geq f \; \forall \; n$, where $\int_\Omega f d\mu > -\infty$, then

$$\lim_{n \to \infty} \inf \int f_n d\mu \geq \int_\Omega (\lim_{n \to \infty} \inf f_n) d\mu. \qquad (A.2.10)$$

(b) If $f_n \leq f \; \forall \; n$, where $\int_\Omega f d\mu < \infty$, then

$$\lim_{n \to \infty} \sup \int_\Omega f_n d\mu \leq \int_\Omega (\lim_{n \to \infty} \sup f_n) d\mu. \qquad (A.2.11)$$

Theorem A.2.3 (*Dominated Convergence Theorem*): If $f_1, f_2, \ldots, g$ are Borel measurable, $|f_n| \leq g \; \forall \; n$, where g is μ-integrable and $f_n \to f$ a.e. $[\mu]$, then f is μ-integrable and $\int_\Omega f_n d\mu \to \int_\Omega f d\mu$.

Corollary A.2.3.1: If $f_1, f_2, \ldots, g$ are Borel measurable, $|f_n| \leq g \; \forall \; n$, where $|g|^p$ is μ-integrable ($p > 0$ and fixed) and $f_n \to f$ a.e.$[\mu]$, then $|f|^p$ is μ-integrable and $\int_\Omega |f_n - f|^p d\mu \to 0$ as $n \to \infty$.

We state two more relevant theorems in this context.

It is well-known from calculus that we can evaluate the double integral $\int \int f(x, y) dx dy$ of a continuous function that vanishes outside a finite rectangle by successively performing the one-dimensional integrals, namely

$$\int \int f(x, y) dx dy = \int dx \int f(x, y) dy.$$

The analogue of this fact for the Lebesgue integral is called Fubini's theorem.

Theorem A.2.4 (*Fubini's Theorem*): Consider the measure space ($\Omega \times \Omega', \mathcal{B} \times \mathcal{B}', \Pi$) which is the product of the measure spaces ($\Omega, \mathcal{B}, P$) and

$(\Omega', \mathcal{B}', P')$(vide Section 3.7). Suppose $f = f(\omega, \omega')$ is a measurable, real-valued function on this product space and that f is integrable with respect to the product measure Π. Then

$$
\begin{aligned}
\int_{\Omega \times \Omega'} f(\omega, \omega') d\Pi &= \int_{\Omega} (\int_{\Omega}' f(\omega, \omega') dP') dP \\
&= \int_{\Omega'} (\int_{\Omega} f(\omega, \omega') dP) dP'.
\end{aligned} \qquad (A.2.12)
$$

If either of these two iterated integrals exist with f replaced by $|f|$, then f is integrable with respect to Π and the equality holds.

Absolute Continuity and Radon-Nikodym Theorem

Suppose that $(\Omega, \mathcal{B})$ is a measurable space and that Q and R are both finite measures on this space. The measure R is said to be absolutely continuous with respect to Q, often denoted as $R << Q$, if $Q(E) = 0$ implies $R(E) = 0$. It can be seen that this must happen if R is defined by

$$
R(E) = \int_E f dQ \qquad (A.2.13)
$$

where f is a non-negative function integrable with respect to Q. The converse is also true.

Theorem A.2.5 (*Radon-Nikodym Theorem*) Suppose that Q and R are finite measures on $(\Omega, \mathcal{B})$ and that R is absolutely continuous with respect to Q. Then there exists a non-negative Q-integrable function f such that (A.2.13) holds for all $E \in \mathcal{B}$.

A.2.1 Integration of a random variable

Since, a random variable is a Borel measurable function, all the results stated above hold for integration of a random variable with respect to a probability measure. However, some of these results are stated explicitly, since they will be used extensively in the later development.

A.2.1.1 Integral of a simple random variable

Consider the probability space $(\mathcal{S}, \mathcal{A}, P)$ and $X(\omega)$, a random variable defined on $\mathcal{A}$. When $X(\omega)$ is a simple function, (i.e. a simple random variable), $\sum_{k=1}^{n} a_k I_{A_k}$, the Lebesgue-Steiltjes (LS) integral of $X(\omega)$ with respect to $(\mathcal{S}, \mathcal{A}, P)$ over $\mathcal{S}$ is given by

$$
\int_{\mathcal{S}} X(\omega) dP(\omega) = \int \sum_{k=1}^{n} a_k I_{A_k} dP(\omega) = \sum_{k=1}^{n} a_k P(A_k). \qquad (A.2.14)
$$

The integral (A.2.14) can be given a simple but important interpretation. We rewrite this as

$$\int_S X dP = \sum_{k=1}^{n} a_k P(X = a_k). \qquad (A.2.15)$$

The sum on the right side is the probability-weighted sum of the values of the function. Since the total probability mass is unity, this sum is the probability-weighted average of the values of the random variable.

In general, for any subset $E \in \mathcal{A}$,

$$\int_E X(\omega) dP(\omega) = \int_S X I_E dP = \sum_{k=1}^{n} a_k P(E \cap A_k). \qquad (A.2.16)$$

A measure μ defined on the class $\mathcal{A}$ is said to be a *totally finite measure* if

$$\mu(A) = \mu(S \cap A) \text{ for each } A \in \mathcal{A}. \qquad (A.2.17)$$

We note below some properties of integral of a simple random variable X. All these properties are easy to prove from the definition (A.2.14).

(i) *Linearity with respect to integrand:* If a and b are constants,

$$\int_E (aX + bY) dP = a \int_E X dP + b \int_E Y dP. \qquad (A.2.18)$$

(ii) *Linearity with respect to measure:* If $\mu(.), \mu'(.), \mu''(.)$ are totally finite measures and a, b are constants, such that $\mu(.) = a\mu'(.) + b\mu''(.)$, then

$$\int_S X d\mu = a \int_S X d\mu' + b \int_S X d\mu''. \qquad (A.2.19)$$

(iii) *Additivity:* If $\{E_i : i \in I\}$ is a finite or countably infinite partition of E, then

$$\int_E X dP = \sum_{i \in I} \int_{E_i} X dP. \qquad (A.2.20)$$

(iv) $|\int_E X dP| \leq \int_E |X| dP.$

(v) If $P(E) = 0$, then $\int_E X dP = 0$.

(vi) (a) $\int_E X dP = 0 \; \forall E$ iff $X(.) = 0[\, P \,]$; (b) $\int_E X dP = \int_E Y dP \; \forall E$ iff $X(.) = Y(.)[\, P \,]$.

(vii) If $X(.) \geq 0[\, P \,]$, then $\int_E X dP \geq 0$ for all events E; equality holds iff $X(.) = 0[\, P \,]$. More generally, if $X(.) \geq |Y(.)|$ [P], then $\int_E X dP \geq \int_Y dP$ for all events E.

(viii) If X, Y are independent random variables, then

$$\int_S XY dP = \int_S X dP \int_S Y dP. \qquad (A.2.21)$$

A.2.1.2 *Integral of a general random variables*

We now extend the definition of integral to the general integrable random variable.

DEFINITION A.2.2: A non-negative random variable $X(.)$ is said to be integrable on the set E with respect to probability measure $P(.)$ *iff* there is a non-decreasing sequence of simple random variables $\{X_n(.) : 1 \le n < \infty\}$, with

$$\lim_{n \to \infty} X_n(.) = X(.) \text{ for each } \omega \in E \qquad (A.2.22)$$

and such that

$$\lim_{n \to \infty} \int_E X_n dP < \infty. \qquad (A.2.23)$$

In this case, we define the integral of $X(.)$ over E with respect to $P(.)$ to be

$$\int_E X dP = \lim_{n \to \infty} \int_E X_n dP. \qquad (A.2.24)$$

If $X(.)$ is any random variable, it is said to be integrable on E with respect to $P(.)$ *iff* both $X_+(.)$ and $X_-(.)$ are integrable. In this case,

$$\int_E X dP = \int_E X_+ dP - \int_E X_- dP.$$

It is obvious that the question of integrability with respect to $P(.)$ arises, only when the random variable is unbounded. For more general measures, which may be infinite on the basic space, even bounded measurable functions may not be integrable.

After defining the integral of a general random variable with respect to $P(.)$ we now investigate whether such integral exists.

We have seen in Theorem 4.6.1 that when $X(\omega)$ is bounded, non-negative and can take infinitely many values, there exists a non-decreasing sequence of simple random variables $\{X_n(.)\}$ such that condition (A.2.20) is satisfied. Here

$$X_n(\omega) = \sum_{i=0}^{N2^n} t_{in} I_{E_{in}}(\omega)$$

where t_{in} is the lowest point of the ith interval M_{in} in which the range-space $R(N)$ of $X(\omega)$ has been divided and $I_{E_{in}}(\omega)$ is the indicator function of the subset $E_{in} = X_n^{-1}(M_{in})$ of $\mathcal{S}$ (vide Section A.1). Clearly, $X_n(.)$ is a simple random variable and $|X_n(\omega) - X(\omega)| < \epsilon$ for every $\epsilon > 0$ for sufficiently large n, uniformly for all $\omega \in \mathcal{S}$. Therefore, the integral exists and

$$
\begin{aligned}
\int_{\mathcal{S}} X \, dP &= \lim_{n \to \infty} \int_{\mathcal{S}} X_n \, dP \\
&= \lim_{n \to \infty} \int_{\mathcal{S}} \sum_{i=0}^{N2^n} t_{in} I_{E_{in}} \, dP \qquad (A.2.25) \\
&= \lim_{n \to \infty} \sum_{i=0}^{N2^n} t_{in} P(E_{in})
\end{aligned}
$$

provided the condition (A.2.23) is satisfied. Also, the limit is independent of the choice of the sequence $\{X_n(.)\}$ satisfying (A.2.22). Lebesgue-integral of $X(\omega)$ with respect to $(\mathcal{S}, \mathcal{A}, P)$ over any set $E \in \mathcal{A}$ is given by (A.2.22).

When $X(\omega)$ is not necessarily bounded, but is non-negative, it is integrable by virtue of the result stated in Remark A.1.2 to Theorem A.1.1 and its integral is given by (A.2.22), provided the condition (A.2.23) is satisfied. However, in this case convergence of $\{X_n(\omega)\}$ to $X(\omega)$ is not uniform in ω. Also, the limit, when it exists, does not depend on the choice of the sequence satisfying (A.2.22) and (A.2.23).

If $X(\omega)$ is arbitrary, it is integrable, if it can be written as $X(\omega) = X_1(\omega) - X_2(\omega)$, where $X_1(\omega), X_2(\omega)$ are both non-negative integrable random variables and in this case,

$$
\int_E X(\omega) dP(\omega) = \int_E X_1(\omega) dP(\omega) - \int_E X_2(\omega) dP(\omega) \qquad (A.2.26)
$$

and its value is independent of choice of $X_1(\omega)$ and $X_2(\omega)$.

The properties of Lebesgue-integral of a random variable $X(\omega)$ with respect to $(\mathcal{S}, \mathcal{A}, P)$ over $E \in \mathcal{A}$, has been stated explicitly in Section 4.7. The rule (4.7.10) follows from the corresponding rule for simple random variables and the fact that approximating simple random variables for $X(.)$ and $Y(.)$ are independent.

A.2.2 The Lebesgue-Stieltjes integral

Let $F(.)$ be a distribution function and let $\mu_F(.)$ be the Lebesgue-Stieltjes Measure induced by it (vide Section 3.4). The Lebesgue-Stieltjes integral of a function $g(x)$ defined on the real line $\mathcal{R}$ is given by

$$
\int_a^b g(t) dF(t) = \int_{[a,b]} g \, d\mu_F \qquad (A.2.27)
$$

where $[a, b]$ is a closed interval $a \le t \le b$.

In the special case, where $F(t)$ is given by $F(t) = t$, the measure induced by it is the ordinary Lebesgue measure, which assigns to each interval its length. The Lebesgue-Stieltjes integral reduces in this case to the Lebesgue integral.

Henceforth in this section, we shall confine ourselves to the probability distribution function $F_X(.)$ and the corresponding probability measure P_X of the random variable X. Recall that $F_X(x) = P_X(-\infty, x]$ and P_X is defined on the measure space $(\mathcal{R}, \mathcal{B}(\mathcal{R}))$. Also, $\mu_F(a, b] = F_X(b) - F_X(a) = P_X(a < X \leq B)$. First we consider three special cases:

(1) $F_X(.)$ *is a continuous function* : In this case, there is no probability mass concentration and $P_X(\{x\}) = 0$ for every $x \in R$. The density function $f_X(.)$ exists and the Lebesgue-Stieltjes integral is expressible in the form

$$\int_a^b g(t)dF_X(t) = \int_a^b g(t)f_X(t)dt \qquad (A.2.28)$$

where the integral is in the sense of Lebesgue. In most cases of practical interest, this is equal to the corresponding Riemann integral and may be computed by ordinary techniques of integration.

(2) F_X *is a step function* : Here probability mass is concentrated at discrete points. In this case the integrand has value $g(t_i)$ at each of the discrete value t_i where the probability mass is concentrated and zero elsewhere. We may consider this function as a simple function which has values t_i's at these points and zero elsewhere. By definition

$$\int g(t)dF_X = \int g(t)dP_X = \sum_i g(t_i)P_X(\{t_i\}). \qquad (A.2.29)$$

(c) *Singular continuous mass distribution* : Munroe (1953, Section 27) has shown that it is possible to construct mass distribution such the probability mass is concentrated at a set of points of Lebesgue measure (length) zero, but the distribution function is continuous. In this case, there is no density function and no discrete mass distribution. For a single real-valued random variable, the situation is very unusual in practice.

(d) *The general case* : The distribution function may be written as a sum of a step function F_{Xd} and a continuous function, F_{Xc}. Thus, $F_X(.) = F_{Xd}(.) + F_{Xc}(.)$. These component functions have the characteristics of distribution functions and define the Lebsgue-Stieltjes measures μ_{Fd} and μ_{Fc} respectively on the Borel sets of the real line. It is obvious that

$$P_X(B) = \mu_{Fd}(B) + \mu_{Fc}(B) \qquad (A.2.30)$$

for any Borel set B.

A decomposition theorem due to Lebesgue shows that it is possible to write

$$\mu_{Fc} = \mu_{Fac} + \mu_{Fsc} \qquad (A.2.31)$$

where μ_{Xac} is an absolutely continuous set function in the sense that it assigns zero probability mass to each Borel set of Lebesgue measure zero. We call this as *absolutely continuous part* of the mass distribution. The other component μ_{Fsc} determines a mass distribution which is continuous and singular in the sense of case (c) above. We refer to this as *singular continuous part* of the mass distribution.

If F_{Xac} and F_{Xsc} are the distribution functions corresponding to μ_{Fac} and μ_{Fsc} respectively and f_{Xac} is the density function corresponding to F_{Xac}, then

$$\int_\infty^\infty g(t)dF_X(t) = \int_{-\infty}^\infty g(t)f_{Xac}(t)dt + \sum_i g(t_i)P_X(\{t_i\}) + \int_{-\infty}^\infty g(t)dF_{Xc}(t)$$

$$(A.2.32)$$

where t_i are the jump points of the step function F_{Xd}.

A.2.3 The Riemann-Stieltjes integral

Let $g(x)$ be a continuous function in $(a, b]$. We divide the interval $(a, b]$ into n subintervals by inserting points $x_1, x_2, \ldots, x_{n-1}$. In order of these intervals, arbitrarily choose points $\xi_1, \xi_2, \ldots, \xi_n$ in these intervals and find the sum

$$S_n = \sum_{i=1}^n g(\xi)[F_X(x_i) - F_X(x_{i-1})], \quad x_0 = a, \ x_n = b. \qquad (A.2.33)$$

If M_i, m_i, respectively, are the upper and lower bound of $g(x)$ in $(x_{i-1}, x_i]$, we define upper and lower sums respectively as

$$S_{nU} = \sum_{i=1}^n M_i[F_X(x_i) - F_X(x_{i-1})], \quad S_{nL} = \sum_{i=1}^n m_i[F_X(x_i) - F_X(x_{i-1}].$$

$$(A.2.34)$$

In the same way as for the ordinary Riemann integral, it can be proved that when $n \to \infty$ in such a way that $\max(x_i - x_{i-1}) \to 0$, S_n tends to a definite limit and is called the Riemann-Stieltjes integral of $g(x)$ with respect to $F(x)$ over $(a, b]$ and is given by

$$\lim_{n\to\infty} S_n = \lim_{n\to\infty} S_{nL} = \lim_\to S_{nU} = \int_a^b g(x)dF_X(x). \qquad (A.2.35)$$

The Riemann-Stieltjes integral (A.2.35) exists in the more general case, when $g(x)$ is bounded in $(a, b]$ and has at most a finite number of discontinuities at r_i, provided $F_X(x)$ is continuous at every r_i.

When $F_X(x) = x$, the Riemann-Stieltjes integral (A.2.35) becomes ordinary Riemann integral.

In case $F_X(x)$ is continuous in $[a, b]$ and has continuous derivative $F_X'(x) = f_X(x)$ throughout, except at most in a finite number of points, Riemann-Stieltjes integral (A.2.35) reduces to the ordinary Riemann integral

$$\int_a^b g(x)f(x)dx. \qquad (A.2.36)$$

If $F_X(x)$ is a step function with jumps $p_i = P_X(\{t_i\})$ at points t_i, Riemann-Stieltjes integral becomes the sum $\sum_i g(t_i)p_i$, which is a finite sum or absolutely convergent infinite series, according as the set of points $\{t_i\}$ is finite or infinite.

The Riemann-Stieltjes integral extends to the case of complex-valued functions $g(x)$ and also to infinite interval (a, b). Here,

$$\lim_{n \to \infty} \sum_1^n g(\xi_i)[F(x_i) - F(x_{i-1})] = \int_{-\infty}^{\infty} g(x)dH(x) \qquad (A.2.37)$$

when $\max(x_i - x_{i-1}) \to 0$ as $n \to \infty$ and $a \to \infty, b \to \infty$ as $n \to \infty$.

Similar results hold if we replace $F_X(.)$ by any other distribution function.

Bibliography

Ash, R.B. and Dade, C.A.B. (2000): *Probability and Measure Theory*, Academic Press, San Diego.

Bailey, N.T.J. (1964): *The Elements of Stochastic Processes*, John Wiley,N.Y.

Bartoszyn'ski, R. and Nicwiadomska-Bugaj, M. (1996): *Probability and Statistical Inference*, John Wiley, N.Y.

Bharucha-Reid, A.T. (1960): *Elements of Theory of Markov Processes and their Applications*, McGraw-Hill, N.Y.

Bhat, B.R. (1991): *Modern Probability Theory*, 2nd edn., Wiley Eastern, New Delhi, India.

Billingsley, P. (1995): *Probability and Measure*, 3rd edn., John Wiley, N.Y.

Bre'maud, P. (1980): *An Introduction to Probability Modelling*, Springer-Verlag, N.Y.

Borel, E. (1909): Sur les probabilitie's de'nombrables et leurs applications arithme'tiques. *Circolo mat.*, **26**, 247 - 271.

Cantelli, P.P. (1917): Sulla probabilita' come limite della frequenza. *Atti Accad. naz. Lincel Rc. Sed. salen.*, s. 5, **26**, 39 - 45.

Castillo, E.; Hadi, A.S.; Balakrishnan, N; Sarabia, J.M. (2005): *Extreme Value and Related Models with Applications in Engineering and Science*, John Wiley, N.Y.

Chung, K.L. (1960): *Markov Process with Stationary Transition Probabilities*, Springer-Verlag, N.Y.

Cramer, H. (1950): *Mathematical Methods of Statistics*, Princeton University Press, 1946 and Asia Publishing House, 1950.

Esse'en, P. (1945): Fourier Analysis of the Distribution Functions, Acta Math., **77**, 1 - 125.

Feller, W. (1968): *An Introduction to Probability Theory and Its Applications*, vol.I, 3rd. edn., Wiley Eastern, New Delhi, India.

Fisz, M. (1963): *Probability Theory and Mathematical Statistics*, John Wiley.

Gnedenko, B. V. (1962): *Theory of Probability*, Chesla Pub. Co. and Mir Publishers.

Halmos, P.R.(1958): *Measure Theory*, Van Hostrand, Princeton.

Harris, B. (1966): *Theory of Probability*, Addison-Wisely, London.

Hodges, J.L. (Jr.) and Lehmann, E.L. (1985): *Elements of Probability*, Holden Day.

Johnson, N.L. and Kotz, S.M. (1977): *Urn Models and Their Applications: An Approach to Modern Discrete Probability Theory*, John Wiley, N.Y.

Johnson, N.L., Kotz, S. and Balakrishnan, N. (1994): *Continuous Univariate Distributions*, John Wiley, N.Y.

Kerich, J.E. (1946): *An Experimental Introduction to the Theory of Probability*, Kimar Munkagaard, Copenhagen.

Kingman, J.F.G. and Taylor, J.J. (1966):*Introduction to Measure and Probability*, Cambridge University Press.

Klenke, A. (2008): *Probability Theory: a comprehensive course*, Springer-Verlag, London.

Kolmogorov, A.N. (1956): *Foundations of Probability*, Chesla Publishing House, N. Y.

Kolmogorov, A.N. and Fomin, S.V. (1975): *Introductory Real Analysis* (translated by R.Silverman), Dover, N.Y.

Laha, R.G. (1982): Characteristic Functions, in *Encyclopedia of Statistical Sciences*, **1**, S. Kotz, N.L. Johnson and C.B. Read (Editors), 415 - 422, New York: Wiley.

Laha, R.G. and Rohatgi, V.K. (1979): *Probability Theory*, John Wiley, N.Y.

Lamperti, J.W. (1996): *Probability: a survey of mathematical theory*, 2nd edn., John Wiley, N.Y.

Loeve, M. (1977): *Probability Theory*, 4th edn., Springer-Verlag, N.Y.

Lomax, K.S. (1954): Business failures: Another example of the analysis of failure data. *Journal of American Statistical Association*, **49**, 847 - 852.

Lukacs, E. (1970): *Characteristic Functions*, 2nd edn., Griffin, London.

Lukacs, E.(1982): *Developments in Characteristic Function Theory*, Griffin, London.

McShane, E.J. (1944): *Integration*, Princeton University.

Mood, A.M., Graybill, F.A. and Boes, D.C. (1974): *An Introduction to Theory of Statistics*, McGraw-Hill and Tata McGraw-Hill.

Moran, P.A.P. (1968): *An Introduction to Probability Theory*, Oxford University Press, London.

Munroe, M.E. (1951): *Theory of Probability*, McGraw Hill, N.Y.

Papoulis, A. (1965): *Probability, Random Variables and Stochastic Process*, McGraw-Hill.

Parzen, E. (1960): *Modern Probability Theory and Its Applications*, John Wiley, N.Y.

Parzen, E. (1962): *Stochastic Processes*, Holden-Day.

Pfeiffer, P.E. (1965): *Concepts of Probability Theory*, McGraw Hill.

Pierce, J.A. (1940): A study of the universe of n finite populations with applications in moment-function adjustments in grouped data. *Annals of Mathematical Statistics*, **11**, 311 - 334.

Pitt, M.R. (1963): *Integration Measure and Probability*, Olliver and Boyd, London.

Port, S. C. (1994): *Theoretical Probability and Applications*, John Wiley,

Prabhu, N.J. (1963): *Stochastic Processes*, Macmillan. N. Y.

Rahaman, N.A. (1967): *Exercises in Probability and Statistics*, Griffin, London.

Rao, M.M. (1984): *Probability Theory with Applications*, Academic Press.

Rohatgi, V.K. (1990): *An Introduction to Probability Theory and Mathematical Statistics*, Wiley Eastern, New Delhi, India.

Sevastyanov, B.A., Chistyakov, V.P. and Zubkov, A.M. (1985): *Problems in the Theory of Probability*, Mir Publishing.

Uspensky, J.V. (1937): *Introduction to Mathematical Probability*, McGraw-Hill and Tata McGraw-Hill.

Whittle, P. (1978): *Probability*, Wiley, John N.Y.

Wilks, S.S. (1976): *Mathematical Statistics*, John Wiley, N.Y.

Subject Index

Table 1. Ordinates of the normal density function

$$\phi(x) = \frac{1}{\sqrt{2\pi}}\, e^{-x^2/2}$$

X	.00	.01	.02	.03	.04	.05	.06	.07	.08	.09
.0	.3989	.3989	.3989	.3988	.3986	.3984	.3982	.3980	.3977	.3973
.1	.3970	.3965	.3961	.3956	.3951	.3945	.3939	.3932	.3925	.3918
.2	.3910	.3902	.3894	.3885	.3876	.3867	.3857	.3847	.3836	.3825
.3	.3814	.3802	.3790	.3778	.3765	.3752	.3739	.3725	.3712	.3697
.4	.3683	.3668	.3653	.3637	.3621	.3605	.3589	.3572	.3555	.3538
.5	.3521	.3503	.3485	.3467	.3448	.3429	.3410	.3391	.3372	.3352
.6	.3332	.3312	.3292	.3271	.3251	.3230	.3209	.3187	.3166	.3144
.7	.3123	.3101	.3079	.3056	.3034	.3011	.2989	.2966	.2943	.2920
.8	.2897	.2874	.2850	.2827	.2803	.2780	.2756	.2732	.2709	.2685
.9	.2661	.2637	.2613	.2589	.2565	.2541	.2516	.2492	.2468	.2444
1.0	.2420	.2396	.2371	.2347	.2323	.2299	.2275	.2251	.2227	.2203
1.1	.2179	.2155	.2131	.2107	.2083	.2059	.2036	.2012	.1989	.1965
1.2	.1942	.1919	.1895	.1872	.1849	.1826	.1804	.1781	.1758	.1736
1.3	.1714	.1691	.1669	.1647	.1626	.1604	.1582	.1561	.1539	.1518
1.4	.1497	.1476	.1456	.1435	.1415	.1394	.1374	.1354	.1334	.1315
1.5	.1295	.1276	.1257	.1238	.1219	.1200	.1182	.1163	.1145	.1127
1.6	.1109	.1092	.1074	.1057	.1040	.1023	.1006	.0989	.0973	.0957
1.7	.0940	.0925	.0909	.0893	.0878	.0863	.0848	.0833	.0818	.0804
1.8	.0790	.0775	.0761	.0748	.0734	.0721	.0707	.0694	.0681	.0669
1.9	.0656	.0644	.0632	.0620	.0608	.0596	.0584	.0573	.0562	.0551
2.0	.0540	.0529	.0519	.0508	.0498	.0488	.0478	.0468	.0459	.0449
2.1	.0440	.0431	.0422	.0413	.0404	.0396	.0387	.0379	.0371	.0363
2.2	.0355	.0347	.0339	.0332	.0325	.0317	.0310	.0303	.0297	.0290
2.3	.0283	.0277	.0270	.0264	.0258	.0252	.0246	.0241	.0235	.0229
2.4	.0224	.0219	.0213	.0208	.0203	.0198	.0194	.0189	.0184	.0180
2.5	.0175	.0171	.0167	.0163	.0158	.0154	.0151	.0147	.0143	.0139
2.6	.0136	.0132	.0129	.0126	.0122	.0119	.0116	.0113	.0110	.0107
2.7	.0104	.0101	.0099	.0096	.0093	.0091	.0088	.0086	.0084	.0081
2.8	.0079	.0077	.0075	.0073	.0071	.0069	.0067	.0065	.0063	.0061
2.9	.0060	.0058	.0056	.0055	.0053	.0051	.0050	.0048	.0047	.0046
3.0	.0044	.0043	.0042	.0040	.0039	.0038	.0037	.0036	.0035	.0034
3.1	.0033	.0032	.0031	.0030	.0029	.0028	.0027	.0026	.0025	.0025
3.2	.0024	.0023	.0022	.0022	.0021	.0020	.0020	.0019	.0018	.0018
3.3	.0017	.0017	.0016	.0016	.0015	.0015	.0014	.0014	.0013	.0013
3.4	.0012	.0012	.0012	.0011	.0011	.0010	.0010	.0010	.0009	.0009
3.5	.0009	.0008	.0008	.0008	.0008	.0007	.0007	.0007	.0007	.0006
3.6	.0006	.0006	.0006	.0005	.0005	.0005	.0005	.0005	.0005	.0004
3.7	.0004	.0004	.0004	.0004	.0004	.0004	.0003	.0003	.0003	.0003
3.8	.0003	.0003	.0003	.0003	.0003	.0002	.0002	.0002	.0002	.0002
3.9	.0002	.0002	.0002	.0002	.0002	.0002	.0002	.0002	.0001	.0001

Table 2. Cumulative normal distribution

$$\phi(x) = \int_{-\infty}^{x} \frac{1}{\sqrt{2\pi}} e^{-t^2/2} dt$$

x	.00	.01	.02	.03	.04	.05	.06	.07	.08	.09
.0	.5000	.5040	.5080	.5120	.5160	.5199	.5239	.5279	.5319	.5359
.1	.5398	.5438	.5478	.5517	.5557	.5596	.5636	.5675	.5714	.5753
.2	.5793	.5832	.5871	.5910	.5948	.5987	.6026	.6064	.6103	.6141
.3	.6179	.6217	.6255	.6293	.6331	.6368	.6406	.6443	.6480	.6517
.4	.6554	.6591	.6628	.6664	.6700	.6736	.6772	.6808	.6844	.6879
.5	.6915	.6950	.6985	.7019	.7054	.7088	.7123	.7157	.7190	.7224
.6	.7257	.7291	.7324	.7357	.7389	.7422	.7454	.7486	.7517	.7549
.7	.7580	.7611	.7642	.7673	.7704	.7734	.7764	.7794	.7823	.7852
.8	.7881	.7910	.7939	.7967	.7995	.8023	.8051	.8078	.8106	.8133
.9	.8159	.8186	.8212	.8238	.8264	.8289	.8315	.8340	.8365	.8389
1.0	.8413	.8438	.8461	.8485	.8508	.8531	.8554	.8577	.8599	.8621
1.1	.8643	.8665	.8686	.8708	.8729	.8749	.8770	.8790	.8810	.8830
1.2	.8849	.8869	.8888	.8907	.8925	.8944	.8962	.8980	.8997	.9015
1.3	.9032	.9049	.9066	.9082	.9099	.9115	.9131	.9147	.9162	.9177
1.4	.9192	.9207	.9222	.9236	.9251	.9265	.9279	.9292	.9306	.9319
1.5	.9332	.9345	.9357	.9370	.9382	.9394	.9406	.9418	.9429	.9441
1.6	.9452	.9463	.9474	.9484	.9495	.9505	.9515	.9525	.9535	.9545
1.7	.9554	.9564	.9573	.9582	.9591	.9599	.9608	.9616	.9625	.9633
1.8	.9641	.9649	.9656	.9664	.9671	.9678	.9686	.9693	.9699	.9706
1.9	.9713	.9719	.9726	.9732	.9738	.9744	.9750	.9756	.9761	.9767
2.0	.9772	.9778	.9783	.9788	.9793	.9798	.9803	.9808	.9812	.9817
2.1	.9821	.9826	.9830	.9834	.9838	.9842	.9846	.9850	.9854	.9857
2.2	.9861	.9864	.9868	.9871	.9875	.9878	.9881	.9884	.9887	.9890
2.3	.9893	.9896	.9898	.9901	.9904	.9906	.9909	.9911	.9913	.9916
2.4	.9918	.9920	.9922	.9925	.9927	.9929	.9931	.9932	.9934	.9936
2.5	.9938	.9940	.9941	.9943	.9945	.9946	.9948	.9949	.9951	.9952
2.6	.9953	.9955	.9956	.9957	.9959	.9960	.9961	.9962	.9963	.9964
2.7	.9965	.9966	.9967	.9968	.9969	.9970	.9971	.9972	.9973	.9974
2.8	.9974	.9975	.9976	.9977	.9977	.9978	.9979	.9979	.9980	.9981
2.9	.9981	.9982	.9982	.9983	.9984	.9984	.9985	.9985	.9986	.9986
3.0	.9987	.9987	.9987	.9988	.9988	.9989	.9989	.9989	.9990	.9990
3.1	.9990	.9991	.9991	.9991	.9992	.9992	.9992	.9992	.9993	.9993
3.2	.9993	.9993	.9994	.9994	.9994	.9994	.9994	.9995	.9995	.9995
3.3	.9995	.9995	.9995	.9996	.9996	.9996	.9996	.9996	.9996	.9997
3.4	.9997	.9997	.9997	.9997	.9997	.9997	.9997	.9997	.9997	.9998

x	1.282	1.645	1.960	2.326	2.576	3.090	3.291	3.891	4.417
$\phi(x)$	.90	.95	.975	.99	.995	.999	.9995	.99995	.999995
$2[1-\phi(x)]$	.20	.10	.05	.02	.01	.002	.001	.0001	.000001